Basic Electronics Theory
4th Edition

Delton T. Horn

TAB Books
Division of McGraw-Hill

New York San Francisco Washington, D.C. Auckland Bogotá
Caracas Lisbon London Madrid Mexico City Milan
Montreal New Delhi San Juan Singapore
Sydney Tokyo Toronto

Notices

Apple	Apple Computer, Inc.
Macintosh	
Heathkit	Heath Co.
IBM	International Business Machines Corp.
Radio Shack	The Tandy Corp.
SQ	Columbia Broadcasting System, Inc.

All other trademarks and registered trademarks are the property of their respective owners.

pbk   13 14 15 DOC/DOC 0 9 8 7 6 5 4 3
hc   1   2   3   4   5   6   7   8   9   10 DOC/DOC 9 9 8 7 6 5 4 3

Library of Congress Cataloging-in-Publication Data

Horn, Delton T.
    Basic electronics theory / by Delton T. Horn. — 4th ed.
        p.   cm.
    Rev. ed. of: Basic electronics theory—with projects and
experiments. 3rd ed. c1989.
    Includes index.
    ISBN 0-8306-4199-8   ISBN 0-8306-4200-5 (pbk.)
    1. Electronics. 2. Electronics—Experiments. I. Horn, Delton T.
Basic electronics theory—with projects and experiments. II. Title.
TK7816.H69   1993
621.381—dc20   93-30749
                                            CIP

Acquisitions editor: Roland S. Phelps
Editorial team: B.J. Peterson, Editor
            Susan Wahlman, Managing Editor
            Joanne Slike, Executive Editor
            Joann Woy, Indexer
Production team: Katherine G. Brown, Director
Design team: Jaclyn J. Boone, Designer
            Brian Allison, Associate Designer   EL1
Cover design: Lori E. Schlosser   4261

Contents

Introduction

Electronics is unquestionably a very rapidly growing field. It is also one of the few industries where prices consistently tend to fall as capabilities increase. A computer system that sells for $1000 now has more memory and computing power than a $5000 system of a few years ago. Not so long ago, VCRs (videocassette recorders) were playthings of the wealthy, costing over $1000 dollars. Today, you can buy a full-featured VCR for under $200, and about half of all American homes have at least one. In addition, new technology makes new applications possible or more practical. Dozens of new electronic components appear on the market almost every month.

To try to keep up with some of the most significant new developments in the field of electronics, and to expand the coverage of certain subjects, this new edition of *Basic Electronics Theory* is written. Very little has been deleted from the earlier editions of this book. However, quite a bit of new material has been added, including expanded and updated information on personal computers. The troubleshooting chapter has been expanded, and new chapters on television and VCRs have been added. There is also some added coverage of controversial related subjects, such as the alleged health aspects of electromagnetic fields.

As with the earlier editions of this book, the goal of this volume is to serve as both an introductory text and as a reference volume. To aid in the learning process, a number of practical experiments are described. I strongly recommend that you actually perform these experiments for yourself, rather than just read about them. Hands-on experience is always the best way to learn. Most of the chapters in this book feature self-tests to help you master the material presented in the text. Most of these tests have been revised and re-written, based on reader response to earlier editions. The answers to these self-tests appear in the appendix.

Of course, electronics is such a wide area, every facet can't possibly be covered in depth in a single book. The emphasis in this book is on the basics and some of the more important of the specialized areas of modern electronics.

The first few chapters briefly outline the relevant aspects of atomic physics and electrical theory. Then a number of chapters are devoted to the most common types of electronic components. Most components you are likely to encounter will be covered somewhere in these chapters. The next few chapters cover basic circuit types such as power supplies, amplifiers, filters, and so on. These basic circuits are used in electronic equipment of almost all types. Finally, the remaining chapters discuss some of the major applications of modern electronics.

I sincerely hope you find this fourth edition of *Basic Electronics Theory* helpful and informative. I also hope you enjoy learning about the exciting, multifaceted field of electronics.

1
What is electronics?

Today, in this increasingly high-tech world, almost anywhere you turn you're likely to find electronics products. Many items you take for granted would have been unthinkable, outside the realm of science fiction, 50 years ago. Some products that are fairly common would have been nearly impossible ten years ago. A few examples of such products are digital watches, pocket TVs, lasers, MIDI (musical instrument digital interface) synthesizers, personal computers, CD players, and VCRs. This list is scarcely comprehensive or complete; there are hundreds of other products that could be included.

There is no denying that electronics is having an increasingly important impact on our lives and society. Because you are reading this book, it's reasonable to assume you are interested in this field, either as a profession or as a hobby.

Just what is included in the field of electronics? As you might have guessed, it is a pretty large area, with many subcategories. But all of these various subcategories have much in common. In the simplest terms, electronics is the study, design, and use of electrical circuits using a variety of components to manipulate electrical signals in some way.

Electronic components

While there are countless variations, there are just a few fundamental components. Most sophisticated, specialized devices are modified versions of some common component type. The basic component categories include:

- Resistors
- Capacitors
- Coils
- Semiconductor junctions (diodes)

- Crystals
- Switches
- Digital gates (actually these are made up of simpler components from the above fist).

Each of these component types and their more common variations will be discussed in this book.

Basic circuits

There might seem to be an infinite variety of electronics circuits. In a way, there are. But all complex circuits are really made up of smaller subcircuits. Once again, there are just a few basic circuit types (with countless variations and modifications) that can be combined in a wide variety of ways. Basic circuit types include:

- Amplifiers
- Power supplies
- Switching circuits
- Filters
- Timers
- Digital gates (Gates can be considered simple subcircuits, or components, depending on the context. This concept is explained in the discussion on digital circuitry).

Don't panic

While electronics can seem like a terribly complex field, the most complex devices, circuits, and concepts can usually be broken down into simpler subunits. If a particular chapter seems difficult to understand, just take it slow. Sometimes, coming back to a tough chapter after reading a later chapter will make some concept a little clearer.

To work with electronics theory, there is no getting around the need for mathematics; math is intimidating for many people. Don't panic, just take it one step at a time. I've made every attempt to make the math and theory in this book as painless as possible. Sometimes things are hard because we expect them to be. In high school I assumed that calculus was beyond me, and didn't take that class. Meanwhile, I was getting involved with electronics as a hobbyist. Some years later, when I started writing in the field, I realized with some surprise that I had been using calculus for some time. Without that intimidating name, it wasn't really so bad after all.

Some of the details of the electronics theory aren't too critical for practical work. If some details here and there seem fuzzy to you, don't worry about it, just try to get a general feel for what is going on. You could discover those troublesome concepts will make a lot more sense to you later, once you are more comfortable with the field.

The most important thing to remember is that you do not have to be a genius or a scientific whiz to understand electronics theory. I believe anyone can learn the basics if they just try and don't get scared off by the seeming complexity, and the unfamiliar terms and concepts.

Interestingly, many of the people who start out most intimidated by the field end up being the most enthusiastic.

Don't expect to learn everything all at once. You will probably have to go through this book two or three times before you have a firm grasp on electronics theory.

The importance of doing the experiments described in the text can't be overemphasized. Hands-on experience is the best way of learning. It will make a lot more sense and stick with you a lot better, if you actually see the principles in operation than to just read about them.

2
Electrons and electricity

To understand electricity and electronics, you need at least a basic grasp of the theoretical structure of atoms. This chapter describes atoms and explains their role in electronics and electricity.

Atoms and their structure

All substances are made up of tiny particles called *atoms*. There are approximately 100 different kinds of atoms (92 occur in nature; others are *synthetic* or created by human beings). A substance that is made up entirely of just one type of atom is called an *element*. Copper, hydrogen, carbon, gold, and oxygen are a few familiar elements.

Two or more elements can be chemically combined into a more complex *compound* substance. For example, water is made up of hydrogen and oxygen atoms. The smallest unit of matter that is recognizable as a compound, rather than as its component elements, is a *molecule*. If you broke a single molecule of water into smaller particles, you'd have two hydrogen atoms and one oxygen atom. Billions of different substances can be formed by various combinations of the basic elements, just as the 26 letters of the alphabet can be arranged into millions of different words.

Although an atom is the smallest particle recognizable as a specific element, atoms themselves are made up of smaller particles. Atoms are made up of particles called *protons*, *neutrons*, and *electrons*. Recent discoveries have indicated the presence of a large number of additional subatomic particles, but you can probably ignore them. The three kinds of subatomic particles are all roughly of equal size, but protons and neutrons are considerably more massive than electrons.

If 250 million hydrogen atoms were laid end to end, they'd span only about one inch. It would take 100,000 electrons (or protons or neutrons) laid side by side to span the width of a single hydrogen atom. Atoms don't contain anywhere nearly that many particles. The hydrogen atom typically consists of only a single proton and a

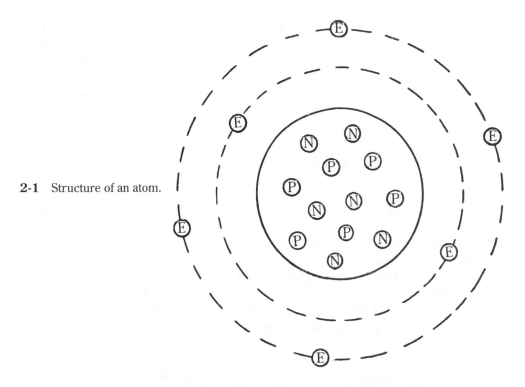

2-1   Structure of an atom.

single electron. Most of the space of an atom is empty. The protons and neutrons are clumped together in the center, forming a structure called the *nucleus*. The electrons revolve around the nucleus. The basic structure of a typical atom (carbon) is usually drawn as shown in Fig. 2-1.

You'll notice that this arrangement is roughly similar to the solar system. The electrons revolve around the nucleus like the various planets revolve around the sun. It is important to realize that a lead proton is exactly the same as, say, a gold proton. What differentiates the elements is simply the number of these particles contained within the atom.

Electrical charge

When an electron is isolated from an atom, it exhibits a tiny *electric charge*. The basic unit for measuring electric charge is the *coulomb*. The combined charge of 6,250,000,000,000,000,000 (6.25×10^{18}) electrons equals a charge of one coulomb.

There are actually two types of electrical charges. The type exhibited by an electron is arbitrarily called a *negative charge*. A proton has the same amount of electrical charge as an electron, but it is the opposite type. It is referred to as a *positive charge*.

Two similarly charged particles (that is, two electrons or two protons) tend to repel each other. Two oppositely charged particles (an electron and a proton) tend to attract each other. This attraction is one of the factors that keeps the electrons in orbit around the oppositely charged protons in the nucleus.

Ordinarily, an atom has an equal number of electrons and protons; therefore, the atom as a whole has no electrical charge. That is, it is electrically *neutral*. But in most cases, an extra electron can be added to an atom, giving the atom (as a unit) a negative charge. Conversely, an electron can be deleted, leaving the atom as a unit with a positive electrical charge.

Neutrons, which are contained within the nucleus along with the protons, have no electrical charge. As their name implies, they are neutral.

Isotopes

The number of particles in an atom determines what kind of atom it is. For instance, an ordinary hydrogen atom (hydrogen is the simplest element) consists of a single proton, a single electron, and no neutrons. Sometimes, however, a hydrogen nucleus does contain one or even two neutrons (heavy hydrogen). In this case, it is still a hydrogen atom, but it has a few different properties (the specifics of the differences are not relevant to your purposes here). This kind of atomic variation is called an *isotope*.

Just because an atom contains neutrons doesn't necessarily mean it is an isotope, of course. Many elements contain a number of neutrons in their basic form. For example, the nucleus of an ordinary lead atom contains 82 protons and 125 neutrons. Its isotopes contain still more neutrons.

Atomic number and atomic weight

Elements are often identified by their *atomic number*, which is simply the number of protons they contain. Each element has a unique atomic number. For example, hydrogen has an atomic number of one. The atomic number for helium is 2, carbon is 6 and lead is 82. *Atomic weight*, on the other hand, is the total number of both the protons and the neutrons.

The atomic number and atomic weight for ordinary hydrogen are identical—1 (this element is the only element where they are equal). However, a hydrogen isotope can have an atomic weight of 2 or 3, but the atomic number is always 1. The atomic number for ordinary lead is 82, but its atomic weight is 207. It is the atomic number that determines what kind of element the atom is.

Electrons and most other subatomic particles can easily be ignored in determining atomic weight, because they have little mass in comparison with protons and neutrons. They do not significantly affect the weight of the atom. They do have some mass, however, because they are matter, and all matter has mass. Remember, an electron is a physical unit, not a unit of energy. Electrons possess potential energy in the form of their negative charge, but they are physical objects. Often people get confused on this point.

Electron rings

Look again at the diagram of the atom in Fig. 2-1. Notice that the electrons circle the atom in a number of fixed, concentric rings. These rings have a definite pattern in the maximum number of electrons each ring can contain.

The first ring, the one closest to the nucleus, can only hold one or two electrons. If the atom has three electrons, two are in this innermost ring; the third is in the second ring, farther out. This second ring can hold up to eight electrons. The third ring can hold 18, the fourth 32, the fifth 50, and the sixth can hold 72 electrons. No known atom has more than six rings. Usually the inner rings are completely filled before the outer rings are started, but there are exceptions.

Obviously, the easiest electrons to remove, giving the atom a positive charge, are those in the outermost ring. They are the easiest to strike with an external force, and they are the farthest from the attraction of the positively charged protons that try to hold the electrons in place.

Conductors and insulators

Certain substances give up an electron (or accept an extra electron) more readily than others. Such substances (typically metals) are called *conductors*, because they can conduct electricity. That is, they allow an electric current to pass through them. This concept is explained in greater depth in the next section of this chapter.

A substance that has strong internal attraction and is thus resistant to releasing or accepting electrons is an *insulator*. Electric current can pass through an insulator, but it takes a far greater amount of force than for a conductor.

Any atom can be made to give up an electron. Conductors are simply those substances that give up electrons without a great deal of external force.

Electrical current

When an electron is knocked free from an atom, it drifts through space until it collides with a second atom that accepts it and throws off one of its own original electrons. This electron then strikes a third atom, and so forth. Each individual electron doesn't travel very far, but the energy of electron movement can be transmitted along the length of the conductor.

A simplified model of this kind of process is shown in Fig. 2-2. A cardboard tube is filled with three ping pong balls. When an extra ball, X, is pushed into the tube, it displaces ball 1, which displaces ball 2, which shoves ball 3 out of the far end of the tube. It all takes place almost instantaneously—when X is pushed in, 3 is shoved out. Each ball moved very little, and fairly slowly, but the energy was quickly transmitted through the tube.

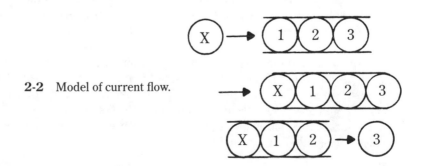

2-2   Model of current flow.

When this process takes place with electrons in a conductor, it is called *electricity*, or *electron current*.

Another way to put it would be to consider current the effective flow of electrons (usually simplified as *electron flow*). If a coulomb (6.25×10^{18} electrons) flows past a given point within one second, the current equals 1 *ampere (A)*.

The ampere is the basic unit of measurement for electric current, but in many practical cases, it is too large a unit to be convenient. In these instances, it is simpler to use the *milliampere* (mA), which equals one thousandth of an ampere, or *microampere* (μA), which equals one millionth of an ampere (one thousandth of a milliampere). For example, 50 mA equals 0.05 A, or 50,000 μA. Generally you can use whichever unit gives you the most manageable figures.

The word *ampere* is sometimes abbreviated as *amp*, or simply the letter *A* as indicated above. Similarly, milliampere can be written as *milliamp* or *mA*. Microampere is usually abbreviated as μA.

In electrical equations, current is usually represented by the letter *I*. The value of *I* is generally assumed to be in amperes unless otherwise stated.

Voltage

Because current specifies the number of electrons moving past a given point within a given time period, it can be considered the speed of the electron flow.

If you visualize electron flow as a toy car, you can draw a clear analogy to the basic elements of electricity. If you set the toy car on a perfectly flat table top, the speed (current) is zero—there is nothing to make it go. But if you tilt up one end of the table, the car starts to roll down the hill. The steeper the slope, the faster it goes.

What is making the car move is the difference between the highest point and the lowest point of the slope. In electricity the "highest point" is a point with a surplus of electrons (negative charge), and the "lowest point" is an electron shortage (positive charge). Because like charges repel and opposite charges attract, a stream of electrons flows from the most negative point to the most positive point.

Remember, each individual electron doesn't move very far, but the disturbance travels through the entire *circuit*. A circuit is a complete path for current flow. If the path is broken at any point, no current flows.

How strongly the current flows depends on the difference in charge between the most negative point and the most positive point of the circuit. This *difference of electrical potential* is also called *voltage* or *electromotive force*.

Voltage is measured in units called *volts* (V). One volt pushes 1 A of current through one *ohm* (Ω) of *resistance*. Resistance is discussed in the next chapter.

If the volt is too large a unit, you can measure voltage in *millivolts* (mV). One volt equals 1000 mV. If the volt is too small a unit, use the *kilovolt* (kV) as the unit of measurement. One kV equals 1000 V. For example, 25 V equals 25,000 mV, or 0.025 kV. Similarly, 2300 V equals 2.3 kV or 2,300,000 mV.

In electrical equations, voltage is usually represented by the letter *E*. *E* is given in volts unless otherwise stated.

Power

If you want to determine how much work the circuit is doing, you need to consider both the voltage and the current. The total energy consumed is called *power* and is measured in *watts* (W). One watt of power is consumed when 1 V pushes 1 A through a circuit.

The relationship between power, voltage, and current is stated in the following formula:

$$P = EI$$ **Equation 2-1**

P is power, *E* is voltage, and *I* is current. So power in watts equals voltage in volts times current in amperes.

This formula can be rearranged if you know the wattage (power) and the voltage and need to find the current. In this case:

$$I = P/E$$ **Equation 2-2**

The meaning of the variables is the same as in Equation 2-1.

The final possibility is if you know the wattage and the current, but not the voltage. Here you can write the equation as:

$$E = P/I$$ **Equation 2-3**

Again, the variables are the same as above. *P* is usually given in units of watts. You can use kilowatts (kW) or milliwatts (mW) if they are more convenient. The conversion is the same as with volts to kilovolts or to millivolts.

Batteries

Now, how can you generate this electrical force to push a current through a circuit? You can convert another form of energy (such as mechanical, chemical, heat, or even light) into electrical energy. One of the most common and simplest of these methods is chemically. Other methods of producing electricity are discussed in other chapters.

Wet cells

If a little hydrochloric acid is poured into a jar of water, the compound starts to break down chemically, producing negative and positive *ions*. An ion is simply an electrically charged particle. This process is called *ionization*, and the acid-water mixture is called an *electrolyte*.

Now, if you put a strip of zinc and a strip of copper into the acid/water solution, and connect the two metal strips with some wire, an electrical current starts to flow between them (Fig. 2-3).

When the zinc strip is placed into such an acid/water solution, it starts to dissolve, emitting positive ions. But every positive ion that leaves the zinc strip leaves behind two electrons, so the strip itself soon has a surplus of electrons. That is, it has a negative charge.

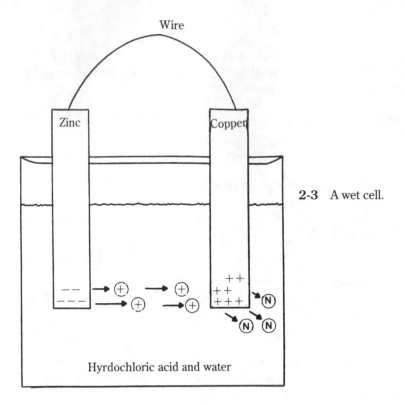

2-3   A wet cell.

Meanwhile, as the positive ions float through the electrolyte, some of them collide with the copper strip. Copper gives up electrons quite easily (it's an excellent conductor), so the positive ions take electrons away from the copper strip and are thus neutralized. Because the copper strip now has a deficiency of electrons, it takes on a positive electrical charge.

The difference between the negative charge on the zinc strip and the positive charge on the copper strip creates an electrical potential, or voltage, between the two strips, or *electrodes*. If the electrodes are connected with a conductive material (such as a piece of wire), an electric current will flow between them from negative to positive. This device is known as an *electric cell*. Because the electric cell I have been describing has an electrolyte in liquid form, it is called a *wet cell*.

This type of cell generates about 1 V. You'll notice that no mention is made about the size of the cell. The size of the cell has no effect on the voltage it produces. A gallon of electrolyte produces no more voltage than a pint. However, a larger cell allows more current to be drawn from it before it is used up. That is, it lasts longer, and can handle a heavier work load.

Different materials can be used for the electrolyte and electrodes. The voltage produced depends on the specific materials used. Wet cells have a number of obvious disadvantages. They have to be rather large to produce a useful amount of current. And, because the electrolyte is a liquid, it can easily be spilled. Also, bubbles can form a sheath around the positive electrode, preventing any further ions from striking it and being neutralized. The sheath stops some of the chemical action, and the voltage quickly drops off to zero. This process is called *polarization*. There are methods of preventing this effect (or, at least, limiting it) but, naturally, they increase the cost and complexity of the cell.

Dry cells

The problems involved with wet cells can be largely avoided by using an electrolyte that is in a pastelike form. Such a cell is called a *dry cell*. A typical dry cell uses a carbon rod for the positive electrode and a pasty electrolyte consisting primarily of ammonia and chlorine. The cell is contained in a zinc can that also serves as the negative electrode. Such a cell is called a *carbon-zinc cell*.

The voltage put out by such a cell is about 1.45 to 1.55 V when fresh—the voltage drops as the cell ages. It is said to have a *nominal voltage* of 1.5 V. The nominal voltage is the value usually used in circuit calculations for such cells.

Once again, the size of the cell affects the amount of current the cell can handle, but the voltage remains constant. Table 2-1 shows typical current-handling capabilities of the four most common cell sizes. Notice that current-handling capability increases with the size of the cell.

Table 2-1. Current-handling capabilities of standard dry-cell sizes—carbon-zinc.	*AAA*	20 mA	1.5 V
	AA	25 mA	1.5 V
	C	80 mA	1.5 V
	D	150 mA	1.5 V

The amount of current that is actually drawn from the cell is determined by the circuit that the battery is connected to (this concept is explained in later chapters). These figures are maximum ratings. Exceeding the current ratings given in Table 2-1 could result in damage or premature failure of the cell.

The primary disadvantage of carbon-zinc cells is their relatively low current-handling capabilities. If they are used to operate even a moderately heavy circuit, they have a short operating life. In addition, these cells have a limited *shelf life*. That is, they can go bad even if they aren't used.

Current is drawn from a battery only when an external circuit, or *load*, is connected between its electrodes, or *terminals*. If the terminals are not connected, there is virtually no path for the current to flow between them (air is a very poor conductor). However, a certain amount of internal current leakage within the cell itself is inevitable, and the leakage discharges (uses up) the cell at a slow, but definite rate, even if no external current is being drawn from it. Also, low-level chemical reactions are occurring constantly between the various substances within the cell, and the reactions can cause eventual deterioration.

Finally, although carbon-zinc cells are carefully sealed, no seal is 100% perfect. The moisture in the electrolyte can evaporate, causing the chemical activity of the cell to cease.

These processes are somewhat affected by temperature, so the way in which a cell is stored affects its shelf life. High temperatures tend to speed up chemical reactions, but lower temperatures slow them down. It's a good idea to store dry cells in a cool, dry place. However, do not subject them to freezing temperatures. If a carbon-zinc battery is frozen, it has an extremely short life when thawed.

The best temperature for storing these cells is about 40 to 50°F (or 4 to 10°C). At these temperatures, a carbon-zinc cell can have a shelf life of up to two or three years, as opposed to about six months when stored at room temperature. If these cells are kept refrigerated, it's usually a good idea to let them warm slowly to room temperature before using them. Some people believe that refrigerating an old dry cell rejuvenates it, but this is not true.

The way these cells generate a voltage is through the use of a chemical that gradually eats away the negative electrode. Clearly, if the electrode is completely destroyed, or even badly damaged, the cell is useless, because there is no way the electrode can be replaced.

However, if a small current is applied to a cell before it is completely dead with *reverse polarity* (that is, negative to positive and positive to negative), then some (not all) of the chemical action within the cell can be partially reversed. This procedure can extend the life of a dry cell somewhat. It won't be as good as new, because much of the chemical activity is irreversible, and the negative electrode is restored unevenly. Also, the process works only a limited number of times for any individual cell.

This process is called *charging*, or, more correctly, *recharging*, the cell. The recharging current must be kept very small, or the reverse polarity destroys the cell. Ordinarily, you must always avoid interconnecting two voltage sources with their polarities reversed.

The low recharging current limitation means the recharging process takes about 12 to 16 hours. It is also important to remove the recharging current when the cell is fully charged.

For the recharging process to be effective, the cell voltage must not be allowed to drop below 1 V (two-thirds of the nominal value), and it must be recharged immediately after it is taken out of service. Once the cell is recharged, it should be used as soon as possible because the shelf life of a recharged cell is quite short.

A number of commercially manufactured recharger units are available today. They can represent a substantial savings if you keep the inherent limitations in mind.

Alkaline cells

By using an alkaline material instead of an acid as the electrolyte, a greater current-handling capability can be achieved. Table 2-2 shows the current-handling capability for the four most popular cell sizes of *alkaline cells*.

AAA	200 mA	1.5 V
AA	300 mA	1.5 V
C	500 mA	1.5 V
D	600 mA	1.5 V

Table 2-2. Current-handling capabilities of standard dry-cell sizes—alkaline.

Alkaline cells come in exactly the same sizes as regular carbon-zinc cells and also have a nominal voltage (when fresh) of 1.5 V. Alkaline cells are directly interchangeable with their carbon-zinc counterparts.

Alkaline cells are usually more expensive than carbon-zinc cells, but their greater current-handling capability means that they last longer in high current-drain applications. A single alkaline cell might last as long as five or six carbon-zinc cells, so they often end up being less expensive to use. However, in very low current-drain applications, alkaline cells might offer no real advantage because of their relatively short shelf life. One major disadvantage of alkaline cells is that they cannot be recharged. Applying a reverse polarity current to an alkaline cell can damage the cell or even cause it to explode.

Nickel-cadmium cells

Another type of electric dry cell that is becoming increasingly popular is the *nickel-cadmium cell* (often shortened to *Ni-Cad*). These cells are specifically designed to handle moderately large current drains and to be fully rechargeable. A typical Ni-Cad cell can usually be recharged 500 to 1000 times before it fails.

Usually, when a Ni-Cad cell fails, the reason is an internal short between the electrodes. Often the cause of this short is allowing the cell voltage to drop down too low before recharging. A Ni-Cad battery voltage should never be allowed to drop below 1.05 V. When a short does develop in a Ni-Cad cell, it can sometimes be "blown away" by briefly applying a moderately large reverse-polarity current. However, this method doesn't always work.

The recharging requirements for Ni-Cad cells made by different manufacturers vary widely. Some are designed to be recharged in just three or four hours. Most require about 14 hours for a full charge. Most Ni-Cad cells are designed so that no damage is done if the charging current is continually applied even after the cell is fully charged. These cells, unlike carbon-zinc cells, can be left in the recharging unit for extended periods of time. Another advantage of Ni-Cad cells is that they are not as temperature-sensitive as other types of dry cells.

Nickel-cadmium cells cost four to eight times as much as most carbon-zinc cells, but because they can be reliably recharged and reused so many times, they are considerably less expensive over time.

One disadvantage of this type cell is that it puts out only 1.25 V when fully charged, as opposed to the 1.5 V produced by a fresh carbon-zinc or alkaline cell. In many applications this difference won't matter, but for many applications, that missing quarter volt can make a significant difference. Fortunately, this deficiency can be overcome by using multiple cells. This problem and its solution are the subject of the next section.

Series and parallel batteries

The cells discussed on the previous few pages are often incorrectly called *batteries*. A true battery consists of two or more interconnected cells. These cells can be wired together either in *series* or in *parallel*.

If you need more voltage than is generated by a single dry cell, you can connect two or more cells in series, that is, one after another. A series is illustrated in Fig. 2-4. The negative electrode, or *terminal*, of cell A is connected to the positive terminal of cell B, and the negative terminal of cell B is connected to the positive terminal of battery C. The load circuit is connected between the positive terminal of cell A and the negative terminal of cell C. The current must therefore pass through each of this series of cells sequentially. The voltages of the cells add. In this example, three 1.5 V cells are connected in a series battery, so the total voltage is 3 × 1.5 V, or 4.5 V.

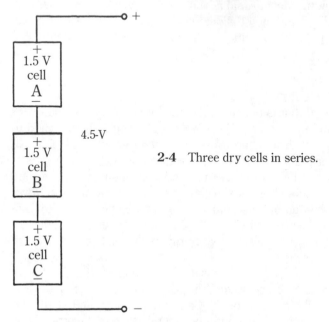

2-4   Three dry cells in series.

Similarly, two 1.5 V cells produce 3 V, eight cells generate a total of 12 V, and so forth. The current-handling capability of the entire battery is the same as that for a single cell. Here is where the quarter volt difference of nickel-cadmium cells can become significant. Eight regular carbon-zinc cells add up to 12 V, but eight Ni-Cad cells generate only 10 V. It would take 10 Ni-Cad cells to produce a voltage of 12.5 V, which would usually be close enough.

In most applications, more nickel-cadmium cells are needed than carbon-zinc or alkaline cells. Some pieces of equipment allow no space for the extra cells, which means nickel-cadmium cells can't be used. This situation is true, for example, with most portable cassette tape recorders.

A lot of modem electronic equipment makes allowances for this difference, however. The battery compartment is made large enough to hold the required number of nickel-cadmium cells, and a spacer is provided to fill the extra space if higher-voltage carbon-zinc or alkaline batteries are used.

Now, look at another possible situation. Suppose your dry cell has enough voltage, but won't handle the required current. Remember that the amount of current drawn from a cell (or battery) is determined by the demands of the load circuit it is powering, but there are definite limits as to how much current a cell can supply. Too high a current drain can prematurely age or destroy a cell. Therefore, it is sometimes necessary to form a battery that increases the current-handling capability rather than the voltage.

Such a battery, shown in Fig. 2-5, is called a *parallel battery*. The positive terminals are all connected together, as are all the negative terminals. The cells can thus work together, each cell providing only part of the current drawn by the load circuit. In a three-cell battery, such as the one shown, each cell provides only one-third of the total current.

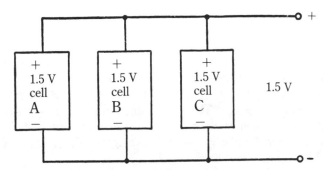

2-5   Three dry cells connected in parallel.

The total voltage of a parallel battery is the same as for a single cell, but the current-handling capability is multiplied by the number of cells in the battery. For example, three size-C batteries in parallel can handle up to 240 mA if they are carbon-zinc, or 1500 mA (1.5 A) if you use alkaline cells. Do not mix cell types.

In actual practice the parallel battery is rarely used. Most battery-operated equipment is designed for low current drain. If a high current drain is necessary, some other type of power source is used. Other power sources are discussed in other chapters. If both higher voltage and higher current are needed, both series and parallel techniques can be used simultaneously.

Schematic symbols

In electronic diagrams, it is generally unnecessary and inefficient to actually draw each of the components, especially in complex circuits. Instead, a form of visual shorthand is used. Drawings using this system are called *schematic diagrams*.

Each kind of electronic component is represented by a specific symbol. In some cases different symbols are used by different technicians, but usually the differences are minor. In this book, as each type of component is discussed, its schematic symbol (or symbols) are also introduced. Schematic diagrams are discussed in detail in chapter 14.

The standard schematic symbol for an electrical cell is shown in Fig. 2-6A, and the symbol for a battery is shown in Fig. 2-6B. Notice that the symbol for a battery is just two (sometimes three) cells together. Usually only two or three cells are shown in a battery on the schematic, regardless of how many cells are actually employed in the circuit. This convention is simply one of convenience. It would be quite awkward to show each individual cell in a 22.5 V or 45 V battery.

2-6   Schematic symbols. An electric cell (A); a battery (B).

A   **B**

Also note that the symbol in Fig. 2-6B is generally used only for series batteries. If parallel batteries are used, the individual cells are indicated.

In some schematics the "+" (plus) sign is not shown at the positive terminal, because it isn't really necessary. The longer line always represents the positive terminal, whether it is marked as such or not. The + sign simply makes the polarity easier to see at a glance. Similarly a "–" (minus) sign might or might not be included at the negative terminal (short line).

Interconnecting wires are generally shown in schematic diagrams as solid lines. These lines never curve, but always bend at sharp angles (usually 90-degree angles). You sometimes can be confused if two lines on a schematic diagram cross each other. Sometimes it means they are electrically connected, other times, the wires simply pass by each other with no electrical contact. Figure 2-7 shows three common systems for showing crossed wires in schematic diagrams. Notice that the no connection symbol in Fig. 2-7A is the same as the connection symbol in Fig. 2-7B. To avoid ambiguity, the symbols in Fig. 2-7C are preferred. Always make sure you know which convention is being used whenever you look at a schematic diagram.

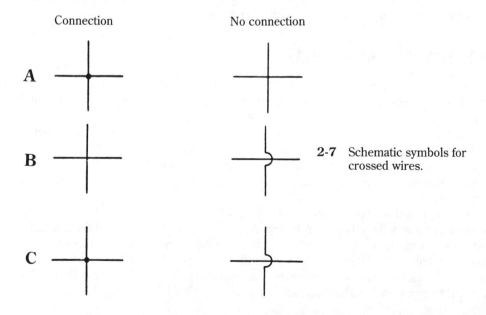

Connection   No connection

A

B

2-7   Schematic symbols for crossed wires.

C

Self-test

1. Which of the following is not a basic part of an atom?

A Coulomb
B Electron
C Proton
D Neutron

2. What flows through a conductor to create an electrical current?

A Protons
B Isotopes
C Molecules
D Electrons

3. What is the basic unit for measuring current flow?

A Volt
B Atomic weight
C Ampere
D Coulomb

4. What is the basic unit for measuring electrical power?

A Watt
B Volt
C Ampere
D Electrolyte

5. An ordinary flashlight battery is which of the following?

A A load
B A dry cell
C A wet cell
D A storage cell

6. Which of the following pairs of metals would make good electrodes in a wet-cell battery?

A Zinc and copper
B Zinc and iron
C Iron and copper
D Two strips of zinc

7. What is the effect of connecting battery cells in series?

A Current increases
B Current decreases
C Voltage increases
D Voltage decreases

8. What is the effect of connecting battery cells in parallel?

A Current increases
B Current decreases
C Voltage increases
D Voltage decreases

9. If the voltage applied to a circuit is 15 V, and the current flow is 2 A, what is the wattage (power) consumed by the circuit?

A 17 W
B 30 W
C 7.5 W
D 0.133 W

10. If 12 V are applied to a circuit that consumes 78 W, what is the current flow through the circuit?

A 936 A
B 0.15 A
C 9.36 A
D 6.5 A

3
Resistance and Ohm's law

What determines how much current a load circuit will draw from a battery, or other voltage source? Return to the analogy of the toy car from chapter 2. It might seem that if you knew the slope of the table, you'd immediately know how fast the toy car would roll down the incline, but other factors also affect the speed. The most important of these factors is friction. The car will roll down a smoothly polished board much faster than it will roll down a coarse, unsanded one. Friction will slow the car—that is, it will impede or resist movement of the car.

The electrical equivalent to friction is *resistance* (represented by the letter R). Resistance impedes, or works against the flow of current. You might think that resistance is something that should always be avoided as much as possible, but it's actually quite a useful factor in practical circuits.

Remember that the higher the current drawn from a battery or cell, the faster the battery or cell will be discharged. Resistance limits the amount of current drawn. It can also, as you'll see, reduce the voltage in certain portions of a circuit.

Ohm's law

The basic unit of resistance is the *ohm*, which is sometimes written as Ω (the Greek letter omega). One volt can push 1 A of current through 1 Ω of resistance. The relationship between these three factors is perhaps the most important concept in electronics. This relationship is defined by a principle called *Ohm's law*. According to this law, voltage equals current times resistance, or:

$$E = IR$$
Equation 3-1

E is the voltage in volts, I is the current in amperes, and R is the resistance in ohms. E will also be in volts if the current is in milliamperes and the resistance in kilohms (kΩ) (see below).

With a little simple algebraic manipulation, you can rearrange the equation to solve for current if voltage and resistance are known:

$$I = E/R \qquad\qquad \textbf{Equation 3-2}$$

Or, solving for resistance with known voltage and current:

$$R = E/I \qquad\qquad \textbf{Equation 3-3}$$

You'll recall that in the last chapter, you learned that power in watts equals voltage times current ($P = EI$). You can combine this equation with Ohm's law to find power consumed when only the resistance and the current are known:

$$P = EI$$

and

$$E = IR$$

$$\text{so } P = (IR)I$$

$$\text{or } P = I^2R \qquad\qquad \textbf{Equation 3-4}$$

Similarly, if you know the resistance and the voltage:

$$P = EI$$

and

$$I = E/R$$

$$\text{so } P = E(E/R)$$

$$\text{or } P = E^2R \qquad\qquad \textbf{Equation 3-5}$$

These equations are quite versatile, and they are absolutely essential to your understanding of electronics.

In most practical circuits the ohm is too small a unit, so the *kilohm* (kΩ—1000 ohms) and the *megohm* (MΩ—1,000,000) are often used.

Resistors

A *resistor* is an electronic component that is designed to introduce a specific amount of resistance into a circuit. Resistors are probably the most commonly used class of components in electronics. A typical resistor is shown in Fig. 3-1. Different size resistors can handle different amounts of power, or wattage. The action of a resistor causes it to heat up. That is, it converts electrical energy into *thermal energy* (heat). If a resistor gets too hot, it can change its resistance value or it can become damaged. To prevent this problem, always use a resistor that will *dissipate* (or handle) the required amount of power. If in doubt, use a larger resistor. A 2 W resistor, for example, will work fine in a ½ W circuit.

Unfortunately, large wattage resistors tend to be more expensive and take up quite a bit of space. So generally, use the smallest resistor that will comfortably handle the required wattage. For most electronic circuits, ½ W resistors are more than sufficient. In circuits built around *integrated circuits* (discussed in another), you would normally use ¼ W resistors.

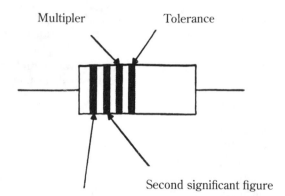

Multipler   Tolerance

Second significant figure

First significant figure

3-1 Color coding bands on a typical resistor.

Color coding resistors

Look again at Fig. 3-1. Notice that there are four bands around the body of each resistor. These bands are color coded, and are used to identify the value of the resistor. The standard resistor color code is given in Table 3-1. Most people who work with electronics use this table so frequently they memorize it. It might look a little complicated at first, but it's really quite simple, once you get used to it.

Table 3-1. Standard resistor color codes.

Color	Band 1	Band 2	Band 3	Band 4
Black	0	0	1	—
Brown	1	1	10	1 %
Red	2	2	100	2 %
Orange	3	3	1000	3 %
Yellow	4	4	10000	4 %
Green	5	5	100,000	—
Blue	6	6	1,000,000	—
Violet	7	7	10,000,000	—
Gray	8	8	100,000,000	—
White	9	9	—	—
Gold	—	—	0.1	5 %
Silver	—	—	0.01	10 %
No color				
	—	—	—	20 %

Suppose you have a resistor with the following markings. The band closest to the end of the resistor (band 1) is red, the next band (band 2) is violet, band 3 is orange and the last band (band 4) is silver. What is the value of the resistor? Because the first band is red, the chart tells you that the first *significant digit* is 2. Similarly, the second band, being violet, indicates the second significant digit is 7. So you now have a value of 27.

The next band (band 3) is the *multiplier*. It is orange on the sample resistor, so you multiply the significant figures by 1000. So the resistor has a total value of 27,000 Ω, or 27 kΩ.

The fourth band tells you the tolerance of the resistor. It is very difficult to manufacture a resistor to an exact value, and it usually isn't necessary to be that exact. The actual value of a resistor might be higher or lower than the *nominal* value indicated by the first three color bands. In the example, the tolerance band is silver, which means the actual value of the resistor is within plus or minus 10% of the marked value (that is—27,000 Ω). That is, this particular resistor might actually be anywhere between 24,300 Ω and 29,700 Ω. For most applications, this value is accurate enough.

In fact, sometimes only 20% accuracy is needed. In this case, there is no fourth band at all. Assuming the resistor had a nominal value of 27,000 Ω, it could actually be anywhere from 21,600 to 32,400 Ω.

In other applications, greater accuracy is sometimes required. If the fourth band of a resistor is gold, that means the device has a 5% tolerance. Again, if the nominal value is 27,000 Ω, the actual value could range from 25,650 to 28,350 Ω.

In some very specialized applications, great precision is sometimes needed. In these cases, 1% (often called *precision*) resistors are used. That is, a nominal value of 1% would require an actual value between 26,730 and 27,270 Ω. The nominal value of a precision resistor is usually printed right on the body of the resistor, and the color code is not used.

Most 1/2 W resistors have a tolerance of 10%. Recently, the trend has been swinging towards more and more 5% units. Not surprisingly, the better the accuracy, the more the resistor is going to cost to manufacture, so if great precision is not required, it generally makes sense to use a wide tolerance resistor. Remember that even a 20% tolerance resistor could be exactly its nominal value, but the manufacturer only guarantees that it will be somewhere between plus or minus 20% of the indicated value.

Here are some additional examples of using the color code:

If the resistor bands are brown, black, green, gold, the value of 1 (brown) 0 (black) × 100,000 (green), plus or minus 5% (gold), or a nominal value of 1,000,000 Ω (1 MΩ).

Another resistor might be marked blue, gray, red, with no fourth band. This translates to 6 (blue) 8 (gray) × 100 (red) or 6800 Ω (6.8 kΩ). The tolerance of this resistor is 20%.

As you grow accustomed to the color code, it will become second nature to you, and you'll be able to read resistor values directly without even thinking about it. Here's a couple more examples for you to work on your own:

If a resistor is marked yellow, violet, yellow, silver, and its actual value is 450,000 Ω, is it within tolerance?

If a resistor is marked red, red, red, gold, and its actual value is 2000 Ω, is it within tolerance?

Types of fixed resistors

If only a few ohms of resistance is needed in a circuit, the resistor can simply be a piece of *nichrome* (nickel-chromium) wire of suitable width and length. Nichrome wire has a much greater resistance than standard copper wire (all conductors have some resistance) so small *wirewound resistors* can be made without the length being unreasonable. This nichrome wire (sometimes called resistance wire) is usually wound around a ceramic core and covered with some insulating material.

Usually the resistances needed in practical electronic circuits are too large for reasonable wirewound resistors, so *composition resistors* are more commonly used. These resistors are usually made of a thin coating of carbon on a ceramic tube. Carbon is only a fairly poor conductor, so a fairly large resistance can be achieved in a relatively small space.

The upper limit for such a *carbon resistor* is generally around 10 MΩ. Of course, the resistor itself is covered with an insulating body. The color coding bands are painted on the outside of the insulation.

Another common type of resistor uses a thin metallic film instead of carbon. *Metal-film resistors* can usually be made to more precise values than carbon composition resistors. Metal-film resistors are also less sensitive to temperature fluctuations (carbon resistors can sometimes change value at temperature extremes) and produce less *internal noise* (random and undesirable voltages and power fluctuations).

All of these devices are called *fixed resistors* because, unless they are in some way damaged, their value is more or less constant (all resistors change value somewhat in response to temperature fluctuations). The schematic symbol for a fixed resistor is shown in Fig. 3-2.

3-2   Schematic symbol for a fixed resistor.

Variable resistors

It is often necessary to be able to alter the amount of resistance in a circuit. In these cases a *variable resistor* is used. Variable resistors are usually called *rheostats* or *potentiometers* (often shortened to *pots*). These terms are more or less interchangeable, but generally the word *rheostat* is used to identify a device that is suitable for heavy-duty ac (alternating-current) circuits (see chapter 5), and potentiometers are generally used in circuits having relatively low power.

Also, potentiometers usually have three terminals. The two outside terminals act like a simple fixed resistance—the resistance between these two terminals does not change. The center terminal, however, is attached to a *slider*, which is controlled by a knob. You can move the slider along the resistance element, which is either wound wire, or a strip of carbon. Depending on the position of the slider, the resistance between it and either of the outside terminals will vary. See Fig. 3-3. Notice that the total of resistance *AB* plus resistance *BC* always equals the constant resistance *AC*. As resistance *AB* increases, resistance *BC* decreases, and vice versa.

A variation on the standard potentiometer is the *slide pot*. It works in exactly

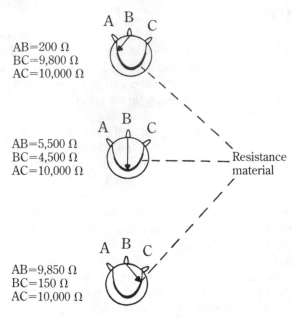

AB=200 Ω
BC=9,800 Ω
AC=10,000 Ω

AB=5,500 Ω
BC=4,500 Ω
AC=10,000 Ω

Resistance
material

AB=9,850 Ω
BC=150 Ω
AC=10,000 Ω

3-3   Resistance of a potentiometer varies as the slider
is moved through its range.

the same way as a regular potentiometer, except the slider moves in a straight line
rather than in circular motion. The only advantage of using a slide pot is that in
certain applications, it is easier to see where the slider is positioned.

The schematic symbol for a potentiometer is shown in Fig. 3-4. The symbol is
the same whether it is for a slide pot or standard round pot.

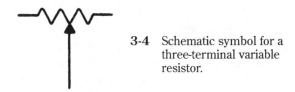

3-4   Schematic symbol for a
three-terminal variable
resistor.

A rheostat, on the other hand, is often a two-terminal device. That is, there is
one fixed terminal and the movable terminal (slider). The second fixed terminal is
simply left off. Figure 3-5 shows the most common schematic symbol for a two terminal variable resistor. Alternatively, the standard symbol for a potentiometer can be
used with one of the outer (fixed) terminals left disconnected.

3-5   Schematic symbol for a
two-terminal variable
resistor.

Potentiometers and rheostats are generally panel controls. To use a potentiometer or rheostat, you rotate a knob (which is connected to the shaft of the variable resistor) that in some way alters the operation of the circuit. For example, in an audio amplifier, potentiometers might be used to adjust the volume and the tone of the sound.

Sometimes, however, a variable resistor is needed to fine tune a circuit, perhaps to compensate for component tolerances that could throw the precise operation of the circuit off. In cases like this, you want a variable resistor you can set and forget. *Trimpots* or *trimmers* are used for this kind of function.

Trimpots are simply miniature potentiometers with sliders that are positioned with a screwdriver, rather than with an external knob. Sometimes, when the correct setting is found, a tiny drop of paint or glue is placed on the screwdriver slot to prevent the slider from being moved out of position accidentally. In other applications, the trimmer might need to be readjusted periodically (perhaps to compensate for changing values as components age), but not often enough to warrant a more expensive and space-consuming front-panel control. Also, there are often critical adjustments that should only be made with special test equipment. To prevent incorrect adjustment by casual users, the adjustments should be inaccessible.

The *taper* of a potentiometer is the way in which the resistance changes in relation to the position of its slider. The two most popular tapers are shown graphically in Fig. 3-6. The *linear taper* in Fig. 3-6A varies the resistance in direct proportion to the position of the slider. You can see that the graph is simply a straight line. The *logarithmic taper* (Fig. 3-6B), on the other hand, has a more complex relationship between its resistance and its slider position. This relationship is based on the mathematical function of logarithms.

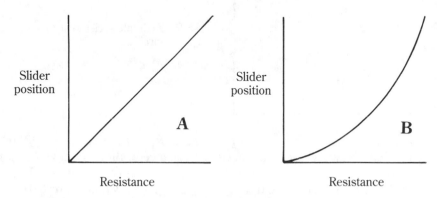

3-6   Graphs of typical potentiometer tapers. Linear taper (A); logarithmic taper (B).

Which type of taper should be used depends on the nature of the specific circuit. For example, volume controls usually have a logarithmic taper because the ear hears in a logarithmic manner. If a linear taper potentiometer were used, most of the apparent range of loudness will be crowded into a relatively small portion of the rotation path of the knob. With a logarithmic-taper potentiometer, the position of the knob will relate more closely to the perceived volume level.

Combinations of resistors

Practical electronic circuits consist of just a single resistance, so you need a way of determining the total value of multiple resistances in various combinations.

Series Resistances

Figure 3-7 shows a simple circuit with two resistors in series. Assume the battery generates three V. The value of R1 is 100 Ω, and R2 is 200 Ω. How do you find the current? Because the total current has to flow through both resistors, it must have the same value for each resistor. The current has to flow through 100 Ω; then it has to flow through an additional 200 Ω more. As you might have guessed, this resistance appears to the current as a single 300 Ω resistor. Resistances in series add. Stated algebraically, the formula for resistance in series is:

$$R_T = R_1 + R_2 \ldots + R_n \qquad \qquad \textbf{Equation 3-6}$$

where R_T is the total resistance. The letter n represents the total number of resistances in the circuit.

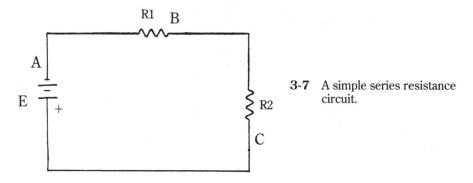

3-7   A simple series resistance circuit.

In the example, you have just two resistances. $R_T = R_1 + R_2 = 100 + 200 = 300 \Omega$. Now that you know the total effective resistance in the circuit, you can use Ohm's law to find the current. You know that $I = E/R$ (current equals voltage divided by resistance), so the current in this circuit equals 3 V/300 Ω, or 0.01 A (10 mA).

Now, what is the voltage dropped by each resistor? Because $E = IR$ (voltage equals current times resistance), the voltage through R1 must equal 0.01 A $\times$ 100 Ω, or 1 V. Similarly, the voltage through R2 equals 0.01 A $\times$ 200 Ω, or 2 V. Notice that adding the voltages dropped across each of the resistors will give you the original source voltage. You can say that all of the voltage is used up by the resistances, which is true of all circuits. As the current passes through each resistor, the resistance causes the voltage to drop. At point A you have the full source voltage, or 3 V. At point B, R1 has dropped 1 V, so there is 3 − 1, or 2 V. R2 drops 2 V, so at point C the voltage is 0. The source voltage is used up.

Finally, you can calculate the total power consumed by the circuit. You'll recall that the formula is $P = EI$ (power or wattage equals voltage times current). In the example, you have 3 V × 0.01 A, or 0.03 W (30 milliwatts—mW).

Try another example, and use a slightly different method of solving it. You'll still be using the circuit shown in Fig. 3-7, but this time the battery generates 12 V, R1 is 1000 Ω, and R2 is 150 Ω. The total resistance in the circuit is 1000 + 150, or 1150 Ω. Because you know that the total power consumed by a circuit can be calculated with the formula $P = E^2/R$, you can insert the known values. $P = 12^2/1150 = 144/1150$ = approximately 0.125 W (125 mW).

Solving for current you can rearrange $P = EI$ to $I = P/E$, or 0.125/12 equals just over 0.01 A (about 10 mA). The voltage drop across R1 is found by Ohm's law. $E = IR = 0.01$ A × 1000 Ω = 10 V. The voltage drop across R2 equals 0.01 A × 150 Ω, or about 1.5 V. You'll notice that the calculated voltage drop is 10 V + 1.5 V, or 11.5 V, rather than the source voltage of 12 V. What happened to that extra half volt? Actually, nothing. You lost it in the calculations because of rounding off. The wattage consumed is actually 0.1252174 W, but rounding this figure off to 0.125 W made the rest of the calculations simpler. There is nothing wrong with rounding off the results of these equations, and usually the calculated values will be close enough. But when a discrepancy does show up, you might find it necessary to go back and work with the exact values.

Here's one last example that you can try solving for yourself. $E = 30$ V, $R_1 = 250,000$ Ω, and $R_2 = 50,000$ Ω. You can use whichever method you prefer to find the current, the wattage, and the voltage drop across each resistor.

Parallel resistances

Now, what if you have a circuit like the one shown in Fig. 3-8? In this case, the electron flow drawn from the battery (the current) is split up between the two resistors. There are two *parallel* paths for the current to follow. Some of it will flow through R1, and some will flow through R2.

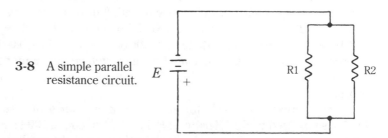

3-8 A simple parallel resistance circuit.

Naturally, more current will flow through the path with less resistance. If both resistors are of equal value, equal currents will flow through them. As far as the voltage is concerned, a parallel circuit looks like two separate circuits, as in Fig. 3-9. The full source voltage is dropped across each resistor.

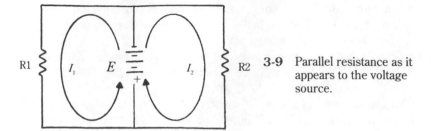

3-9 Parallel resistance as it appears to the voltage source.

Suppose the circuit is powered by a 6 V battery. R_1 is 1000 Ω and R_2 is 3000 Ω. You already know that the full source voltage (6 V) will be dropped across each resistor, so you can use Ohm's law and solve for the currents separately. For R1, $I = E/R = 6\ V/1000\ Ω = 0.006\ A$ (6 mA). The current through R2 is 6 V/3000 Ω, or 0.002 A (2 mA). R1 draws 6 mA from the battery, and R2 draws an additional 2 mA, so the total current drawn by the parallel circuit is 8 mA.

Solving for the equivalent resistance of the entire circuit, you can use the formula $R = E/I = 6\ V/0.008\ A = 750\ Ω$. Notice that this total equivalent resistance is less than the value of the smallest resistor. This relationship is always true in a parallel circuit. If the two resistors are equal, the equivalent resistance will be exactly one half their individual value.

Another formula for determining the equivalent resistance of n resistors in a parallel circuit is:

$$\frac{1}{R_1} + \frac{1}{R_2} \cdots + \frac{1}{R_n} = \frac{1}{R_T}$$

Equation 3-7

R_T is the total effective resistance of the parallel circuit. Using the example above, you find 1/1000 + 1/3000 = 0.001 + 0.00033333 = 0.00133333. Taking the reciprocal (1/0.00133333), you get 750 Ω. You get the same answer, no matter which method you use.

The power consumed by the circuit is solved in the usual way. That is, $P = EI$. In this circuit, 6 V × 0.008 A = 0.048 W. This value can be rounded off to about 50 mW.

As you can tell from Equation 3-7, any number of resistances can be combined in parallel. For example, imagine a circuit with four resistors in parallel. Their values are 1000 Ω, 2200 Ω, 6800 Ω and 10,000 Ω. So you have 1/1000 + 1/2200 + 1/6800 + 1/10,000 = 0.001 + 0.0004545 + 0.0001471 + 0.0001 = 0.0017016. Taking the reciprocal to find the total effective resistance, you get just under 590 Ω. Notice that this equivalent value is lower than the individual value of any of the separate resistors.

Here is another example for you to work on your own. Assume the circuit in Fig. 3-8 is powered by a 12 V battery. R_1 is 470 Ω, and R_2 is 100 Ω. Find the equivalent resistance, the current drawn by each resistor, and the total power consumed by the circuit.

Series-parallel combinations

In actual practice, you'll rarely come across a circuit with just series resistances or just parallel resistances. Usually you'll find a combination of the two forms. Take a look at Fig. 3-10. Here you have a circuit with four resistors both in series and in parallel. It might look complicated to solve for such a combination circuit, but it's easy enough if you go one step at a time.

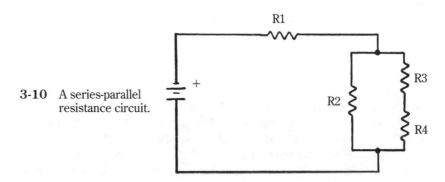

3-10 A series-parallel resistance circuit.

Assume the source voltage is 15 V. R_1 equals 1500 Ω, R_2 equals 2200 Ω, R_3 equals 470 Ω, and R_4 equals 1000 Ω. First you should find the equivalent resistance of the R_3-R_4 combination. Because these two resistances are in series, their values simply add. That is, $470 + 1000$ Ω $= 1,470$. For simplicity, you can consider this combination as a single resistor, R_A. See Fig. 3-11.

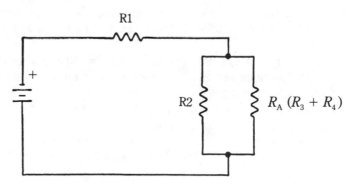

3-11 Simplification of Fig. 3-10.

Now solve for the R_2-R_A parallel combination. $1/2200 + 1/1470 = 0.0004545 + 0.0006803 = 0.0011348$. So the equivalent resistance is just over 880 Ω. Call this value R_B. See Fig. 3-12.

R1

R_B $(R_2 + R_A)$

3-12   Further simplification of Fig. 3-10.

Now, you just have R_1 in series with R_B. $R_1 + R_B = 1500 + 880 = 2380$ Ω, which is the total equivalent resistance for the entire circuit. Solving for the total circuit current, you find that because $I = E/R$, the current equals 15 V/2380 Ω, or about 0.0063 A (6.3 mA). The full current flows through R1, because there is no other path for it to take to bypass this resistor. You can solve for the voltage drop across R1. $E = IR = 0.0063$ A $\times$ 1500 $\Omega = 9.45$ V.

Looking at the RB combination as a single resistor, you find the total voltage drop is about 0.0063 $\times$ 880, or 5.55 V (9.45 V across R1, and 5.55 V across RB equals the source voltage—15 V). Because RB consists of R2 and RA in parallel, you know the voltage dropped across each of these resistances is equal. Specifically, 5.55 V. The current through R2 equals 5.55 V/2200 Ω, or approximately 0.0025 A (2.5 mA). The current through the R_A combination is 5.55 V/1470 Ω, or about 0.0038 A (3.8 mA). R_A is actually R3 and R4 in series. Because these two resistors are in series, they pass the same current—3.8 mA. The voltage drop through R3 is 0.0038 A $\times$ 470 Ω, or 1.77 V. Across R4 it is 0.0038 A $\times$ 1000 Ω, or 3.78 V. Notice that 3.78 V + 1.77 V equals 5.55 V.

Further, the current through R2 (2.5 mA) plus the current through RA (3.8 mA) equals 6.3 mA—the same value you got for the entire circuit. You can see how all these equations are interconnected.

Finally, the power through the entire circuit is 15 V times 0.0063 A, or 0.0945 W. You can round this value off to about 95 mW.

Now solve the various circuit values for the same circuit if the battery generates 12 V, R_1 is 4700 Ω, R_2 is 100,000 Ω, R_3 is 33,000 Ω, and R_4 is 27,000 Ω.

Self-test

1. What is the basic unit of resistance?

A Ohm
B Coulomb
C Mho
D Pot

2. If the bands on a resistor are yellow, violet, red, and gold, what is the resistance value?

A 470 Ω, 5%
B 470 Ω, 10%
C 4700 Ω, 5%
D 47,000 Ω, 5%

3. If the bands on a resistor are red, red, orange, and silver, what is the resistance value?

A 223 Ω, 10%
B 220 Ω 5%
C 2,200 Ω 20%
D 22,000 Ω, 10%

4. Which of the following is not a valid expression of Ohm's law?

A $R = E/I$
B $E = IR$
C $I = E/R$
D $R = E^2/I$
E None of the above

5. If a 3300 Ω resistor and a 22,000 Ω resistor are connected in series, what is the total resistance?

A 18,700 Ω
B 2870 Ω
C 25,300 Ω
D 5500 Ω
E None of the above

6. If three resistors, each with a value of 560 Ω, are connected in parallel, what is the total resistance of the combination?

A 187 Ω
B 1867 Ω
C 560 Ω
D 1680 Ω
E None of the above

7. Assume 15 V is applied to a simple series circuit consisting of a 2200 Ω resistor and a 4700 ohm resistor. What is the value of the current flowing through this circuit?

A 0.010 A
B 0.0022 A
C 22 A
D 0.460 A

8 A 33 kΩ resistor is connected in series with a parallel combination made up of a 56 kΩ resistor and a 6.8 kΩ resistor. What is the total combined resistance of these three resistors?

A 95,800 Ω
B 49,069 Ω
C 39,067 Ω
D 63,769 Ω
E None of the above

9. When resistors are connected in series, what happens?

A Nothing
B The effective resistance is decreased
C The effective resistance is increased
D The tolerance is decreased
E None of the above

10. When resistors are connected in parallel, what happens?

A Nothing
B The effective resistance is decreased
C The effective resistance is increased
D The tolerance is decreased
E None of the above

4

Kirchhoff's laws

Kirchhoff's laws are a handy set of tools for analyzing what's going on within an electrical circuit. You have a choice of whether to use Kirchhoff's voltage law or Kirchhoff's current law. They are just two different paths to the same ends. Ultimately, they give the same results.

Kirchhoff's laws are especially useful in analyzing circuits that cannot be broken down into simple series, parallel, or series-parallel combinations of resistances. Such a circuit is shown in Fig. 4-1.

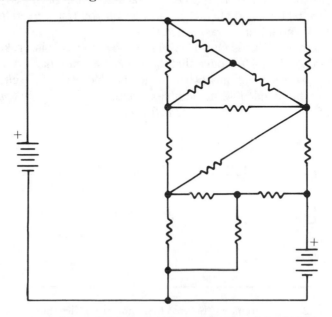

4-1   Some circuits cannot be broken down into simple series-parallel combinations.

For those of you who hate math, there is no way to get around using at least some math in analyzing electronic circuits. Fortunately, the math required for Kirchhoff's laws is not too advanced or complicated. The math used here is as painless as possible. Once you get used to working with these equations, you probably won't have any trouble with them at all, even though they do seem intimidating at first.

If you ever took a course in algebra, you're more than halfway home already. If you've never had any algebra (or if you've forgotten what you learned), just go through each of the steps described in the text slowly and carefully. The math should start making sense to you. Just don't let *mathaphobia* get the best of you. Don't panic. It's really not as hard as it looks.

Kirchhoff's voltage law

According to Kirchhoff's voltage law, "the algebraic sum of the voltage sources in any loop is equal to the algebraic sum of the voltage drops around the loop." If you don't understand this statement, don't worry about it. In somewhat simpler terms, it means—the total amount of voltage put into the loop must be exactly cancelled out to the voltage used up (dropped) within the loop. This idea will become clearer as you go on. To understand Kirchhoff's voltage law, you first need to know what is meant by a *loop*.

Loops

In the Kirchhoff system, a loop is any closed conducting path within the circuit. Loops can be made up of any combination of conductors, resistances, reactances, or voltage sources (but not current sources). (Reactances are discussed in other chapters, especially chapter 6 and chapter 8. For now all you really need to know is that a reactance is essentially an ac resistance.)

A fairly simple circuit is shown in Fig. 4-2. This circuit is broken up into its component loops in Fig. 4-3. Notice that the loops are redundant. Although the circuit is made up of three loops, any two of the loops include all of the circuit elements. The third loop is not needed. More complex circuits will break down into more than three loops, of course, but the principle is the same.

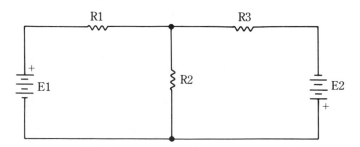

4-2   This circuit will be used to demonstrate Kirchhoff's laws.

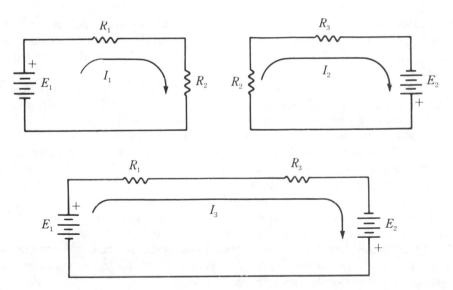

4-3   The circuit of Fig. 4-2 shown in component loops.

In working with Kirchhoff's voltage law, you will use the minimum number of loops that will include all of the circuit elements. To use any more loops would merely be redundant and make unnecessary extra work for you.

Loop Currents

The next expression you need to understand is *loop current*. The definition is pretty obvious. A loop current is simply the current assumed to flow through a given loop. The loop currents are indicated in Fig. 4-3.

It is important to realize that you realize that loop currents are a theoretical concept. They exist only mathematically. If you actually measured the current at a given point in the circuit, you might or might not get a value close to the assumed loop current. There is a perfectly good reason for this. The tested point can be (and probably is) part of more than one loop. There might well be multiple loop currents flowing through that circuit point. For example, in Fig. 4-3, loop currents I_1 and I_2 both flow through resistance element R_1.

For purposes of Kirchhoff circuit analysis, you simply artificially separate these simultaneous currents into independent loop currents. This concept is somewhat confusing, so it might be worthwhile to go back and reread the brief discussion on loop currents before going any further.

Sign conventions

Kirchhoff's voltage law deals with *algebraic sums*. Using algebraic sums, you will be dealing with positive (plus) and negative (minus) quantities. Various values will add or subtract, depending on what *algebraic sign* they have. Obviously you need a set of conventions or rules to determine whether a given value should be given a plus or minus sign. The procedure for determining the proper signs is: pick a direction, either clockwise or counterclockwise. It doesn't really matter which one you pick, as long as your choice remains consistent throughout the analysis. A negative sign simply indicates that the current is moving in the opposite direction from the one you selected.

All of the loop currents must be assumed to flow in the same direction (all clockwise, or all counterclockwise). It is important to remember that the use of a plus or minus sign here does not refer to ordinary electrical polarity. It is just a mathematical convention for your convenience in analyzing the circuit.

If a current flows through a resistor in the same direction as the loop current for that loop, the voltage drop across the resistance is positive. Of course, if the current flowing through the resistor were in the opposite direction, you will consider the voltage drop across the resistance to be negative.

There are two kinds of current flowing through the loop. There is the loop current for the loop presently under consideration, and there are loop currents from other loops in the circuit. As a rule of thumb, voltage drops caused by the present loop current will almost always be positive. Voltage drops due to currents from other loops can be either positive or negative.

Loops can contain voltage sources. If a loop current passes through a voltage source from the negative terminal to the positive terminal, that current has a positive value. If it flows in the opposite direction (from the voltage source positive terminal to the negative terminal), that current is negative. By following these rules, you can determine the proper sign for any current or voltage drop in a loop.

Kirchhoff's voltage law in action

You might be scratching your head at this point. You might be starting to believe this is all beyond you, but don't worry. Kirchhoff's voltage law is a little difficult to explain abstractly, but once you go step by step through a practical example, you should have a better grasp of the concepts described.

You will analyze the simple circuit shown in Fig. 4-2. Of course, you can't do that without knowing the voltages and resistance values in the circuit. You will assume the following values:

E_1 12 V
E_2 6 V
R_1 100 Ω
R_2 47 Ω
R_3 220 Ω

In analyzing the circuit, you can use either two of the loops shown in Fig. 4-3. Use loop A and loop B. These loops are shown in Fig. 4-4. The loop currents are assumed to flow in a clockwise direction. Loop A consists of E1, R1, and R2. The loop current for loop A is labelled I1. Loop B consists of E2, R2, and R3.

The loop current for loop B is labelled I2. Notice that resistance R2 is part of both loops. Therefore, both the I_1 and I_2 currents flow through this component.

Kirchhoff's voltage law states that the algebraic sum of all voltage sources in the loop equals the algebraic sum of all voltage drops in the circuit. In other words,

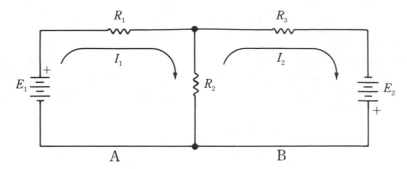

4-4   Use loop A and loop B from the circuit of Fig. 4-2.

the total amount of voltage put into the loop must be exactly cancelled out to the voltage used up (dropped) within the loop.

The only voltage source in loop A is E1. There is just one voltage drop across resistor R1, because only I1 flows through it. There are two voltage drops across resistor R2, however. One is due to loop current I_1, and the other is due to the external current (from loop B) I_2. These three voltage drops must equal the voltage supplied by E1. In algebraic terms:

$$E_1 = (I_1 \times R_1) + (I_1 \times R_2) - (I_2 \times R_2)$$

Each voltage drop is calculated simply by Ohm's law:

$$E = I \times R$$

The $(I_2 \times R_2)$ voltage drop is negative because current I_2 is flowing through R_2 in the opposite direction as the loop current I_1. The same approach is used to find the algebraic formula for loop B. Its only voltage source is E_2. Only loop current flows through resistor, so it contributes a single voltage drop. Both I_1 and I_2 flow through R_2, so it has two voltage drops. I_1 (from loop A) flows in the opposite direction of I_2, so it causes a negative voltage drop. In algebraic terms, you have:

$$E_2 = (I_2 \times R_3) + (I_2 \times R_2) - (I_1 \times R_2)$$

Plugging in the component values for the circuit as listed above, the equation for loop A becomes:

$$E_1 = (I_1 \times R_1) + (I_1 \times R_2) - (I_2 \times R_2)$$
$$12 = 100(I_1) + 47(I_1) - 47(I_2)$$

Combining the like terms (I_1), this equation can be simplified to:

$$12 = 147(I_1) - 47(I_2)$$

Similarly, plugging the component values into the equation for loop B, you get:

$$E_2 = (I_2 \times R_2) + (I_2 \times R_3) - (I_1 \times R_2)$$
$$6 = 47(I_2) + 220(I_2) - 47(I_1)$$

Once again you combine the like terms (I_2 in this case) and simplify the equation to:

$$6 = 267(I_2) - 47(I_1)$$

Many people hate math, but there's no getting around the need to use simple algebra to solve Kirchhoff's laws. Try to follow each step of the process. You need to solve for the currents (I_1 and I_2). Because you have two variables, you will have to define one in terms of the other. You can rearrange the equation for loop A to solve for current I_1:

$$12 = 147(I_1) - 47(I_2)$$
$$147(I_1) - 47(I_2) = I_2$$

(First you just reverse the order of the two halves of the equation. The reversal is just for convenience in writing the steps in the changing equation down. No values are changed here.)

$$147(I_1) - 47(I_2) + 47(I_2) = 12 + 47(I_2)$$

(Any changes you make on one side of the equation you must also duplicate on the other side of the equation. Adding $47(I_2)$ to the left side of the equation cancels this factor out, but you must add it to the right side to keep the equation consistent.)

$$147(I_1) = 12 + 47(I_2)$$

(The $47(I_2) - 47(I_2)$ in the last step cancels out, leaving a value of zero, which can be dropped from the equation.)

$$147(I_1)/147 = (12 + 47(I_2))/147$$

(You divide both halves of the equation by 147 to remove this element from the left side of the equation. This step moves the 147 over to the right half of the equation.)

$$I_1 = (12 + 47(I_2))/147$$

(The two 147s in the left hand of the equation cancel each other out and are dropped. Notice that you haven't really changed the equation. The values remain the same, you've just moved the various elements around at your convenience.)

You now can define current I_1 in terms of I_2. Moving on to the equation for loop B, you can substitute the right half of the last equation anywhere the term I_1 appears in the second equation. That is:

$$6 = 267(I_2) - 47(I_1)$$
$$6 = 267(I_2) - 47((12 + 47(I_2))/147)$$

You now have just one variable to solve for in this equation (I_2). You can find the value of I_2 through simple rearrangement of the terms:

$$6 = 267(I_2) - ((47 \times 12) + (47 \times 47)(I_2))/147$$
$$6 = 267(I_2) - ((47 \times 12)/147) + ((47 \times 47(I_2))/147)$$
$$6 = 267(I_2) - (564/147) - (2209(I_2)/147)$$
$$6 = 267(I_2) - 3.8 - 15(I_2)$$

Note—values have been rounded off in the division. This rounding will cause a slight error in the final results, but the errors due to rounding off will be negligible.

$$6 = 267(I_2) - 15(I_2) - 3.8$$
$$6 = 252(I_2) - 3.8$$
$$6 + 3.8 = 252(I_2) - 3.8 + 3.8$$

(3.8 is added to both sides of the equation to maintain the proper balance.)

$$9.8 = 252(I_2)$$
$$9.8/252 = 252(I_2)/252$$

(252 is divided into both halves of the equation.)

$$0.04 = I_2$$

(Note that you have rounded off the division again.)

Current I_2 has a value of approximately 0.04 A (or 40 mA). Earlier, defined I_1, in terms of I_2:

$$I_1 = (12 + 47(I_2))/147$$

Because you now know the value of I_2 (0.04), you can simply plug in this value and solve for I_1:

$$I_1 = (12 + (47 \times 0.04))/147$$
$$= (12 + 1.8)/147$$
$$= 13.8/147$$
$$I_1 = 0.09$$

A slight error has been introduced due to rounding off the values during the math. Usually this won't matter, but you should be aware of round-off errors, or you'll go nuts trying to figure out why the final mathematical results don't match up perfectly.

Please remember that you are working with loop currents here (I_1 and I_2). Loop currents are mathematical fictions. If you insert an ammeter into the circuit you will not always be able to measure 0.04 A or 0.09 A at any point. The loop currents are assumed to exist only for the purposes of Kirchhoff's voltage law.

Knowing the currents, you can now go back and find the voltage drops for each of the resistances in the circuit. Remember that there are four voltage drops in this circuit, because both I_1 and I_2 flow through R_2, but only I_1 flows through R_1, and only I_2 flows through R_3. The four voltage drops (stated via Ohm's law), are as follows:

$$I_1 \times R_1, I_1 \times R_2, I_2 \times R_2, I_2 \times R_3$$

You know the resistance values from the original parts list, and you have just solved for the current values. Now, to find the voltage drops, you just have to plug in the appropriate values:

(a) $I_1 \times R_1 = 0.09 \times 100 = 9$ V
(b) $I_1 \times R_2 = 0.09 \times 47 = 4.2$ V
(c) $I_2 \times R_2 = 0.04 \times 47 = 1.9$ V
(d) $I_2 \times R_3 = 0.04 \times 220 = 8.8$ V

To make things a little more convenient, Each voltage drop has been assigned an identifying letter from a to d. Loop A includes three of the voltage drops, a, b, c.

$$E_1 = (I_1 \times R_1) + (I_1 \times R_2) - (I_2 \times R_2)$$
$$E_1 = a + b - c$$

Plugging in the values you now know:

$$12 = 9 + 4.2 - 1.9$$
$$12 = 11.3$$

Don't be thrown by the apparent 0.7 V error. It is simply the cumulative result of the rounding errors throughout the process. If you go back and use a calculator to determine the various values exactly, the error should disappear (or shrink significantly—the calculator might do some minor rounding off of its own).

Allowing for rounding off errors, the total voltage sources in the loop equal the total voltage drops in the loop, as Kirchhoff's voltage law predicts. Similarly, loop B also includes three of these voltage drops, b, c, d.

$$E_2 = (I_2 \times R_2) + (I_2 \times R_3) - (I_1 \times R_2)$$
$$E_2 = (c) + (d) - (b)$$

Plugging in the known values:

$$6 = 8.8 + 1.9 - 4.2$$
$$6 = 6.5$$

Once again, you have a slight imbalance due to cumulative rounding errors, but the total voltage sources in the loop again is approximately equal to the total voltage drops in the loop. If you recalculate all the values exactly (without any rounding off), you will find that Kirchhoff's voltage law accurately predicts what is happening here. The results of the analysis of this circuit using Kirchhoff's voltage law are shown in Fig. 4-5.

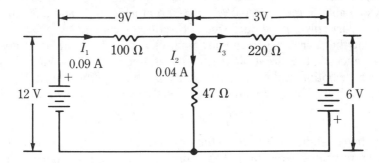

4-5   The results of Kirchhoff's voltage law analysis of Fig. 4-2.

For simplicity in the discussion to this point, you have rounded off the calculated values as much as possible. Inevitably this comes at a cost in accuracy of results. It is okay for quick-and-dirty calculations. For more precision work, you probably won't want to round off quite this much. Rounding errors are often (but not always) somewhat cumulative.

Some technicians argue that all calculations should be performed with a minimum of three significant figures for adequate accuracy. To illustrate the difference this makes, you will quickly work through the same major equations from the preceding discussion, except this time you will round off all values to three significant digits.

$$6 = 267(I_2) - ((47 \times _{12}) + (47 \times 47)\ (I_2)/147)$$
$$= 267(I_2) - ((47 \times 12)/147) + ((47 \times 47(I_2))/147)$$
$$= 267(I_2) - (564/147) - (2209(I_2)/147)$$

$$6 = 267(I_2) - 3.34 - 15.0(I_2)$$
$$= 267(I_2) - 15(I_2) - 3.34$$
$$= 252(I_2) - 3.34$$

$$6 + 3.34 = 252(I_2) - 3.34 + 3.34$$
$$9.34 = 252(I_2)$$
$$9.34/252 = 252(I_2)/252$$
$$0.037 = I_2$$

$$I_1 = (12 + 47(I_2))/147$$
$$= (12 + (47 \times 0.037))/147$$
$$= (12 + 1.739)/147$$
$$= 13.739/147$$
$$= 0.093$$

(a) $I_1 \times R_1 = 0.093 \times 100 = 9.3$ V
(b) $I_1 \times R_2 = 0.093 \times 47 = 4.37$ V
(c) $I_2 \times R_2 = 0.037 \times 47 = 1.74$ V
(d) $I_2 \times R_3 = 0.0_37 \times 220 = 8.14$ V

$$E_1 = (I_1 \times R_1) + (I_1 \times R_2) - (I_2 \times R_2)$$
$$= a + b - c$$
$$= 9.3 + 4.37 - 1.74$$
$$= 11.93 \text{ V}$$

(Notice this answer is much closer to the expected nominal value of 12 V than the 11.3 V you got in the original equations. It is still not exact, because you are still rounding off the calculation values, just not as much as before.)

$$E_2 = (I_2 \times R_2) + (I_2 \times R_3) - (I_1 \times R_2)$$
$$= c + d - b$$
$$= 1.74 + 8.14 - 4.37$$
$$= 5.51 \text{ V}$$

In this case, you have almost as much rounding off error as before (6.5 V instead of the nominal 6 V expected), but this time the error is in the opposite direction. Such mathematical oddities will turn up from time to time. In most practical circuits such errors probably won't be of too much significance. Component tolerances can often result in as much deviation from the ideal, nominal values anyway. You are just off by 8%. Most practical, real-world resistors have tolerances of 5% or

10%. In noncritical applications, this much error would be quite acceptable. In a circuit intended for a high precision application, there will probably be one or more potentiometer (manually adjustable resistors) to permit you to precisely calibrate the circuit. Again, it probably won't matter too much if these equations are slightly off during the design of the circuit. The output of the circuit can be adjusted to match some calibration standard. The calculations just put you in the correct ballpark, so as little adjustment as possible will be necessary to simplify circuit calibration procedures.

Kirchhoff's current law

There is an alternate to Kirchhoff's voltage law, and it is Kirchhoff's current law. This second Kirchhoff law permits you to deal with actual currents rather than the mathematical fictions (loop currents) of Kirchhoff's voltage law.

According to Kirchhoff's current law, "the amount of current flowing into a node always exactly equals the current flowing out of that node." That idea certainly makes sense when you think about it. What you put into a node is what you get out of it. In more mathematical terms, the algebraic sum of all currents flowing through a node is zero. Obviously, before you go any further, you should understand just what you mean by the term *node*.

Nodes

A node is simply the connection point between two or more conductors. The simple example circuit is shown in Fig. 4-6, with the nodes indicated. This particular circuit has just two nodes (A and B).

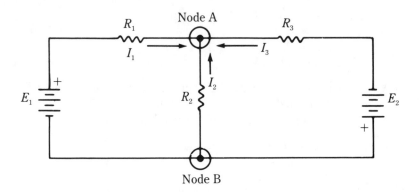

4-6   The circuit of Fig. 4-2 with the current nodes indicated.

Current flowing into any node is always assumed to be positive. Current flowing out of the node is assumed to be negative. For voltage drops across any resistance elements, the terminal where the current enters is assumed to be at a higher potential (more positive) than the terminal where the current exits. This concept is illustrated in Fig. 4-7.

4-7 The terminal of a resistance element where the current enters is assumed to be more positive than the terminal where the current exits.

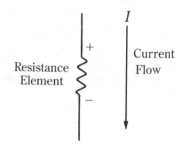

When using Kirchhoff's current law, the first step is to count the number of nodes in the circuit to be analyzed. If the circuit has N nodes, you will need to examine N–1 nodes to completely analyze the circuit. The required number of node equations is always one less than the total number of nodes in the circuit.

Kirchhoff's current law in action

In the sample circuit of Fig. 4-6, you have two nodes, so you only need to look at one to analyze the circuit. Use node A. You could just as easily use node B. The choice of which one to leave out is purely arbitrary.

There are three current paths into node A. These are marked in Fig. 4-6 as follows:

$$I_1, I_2, I_3$$

According to Kirchhoff's current law, the algebraic sum of these three currents must be equal to zero. That is:

$$I_1 + I_2 + I_3 = 0$$

Note that at this point you don't know which of these currents are positive and which are negative. This basic equation can't do you much good until you relate the currents to the voltages and resistances within the circuit.

Current I_1 flows through resistor R_1. Thanks to Ohm's law $(I = E/R)$, you know that current I_1 must be equal to the voltage drop across R_1, divided by the value of the resistance.

The voltage drop across R1 must be equal to the voltage going into the resistance element at the positive terminal (which is E1, in this case) minus the voltage at the negative terminal of the resistance element. You will call the voltage at the negative terminal Ea. The current direction of I_2 tells you that node A is less positive (more negative) than node B, so voltage E_a takes on a negative sign. Putting all of this together, you can create a simple Ohm's law equation for current I_1:

$$I_1 = (E_1 - (-E_a))/R_1$$

The two negative signs in front of E_a cancel each other out, which leaves you with:

$$I_1 = (E_1 + E_a)/R_1$$

Current I_2 is defined by the voltage drop across R_2, which is simply equal to E_a, so:

$$I_2 = E_a/R_2$$

Finally, current I_3 is determined by the voltage drop across resistor R3. The input voltage for this resistance element is equal to the voltage put out by voltage source E_2. The output voltage of E3 is again E_a. E_2 is negative because of the polarity of the voltage source. E_a is negative because of the direction of the I_2 current flow. Putting all these elements together, you find that current I_3 works out to:

$$I_3 = (-E_2 - (-E_a))/R_3$$
$$= (-E_2 + E_a)/R_3$$
$$= (E_a - E_2)/R_3$$

The next step is to substitute these individual current formulas into the original node equation:

$$I_1 + I_2 + I_3 = 0$$
$$((E_1 + E_a)/R_1) + (E_a/R_2) + ((E_a - E_2)/R_3) = 0$$

You can simplify and rearrange the equation as follows:

$$(E_1/R_1) + (E_a/R_1) + (E_a/R_2) + (E_a/R_3) - (E_2/R_3) = 0$$
$$(E_a/R_1) + (E_a/R_2) + (E_a/R_3) + (E_1/R_1) - (E_2/R_3) = 0$$

(You've just rearranged the order of the items in the equation to make the following steps clearer.)

$$(E_a/R_1) + (E_a/R_2) + (E_a/R_3) + (E_1/R_1) - (E_2/R_3) -$$
$$((E_1/R_1) - (E_2/R_3)) = 0 - ((E_1/R_1) - (E_2/R_3))$$

(The same factor is subtracted from both sides of the equation to maintain equality.)

$$(E_a/R_1) + (E_a/R_2) + (E_a/R_3) = - ((E_1/R_1) - (E_2/R_3))$$
$$(E_a/R_1) + (E_a/R_2) + (E_a/R_3) = - (E_1/R_1) + (E_2/R_3)$$
$$(E_a/R_1) + (E_a/R_2) + (E_a/R_3) = (E_2/R_3) - (E_1/R_1)$$

(You've just rewritten the equation in slightly simpler form, without really changing anything.)

$$E_a \times ((1/R_1) + (1/R_2) + (1/R_3)) = (E_2/R_3) - (E_1/R_1)$$

(Here you have just factored the left side of the equation. What the equation expresses has not been changed in any way; you've just modified the way you are expressing it.) Before you can go any further, you will need some specific component values to work with. You will assign the following component values to the circuit:

E_1 9 V
E_2 12 V
R_1 10 Ω
R_2 20 Ω
R_3 50 Ω

Notice that all of the elements in the equation except E_a are defined by the parts list. In a more complex circuit, there can be more than one unknown voltage element. Multiple unknowns can be solved by combining multiple node equations. In this simple circuit, you have only one node and one unknown voltage value to worry about.

Plugging the known values from the parts list into the last form of the equation, you get:

$$E_a \times ((1/R_1) + (1/R_2) + (1/R_3)) = (E_2/R_3) - (E_1/R_1)$$
$$E_a \times ((1/10) + (1/50) + (1/20)) = (12/50) - (9/10)$$

Now you must start solving and simplify the values:

$$E_a \times (0.1 + 0.02 + 0.05) = 0.24 - 0.9$$
$$E_a \times 0.17 = -0.66$$

(You can divide both sides of the equation by 0.17 to remove this factor from the left side and solve for E_a.

$$(E_a \times 0.17)/0.17 = -0.66/0.17$$
$$E_a = -0.66/0.17$$
$$E_a = -3.88 \text{ V}$$

The negative sign here simply means that the actual polarity is the opposite of the one you assumed. As you can see, you really can't get the polarity "wrong," as long as you're consistent. The equations will work out with the proper signs.

You now have enough information to go back and solve for each of the currents:

$$I_1 = (E_1 + E_{a)}/R_1$$
$$= (9 + (-3.88))/10$$
$$= (9 - 3.88)/10$$
$$= 5.12/10$$
$$= 0.512 \text{ A}$$
$$= 512 \text{ mA} = I_1$$

$$I_2 = E_a/R_2$$
$$= -3.88/20$$
$$= -0.194 \text{ A}$$
$$= -194 \text{ mA} = 12$$

$$I_3 = (E_a - E_2)/R_3$$
$$= (-3.88 - 12)/50$$
$$= -15.88/50$$
$$= -0.318 \text{ A}$$
$$= -318 \text{ mA} = I_3$$

The negative values for I_2 and I_3 simply mean that the actual direction of current flow is the opposite of that shown in Fig. 4-6. These currents flow out of, rather than into, node A. These are the actual currents in the circuit, not mathematical fictions like the loop currents used in Kirchhoff's voltage law.

You can double-check your work by plugging these derived current values back into the original node equation:

$$I_1 + I_2 + I_3 = 0$$
$$0.512 + (-0.194) + (-0.318) = 0$$
$$0.512 - 0.194 - 0.318 = 0$$

Yes, it works. Kirchhoff's Current Law gave an accurate prediction of how the currents in the circuit interact at the examined node.

Other circuits will end up with slightly different equations. Naturally, the more nodes there are in the circuit, the more equations you will have to work with.

Self-test

1. Which of the following is not one of Kirchhoff's laws?

A Kirchhoff's voltage law
B Kirchhoff's resistance law
C Kirchhoff's current law
D None of the above

2 For Kirchhoff's voltage law, circuits are divided into which of the following?

A Nodes
B Ohms
C Series-parallel circuits
D Loops

3. What is a node?

A A mathematical fiction
B A terminal point for a loop current
C A connection point between two or more conductors

4. Which of the following correctly describes Kirchhoff's voltage law?

A The algebraic sum of all the voltage sources in a loop equals the algebraic sum of all voltage drops in that loop.
B The algebraic sum of all the currents entering a node equals the algebraic sum of all currents exiting that node.
C A voltage drop equals the current multiplied by the resistance.
D The algebraic sum of all loop currents equals the algebraic sum of all voltage drops in that loop.

5. Which of the following cannot be included in a loop for Kirchhoff's voltage law?

A Resistances
B Current sources
C Voltage sources
D Reactances

6. How many nodes are needed to completely analyze a circuit according to Kirchhoff's current law?

A One
B Two

C All nodes in the circuit

D One less than the total number of nodes in the circuit.

7. Loop currents should be assumed to flow in which direction?

A Clockwise

B Counterclockwise

C Either A or B can be arbitrarily selected

8. According to Kirchhoff's current law, what is the algebraic sum of all currents entering and exiting a node?

A A positive value

B A negative value

C The algebraic sum of all loop currents

D Zero

9. If a resistance element is part of two loops, how many voltage drops must be calculated for that component?

A None

B One

C Two

D Three

10. In Kirchhoff's current law, which terminal of a resistance element is assumed to be at a higher potential (more positive) than the other?

A The terminal where the current enters the resistance element

B The terminal where the current exits the resistance element

C The terminal closest to the node being analyzed

D Either A or B can be arbitrarily selected

5

Alternating current

So far you have been working with circuits where the current flows in only one direction. This form of electricity is called *direct current*, or *dc*. Many other circuits, however, operate on a voltage and current that continuously vary in value, according to a repetitive, periodic pattern. Electricity in this form is called *alternating current*, or *ac*.

Varying voltage and current

The polarity of a voltage source determines the direction of current flow. The current will flow from the negative terminal of the voltage source to the positive terminal. If the polarity of the voltage source is reversed, the current will flow in the opposite direction.

Take a look at the circuit in Fig. 5-1. Here you have a circuit with two voltage sources, each a 3 V battery. Because these batteries are connected with opposing polarities, if they were allowed to have an equal effect on the circuit, they would simply cancel each other out. No current would flow through the circuit, and the power consumed would be zero.

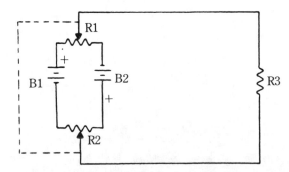

5-1 A circuit powered by a voltage with reversible polarity.

The two potentiometers labeled R1 and R2 control the relative effect of the two batteries in the main circuit. The dotted line between these two schematic symbols indicates that they are mechanically tied together. That is, one knob controls both potentiometers simultaneously. Such multiple components are said to be *ganged*. Potentiometers can be dual, triple, or even quadruple ganged. Dual ganged pots are not uncommon, but larger combinations tend to be fairly rare.

For convenience in this discussion, refer to R1 and R2 as if they were a single potentiometer, because they always work in unison. If the potentiometer slider is in the exact center of its path of rotation, equal resistances will be seen by each of the source voltages. So both batteries will present an equal, but opposite voltage to the main circuit. This results in the voltages canceling each other out, and no current flows through the circuit.

If, however, the slider is moved all the way towards battery 1, that battery will see a minimum resistance, and battery 2 will see a maximum resistance. In other words, most of battery 2 voltage is dropped across the resistance of the potentiometer. But most of the voltage from battery 1 makes its way through to the external circuit. Therefore, as far as the load circuit is concerned, battery 2 doesn't exist. Battery 1 provides the power to operate the circuit.

At the other extreme of slider path, the situation is reversed—battery 2 is dominant, and battery 1 is ignored. At intermediate positions of the slider, the two voltages will interact in a subtractive manner. The battery closest to the slider will have the greater effect, but the other battery will cancel out some of the voltage.

For example, if the potentiometer is set so that 2.5 V is passed from battery 1, but battery 2 is allowed to put out only 0.5 V, the load circuit will see a voltage source of 2 V (battery 1 − battery 2). This voltage will have the same polarity as the larger of the opposing voltages. In this case, battery 1 determines the polarity.

If you draw a graph of the effective voltage seen by the circuit, as the potentiometer slider is rotated through its entire range, it would look like Fig. 5-2.

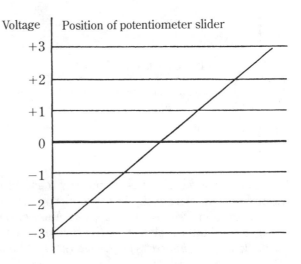

5-2   The voltage through the circuit of Fig. 5-1 as the potentiometers are moved through their entire range.

Now, suppose you start with the slider in its center position, and then smoothly rotate the knob back and forth. A graph of the effective voltage under these circumstances would resemble the one in Fig. 5-3.

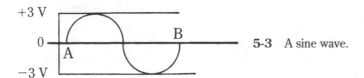

5-3   A sine wave.

Notice that there is a repeating pattern in this graph. If you happen to be familiar with trigonometry, you might recognize this graph as a series of representations of the function called the *sine* of an angle. For this reason, this *wave shape* is called a *sinusoidal*, or *sine wave*. Each complete pattern, without repetition (as from point A to point B in Fig. 5-3) is called a *cycle*, or a *wave*.

If you try a few Ohm's law equations, you'll find that the current drawn by the circuit varies in step with the fluctuations of the applied voltage. When the voltage goes up, the current goes up, and vice versa. The current is said to be *in phase* with the voltage. A purely resistive circuit does not alter the phase relationship. When the voltage and current in a circuit fluctuate in this manner, you have an *alternating current*.

ac voltage sources

Figure 5-4 shows the schematic symbol for any ac voltage source that generates a sine wave (other wave shapes are discussed in other chapters). Notice that because the current is constantly reversing itself, there is no fixed polarity for such a voltage source.

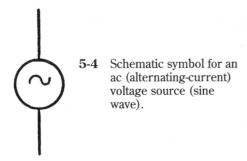

5-4   Schematic symbol for an ac (alternating-current) voltage source (sine wave).

An ac source changes its polarity many times each second. By counting the number of complete cycles in a second, you get the *frequency* of the wave. Frequency is measured in *cycles per second* (cps). Another name for a cycle per second is *hertz* (Hz), named after a pioneer in the field of electricity. One thousand hertz is a *kilohertz* (kHz), or a *kilocycles per second* (kcps). Similarly, a million cps is one *megahertz* (MHz), or *megacycle*.

Typically, ac power sources operate at a fairly low frequency. In the United States, house current alternates at a 60 Hz rate. Other countries use a 50 Hz standard. It might seem that using ac would just complicate matters, but actually, the reverse is true. For low-power applications, it is easier to make dc devices (that is, batteries) than ac devices. In heavy-wattage installations, ac is much more practical to generate. It is also easier to transmit ac over long lines. Ease of transmission is why the electric power companies operate on ac. Also, as you shall see in other chapters, many circuits will operate quite differently depending on whether the electron flow through them is ac or dc.

The most readily available source of ac for powering electronic circuits is the house current provided by the electric company through ordinary wall sockets (see Fig. 5-5). The wall socket is a power source of about 110 to 120 Vac (the level fluctuates somewhat) with a nominal frequency of 60 Hz. This frequency also fluctuates slightly, but the average is usually very close to 60 Hz.

5-5   Wall socket for ac voltage.

Many electronic circuits are designed to operate directly from an ac wall socket, so schematics often include a line cord. The symbol is shown in Fig. 5-6A. In some circuits, power is first passed through a *transformer* (see chapter 9) to change the voltage or through a power supply (see chapter 32) to convert the ac to a dc voltage.

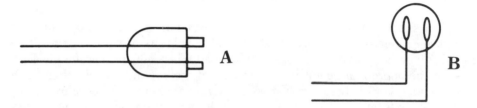

5-6   Schematic symbols. ac line cord (A); ac socket (B).

Many devices that operate off standard ac also include an extra socket to carry the power on to other equipment. The schematic symbol for an ac socket is shown in Fig. 5-6B.

Notice that an ac line cord contains two wires. One is a reference line, and the other carries the actual current. With many circuits, it doesn't matter which is which, but some circuits require that the power must come through the wires in a specific way. To ensure that the cord is plugged into the wall socket properly, such equipment uses a *polarized plug*. Such a plug can be plugged into the wall socket in only one way. If its prongs are reversed, it will not fit into the socket. A standard *nonpolarized plug* can be reversed.

Sometimes, especially with equipment that draws heavy current levels (high wattage), extra protection is necessary—both for the circuit itself and to prevent dangerous shock hazards to the user. In such equipment a third wire is needed to *ground* the circuit. The word *ground* actually has two meanings in electronics. An *earth ground* offers protection by returning the voltage to the earth itself, via a metal post of some kind that makes a good contact with the ground (see Fig. 5-7). In this manner, any potentially dangerous voltage is shunted into the earth, rather than into the components of the circuit or into the body of the operator. A cold water pipe is often a good place to make an earth ground connection.

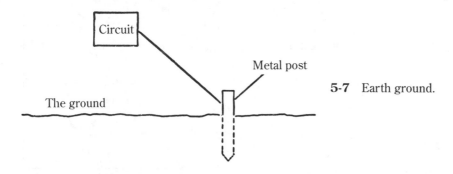

5-7   Earth ground.

For standard ac power, the screw in the center of the wall socket plate is connected to earth ground. In a three-prong socket, the third prong is also connected to earth ground. When a device with a three prong plug must be used with a two-prong socket, an *adaptor*, is used. The wire must be connected to the center screw of the socket plate. Leaving this wire unconnected defeats the purpose of the third prong, and this practice could be extremely dangerous!

The second kind of ground found in electronics work is also called a *common ground*, or simply *common*. This ground, as the name implies, is a point common to all signals in a circuit, so it is used as a reference point to measure voltages in the circuit. A common, by definition, is at 0 V, so all other voltages can be measured between the appropriate point in the circuit and the common. In dc circuits, the common point is usually either the negative terminal (somewhat more frequently) or the positive terminal of the voltage source. Either can be used without affecting circuit operation. It is just a reference point for measurements.

The type of common used will affect the polarity of the voltages measured. For example, in a circuit powered by a 3 V battery with a negative ground, the source voltage will be +3 V. If a positive ground is used, the source voltage is −3 V. This difference will not affect any circuit calculations as long as you are consistent throughout the circuit. If all voltages are of the same polarity, their signs can be ignored for calculations. If the voltage is negative, the current must be negative too, which means resistance will always work out to be positive ($R = E/I$).

In quite a few circuits, the common ground is attached to the metal *chassis*, or case of the circuit. When the common point covers a large area, it is called a *ground plane*. Some circuits (especially *digital* circuits—discussed in other chapters) require a ground plane to avoid erratic operation.

Figure 5-8 shows the three most common schematic symbols for a ground point. The symbol at Fig. 5-8A is generally only used to indicate an earth ground, but the other two can represent either earth ground or common ground (sometimes also called *chassis ground*). In most cases when both types of ground are used, they are identical; that is, they are at the same point. But this situation is not always the case.

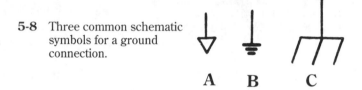

5-8   Three common schematic symbols for a ground connection.

A   **B**   **C**

All points in a circuit with a ground symbol are electrically connected with all other similarly marked points. That is, the circuits in Figs. 5-9A and 5-9B are actually identical.

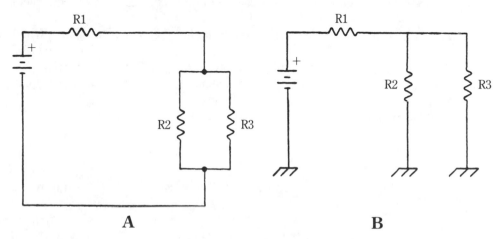

A   **B**

5-9   Identical circuits showing how the ground symbol is used.

ac values

Because the levels in an ac circuit are continuously changing, determining voltage and current values isn't the straightforward matter it is with dc circuits. As an example, assume you have a sine wave that reaches a maximum, or *peak* value of 10 V (it would also reach a negative peak of −10 V). To say you have 10 Vac is rather misleading, because the voltage actually reaches the full 10 V for a brief instant during each cycle. The rest of the time, the voltage is lower.

Because the voltage varies between a 10 V positive peak and a 10 V negative peak, there is a 20 V difference between the two peaks. That is, the voltage varies 20 V *peak-to-peak*. Although it is frequently useful to know this value, it would obviously be quite misleading to say that the voltage is 20 Vac.

It might seem that you could just take an average of the various instantaneous values passed through during the complete cycle, and come up with an average voltage. Unfortunately, because the positive portion of the cycle is a mirror image of the negative portion, all of the values are canceled out, and you are left with an average voltage of zero.

A reasonable solution would be to take only half of the cycle—either just the positive portion, or just the negative portion—and take an average of that. This process could be rather tedious to work out, but it has been mathematically proven that the average of half a cycle of a sine wave is always equal to 0.636 times the peak value. So, in this example, if the peak voltage is 10 V, then the *average voltage* is 6.36 Vac.

Conversely, if you know the average value, and need to find the peak voltage, you can multiply the average voltage times 1.572327. This value can usually be rounded off to 1.57, or even 1.6.

Although the average value can give you a fair idea of how much voltage is being passed through a circuit, a major disadvantage of using this kind of value is that the relationships stated in Ohm's law no longer hold true. What you need is a way to express ac voltage in terms that can be directly compared to an equivalent dc voltage.

Such an equivalent value can be found by taking the *root-mean-square* (rms) value of the sine wave. The mathematics are fairly complex, but it works out to 0.707 times the peak value. In this example, you have 10 × 0.707, or 7.07 V rms. This value is most commonly used for ac measurements. By using rms values, you can use Ohm's law in the same way it is used with dc circuits.

Here is a summary of the basic ac equations.

$$\text{rms} = 0.707 \times \text{peak} \qquad \textbf{Equation 5-1}$$

$$\text{rms} = 1.11 \times \text{average} \qquad \textbf{Equation 5-2}$$

$$\text{Average} = 0.9 \times \text{rms} \qquad \textbf{Equation 5-3}$$

$$\text{Average} = 0.636 \times \text{peak} \qquad \textbf{Equation 5-4}$$

$$\text{Peak} = 1.41 \times \text{rms} \qquad \textbf{Equation 5-5}$$

$$\text{Peak} = 1.57 \times \text{average} \qquad \textbf{Equation 5-6}$$

$$\text{Peak-to-peak } 2 \times \text{peak} \qquad \textbf{Equation 5-7}$$

Equations 5-1, 5-5 and 5-7 are the most frequently used formulas. The average value is rarely of practical importance. The same equations are used for current and for power in ac circuits.

Phase

The current from an ac voltage source is ordinarily in step with the voltage. That is, when the voltage increases, so does the current, and when the voltage decreases, the current also decreases. The voltage and the current are *in phase*. Their cycles start at the same instant. In other chapters, you'll learn how some components can throw the voltage and current *out-of-phase*. That is, one is delayed so that the two are no longer in step with each other.

Multiple ac sources

Remember that when you have two or more dc voltages in series (such as cells in a battery), you can find the total voltage simply by adding the component voltages from each individual source. Or, if you have two voltages of opposite polarity, you can simply subtract the smaller from the larger to find the total effective voltage in the circuit. With ac sources, however, the situation is much more complex. If the two voltage sources in a circuit like in Fig. 5-10 are in phase with each other, there

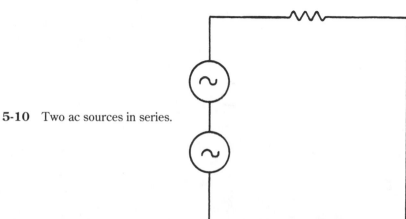

5-10   Two ac sources in series.

is no problem. you can simply add the voltages, just as with dc. Or, if the voltage sources are 180 degrees out of phase with each other (one full cycle equals 360 degrees—see Fig. 5-11), you can just subtract the smaller from the larger voltage.

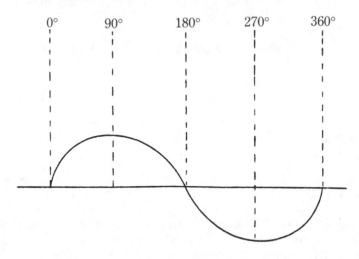

5-11   The degrees of a cycle.

Of course, if two equal ac voltages are 180 degrees out of phase with each other, they will cancel each other out, leaving a net voltage of zero.

To calculate any other phase relationship requires fairly complicated mathematics. For example, consider the graphs in Fig. 5-12. The voltage in graph B is 60 degrees out of phase with the voltage in graph A. Mathematically calculating the effective voltage of A and B would require numerous equations.

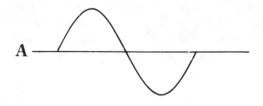

5-12   Out-of-phase ac voltages.

Vector diagrams

Fortunately, there is an easier way to solve combinations of out-of-phase ac signals. The easier solution is a method using *vector diagrams*. Vector diagrams might look and sound rather complicated, and the mathematical theory behind them is rather complex, but actually using these diagrams in practical situations is really quite simple. If you know how to use a ruler and a protractor, you should have no trouble.

Use a vector diagram to solve for the resultant voltage in Fig. 5-12. Assume voltage A is 5 V and voltage B is 3 V. Voltage B is 60 degrees out of phase with voltage A. First draw a straight line to represent voltage A. You can use any scale you find convenient. For example, if you are using a scale of one inch equals 1 V, draw line A five inches long. This step is shown in Fig. 5-13.

5-13 Step 1 of a vector diagram.
Line A

Starting at the point of origin of line A, draw a second line to represent voltage B. Draw this line to the same scale as line A. In the example, line B should be three inches long. You also have to use a protractor to draw line B at an angle equal to the phase difference. Naturally, line A is at 0 degrees. In the example, line B should be at a 60-degree angle to line A. This angle is shown in Fig. 5-14.

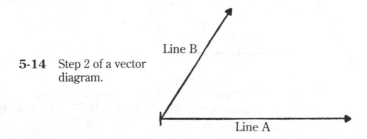

5-14 Step 2 of a vector diagram.
Line B
Line A

Draw a dotted line parallel with line B from the open end of line A. This new line (line C) should be the same length, or longer than, line B. Line C is added to the vector diagram in Fig. 5-15. Draw a fourth line (dotted line D in Fig. 5-16) from the open end of line B to line C. This line will be parallel with line A.

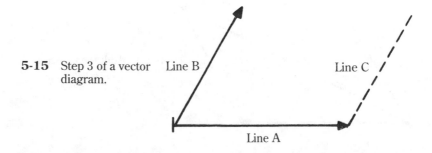

5-15 Step 3 of a vector diagram.
Line B
Line C
Line A

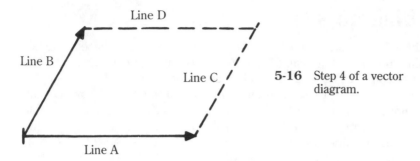

5-16   Step 4 of a vector diagram.

Draw one last straight line from the original point of origin (the junction of line A and line B) to the junction of line C and line D. This is the line labeled line E in Fig. 5-17. If you now measure line E, its length will represent the resultant voltage, drawn to the same scale as line A and line B. In the example, it comes out to about 8.5 inches, or 8.5 V. The angle of line E with respect to line A is the same as the phase difference of the resultant voltage (compared to voltage A).

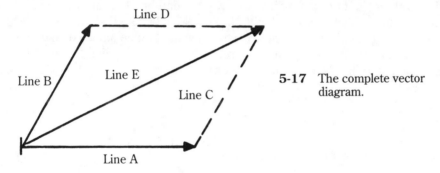

5-17   The complete vector diagram.

Additional examples of vector diagrams are shown in Figs. 5-18 and 5-19. Also try drawing your own vector diagrams to solve for the resultant voltage for each of the combinations listed in Table 5-1.

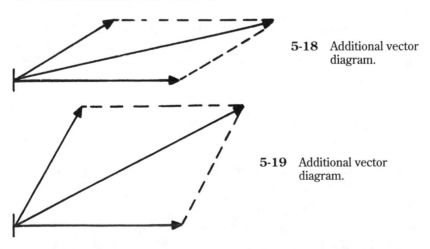

5-18   Additional vector diagram.

5-19   Additional vector diagram.

Table 5-1.
Values for practice vector diagrams.

Voltage A	Voltage B	Phase
6	6	45 degrees
8	3	90 degrees
7	5	30 degrees

Combining ac and dc

What happens when you have a circuit with both a dc voltage source and an ac voltage source, like the one illustrated in Fig. 5-20? Assume the ac puts out a 5 V peak sine wave, and the dc source is 9 V. At any given instant, the ac signal has a specific value. For instance, whenever the signal crosses the 0 line, it has an instantaneous value of 0 V. At this instant, the ac source has no effect on the circuit—the total circuit voltage at this brief point in time is simply the 9 V from the dc source.

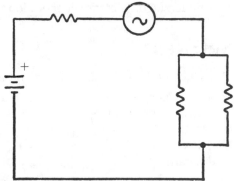

5-20   A circuit with ac and dc
voltage sources.

However, at another instant in its cycle, the ac source is putting out its peak value of 5 V. Meanwhile, the dc source is still producing 9 V. These values are simply added, and the instantaneous value is 14 V. When the ac voltage reaches its negative peak (−5 V), the polarity opposes that of the 9 Vdc source. In this case you have −5 V and +9 V, so the instantaneous value is +4 V.

If you continue figuring the instantaneous values at various points throughout the cycle, you will find you have a sine wave that varies between +4 V and +14 V. The graph of the resultant voltage is shown in Fig. 5-21. Notice that it is identical to an ordinary sine wave, except it varies above and below 9 V, rather than 0 V. You say you have 3.535 Vac rms (5 V peak) *superimposed* on 9 Vdc.

If the ac voltage peak is less than the dc voltage, circuit polarity will not be reversed at any time in the cycle. If, on the other hand, the ac voltage is larger than the dc voltage, then polarity will be reversed during at least some of the cycle. For example, 10 Vac peak superimposed on 6 Vdc will produce a combined voltage that fluctuates between −4 V and +16 V.

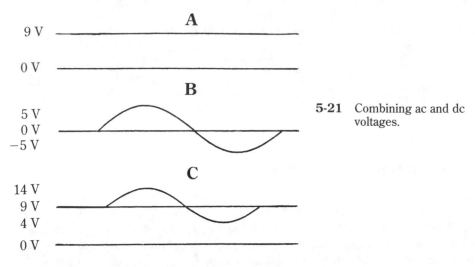

5-21 Combining ac and dc voltages.

In solving most circuit equations, the ac and dc elements are handled separately. In many cases, one or the other can be ignored without adversely affecting your understanding of circuit operation.

Self-test

1. Which of the following is not a unit for measuring ac frequencies?

A Cycles per second
B Waves per second
C Hertz
D Kilocycles
E Megahertz

2. If an ac signal has an rms voltage of 75 V, what is the peak voltage?

A 83.25 V
B 53.025 V
C 105.75 V
D 117.75 V
E None of the above

3. If an ac signal has a peak voltage of 55 V, what is the average voltage?

A 34.98 V
B 61.05 V
C 86.35 V
D 38.885 V
E None of the above

4. If an ac signal has an average voltage of 18 V, what is the rms voltage?

A 16.2 V
B 12.726 V

C 25.38 V
D 19.98 V
E None of the above

5. How many degrees are there in one complete wave cycle?

A 180 degrees
B 90 degrees
C 360 degrees
D 720 degrees
E None of the above

6. When comparing rms voltages and average voltages, which of the following statements is true, assuming sine waves?

A The average voltage is always greater than the rms voltage.
B The rms voltage is always greater than the average voltage.
C Either the rms voltage or the average voltage might be larger.
D There will always be a very large difference between the rms voltage and the average voltage.

7. If two equal frequency ac signals of exactly 5 V each are combined with one the signals 180 degrees out of phase with the other, what will be the value of the resultant voltage?

A 0 V
B 5 V
C 10 V
D 2.25 V
E None of the above

8. If the combination of an ac voltage and a dc voltage has an instantaneous voltage that varies through a range from −2 V to +10 V, what is the peak ac voltage of the combination?

A 12 V
B 10 V
C 6 V
D 16 V
E None of the above

9. What is the rms value of the ac voltage of the signal described in question 8?

A 8.48 V
B 4.242 V.
C 7.07 V
D 14.1 V
E None of the above

10. In the signal described in question 7, what is the value of the superimposed dc voltage?

A 0 V
B −2 V
C +6 V
D +4 V
E None of the above

6
Capacitance

It is mentioned in chapter 3 that the most commonly used component in electronic circuits is the resistor. Probably the second most commonly used component is the *capacitor*.

What is capacitance?

If two metal plates are separated by an insulator (or *dielectric*), and a dc voltage is applied between the plates, current will not be able to cross the dielectric. But a surplus of electrons will be built up on the plate connected to the negative terminal of the voltage source, and there will be a shortage of electrons on the plate connected to the positive terminal. The voltage source will try to force electrons into one plate (negative terminal) and draw them out of the other (positive terminal).

At some definite point, these plates will be completely saturated. No further electrons can be forced into the negative plate, and no more electrons can be drawn from the positive plate. At this point, the plates have an electrical potential equal to that of the voltage source. In fact, the plates now act like a second voltage source in parallel with the first, and with the opposite polarity. Figure 6-1 shows the equivalent circuit. Naturally, because these opposing voltages are equal, they cancel each other out and no current can flow between the voltage source and the plates in either direction. The plates are said to be *charged*.

Now if the voltage source is removed from the circuit, as in Fig. 6-2, the plates will stay charged, because there is no place for the electrons on the negative plate to go. Similarly, there is no place for the positive plate to draw electrons from. The voltage is stored by the plates.

Replacing the missing voltage source with a resistor, as shown in Fig. 6-3, provides a current path for the excess electrons stored on the negative plate to flow to

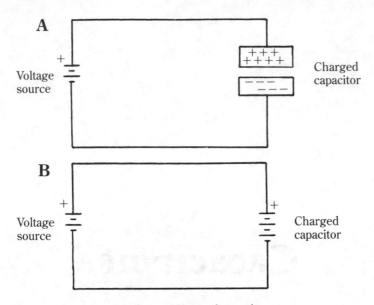

A

Voltage source

+++
++++

− − −

Charged capacitor

B

Voltage source

+

+

Charged capacitor

6-1   Equivalent circuits for a charged capacitor.

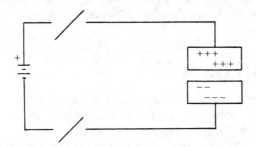

+

+++
+++

− −
− − −

6-2   Charged capacitor removed from the power source.

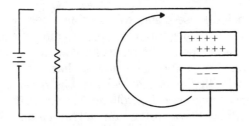

++++
++++

− − −
− − − −

6-3   Discharging a capacitor.

the positively charged plates. This flow will continue until both plates are returned to an electrically neutral state. This process is called *discharging* the plates.

Such a device (two conductive plates, separated by insulator) is called a *capacitor*. A capacitor is used to store electrical energy. At one time, capacitors were known as *condensers*, but this term is somewhat misleading and has fallen into disuse. You might run across it once in a while. Just remember, it is simply another name for a capacitor.

A capacitor cannot hold a charge indefinitely. Even air can conduct some current, so the charge will slowly seep off into the air. This action is a form of *leakage*. There will also be some leakage through the insulating dielectric. Of course, if all other factors are equal, the lower the internal leakage, the better the capacitor.

Now, consider what happens in a capacitor when an alternating current is applied to it. See Fig. 6-4. During the first part of the cycle, as the source voltage increases from zero, it will charge the plates of the capacitor in a manner similar to the dc circuit described above. The polarity of the charged capacitor opposes that of the source voltage.

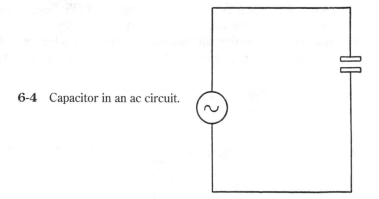

6-4   Capacitor in an ac circuit.

The capacitor might or might not be completely charged by the time the applied voltage passes its peak and starts to decrease again (depending on the size of the plates, how much voltage is applied, and the frequency of the ac signal). In either case, as the applied voltage decreases, a point will be reached when it is less than the charge stored in the capacitor. This situation will allow the capacitor to start discharging through the ac voltage source.

The capacitor might or might not be completely discharged when the ac voltage reverses polarity, but because the source polarity is the same as the capacitor polarity, the voltages aid, quickly discharging the capacitor the rest of the way, then charging it with the opposite polarity from the original charge. When the ac source voltage reverses direction, the capacitor is discharged again, and the entire process is repeated with the next cycle of the ac waveform.

If you constructed the circuit illustrated in Fig. 6-5 with a dc voltage source, the lamp would not light, because the dc current cannot flow through the circuit—it is blocked by the dielectric. The capacitor acts like an open circuit as far as direct current is concerned.

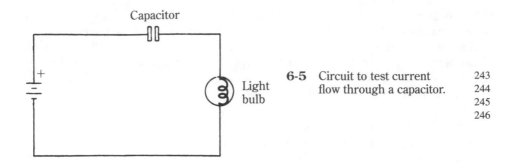

Capacitor

Light
bulb

If, however, the same circuit were built with an ac voltage source, the lamp will light (see Fig. 6-6). The light indicates that alternating current is flowing through the circuit. Of course, virtually no current (except the tiny leakage current) will flow across the dielectric itself. Remember, any given electron doesn't travel very far in an electric circuit. It merely moves far enough to disturb its neighbor. The process of charging, discharging, and recharging a capacitor from an ac voltage source, gives the same effect as if the current was actually flowing through the capacitor itself. Moreover, if you decrease the frequency of the ac source, the lamp will dim. Increasing the frequency will cause the lamp to burn brighter. A capacitor lets more current flow as the frequency of the source voltage is increased.

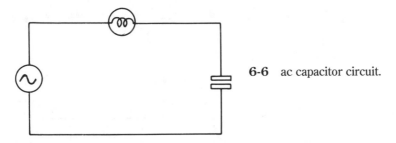

6-6   ac capacitor circuit.

If you measured the dc resistance of a discharged capacitor, the meter needle would show a sharp kick down to a moderately low resistance as the capacitor is being charged. Then it will settle down to a very high resistance value. In an ideal capacitor you would have an infinite resistance; that is, you would have a completely open circuit. However, you've already learned that ac can flow through a circuit with a capacitor. The apparent resistance of a capacitor in an ac circuit is less than its dc resistance. This apparent ac resistance is called *capacitive reactance*. Its value decreases as the applied frequency increases. A capacitive reactance slows down voltage more than it does current, so the voltage lags the current by 90 degrees (assuming a purely capacitive circuit).

Before you can understand the formula for determining capacitive reactance, you need to know how capacitance is determined. The basic unit of capacitance is the *farad*. If one ampere of current flows when the applied voltage changes at a rate of one volt per second, you have one farad of capacitance (1 F). In actual circuits, the farad is far too large a value. Instead, you generally use the *microfarad* (μF), which is one millionth of a farad. The abbreviation μF is generally preferred today.

A still smaller unit is the *picofarad* (pF), which is one millionth of a microfarad. Sometimes a picofarad is referred to as a *micromicrofarad* (μμF).

Capacitance can be increased by making the metal plates larger, or by bringing them closer together or, in other words, making the dielectric thinner. The specific dielectric used also has an effect on the capacitance of the unit. Air, for example, has a *dielectric constant* (K) of one. Mica, however, has a dielectric constant of 6. Paper is between 2 and 3, and titanium oxide can have a dielectric constant as high as 170.

Capacitance (in picofarads) can be calculated with the following formula:

$$C = \frac{0.0885 \times K \times A}{T}$$

Equation 6-1

A represents the area of the side of one of the plates that is actually in physical contact with the dielectric. This area is measured in square centimeters for this equation. *T* represents the thickness of the dielectric (or the space between the plates), and is also measured in centimeters. *K*, of course, is the dielectric constant.

Assume you have a capacitor where *A* is 35 square centimeters, *T* is 2 centimeters, and the dielectric is air ($K = 1$). Algebraically, you have $C = (0.0885 \times K \times A)/T = 0.0885 \times 1 \times 35)/2 = 3.0975/2 = 1.54875$, or about 1.5 pF.

If, however, the dielectric is changed to mica ($K = 6$) and all other variables are the same, you find that $C = (0.0885 \times A/T = 0.08854 \times 6 \times 35)/2 = 18.585/2$ 9.2925, or approximately 9 pF.

Now, try increasing *T* to a value of 5. $C = (0.0885 \times 6 \times 35)/5 = 18.585/5 = 3.717$, or just under 4 pF. Increasing *T* decreases the capacitance, but increasing *K* increased the capacitance. The capacitance will also increase with an increase in *A*. For example, you can change *A* to 50 square centimeters. Then, $C = (0.0885 \times 6 \times 50)/5 = 26.55/5 = 5.31$, or just over 5 pF.

As one final example, see what happens when the dielectric constant is increased by a very large amount. Replace the mica ($K = 6$) with titanium oxide ($K = 170$). Now the equation comes out to $C = (0.0885 \times K \times A)/A/T = (0.0885 \times 170 \times 50)/5 = 752.25/5 = 150.45$, or approximately 150 pF.

Now, assume you have a capacitor that uses paper as a dielectric ($K = 2.5$). $A = 90$ square centimeters, and $T = 3$ centimeters. What is the capacitance?

In actual practice, you will rarely need to use this formula, unless you are manufacturing capacitors, but it is helpful to understand how the variables interrelate in determining capacitance.

Returning to capacitive reactance, its value is determined by the capacitance and the frequency of the applied ac voltage. Reactance is stated in ohms, like ordinary dc resistance. The formula is as follows,

$$X_c = \frac{1}{2\pi FC}$$

Equation 6-2

In this equation, *F* stands for the frequency (in hertz), and *C* represents the capacitance in farads (not in microfarads or picofarads). The symbol π is the Greek letter *pi*, and it is used in many formulas to represent a universal constant. The value of pi is always approximately 3.14, so $2 \times \pi = $ about 6.28. This means the formula can be written as $X_c = 1/(6.28 \times F \times C)$.

Notice that if the applied frequency is 0 (that is, dc) the value of the denominator becomes 0, regardless of the capacitance value, because zero multiplied by any number always equals zero. 1/0 is unsolvable—it equals theoretical infinity. Infinite resistance in a circuit, of course, acts like an open or incomplete circuit. This is exactly the way a capacitor behaves in a dc circuit.

Try a few examples with ac frequencies. Start with a 1 μF (1×10^{-6} farad) capacitor. The denominator equals $2\pi FC$, which you can rewrite as $6.28 \times (1 \times 10-6) \times F$, or $6.28 \times 10^{-6} \times F$, so the reactance equation for a 1 μF capacitor is $X_C = 1/(6.28 \times 10^{-6} \times F)$.

If F is 10 Hz, you find that $X_C = 1/(6.28 \times 10^{-6} \times 10) = 1/(6.28 \times 10^{-5})$, or about 16,000 Ω (16 kΩ. If F is increased to 50 Hz, $X_C = 1/(6.28 \times 10^{-6} \times 50) = 1/(3.14 \times 10^{-1})$, or approximately 3,200 Ω (3.2 k). If F is 600 Hz, then $X_C = 1/(6.28 \times 10^{-6} \times 600) = 1/(3.768 \times 10^{-3})$, or about 265 Ω.

You can see that when the capacitance remains constant, and the frequency increases, the reactance decreases. Additional examples are given in Table 6-1.

Table 6-1. Capacitive-reactance samples.

Capacitance	Frequency		
	10 Hz	100 Hz	1000 Hz
0.001 μF	16 M Ω	1.6 M Ω	160 kΩ
0.01 μF	1.6M Ω	160 kΩ	16 kΩ
0.1 μF	160 kΩ	16 kΩ	1600 Ω
1 μF	16 kΩ	1600 Ω	160 Ω
10 μF	1600 Ω	160 Ω	16 Ω
100 μF	160 Ω	16 Ω	1.6 Ω

Ohm's law and reactance

Ohm's law works with reactance in exactly the same way it does with regular dc resistance. The only difference that you must keep in mind is that the result of any specific equation is true only for a single, specific frequency.

For an example, consider the circuit shown in Fig. 6-6, which consists of just a capacitor and an ac voltage source. For the sake of simplicity, consider only reactance. In this simple case, the effects of regular resistance can be considered minimal. In most practical circuits, however, you will have to consider both resistance and reactance. You'll see how this is done later in the book.

Assume the capacitor has a value of 1 μF, so you can use the reactance values found in the examples in the last section. Also assume the ac voltage source generates 100 V rms. According to Ohm's law, the current equals the voltage divided by the resistance, or, in this case, the voltage divided by the reactance. At 10 Hz, you know that a 1 μF capacitor has a reactance of about 16,000 Ω. $I = E/X = 100/16,000 = 6.25 \times 10^{-1}>$ A, or 0.625 mA. This value could also be written as 625 μA. When the frequency is increased to 50 Hz, the capacitive reactance drops to about 3200 Ω. $100/3200 = 3.125 \times 10^{-3}$ A, or about 3 mA. At 600 Hz, the reactance is a mere 265 Ω, so $I = 100/265 = 0.038$ A, or 38 mA. Notice that as the frequency goes up, the current also increases, while the reactance value goes down.

According to Ohm's law, current and resistance are always in an inverse relationship to each other. When one increases, the other decreases. Now, assume the voltage source generates 24 V rms, and the capacitor is 0.2 μF (2×10^{-7}). How much current flows in the circuit at 10 Hz, 50 Hz, and 600 Hz?

Types of capacitors

In the discussion of dielectric constants, you learned that a number of different materials can be used as a dielectric in a capacitor, and the dielectric constant of each material helps to determine the capacitance of a device of a given size. Capacitors are generally classified according to their dielectric material. Some of the more common types of capacitors are discussed in this section.

Regardless of the type of capacitor, the schematic symbol is generally the same. The two most common schematic symbols for capacitors are shown in Fig. 6-7. Most capacitor types are basically interchangeable.

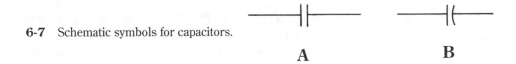

6-7   Schematic symbols for capacitors.

A   **B**

Ceramic capacitors

Perhaps the most commonly used type of capacitor consists of a wafer of ceramic material between two silver plates. Leads are connected to the plates, and the entire assembly (except for the leads, of course) is encased in a protective plastic shield. See Fig. 6-8. This type is called a *ceramic capacitor*, or, because of its most common shape, a *ceramic disc*. Some ceramic capacitors, however, are enclosed in rectangular cases. There is no electrical difference, and often the rectangular units are called ceramic discs anyway.

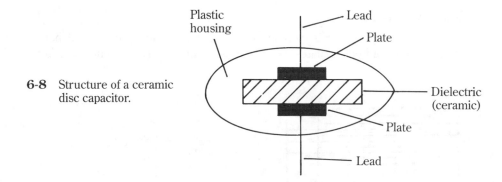

6-8   Structure of a ceramic disc capacitor.

The values for ceramic capacitors usually range from about 10 pF to about 0.05 μF. Occasionally you'll also find ceramic capacitors with values up to 0.5 μF.

There are two important voltage ratings for capacitors. One is the *breakdown voltage*. This voltage is the absolute maximum voltage the dielectric material can withstand without breaking down and rupturing. For most dielectrics, the breakdown voltage is quite high (typically, more than 10 kV), so you generally don't have to worry about it in practical circuits. The second important voltage rating in a capacitor is the *working voltage*. This voltage is the maximum dc voltage the manufacturer recommends to be placed across the plates safely. Exceeding this voltage can damage or destroy the capacitor, and perhaps other nearby components. The working voltage is usually much lower than the breakdown voltage.

One thing that is important to remember is that the working voltage is given in dc volts. Because ac ratings are usually given in rms units, you must find the peak voltage to determine if it can safely be applied to a given capacitor. For example, if a capacitor has a working voltage rating of 100 V, you should not apply an ac voltage of more than 100 V peak. In terms of rms voltage, this is 100 × 0.707, or no more than 70.7 V rms. If you put 100 V rms across this device, the peak voltage would actually be 100 × 1.41, or 141 V. This much overvoltage could easily damage the capacitor.

Working voltages for standard ceramic capacitors generally range from 50 to about 1600 V. If a lower voltage unit can do the job, it will generally be less expensive, but a higher voltage rating can always be substituted. Capacitors with low working voltages also tend to be physically smaller.

Special high capacitance ceramic capacitors are also available with capacitances up to about 0.5 μF. Often this will be marked as 0.47 μF—capacitor values aren't precise enough for this small difference to matter. The working voltage for these large capacitance units is generally only about 10 to 100 V.

The values are usually stamped right on the case of ceramic capacitors. This type of capacitor is very widely used because it is inexpensive, comparatively small, and has a low degree of *power loss* (leakage resistance). In radio frequency tuned circuits (very high frequencies) ceramic capacitors often prove to be unsatisfactory because of poor stability. But at frequencies up to about 100 kHz, they are generally an excellent choice.

Mica capacitors

Another popular type of capacitor uses thin slices of mica sandwiched between a number of interconnected plates. See Fig. 6-9. Again, this type of capacitor is named

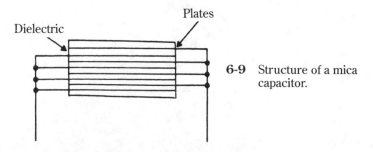

6-9   Structure of a mica capacitor.

after its dielectric. It is called a *mica capacitor*. As a rule, mica capacitors are some-what more expensive than ceramic discs, but they tend to be much more stable at radio frequencies. Their values typically range from 5 pF to 0.01 μF, and their working voltages are generally between 200 and 50,000 V.

A special type of mica capacitor is the *silver-mica capacitor*. This capacitor is also sometimes called a *silvered-mica capacitor*. In this type of capacitor a very thin coating of silver is applied directly onto the mica sheets, instead of the usual foil plates. These units are more expensive, naturally, but they can have capacitance tolerances as close as 1%. Most ordinary capacitors have a tolerance of only 10 to 20%. In most circuits, a plus or minus error of 20% in the capacitance value really won't matter much, or it can be compensated for by fine tuning a variable component. But in some special purpose circuits, precise values are needed, so silver-mica capacitors are used.

Sometimes a mica capacitor has its value stamped directly on its body, but other units are identified via a color code similar to the one used with resistors. Figure 6-10 shows the arrangement of color coding dots on a typical mica capacitor. Some indication, such as the arrow in the figure, will be given so that you will know which dot is which.

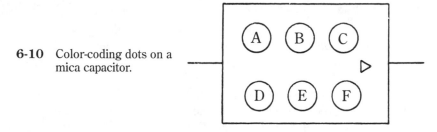

6-10 Color-coding dots on a mica capacitor.

The first dot (dot A) is generally either white or black. This dot simply indicates that the unit is a mica capacitor. Dot B is the first significant figure of the capacitance value, and dot C is the second significant figure. The value for each color is shown in Table 6-2. Notice that the values for dot B and dot C are identical to the resistor color code.

The multiplier is given by dot F, directly below dot C—not dot D at the beginning of the next row, as might be expected. Dot F corresponds to the third band on a resistor. The color values are basically the same as in the resistor color code, but capacitors don't have as wide a value range as resistors, so some colors aren't used as multipliers. Multiplying the significant figures by the multiplier gives the capacitance value in picofarads (pF). To convert to microfarads (μF), you must divide the result by 1,000,000.

Dot D in the lower left hand corner gives the *temperature coefficient* for the capacitor. The capacitance will vary somewhat with changes in temperature. The temperature coefficient tells you the maximum extent of the fluctuation for a given change in temperature. It is given in parts per million per degree centigrade. For example, if dot D is red, the temperature coefficient is ±200. This means if the temperature changes one degree centigrade, the value might increase or decrease by no more than 200 parts per million or 0.02%.

Table 6-2. Capacitor color codes.

Color	Dot A	Dot B	Dot C
Black	Mica	0	0
Brown	—	1	1
Red	—	2	2
Orange	—	3	3
Yellow	—	4	4
Green	—	5	5
Blue	—	6	6
Violet	—	7	7
Gray	—	8	8
White	Mica	9	9
Silver	Paper	—	—
Color	**Dot D**	**Dot E**	**Dot F**
Black	±1000	±20%	×1
Brown	±500	±1%	×10
Red	±200	±2%	×100
Orange	±100	±3%	×1000
Yellow	−20 to +100	—	×10000
Green	0 to +70	±5%	—
Gold	—	±5%	×0.1
Silver	—	±10%	×0.01

Notice that some capacitors can increase in value more than they decrease. In fact if dot D is green, the capacitance can only increase. The value marked is said to be the *minimum guaranteed value*.

Dot E simply gives the manufacturing tolerance of the capacitor. It corresponds to the fourth band on a resistor.

Suppose you have a capacitor with the following markings: dot A is white, dot B is brown, dot C is black, dot D is orange, dot E is black, and dot F is brown. What is the value of this capacitor?

White at dot A simply indicates that this is a mica capacitor. Dot B is brown, which means the first significant figure is 1. Black at dot C means the second significant figure is 0. The value of the capacitor is 10 times the multiplier value. Dot D is orange so the capacitor has a temperature coefficient of plus or minus 100 parts per million (0.01%). The tolerance is given by dot E, which is black. This tolerance guarantees the actual value is within plus or minus 20% of the value marked. Dot F is the multiplier. It is brown on the sample resistor, so the multiplier is 10. The value of this capacitor is 10 × 10, or 100 pF plus or minus 20%, with a temperature coefficient of plus or minus 0.01%. Because the marked value has a tolerance of 20%, the actual value of the capacitor can range from 80 to 120 pF and still be correctly marked.

Assume the capacitance value is exactly as marked—100 pF—and the temperature changes 3°C. This temperature change will cause the value of the capacitor to change as much as 0.03 pF in either direction.

Now, suppose the temperature changes 50°C. The value can change by plus or minus 0.5 pF. That is still not very much. You can see that the temperature coefficient can usually be ignored in practical circuits unless very wide fluctuations of temperature are expected. If dot A is black, it simply means the capacitor was manufactured to military specifications.

Paper capacitors

Sometimes a higher capacitance is needed in a circuit than a ceramic disc or a mica capacitor can provide. Another type of capacitor can be formed by placing a strip of waxed paper (dielectric) between two long strips of tin foil (plates). These foil strips are often several feet long, but only about an inch wide. This assembly can be tightly rolled up to save space, and it is usually enclosed in a cardboard tube or plastic container. Such a device is called a paper capacitor or a tubular capacitor. See Fig. 6-11. Of course, leads are connected to the foil strips and brought out through the ends of the cardboard tube. In some units, the capacitor is sealed in a metal or plastic case to prevent moisture from getting into the capacitor.

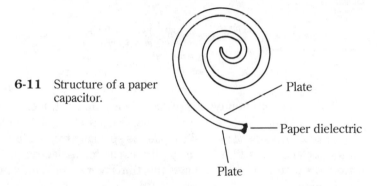

6-11 Structure of a paper capacitor.

Plate

Paper dielectric

Plate

The capacitance range for paper capacitors is from about 0.0001 μF to about 1 μF, with working voltages from 200 to 5000 V. Paper capacitors are relatively inexpensive, but they have a fairly low leakage resistance (a paper capacitor has a lower leakage resistance than, say, a mica capacitor), and they have some trouble giving optimum performance at the upper radio frequencies because of high dielectric losses. However, they are an excellent choice when a fairly large amount of capacitance is needed in a low or medium frequency circuit, especially when space is at a premium.

Paper capacitors generally have their values stamped directly onto their bodies, but sometimes they are color coded with the same system used for mica capacitors. For a paper capacitor, dot A is always silver.

Synthetic film capacitors

A similar type of capacitor can be made using a thin coating of synthetic film instead of waxed paper as the dielectric. Such a capacitor is usually identified by the specific type of film it uses. For example, there are *polystyrene capacitors*, *Mylar capacitors*, and *polyester capacitors*. These synthetic films are usually thinner than the paper used in paper capacitors, so a film capacitor of a given value will tend to be somewhat smaller than a paper capacitor of the same capacitance.

Another advantage of these film capacitors is that they can operate over a wider temperature range than the paper units. In other words, they have a lower temperature coefficient. They also tend to be more precise, and closer to the marked value. The capacitance of synthetic film capacitors ranges from 0.001 μF to about 2 μF. The working voltages are generally between 50 and 1000 V.

Electrolytic capacitors

The capacitors you read about so far are *nonpolarized*. That is, they can be placed in a circuit with the leads connected in either direction. Other capacitors are designed to work only in dc circuits and can only be hooked up one way. Such a component is said to be *polarized*.

The schematic symbol for a polarized capacitor is basically the same as for a regular capacitor, but a plus symbol (+) is added at the positive terminal to indicate polarity. See Fig. 6-12. This terminal must be kept positive with respect to the other terminal.

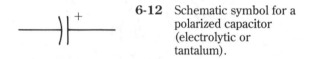

6-12 Schematic symbol for a polarized capacitor (electrolytic or tantalum).

The most common type of polarized capacitor is the electrolytic capacitor. This type of capacitor consists of a pasty, semiliquid electrolyte between aluminum foil electrodes, or plates. The positive plate is specially treated to form a thin oxide film on its surface. This film serves as the capacitor dielectric. If the capacitor is subjected to a reverse polarity voltage, this thin layer of film could be punctured, ruining the capacitor.

Because the dielectric is such a thin film, very high capacitance values can be achieved in a reasonably small space. Typical values range from 0.47 μF to 10,000 μF, and occasionally even higher. Working voltage ratings can be anywhere from 3 to 700 V. The leads on an electrolytic capacitor can be placed in either an *axial* or a *radial* arrangement.

Internally, an electrolytic capacitor resembles the construction of a paper capacitor except, of course, the dielectric is considerably thinner. One problem with electrolytic capacitors is that the semiliquid electrolyte can dry out, rendering the capacitor useless. This failure generally occurs when the capacitor is not in use. Applying a voltage across it seems to prevent the electrolyte from drying out—so it is not a good idea to stock pile electrolytic capacitors in large quantities for possible future use.

Similarly, drying out can sometimes occur if the electrolytic capacitor is operated from too low a voltage. If the voltage is extremely low, then, as far as the capacitor is concerned, it is still just sitting on the shelf. Certainly a 15 V capacitor could be used in a 5 V circuit, but you wouldn't want to use a 500 V unit. Besides, capacitors with higher working voltages are much larger, heavier, and more expensive than smaller units.

Preferably, an electrolytic capacitor should be operated at a voltage between one-third and two-thirds of its rated maximum. This voltage is enough to keep the electrolyte from drying out, but allows for unexpected overvoltage surges.

It might seem that the electrolytic capacitor would be of rather limited value because of its inability to function in an ac circuit, but actually it has a large number of important applications. These include power supply filtering, circuit coupling, audio frequency bypassing and achieving large time constants in RC (resistive-capacitive) circuits (described elsewhere in this chapter).

Electrolytic capacitors usually have a very wide tolerance. They might vary from their normal value by as much as plus or minus 50 to 70%. Except in critical timing applications, the precise value is rarely important, so this loose tolerance is acceptable.

Tantalum capacitors

Closely related to the electrolytic capacitor is the *tantalum capacitor* (see Fig. 6-13). Like the electrolytic capacitor, it can only be used in dc circuits. A dot on the body of the capacitor indicates the positive lead.

6-13   A tantalum capacitor.

Tantalum capacitors offer a number of advantages over their electrolytic counterparts. They tend to be considerably smaller for a given capacitance value. They aren't as prone to drying out, and they have less leakage. Their values tend to be more precise (lower tolerance) and they are much less susceptible to noise. Especially because of their noise resistance, they are generally preferred in computer applications. On the other hand, they tend to be more expensive than electrolytic capacitors, and they are limited to a much smaller range of capacitances and working voltages. Usually they range only from 0.5 μF to about 50 μF with working voltages that are rarely higher than 50 V.

Variable capacitors

Most capacitors have a single fixed value, but, just as there are variable resistors, there are *variable capacitors* too. One type uses a springy material for the plates. If they weren't held in place, the plates would fly apar' Between the plates is a piece of dielectric material, such as mica. The assembly is held together by a screw. By tightening or loosening this screw, the distance between the plates can be changed, thus altering the effective capacitance of the device. Figure 6-14 illustrates the structure of this type of capacitor. This kind of variable capacitor is called a *padder capacitor*, or a *trimmer capacitor*.

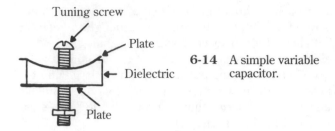

6-14 A simple variable
capacitor.

Another type of variable capacitor consists of a series of interwoven metal plates. One set of plates (called the *stator*) is stationary. The other set of plates (called the *rotor*) can be moved by a knob. The amount of overlap between the plates determines their effective area. Moving the rotor electrically acts like the plates are being increased or decreased in area. Figure 6-15 illustrates the way this works. The dielectric in this type of capacitor is simply air. It is called a *variable- air capacitor*. The schematic symbol for any type of variable capacitor is shown in Fig. 6-16.

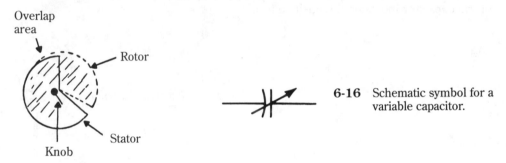

6-16 Schematic symbol for a
variable capacitor.

6-15 A section of a variable
air capacitor.

RC time constants

If a resistor and a capacitor are connected in series across a voltage, as in Fig. 6-17A, the capacitor will be charged through the resistor at a specific rate, deter-

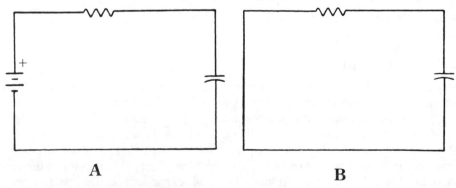

A   **B**

6-17 A series RC (resistive-capacitive) circuit.

mined by both the capacitance and the resistance. The capacitance, of course, determines how many electrons the negative plate can hold when fully charged, and the resistance slows down the flow of electrons.

The time it takes for the capacitance to be charged to 63% of its full potential charge level is called the *time constant* of the combination. For obvious reasons, such a combination of a resistor and a capacitor is called an *RC circuit*.

Similarly, if the voltage source is removed, as in Fig. 6-17B, the capacitor will be discharged through the resistor. In this case, the time constant is defined as the time it takes the capacitor to drop down to 37% of its fully charged value. The charging time constant and the discharging time constant are always equal.

The time constant of a specific RC circuit can be found simply by multiplying the resistance and the capacitance:

$$T = RC \hspace{4cm} \textbf{Equation 6-3}$$

T is the time constant in seconds, R is the resistance in MΩ (1 MΩ = 1,000,000 Ω), and C is the capacitance in microfarads (μF). As an example, if the resistor has a value of 100,000 Ω (0.1 MΩ) and the capacitor is 10 μF, the time constant will be equal to 0.1 MΩ $\times$ 10 μF, 1 s (second). The same time constant could be achieved with other RC combinations. For example, 1 MΩ and 1 μF, or 10,000 Ω and 100 μF would also result in a one second time constant.

As another example, take a 470,000 Ω (0.47 MΩ) resistor and a 2.2 μF capacitor. The time constant in this circuit would equal 0.47 MΩ $\times$ 2.2 μF, or 1.034 s. If the resistor is 22,000 Ω (0.022 MΩ), and the capacitor is 0.3 μF, the time constant is 0.022 $\times$ 0.3, or 0.0066 s (6.6 ms—milliseconds).

Here are two more examples for you to work out on your own. What is the time constant for an RC circuit consisting of a 3.9 MΩ resistor and a 0.68 μF capacitor? How about a 910,000 Ω resistor and a 15 μF capacitor?

Filters

A *filter* is a circuit that allows some frequencies to pass through it but blocks other frequencies. A capacitor is automatically a sort of filter by definition, because it allows higher frequencies to pass through it easily, but it blocks a dc signal (that is, a signal of 0 Hz). A filter that passes high frequencies, but blocks low frequencies is called a highpass filter.

The circuit for a practical, but simple highpass filter is shown in Fig. 6-18. High frequencies are passed with little or no *attenuation*. That is, they are not reduced in

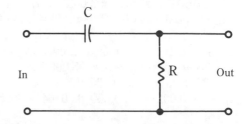

6-18 Passive high-pass filter.

level. But, at some specific frequency, the filter starts to impede the passage of the signal. This is the *cutoff frequency*.

It would be impossible to design a practical filter that would completely block all signals below the cutoff frequency and completely pass all frequencies above it. Instead, the level of the signal allowed to pass through the filter begins to drop off at a regular rate. This dropping off is called the *slope* of the filter. Obviously, the steeper the slope, the better the filter. For a simple filter, such as the one in Fig. 6-18, the slope is quite broad. A graph of the filter frequency response is shown in Fig. 6-19.

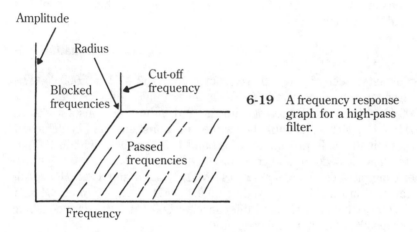

6-19 A frequency response graph for a high-pass filter.

The formula for finding the cutoff frequency of this type of filter is:

$$F_c = \frac{159,000}{RC}$$ **Equation 6-4**

Resistance (R) is in MΩ, and capacitance (C) is in microfarads, so the resulting cutoff frequency (F_c) is in hertz. Notice that the denominator in this equation is the formula for finding the time constant of an RC circuit. The formula could be rewritten as:

$$F_c = \frac{159,000}{T}$$ **Equation 6-5**

T is the time constant in seconds.

Try a few examples with Equation 6-4. Suppose the circuit in Fig. 6-18 consists of a 20 MΩ resistor (20,000,000 Ω) and a 0.68 μF capacitor. The cutoff frequency would equal 159,000/(20 × 0.68) = 159,000/13.6 or approximately 11,600 Hz.

If you replace the resistor with an 820,000 Ω (0.82 MΩ) unit, and keep the same capacitor, the cutoff frequency now becomes 159,000/(0.82 × 0.68) = 159,000/0.5576 or approximately 285,000 Hz.

Now keep the 0.82 MΩ resistor and use a 0.002 μF capacitor. F_c = 159,000/(0.82 × 0.002) = 159,000/0.00164 = 96951220 Hz, or just under 97 MHz. Notice that if either the capacitance or the resistance is decreased in value, the time constant decreases, and the cutoff frequency increases. Of course, it also works in the other direction. Increasing either the resistance or the capacitance will increase the time constant and decrease the cutoff frequency.

Try finding the cutoff frequency for the following combinations.

R	C
470,000 Ω	0.1 µF
3.9 MΩ	2 µF
22 kΩ	150 µF

Now, what would happen if you reversed the positions of the resistor and the capacitor, as shown in Fig. 6-20?

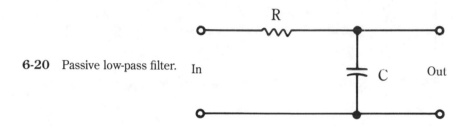

6-20   Passive low-pass filter.

In this case, the capacitor provides a *short circuit*, or *shunt* across the signal source for high frequencies, but not for low frequencies and dc, which can't pass through the capacitor. Because current will tend to flow through the path with the least resistance (or reactance), very little high-frequency current will get through to the output.

Low frequencies, however, face a high reactance if they try to flow through the capacitor to ground, so they will appear at the output. This filter, then, passes low frequencies, but blocks high frequencies. Not surprisingly, this circuit is called a *lowpass filter*. It operates as an exact mirror image of the highpass version, as is indicated by its frequency response graph (Fig. 6-21). The formula for finding the cutoff frequency is identical for both circuits.

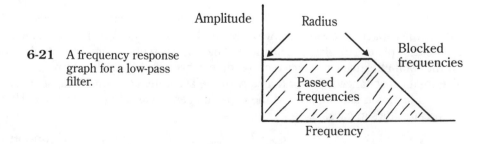

6-21   A frequency response graph for a low-pass filter.

Combinations of capacitors

Now, examine what happens when you have more than one capacitor in a circuit.

Capacitors in parallel

A circuit with two capacitors in parallel, as shown in Fig. 6-22, can be drawn more pictorially, as in Fig. 6-23. Because plates A and B are tied together, they are at the same electrical potential. you can think of them as a single plate. Similarly, plates C and D are electrically combined into an apparent single plate.

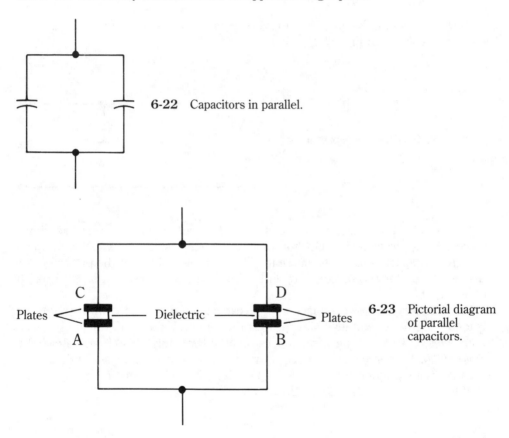

6-22   Capacitors in parallel.

6-23   Pictorial diagram of parallel capacitors.

Remember that the larger the surface area of the plates in a capacitor, the higher the capacitance will be. Obviously, combination plate A-B is going to be larger than either plate A or plate B separately. The same is true of combination plate C-D. So the total effective capacitance of multiple capacitors in parallel always increases. The total capacitance is larger than any of the separate, component capacitances.

In fact, you can simply add the capacitances of capacitors in parallel. That is, for n capacitors in parallel:

$$C_T = C_1 + C_2 + \ldots C_n$$ **Equation 6-6**

Notice that this formula is the same as the formula for finding the total resistance of multiple resistors in series.

Capacitors in series

Similarly, capacitors connected in series, as in Fig. 6-24, work against each other, reducing the total effective capacitance of the circuit. The formula for capacitors in series mirrors the formula for multiple resistors in parallel:

$$\frac{1}{C_T} = \frac{1}{C_1} + \frac{1}{C_2} + \ldots \frac{1}{C_n}$$

Equation 6-7

6-24 Capacitors in series.

Therefore, two 0.1 μF capacitors in series would act like a single 0.05 μF capacitor. If the same capacitors were connected in parallel they would equal 0.2 μF. Of course, both series and parallel combinations of capacitances can be included within a single circuit, just as with resistances.

For example, consider the string of capacitors in Fig. 6-25. Assume C_1 is 0.1 μF, C_2 is 0.033 μF, C_3 is 0.0015 μF, and C_4 is 0.22 μF. First solve for the series combination of C_1 and C_2. $1/C_T = 1/C_1 + 1/C_2 = 1/0.1 + 1/0.033 = 10 + 30.30303 =$ about 40. Taking the reciprocal, you find the series combination of C_1 and C_2 is approximately 0.025 μF. This capacitance is in parallel with C_3, so $C_T = 0.025 + 0.0015 = 0.0265$.

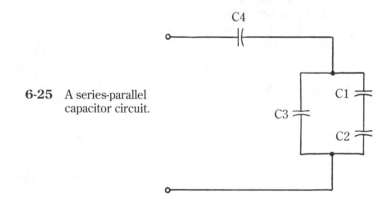

6-25 A series-parallel capacitor circuit.

This effective capacitance is in series with C_4, so the effective capacitance for the entire circuit equals $1/0.0265 + 1/0.22 = 37.7 + 4.5 = 42.2$. Taking the reciprocal, you conclude that the total effective capacitance is just under 0.024 μF.

Now, find the total effective capacitance for the circuit if $C_1 = 15$ μF, $C_2 - 0.47$ μF, $C_3 = 3.3$ μF, and $C_4 = 2.2$ μF.

Stray capacitances

Because a capacitor is simply two conducting surfaces, separated by an insulator, small, unintentional capacitances can be formed by adjacent wires, or component leads. Generally, these *stray capacitances* are far too small to be of any real significance, but in some very high frequency circuits (such as radio circuits) they can be very troublesome. These undesirable capacitances can allow signals to pass into portions of the circuit where they could hinder proper operation.

To prevent such stray capacitances in high frequency circuits, leads should be as short as possible reducing the effective plate area. Leads should also be *shielded* (that is, enclosed in a conductor that is connected to ground—either earth ground or common ground) if more than just a few inches long.

Self-test

1. What is a dielectric?

A A conductive plate in a capacitor
B A measurement of capacitance
C A charged particle
D An insulator between two metal plates in a capacitor
E None of the above

2. Which of the following describes the action of a capacitor?

A Opposes changes in current flow
B Converts ac into dc
C Creates a dc resistance
D Stores electrical energy
E None of the above

3. What type of signal experiences the greatest resistance through a capacitor?

A Low-frequency signals
B ac signals
C High-frequency signals
D Out-of-phase signals
E None of the above

4. Assuming an ideal capacitor, with no leakage, what is the capacitive reactance of a 10 µF capacitance at dc (0 Hz)?

A $0 \, \Omega$
B $1,000,000 \, \Omega$
C $16,00 \, \Omega$
D $63 \, \Omega$
E None of the above

5. What is the capacitive reactance of a 33 µF capacitor at 500 Hz?

A $0 \, \Omega$
B $9.55 \, \Omega$

C 144 Ω
D 1,000,000 Ω
E None of the above

6. What is the capacitive reactance of a 33 μF capacitor at 6500 Hz?

A 0.74 Ω
B 7.4 Ω
C 96 Ω
D 1122 Ω
E None of the above

7. What is the capacitive reactance of a 0.47 μF capacitor at 4300 Hz?

A 8 Ω
B 65 Ω
C 79 Ω
D 2121 Ω
E None of the above

8. If two 0.25 μF capacitors are connected in series, what will be the total effective capacitance?

A 0.50 μF
B 0.0625 μF
C 0.125 μF
D 2.5 μF
E None of the above

9. If two 0.25 μF capacitors are connected in parallel, what will be the total effective capacitance?

A 0.50 μF
B 0.0625 μF
C 0.125 μF
D 2.5 μF
E None of the above

10. What is the cutoff frequency of a passive lowpass filter made up of a 0.56 μF capacitor and a 330 kΩ resistor?

A 860 kHz
B 5.4 kHz
C 54 Hz
D 860 Hz
E None of the above

7

Magnetism and electricity

Very closely related to the concept of electricity is the concept of magnetism. In this chapter, you study how these two phenomena interact.

What is a magnet?

Magnetism has been known to humans for well over 2000 years. The ancient Greeks discovered a peculiar lead-colored stone that had the mysterious ability to attract small particles of iron ore. Some time later, the Chinese found a practical use for this seemingly magical stone. They learned that if a piece of this stone is suspended on a string or floated on a liquid it always tries to point in one specific direction (north). Because they used this device to lead them through the desert, the stone came to be called *lodestone* (that is, the leading stone).

You know now that the lodestone is a natural *magnet*. Although in some ways, magnetism is still rather mysterious, much is now known about its properties. Magic is not involved. You can make magnets out of certain other materials, even though they aren't naturally magnetized. Lodestone is a fairly weak magnet, but stronger magnets can be made of iron, nickel, cobalt, or steel.

The two opposite ends of a magnet are called the *poles*. See Fig. 7-1. One pole will tend to point towards the earth's north pole if the magnet is floated or freely suspended. This north-seeking pole is called the *north pole* of the magnet. The other pole is referred to as the *south pole*.

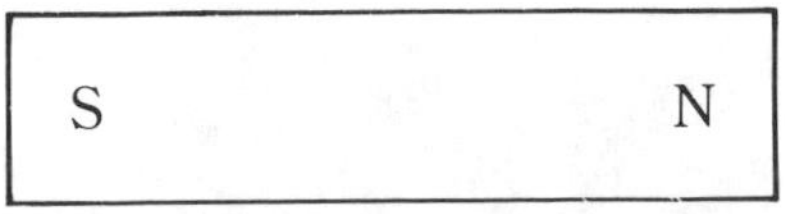

S   N

7-1 Poles of a magnet.

Remember that in an electrical circuit, like charges repel and opposite charges attract. The same effect occurs with magnetic poles. If two magnets are brought together, north pole to north pole, they will try to repel each other. If, however, one of the magnets is turned around so that the north pole of one magnet is facing the south pole of the other, the magnets will exhibit a strong attraction towards each other.

If you place a bar-shaped magnet under a sheet of paper, sprinkle some iron filings on top of the paper, and shake the paper gently, the filings will tend to arrange themselves into a pattern like the one shown in Fig. 7-2.

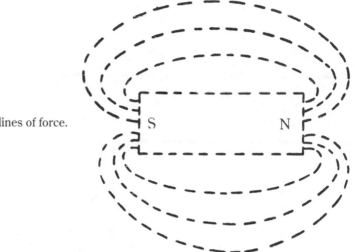

7-2   Magnetic lines of force.

Notice that the iron filings arrange themselves in a set of parallel lines arcing from one pole to the other. These lines never cross or unite. They are an indication of the *magnetic lines of force*, or *flux*. The area they cover is the *magnetic field*.

The flux flows from the north pole to the south pole of the magnet, just as electrical current flows from the negative terminal to the positive terminal of a voltage source. The flux is produced by a force called *magnetomotive force*. Magnetomotive force is somewhat analogous to electrical voltage (which is also sometimes called *electromotive force*).

Just as certain substances conduct electrical current better than others, certain substances allow magnetic lines of flux to pass through them more readily than other substances. In other words, some materials present a greater resistance to the flux. The magnetic equivalent of resistance is called *reluctance*.

The similarities between magnetism and electricity are so strong that Ohm's law applies to magnets too. In magnetic circuits, flux equals magnetomotive force divided by reluctance. This relationship directly reflects the electrical formula, current equals voltage divided by resistance ($I = E/R$).

Producing magnetism with electricity

When an electric current passes through a conductor, such as a piece of copper wire, a weak magnetic field is produced. The magnetic lines of force encircle the wire at right angles to the current flow, and are evenly spaced along the length of the conductor. See Fig. 7-3. The strength of the magnetic field decreases at greater distances from the conductor. The size and overall strength of the magnetic field is dependent on the amount of power flowing through the electrical circuit, but it is always fairly weak. The magnetic force surrounding the conductor can, however, be dramatically increased by winding the wire into a coil, so the lines of force can interact and reinforce each other.

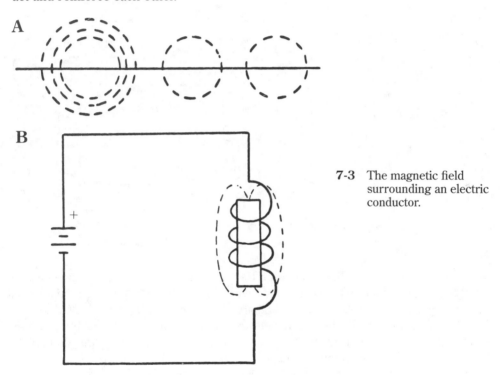

A

B

7-3 The magnetic field surrounding an electric conductor.

An even greater magnetomotive force can be generated if the coil is wound around a piece of low reluctance material, such as soft iron. Because the magnetomotive force vanishes as soon as the current stops flowing in the wire, you have a magnet that can be turned on and off. The strength of the magnet is also electrically controllable. Such a device is called an *electromagnet*.

Producing electricity with magnetism

Because you can produce magnetism with an electrical current, it shouldn't be surprising that you can also produce electricity with a magnet. Look at Fig. 7-4. It is basically the same as Fig. 7-3, but there is no electrical voltage source, and the material in the center of the coil (the *core*) is a permanent magnet.

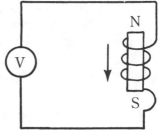

7-4   Producing electricity with a magnet.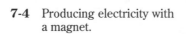

If you move the magnet up through the coil of wire, an electric current will start to flow through the wire. The strength of this induced current depends on a number of factors. These include the intensity of the magnetic field, how many lines of force are cut by the conductor, the number of conductors (each turn of the coil acts like a separate conductor in this case) cutting across the lines of force, the angle at which the lines are cut, and the speed of the relative motion between the magnet and the conductor.

This current will continue to flow until either the magnet is too far away for any of its lines of force to cut across the conductor, or the magnet stops moving.

If the magnet and the coil are stationary with respect to each other, no current is induced. Then, if you push the magnet back down through the coil (the direction of the movement is reversed) current will also flow, but it has the opposite polarity. That is, it flows in the other direction. The exact same effect can be achieved if the magnet is stationary and the conductor is moved. It is the relative motion between the components that is important.

All this might not seem terribly useful, because you have to keep moving the magnet or the coil back and forth to produce a continuing current. The current will keep reversing polarity each time the direction of movement is changed, but this method is actually a very efficient way of producing electricity.

This concept is used by power companies to produce their high-wattage ac power. Any of a number of mechanical means can be used to rotate a conductor between a magnetic north pole and south pole (see Fig. 7-5). It is usually more

7-5   Producing ac electricity with a magnet.

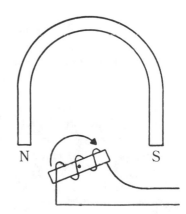

practical to rotate the conductor rather than the magnet. Because the conductor is rotating between the magnetic poles, the direction of its relative movement between the poles appears to alternate, so the induced current, as mentioned above, is an alternating current. Very large amounts of electrical power can be produced in this manner.

Health concerns and electromagnetic fields

These are times of health and safety concerns. Almost every week, the news tells you of something new to worry about. Sometimes these reports are based on valid, even serious concerns. Other times they are based almost entirely on mistakes and misinterpretations, or even deliberate hoaxes. In almost all cases, the practical risk in the real world is greatly exaggerated in general news reports. There is a crying need for intelligent perspective in reporting ecology and health issues to the general public.

It is no longer possible or reasonable to write a book on technological theory without considering such issues. These issues are inescapable in the modern society. You can't afford to ignore them. Unfortunately, but inevitably, the concern forces the book a bit of a detour from the discussion of electronics theory. But the issues at hand are important enough to make such a detour worthwhile.

At the risk of going even further astray from the main subject of this book, it is probably useful to first offer some general thoughts about such issues and how they are reported to and perceived by the general public. Then you can take a more informed look at the current concerns over the alleged health-related effects of electromagnetic fields, and other specific issues that relate to electronic technology.

The problem of proof versus disproof

It is a sad fact that overall the reporting of scientific issues in the popular press (television, newspapers, general news magazines, etc.) has been deplorable. There are very few news reports on scientific issues that do not suffer from major errors of fact. This is not to say such misstatements are due to deliberate or conscious bias. They are the results of poor (or non-existent) scientific training on the part of the reporters. Generally, a solid science background is not considered a requirement for science news reporting. More often than not, such assignments are made solely on journalistic qualifications. Often, any related technical experience or schooling (or lack thereof) is considered irrelevant. But in today's culture, scientific and technological issues are vitally important. How can they adequately be reported by someone who doesn't really understand them? It's like sending someone who has never heard of football out to cover the Super Bowl. No matter how good a journalist the person is, he or she will turn in a very distorted story, filled with errors and misinterpretations. The reporter will not know which facts are important and need emphasis and which are more or less irrelevant.

Scientific and technological issues are often complex, and most professionals in this area don't have very good communications skills. They can write reports that are understandable by their equally trained colleagues, but they all too often don't really know how to break down the relevant information so it is understood by the general public. So the scientifically untrained reporter interviews a scientist, and even though he or she doesn't really understand the underlying theories and principles, the reporter must simplify and condense what the scientist said for the readers. But if the reporter doesn't fully understand the key principles involved, how can he or she avoid introducing errors and misconceptions when simplifying or even rewording. Even using the scientist's exact words might lead to misinformation if quotes are taken out of context.

For example, scientific caution often leads to public paranoia over undefined risks. When a scientist says, for instance, that there is no 100% safe level of radioactivity, that is not necessarily grounds for concern. It certainly doesn't mean that all radioactivity is inherently bad. There is some scientific evidence that indicates that life would not be possible without some minimal amount of radioactivity. Even a level of zero would be unsafe in the sense the scientist means.

Radioactivity occurs when an atom emits subatomic particles and energy. When certain subatomic particles hit a chromosome, they can cause genetic damage. The more subatomic particles there are in a given area (that is, the higher the level of radioactivity), the better the odds that a particle will hit a chromosome and cause damage. As long as there is at least one subatomic particle, there is always some chance that it might make a lucky hit. The odds would be overwhelmingly against a hit under such circumstances but not impossible. A scientist would not say the situation was completely risk free, unless the chances of the negative event occurring are true zero, without rounding off. If there is 1 chance in 1,000,000,000,000, it still counts as a risk to the scientist, even though in practical terms, there is virtually no risk at all. So when the scientist says no level of radioactivity is completely safe, he just means the risk is never true zero, but the lay person interprets the statement as meaning that any radioactivity at all is actively dangerous.

Often the odds of being harmed by a technological risk that many people worry about are considerably poorer than the odds of being killed by a herd of stampeding zebras on Main Street. It could happen, but it's not at all likely.

Another example is that some reports stated that scientists had not ruled out the possibility of AIDS being contracted by kissing. When scientists were asked if it was possible to catch AIDS by kissing, they replied that such a possibility had not being conclusively disproven, even though there was no scientific evidence to suggest it was possible. The odds were very strongly against such a connection, but the scientists didn't want to claim to be all knowing. It is always extremely difficult to conclusively prove that there is a true 0% chance of contracting any disease by any given means. It has not been conclusively proven that making funny faces in a mirror can't cause AIDS, but there is no realistic reason to worry about making funny faces in the mirror. The specific connection has been neither proven nor disproven, and existing evidence suggests it is an unlikely connection.

Just because something has not been conclusively disproven does not mean it has been proven. Similarly, just because something has not been conclusively proven does not mean it has been disproven. It seems that many, if not most lay people, and even some scientists and technical professionals with excellent academic credentials, have great difficulty making this critical distinction. It is vital to understand this to assess any technological risk intelligently.

There is a critical difference between proof and evidence that is often ignored, even by professionals and experts. Inevitably this leads to foolish dogmatism and misinterpretation of facts. Proof is absolute and unquestionable. In the real-world, true scientific proof is very rare and perhaps even nonexistent. There is never any intelligent controversy over proof. Evidence, on the other hand, is subjective by definition. It requires interpretation. Two equally qualified experts might interpret the exact same piece of evidence in two very different, and perhaps even totally opposite, ways. To complicate matters further, they might not agree on just what is or isn't valid, meaningful evidence for the question at hand. One expert's "conclusive evidence" might be unconvincing, or even totally irrelevant to another expert with equal credentials.

Just because someone is an expert in a given field, does not mean he or she is automatically always right. An experiment that gives very impressive results, which might look like conclusive proof to the lay person (and some experts) might later be found (probably by other experts) to be seriously flawed in its procedures, perhaps in a very subtle and unexpected way, so the results are not truly meaningful at all. There is a tendency to say "such-and-such an experiment (or series of experiments) proves this-and-that conclusion." In almost all cases it would be more accurate to say that "such-and-such an experiment (or series of experiments) indicates this-and-that conclusion."

Lay people must always bear in mind that all experts in every field are human beings too, with the same sorts of failings, unconscious biases, and blind spots as the rest of humanity. Every expert is wrong in his field of expertise at least some of the time.

Any time there is a controversy on any scientific issue, it means that the experts are in disagreement over the interpretation of the available evidence. For every expert who says yes, there is another expert with equally impressive credentials who says no. If the issue had been proven, there could be no scientific controversy.

Curiously, and perhaps surprisingly, the lay person is more likely to be exposed to the radical, minority viewpoint in scientific controversies. An interested, intelligent lay person might even find it difficult to find nonprofessional materials presented the majority viewpoint. This is because a radical, sensationalist theory will sell a lot more books at the newsstand, and a book covering a conservative, traditional theory will appear dry and dull to the lay person, who will probably leave it on the racks. So the conservative, mainstream scientific viewpoint is not as potentially profitable to publishers, so such general-level books are less frequently written and published. The mainstream scientific viewpoint is usually covered in some detail in the technical and professional journals for the field in question. Such publications might be hard for the general public to get hold of, or to understand them once found. But the wild, minority theories are covered in the news, in general magazines, and in popular books. The wilder and the more sensational the theory, the better it will sell to the general public. A viewpoint with little scientific support might appear to be scientific truth to the general public.

Notice that such theories are not automatically wrong. The mainstream scientific community has been dead wrong many times in the past and will similarly err many times in the future. The point is, such sensationalist theories should not be accepted as gospel. The more media hoopla there is over any scientific idea, the more controversial it must be, and the more strongly it goes against mainstream scientific thinking.

It is rarely, if ever, valid or reasonable to conclude that the experts who oppose the sensationalist theory have a vested interest, or in the pay of those who do. In fact, such a conclusion is almost always a sure sign of paranoid thinking. If there wasn't good scientific reason doubt the new theory, there would be no real controversy over it. The theory might ultimately turn out to be completely true, but as long as the controversy exists, there is room for doubt, and it is unscientific and even foolish to insist "such-and-such is absolutely true" or "such-and-such is absolutely false" while the controversy continues, indicating that not all qualified individuals have drawn the same conclusions from the inevitably subjective and incomplete evidence.

Often the new, controversial theory ultimately turns out to be only partially true. For example there might be some risk involved with "such-and-such," but not nearly as much as was claimed in the sensationalist literature.

Many authors, even those with strong academic credentials in the field will write a sensationalist book more for monetary than scientific reasons. They might (perhaps not completely consciously) exaggerate the case to sell more books. Read any scientific literature oriented toward the general public cautiously, and never let any one author fully convince you on the issue. (Yes, that includes the author of this book.)

Often the title will give you a clue about the author's intent in writing the book. For example, one book on the subject of the alleged health risks of electromagnetic fields, which you will read about, has the lurid, sensational and alarmist title *Currents Of Death*. Such a wild, almost paranoid title should raise some doubt in the intelligent reader. To be fair, sometimes publishers change the author's title to help the book sell more copies, so check out the text before drawing conclusions. If a book has a wild title but cautious, well-reasoned arguments in the text, the title probably wasn't the author's. But often, the text will be just as wild-eyed and paranoid as the title. Take such an author's arguments with a grain of salt. The author might be sincere, but is lacking in perspective on the issue. Many fully qualified experts have an axe to grind that might get in the way of their scientific judgment.

Just because an author has a string of degrees, it does not mean he or she is automatically correct in his or her conclusions. In the book, the author might slant the evidence to support conclusions and ignore or misrepresent conflicting evidence. This might not be deliberate bias or fraudulent intent. Strong personal beliefs as a human being might be clouding his or her judgment.

Do electromagnetic fields affect human health?

The long detour before discussion of the specifics of the current electromagnetic field controversy is for a reason. Virtually all of the published information on this subject has been from the minority viewpoint that there are serious health risks to the general public from artificially generated electromagnetic fields. The main-stream scientific viewpoint is that there is little or no real risk to the general public from artificial electromagnetic fields. That does not necessarily mean that the minority viewpoint is wrong. They might be right, or they might be partially right. The problem is that the lay public is being given the false impression that the minority viewpoint is the mainstream, majority conclusion of all "honest" experts. This idea is simply not true.

This issue is discussed as honestly and fairly as the author can in the limited space available. Personal opinions do not intentionally color the presentation of the evidence. In fairness to you, the author's opinions will be revealed before you complete the subject. No one is obligated to share the author's viewpoint. It is presented only so that you can compensate for any unintended bias in the discussion of the controversy.

Natural electromagnetic fields

First, be aware that you are always surrounded by electromagnetic fields. They occur naturally, and there is some fairly strong scientific evidence that if natural electromagnetic fields did not exist, life itself could not be sustained. In many respects, the Earth itself is a giant magnet. An reasonable discussion of the health aspects of electromagnetic fields must begin with an understanding of the natural electromagnetic background you are all continuously exposed to.

Because electromagnetic fields are completely natural phenomena, any risks associated with them must be of quantity, not of kind. Despite some of the more paranoid and sensationalistic of some of the writings on this subject, electromagnetic fields are not inherently harmful or bad.

All electromagnetic fields are, by definition, force fields. That is, they carry energy, and can produce an action at a distance. For example, a permanent magnet can move a small metallic object some distance away. The farther from the source, the weaker the electromagnetic field gets. The field strength drops off rather quickly, following a logarithmic, rather than a linear pattern.

It is also important to realize that there are two basic types of radiation. Not all radiation is harmful. For example, ordinary light and heat are both types of radiation. Some forms of radiation are *ionizing*, and others are *non-ionizing*. Ionizing energy, such as X rays, are known to be biologically harmful and can cause a direct chemical reaction (ionization). Traditionally, non-ionizing radiation was considered to be mostly or entirely harmless. This is the cornerstone over the current controversy over the alleged health effects of electromagnetic fields, which are non-ionizing. If they have any effect, it is apparently indirect, and it is certainly less than that of ionizing radiation. If non-ionizing radiation was more potentially harmful than ionizing radiations, its effects would been discovered earlier, or at least at about the same time. The effects of ionizing radiation have been known for some time. The effects of non-ionizing radiation are still unclear and scientifically questionable.

In a sense, the planet Earth is a gigantic bar magnet, as shown in Fig. 7-6. The north and south poles of magnets are named for their electromagnetic similarity to the Earth's North Pole and South Pole. The true magnetic poles of the Earth are not entirely stationary. They move around slightly, and are usually not located precisely at the true geographic poles.

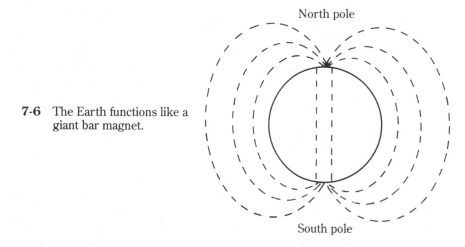

7-6 The Earth functions like a giant bar magnet.

There is some scientific evidence that from time to time in the past, the electromagnetic field of the Earth has reversed polarity. That is, the North magnetic pole became the South magnetic pole, and vice versa. Some scientists hypothesize that such large-scale magnetic field reversals were responsible for the periodic mass deaths of species that have mysteriously occurred in the Earth's past. This theory is quite intriguing, but it is very, controversial, and must not be accepted as proven fact. Many, perhaps most scientists today do not believe any such magnetic pole reversals have ever occurred, or even that they could occur. The existing evidence, although somewhat impressive, is still highly questionable.

Many (not all) authors who believe in the harmful effects of electromagnetic fields tend to accept such controversial theories as given facts. This acceptance places their hypotheses on shaky ground right from the start.

The core of the Earth is molten rock, very heavy in iron. The spinning core of the Earth creates a dipole magnetic field with a magnetic north pole and south pole. As with any permanent magnet, force lines extend from pole to pole, as shown in Fig. 7-6. This illustration is not accurate, however. It shows what the magnetic fields of the Earth would look like if it was alone in the universe. But the Earth is far from the only electromagnetic object in the universe. It's electromagnetic fields are inevitably acted upon by other cosmic objects. The primary influence is the Sun, because it is so large and relatively close. For your purposes, you can assume that only the electromagnetic interference from the Sun is of significance. The effects from other planets, the moon, and nearby stars are real, but far weaker and more subtle than the effects from the Sun. They don't change the overall picture significantly.

The Sun constantly emits a force known as the *solar wind*. Now, this is not a true wind, in the way you normally think of it here on Earth. An ordinary wind is a movement in air, and there is no air in space. The solar wind is a flow of high-energy atomic particles emitted from the surface of the Sun. In many respects, a solar wind acts rather like an ordinary Earth wind, so the name is appropriate, as long as you don't take it too literally.

The solar wind contains particles with very high energy and moving at high speeds. Some of these high energy particles are of the ionizing type, but others are non-ionizing. There is still plenty of energy left in the solar wind by the time the particles travel the distance from the Sun to the Earth. The solar wind therefore interacts with the Earth's natural electromagnetic fields. On the side facing the Sun, the solar wind particles push against the Earth's magnetic fields, compressing them. Meanwhile, the fields on the far side of the planet are "blown" outward by the solar wind to form a long magnetotail. These effects are shown in Fig. 7-7.

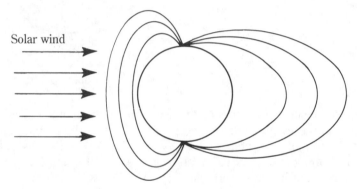

Solar wind

7-7   The Earth's magnetic fields are distorted by the effects of solar wind.

The collision of the solar wind particles with the Earth's magnetic fields creates a *bow shock region* in which these forces interact. Two special areas within this bow shock region are known as the Van Allen belts. Some of the solar wind high-energy particles are trapped within these belts, where they constantly spiral between the north and south ends of the *ducts*, as shown in Fig. 7-8.

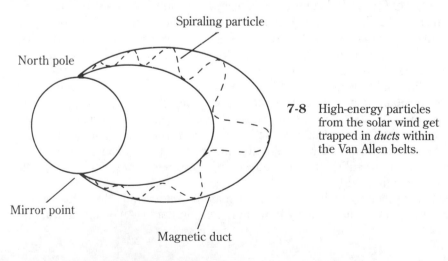

Spiraling particle

North pole

7-8   High-energy particles from the solar wind get trapped in *ducts* within the Van Allen belts.

Mirror point

Magnetic duct

The magnetosphere (the distorted magnetic fields surrounding the Earth) shields the planet from much of the Sun's radiation, especially the potentially harmful ionizing rays. If the magnetosphere was destroyed or removed, all life on Earth would cease to exist. Space flights beyond the Earth's magnetosphere must be of limited duration to prevent harmful effects to the astronauts exposed to the Sun's powerful radiation without this natural shield. (Theoretically, an artificial magnetic shield could be designed and incorporated into future spacecraft, but such technology does not yet exist.) Such space flight missions beyond the Earth's magnetosphere must also be timed carefully so they occur during quiet periods in the Earth's cycle. During a solar storm, the ionizing radiation emitted from the Sun is much greater.

The Earth rotates within the magnetosphere, which remains stationary. The magnetosphere itself does not rotate. The same side always faces the Sun. As the Earth rotates under the magnetosphere, a given spot on the planet's surface will experience a daily pattern of up and down fluctuations in the strength of the natural electromagnetic field. At certain times of day, the electromagnetic field is stronger, and at other times it is weaker. Some scientists believe these electromagnetic level fluctuations might help explain the daily biological rhythms that occur in many species. For example, people who are placed in caves or enclosed buildings for extended periods with no way to tell time will still tend to synchronize their sleep-waking patterns with the Earth's night-day cycle, at least, approximately. Actually, there seems to be a tendency to act as if the day was a little shorter than the Earth's actual 24 rotation speed.

This theory has a lot to recommend it. However, before ascribing too much importance to this still controversial theory, remind yourself that there are almost certainly other physical phenomena that has a daily fluctuation pattern synchronized to the Earth's rotation. The fluctuations in the natural electromagnetic field might have nothing at all to do with biological rhythms, despite the suggestive evidence. It seems like that they do have at least some effect, but this is far from proven to date.

Magnetic storms

In the Earth's upper atmosphere, the interaction between the solar wind particles and the magnetosphere generates very large electrical currents, often with power levels in the billions of watts range. This level is much larger than virtually all electrical sources created by humans. This natural atmospheric electricity creates ionizing radiation and a number of electromagnetic waves in the ELF (extra-low frequency) and the VLF (very-low frequency) range. The VLF range runs from 100 to 1000 Hz, and the ELF range covers everything from 100 Hz on down to 0 Hz (dc). These effects are all normal in a quiet field resulting from a steady flow of solar wind.

Obviously, because all life on Earth is exposed to these effects, they can not be considered inherently harmful, at least, not at their naturally occurring levels. You must keep in mind that the upper atmospheric layers shield much of this energy from reaching the surface of the Earth itself. Much of the shielding is performed by the natural ozone layer, which is why detected holes in the ozone layer are worrisome.

The picture is complicated by the fact that the Sun is not at all a steady energy source. Solar activity rises and falls over an eleven year sunspot cycle. During periods of high solar activity, solar storms occur with some frequency. These so- called solar storms result from *solar flares*, or huge energy eruptions on the surface of the Sun. Some solar flares can shoot up hundreds of miles above the surface of the Sun. The high energy of a solar flare increases the number of high-energy particles emitted from the Sun. In effect, you have a higher gust of solar wind. There are more X rays, proton streams, and electromagnetic waves in the RF (radio-frequency) region hitting the Earth's magnetosphere. Not surprisingly, these high energy bursts of solar wind can cause significant magnetic field disturbances in the Earth's vicinity. Such disturbances are called *magnetic storms*. They are not necessarily associated with meteorological storms (rain, thunderstorms, wind storms, etc.). Unlike Earth-bound meteorological storms, which are usually geographically localized, magnetic storms often affect much of the Earth's surface area at the same time.

During a magnetic storm, the strength of the Earth's electromagnetic field fluctuates wildly, exhibiting great increases in strength. In some magnetic storms, these disturbances can induce very high power currents in electric power transmission lines and telephone lines. This induced power can sometimes cause a breakdown of the system. At the same time, similar disturbances in the ionosphere can cause significant interference or even complete breakdown in radio and television signals. These effects are relatively uncommon, but they are not rare or unusual in scientific terms. Again, such magnetic storms are a purely natural phenomenon, even though you notice their effects mostly in the disruption of communications and power systems.

There is some (mostly statistical) evidence that animal and possibly even human behavior patterns might be affected during a magnetic storm. Such a connection is highly controversial, however, and far from proven to the satisfaction of most scientists today. Even among those who are convinced of such effects, there seems to be considerable disagreement over just what the effects are—that is, what changes supposedly appear in animal or human behavior. It is inherently difficult to scientifically study such effects (if any), because magnetic storms are not frequent, and they are usually of rather brief duration. It is also difficult to predict just when such natural magnetic disturbances will occur.

Electromagnetic fields created by humans

In the modern, technological world, there are many artificial sources of electromagnetic fields. Any radio or television transmitter is, by definition, a high-frequency electromagnetic field generator. The broadcast signal being transmitted is nothing but a modulated electromagnetic wave. Few scientists today believe that these signals are potentially harmful, unless, perhaps, you are very close to a very powerful transmitter. Even then, inherent risks seem rather doubtful. The largest risk close to a radio or television transmitter antenna is the large ac power levels themselves. If you accidentally complete a circuit to ground, you could suffer a serious electrical shock or perhaps even electrocution. There isn't much controversy there.

The present controversy over the alleged health effects of electromagnetic fields concerns primarily low-frequency fields, generally in the ELF (0 Hz to 100 Hz) range. Through most of the world, ac power plants generate ELF signals with a frequency of either 50 Hz or 60 Hz. As it happens, these particular frequencies do not naturally occur to any appreciable degree in the normal electromagnetic spectrum of the Earth, although they might briefly show up during a magnetic storm.

There are a lot of ac power plants operating in the world today to meet society's increasing energy demands. In the course of the last half century or so, the electromagnetic fields created by humans have more than duplicated the hypothetical changes in strength and frequency that are assumed to have been responsible to past species die outs. Although this fact has been used to fuel paranoia and sensationalism on the electromagnetic field issue, it actually does not indicate a cause for major concern at all. Instead, it seems to call the worrisome theories into serious question. If you have already artificially exceeded the electromagnetic levels that supposedly caused mass species die outs, why haven't there been widespread biological disasters? There have been a lot of species extinctions in recent times. Some have been entirely natural (species became extinct long before mankind and its technology came along), but most can be directly explained by other human causes—mainly pollution, hunting, and destruction of natural habitats. There haven't been any mysterious extinctions that can be reasonably attributed to electromagnetic field disruptions caused by modern technology.

These man-made electromagnetic fields have been around for decades and have covered most of the globe. Any health effects therefore must be very subtle in nature, otherwise there would be no room for controversy. This supposition does not mean that all concern over health effects of electromagnetic fields is trivial and inappropriate, but it does indicate a need to keep perspective. Subtle health effects can be cumulatively important, and could conceivably lead to disastrous effects in the future. But lurid sensationalism such as *Currents of Death* is unquestionably unscientific and irresponsible. The more extreme claims being made in this controversy have more to do with superstitious paranoia and *technophobia* than science. The ridiculous claims made by some paranoids could ultimately worsen the problem if it exists. If there is a subtle health risk from man-made electromagnetic fields, the extreme claims are so ridiculous that the mainstream of science is likely to dismiss the entire question altogether as unworthy of investigation. The legitimate grounds for concern gets covered up and masked by the nonsense, so the necessary research is delayed, if not put off altogether.

Those with serious, scientifically valid concerns about the potential health effects of electromagnetic fields should speak out the loudest against the sensationalist nonsense spewed out by the paranoid extremists, which can only hurt the overall cause in the long run. The extremists discredit the entire question, which is generally the case in almost all such controversies.

Electromagnetism and biology

Electromagnetic phenomena have been known since ancient times. The mysterious and amazing ability of a magnet to act upon other objects with no direct physical connection between them has stimulated the human imagination. Until relatively recently, only natural magnetism and static electricity were familiar. But in the last couple of centuries, electricity has gone from a curiosity to an all-persuasive fact of life.

The link between electricity and magnetism was discovered early. Another early discovery was the presence of electrical currents within living things. For example, nerve impulses are both chemical and electrical in nature. Because electricity and magnetism seemed so magical to begin with, and there was definitely some biological connection, many fanciful theories and devices were created in massive quantities, especially in the 19th century. Hundreds of electrical or magnetic cures were touted for almost every imaginable ailment from broken bones to arthritis to general fatigue and listlessness. Some of these electromagnetic cures were concocted by sincere scientists with an incomplete understanding of the phenomena involved, but many were the work of out and out quacks and con artists. None of them were accepted by mainstream medicine because none of them worked. In fact, many of the so-called electrical cures were actually quite dangerous, subjecting the patient to potentially severe shock risks.

Perhaps the most "successful" electrical cure was shock therapy used for mental patients. A controlled electrical shock could calm down a violent or seriously disturbed patient. It also rattled their brains, damaged memories, and often affected normal nerve functioning. This rather barbaric treatment enjoyed quite a startling popularity in mainstream medicine for a while, but is now almost universally rejected as far more harmful than helpful. In fact, it is doubtful that it ever did much more good than simply hitting the troublesome patient in the head.

When hypnotism was first discovered, it was often called *animal magnetism*. It was imagined that some sort of mysterious electromagnetic rays were transmitted from the hypnotist to the subject. Of course, there were never any such rays. Neither electricity nor magnetism have anything at all to do with the altered state of consciousness known as hypnotism. In fact, self-hypnosis demonstrates that a separate hypnotist is not required at all (although a hypnotist can be helpful in guiding the session). There is good reason to suspect that all hypnosis is actually self-hypnosis, even when there is a separate hypnotist offering suggestions to the hypnotized subject. The subject is never really under the hypnotist's control. Hypnotized subjects can and do reject hypnotic suggestions that go against their moral code and values.

The ultimate electromagnetic "cure" was suggested in Mary Shelley's famous novel, *Frankenstein*. In this novel, Dr. Frankenstein revived a reassembled corpse by applying electricity from captured lightening. Shelley probably got the idea from experiments that showed dismembered frog legs would jerk if an electrical voltage was applied to them. Of course, the voltage was simply stimulating the electrochemical nerve endings in the frog legs. They were completely dead, and the applied electricity made no difference in that fact, despite the resulting lifelike movement. Such experiments have no more to do with regenerating life by electrical means than using electricity to cause a motor to move. It's just a natural electromechanical phenomena.

Electromagnetic cures are still advertised from time to time, but to date there is no solid evidence at all that any such gadgetry works, or is even based on scientific principles.

Of course, electricity can have definite harmful biological effects. The effects are especially present in high power (large current capacity) ac voltages. Low-power electrical shocks can cause some of the body's nerve synapses to fire or to lock up, more or less randomly. The shock can cause bodily jerks, and temporary paralysis. A shock victim might be unable to control muscles sufficiently to let go of the hot wire giving the shock.

The uncontrollable physical jerks can lead to falling or hitting things, or other potentially dangerous accidents. If the applied current is increased, seizures, heart failure or other serious medical conditions could be caused. An electrical shock of enough force can be fatal.

Of course, electricity and magnetism are not the same thing, although they are closely related. Biological processes do utilize tiny electrical currents, though there doesn't appear to be any way to positively stimulate biological processes by applying external electrical currents. But what of magnetism?

Here the biological connection is even more tentative and vague. It is true that all living things are surrounded by a magnetic field extending out into space. Theoretically it would seem that living organisms possibly could be influenced by external electromagnetic fields. After all, there would have to be a physical interaction of the external electromagnetic fields and the dc electrical currents flowing through the organisms. But in practical terms, the magnetic field strength would almost certainly have to be extremely large to make much difference. Any external electromagnetic field would have to be stronger than the Earth's natural electromagnetic field to counteract its natural effects (if any). If a compass works well in the vicinity, the primary magnetic field in the area must be the Earth's natural magnetic field, so it is doubtful that any other localized electromagnetic field is having a very significant influence. The natural force field is stronger, so it must be the primary controlling force.

Of course, some artificial electromagnetic field sources are strong enough to cause a compass to malfunction and give an incorrect reading. If there are any biological effects from external electromagnetic fields, they would occur almost exclusively under such conditions. A compass is surely more sensitive to magnetic fields than any biological organism.

There is some controversial evidence that some biological organisms might sense and respond to relatively minor fluctuations in the electromagnetic field of the Earth. But all of the existing evidence indicates that such effects are very minor and secondary. No known living thing responds to magnetism as a primary sense.

Biological rhythms or cycles have been well established since the 1960s. Many living things follow daily and annual patterns in certain ways. The sleep-waking cycle is an obvious example. Some creatures are *diurnal*—that is, they sleep at night and are active during the day. Other creatures are *nocturnal*, sleeping during the day, and becoming active at night. They will more or less maintain these daily patterns, even if ordinary sense clues are made unavailable. For example, in a cave or an enclosed room, the creature can not see if the Sun is up or not. So how do they know when it is night and when it is day?

These biological cycles are still largely a mystery. Anyone who claims to definitely know exactly how they work is a fool or a charlatan. A number of hypotheses have been suggested. One popular hypothetical explanation is that the creatures in some way sense the fluctuations in the Earth's magnetic field and adjust the timing of their biological cycles accordingly.

Notice that even if this hypothesis is correct, the magnetic influence still must be a subtle one. If a dozen subjects are studied under the same conditions (including the same exposure to the same magnetic fields), their biological cycles will not be perfectly synchronized. In some cases they might be significantly out of phase with one another. This means something else must be playing a part in the equation.

Some experiments have indicated more or less consistent response to controlled magnetic field changes in some biological organisms. The most successful of these experiments have been with very, simple organisms, such as bacteria. It is much, more difficult to establish a definite magnetic response in more complex biological organisms.

The most intriguing experiments along these lines with developed creatures have been some studies of homing pigeons that suggest the birds might use magnetic cues as a secondary, back-up method of finding their way home. Of course, the more obvious senses are far more important. There is no question that homing pigeons use visual cues whenever they can. But when the pigeons are fitted with special contact lenses that make it impossible to navigate visually, they can still find their way home. If a pigeon is fitted with the contact lenses and a small permanent magnet, they often seem to get confused or even lost. This result suggests they are somehow using magnetic cues which are blocked by the magnet they are carrying. On the other hand, if a pigeon is not visually impaired, attaching a permanent magnet to it doesn't seem to make much difference in its ability to find its way home. The magnetic cues are obviously secondary in the pigeons' navigation processes.

Dr. Robin Baker of the University of Manchester in England has conducted some controversial experiments suggesting that even humans have an innate ability to sense the direction of magnetic north. Placing a bar magnet against a subject's forehead for about 15 minutes can disrupt this magnetic sense for as long as two hours after the magnet is removed. These results are far from universally accepted in the scientific community. Most scientists doubt that any such magnetic sense exists in humans. After all, if such a sense existed to any particular degree, why were such direction finding devices as the compass and the sextant ever invented? Human beings can get lost too easily to make an ability to sense magnetic north a very likely suggestion. It's not just a matter of modern artificial electromagnetic fields confusing a natural magnetic sense. People were getting lost long before they started using electricity.

Once again, there might be something to Baker's proposed magnetic sense, but it is obviously a very weak, secondary sense that is normally almost completely unused by humans. (It would probably be reasonable to delete the word *almost* from that last sentence, but there is always some room for doubt.)

Until fairly recently, it was assumed that any biological effects from artificial electromagnetic fields were theoretically impossible. Modern science has indicated there is a real possibility of some subtle effects. Notice that these effects are not proven, they are just theoretically possible. This implies a need for further research, not for fear or paranoid concern. Any such effects must logically be quite subtle and secondary.

Most of the current links between health problems and artificial electromagnetic fields have been statistical. This means there is considerable room for error.

There might be some unaccounted for factors. The apparent statistical link might be coincidental, or the result of a common cause. For example, assume say condition A appears to be statistically linked with condition B. Does this mean condition B causes condition A? Possibly, but not necessarily. There might be an unaccounted condition C that causes both condition A and condition B. Or condition C might simply cause condition A, and condition B is simply an unrelated red herring. Statistical links can be useful scientific tools, but they never make up adequate scientific proof by themselves. There is always some built-in margin of error in any set of statistics. Real life always has too many variables to permit simple, unambiguous statistical links.

There have been a number of cases of workers employed around strong electromagnetic field sources having problems—usually just a vague feeling of being slightly ill, or abnormally tired. A few statistical studies of populations living near power plants or other large artificial ELF sources have greater incidence of certain psychological or medical effects. However, these studies have been far from conclusive. Different studies as often as not indicate entirely different effects under apparently similar conditions. You must also remember that any power plant or other large-scale ELF source will almost certainly be putting out larger than normal pollution and heating levels, adding more variables to the cause-and-effect question. Such statistical studies are suggestive, but they are not valid scientific proof themselves.

High frequency electromagnetic signals, especially microwaves, have been demonstrated to cause unnatural heating or feverlike effects under some conditions. Such high-frequency signals can also cause chromosomal damage under some conditions, especially in ionizing radiation is involved. This happens because the chromosomes or other cellular level components can physically resonate with the wavelength of the high frequency signal. Such resonance effects could not occur with ELF (extra- low frequency) fields, of course. The wavelengths are far, too long. This fact does not automatically rule out any biological effect from ELF fields—only direct resonance effects.

It is true that ELF fields are effectively stronger than comparable high frequency fields. That is, a ELF wave of a given wattage will tend to travel further than a high frequency signal of the same wattage.

ELF fields are more penetrating than higher frequency fields. An ordinary RF signal can be blocked by the ground or the ocean, so regular radio won't work in a submarine. But signals can be transmitted too and from a submerged submarine if an ELF carrier frequency is used. Because the higher the frequency of an electromagnetic field the more easily it is blocked, it is logical to assume that ELF signals can get places that higher-frequency signals will never reach. Does this make a difference biologically? Possibly, but it has not been scientifically proven as yet.

Some experiments have indicated that prolonged exposure to strong ELF fields might (but not always) lead to an increase in certain biochemicals in the body. These biochemicals are normally associated with stress. If this connection is true, it implies that strong ELF fields could, under some conditions, cause stress responses. Of course, a great many different things can cause stress in humans. There is no reason to assume that ELF fields would be inherently any worse than other stress stimulator.

ELF fields have been experimentally linked to cancer, but then, what hasn't been? If everything that supposedly causes cancer really did, it is truly remarkable that the entire human race didn't die out years ago.

There have been other experiments with rats that indicated prolonged exposure to strong ELF fields might be related to increased learning difficulties, and genetic damage. There might also be some effect on fetal development. These experimental results are all still controversial. It is also worthwhile to remember that virtually all such experiments involve subjecting the subject rats to much stronger ELF fields than the average human being is likely to encounter. Once again, you have some cause for concern and further research, but there are hardly any reasonable grounds for paranoia or fear of a massive epidemic.

ELF fields in the average home

The strength of the Earth's natural magnetic field at any given location typically has an average value of about 0.5 gauss. (*Gauss* is measure of magnetic flux density.) The daily fluctuation is only about 0.1 gauss. This is a seemingly small level. A small permanent magnet like the one that holds your refrigerator door closed is typically about 200 gauss. Such facts are used by sensationalist authors to stir up inappropriate fear. It sounds like an impressive difference between the Earth's natural magnetic field and the refrigerator door magnet. But the magnetic field surrounding the permanent magnet drops off very rapidly with distance. The 0.5 gauss average value for the Earth's magnetic field already takes these drop off factors into account. The actual magnetic field emitted by the Earth is much larger than this. A few inches away from the refrigerator door magnet, its magnetic field will be negligible in comparison with the Earth's natural magnetic field. You can easily prove this yourself. Get a small compass. If you place it right next to the magnet, it will give an incorrect reading for north. Move a few inches away, and the compass will start working normally.

This apparently dramatic comparison—0.5 gauss to 200 gauss is misleading and even a little dishonest, because the 0.5 gauss value for the Earth's magnetic field is distance compensated, and the 200 gauss rating for the magnet is not. Such logical errors apply to many, if not most, of the claims for electromagnetic health risks from household appliances.

At least one author claims that some devices, such as electrical blankets exhibit significant electromagnetic fields, even when the switch is off. This is not only incorrect, it is absolutely ridiculous. When the switch is off, there is no current flow, which means there is nothing to generate any electromagnetic field at all. If the author in question measured such fields, as claimed in the book, they obviously came from some other source. Such silly arguments should be ignored.

Some electromagnetic alarmists express concern over the electromagnetic fields surrounding such lower power devices as electric clocks. Again, this concern doesn't hold up to common sense. The electric clock on my desk is rated for 2.5 W. About half of this input power is used by a small lamp to illuminate the clock face. The remaining power must overcome the physical inertia and friction of the motor. In any electrical circuit most of the unused power is wasted as heat dissipation. Sure, the coils in the motor are surrounding by electromagnetic fields, as explained elsewhere in this book. But these fields are quite weak unless very close. The self-induction effects of a coil (see chapter 8) normally occur over distances of just a tiny fraction of an inch. If there is too much spacing between the coil loops, it will not work. In conducting some informal experiments, it appears that at a distance of about two feet, the gauss meter couldn't detect the presence of an electric clock at all. To be worried about the electromagnetic fields surrounding an electric clock is pure superstitious paranoia.

The most serious concern of electromagnetic radiation in the average home would be in a television set. Virtually all the energy radiated from any television set is in the form of simple light (from the picture tube) and heat. Old tube television receivers sometimes generated fairly large amounts of X rays, which could possibly be harmful. Extensive shielding was legally required. These legal requirements for X-ray shielding have recently been relaxed considerably. This is not something for anyone to worry about. It is not a matter of trusting the manufacturers to keep the X-ray emissions down. The old problem was from high voltage tube circuits which are now totally obsolete. The modern solid-state circuitry used in all of today's television sets do not act as X-ray sources to any meaningful extent. (Remember, your own body emits some X rays.) The old required shielding serves no particular purpose, because what it shielded against is no longer being generated in the first place.

A television set does emit some RF electromagnetic fields. These drop off quite quickly with distance. It probably wouldn't hurt to avoid sitting too close to the television set, though it is questionable as to just how harmful such energy is. It is essentially the same as the broadcast signals that are always all around you whether you even own a television set or not. Proof that such emissions from the television receiver drop off quickly with distance are easy to come by. Place two television sets near each other, and tune them to different channels. They don't interfere with each other, do they?

A lot has been written about computer monitors and radiation. Much of what has been said on this subject is pure nonsense. A computer monitor certainly is no more harmful than an ordinary television set, because electronically, they are virtually the same thing. The biggest difference is that a dedicated computer monitor doesn't have any television tuner circuitry. Computer monitors are usually designed with a somewhat wider frequency response than ordinary television receivers, but this isn't of any particular significance for the issue at hand.

The legitimate concern over computer monitors is that the user normally sits so much closer to the screen than someone watching television. Documented risks mostly concern eyestrain, back pains, and neck pains from holding an unnatural posture for extended times.

There might be some conceivable risk from the computer itself, as opposed to the monitor. The circuitry in a computer uses very high frequencies, and some high-frequency electromagnetic fields are emitted. The power levels are usually quite low, however. There might be some risk, but it is probably not severe.

To be on the safe side, it might be a good idea to alternate computer work with some other tasks away from its electromagnetic fields. The risk, if any, seems to be the result of prolonged rather than momentary exposure. Even the most rabid sensationalists seem to agree that exposure over time is the critical factor. Brief exposure to most electromagnetic fields is almost certainly of negligible importance.

Some tentative conclusions

The author's opinions on the subject follow so that you can compensate for any inadvertent bias in the discussion. There is little or no realistic risk from electromagnetic fields to the average citizen. There might be some subtle effects, but there are many far greater risks that you are constantly exposed to that this one seems rather negligible. Common household appliances are almost certainly not a significant threat. If you have any concern, the most reasonable response is to minimize usage and to keep a few feet away during normal use, if possible. Coming close to operate controls would not substantially increase any risk—it is prolonged exposure that counts, if anything.

The more or less constant electromagnetic fields from your ac electrical wiring in the walls will almost certainly swamp out the electromagnetic fields emitted by almost any home-size appliance unless you remain in abnormally close contact with it for extended periods. The electromagnetic fields generated by the house wiring are always there, whether you are using any appliances or not.

However, if you live directly under a power-line transfer point or work in a power plant or some other installation with a lot of heavy-duty electrical equipment, the electromagnetic field levels you are exposed to might be abnormally high. Greater research is needed to determine exactly what the true effects are, if any, and what can be down about the problem. What type of shielding would be effective? How much is too much?

If you are concerned about nearby power lines, most electrical companies will do a electromagnetic field reading for you at no cost, or for only a small fee. This could help set your mind at ease.

The most likely effects, if any, from prolonged exposure to electromagnetic fields seem to be in the form of increased stress stimulation. A great many factors can contribute to stress. Stress can cause many health problems, some even fatal, and its effects are often cumulative, so this is not an entirely trivial concern.

There have been experiments showing that sounds just outside the audible spectrum (either very low or very high) can cause stress effects. Perhaps the risk element is not really ELF fields at all, but an irritating sub-audible tone. This is just a casual top-of-the-head suggestion. It is not intended as a serious scientific hypothesis, just a vague possibility. Perhaps you might want to try to devise some small experiments to test this idea.

At the very worst, the health risks from electromagnetic fields (if they exist at all) are subtle and indirect. The issue calls for greater examination and research, but it is hardly grounds for panic, or even widespread concern among the general population. There are many far worse problems and risks facing us. The potential health risks of electromagnetic fields should not be ignored, but they shouldn't be blown out of all realistic proportion either.

It is not a major threat or critical health crisis. It is been the author's experience that the people who make the most noise about the alleged health risks of electromagnetic fields tend toward being a bit technophobic overall, mistrusting all technology in general. This does not prove that there is no risk at all, of course. But it does suggest that the potential risks are probably being exaggerated.

Self-test

1. Which of the following cannot be used to make a magnet?

A Lodestone
B Cobalt
C Carbon
D Iron
E Nickel

2. What are the ends of a magnet called?

A Poles
B Lodestones
C Ions
D Armatures
E None of the above

3. What is another name for magnetic lines of force?

A Armature
B Flux
C Magnetic pole
D Lodestone
E None of the above

4. If like poles of two magnets are brought near each other, what will happen?

A They will attract each other
B They will be damaged
C They will repel each other
D An electrical current will be generated
E None of the above

5. What is the magnetic equivalent to electrical voltage?

A Flux
B Magnetomotive force
C Reluctance
D Magnetic field
E None of the above

6. What is the magnetic equivalent of electrical current?

A Flux
B Magnetomotive force
C Reluctance
D Magnetic field
E None of the above

7. What is the magnetic equivalent to electrical resistance?

A Flux
B Magnetomotive force
C Reluctance
D Magnetic field
E None of the above

8. How can an electrical current be induced with a coil and a magnet?

A Placing the coil at right angles to the magnetic field
B Placing the coil parallel to the magnetic field
C Holding both the magnet and the coil perfectly stationary
D Moving either the magnet or the coil
E None of the above; it can't be done

9. Rotating an armature in a magnetic field produces what type of electricity?

A Static
B ac
C dc
D Pulsating dc
E None of the above

10. What is the frequency range for an ELF field?

A 0–100 Hz
B 100–1000 Hz
C 500–1000 Hz
D 100–5000 Hz
E None of the above

8

Inductance

In the last chapter, you learned that winding a wire carrying an electric current into a coil will increase the electromagnetic effect. Another result of passing current through a coil of wire is a phenomenon called *inductance*. Inductance is another important factor in electronic circuits.

What is inductance?

Because an electric current flowing through a coil of wire can create a magnetic field, and a magnetic field moving relative to a coil of wire can create an electric current, what happens when the current flowing through a coil changes? As long as current flows through the coil at a steady, constant level, and in just one direction (dc), a nonmoving magnetic field is generated. As long as the magnetic field and the coil are stationary in relation to each other, the magnetic field will have no particular effect on the current flow through the coil. But if the current through the coil starts to drop, the magnetomotive force generated by the coil will also be decreased, causing the magnetic lines of force to move in closer. Some of these moving lines of force will cut through some of the turns of the coils, inducing an electric current in the coil. This *induced current* will flow in the same direction (same polarity) as the original current.

Of course, this induced current passing through the coil will produce a magnetic field of its own. A finite time is required for this back-and-forth effect to die down. Current through a coil cannot be stopped or reversed in polarity instantly. Inductance tends to oppose any change in the current flow.

In some ways inductance is the opposite of capacitance. Capacitance offers very little resistance to high frequencies, but opposes low frequencies, or dc (constant current). Inductance, on the other hand, passes dc with practically no resistance, but opposes higher ac frequencies (changing current). This opposition to high frequencies is called *inductive reactance*.

Inductance is measured in *henries* (H). One henry is the inductance in a circuit in which the current changes its rate of flow by one ampere per second and induces one volt in the coil. The henry is too large a unit for practical electronic circuits, so the millihenry (one thousandth of a henry—mH) is more commonly used.

Inductive reactance

The formula for inductive reactance is:

$$X_{L} = 2\pi FL$$ **Equation 8-1**

where X_{L} is the inductive reactance in ohms, L is the inductance in henries (not millihenries), and F is the frequency in hertz. 2π, of course, is a constant, equalling approximately 6.28.

Suppose you have a circuit with 100 mH of inductance (0.1 H). If the frequency of the source voltage is 60 Hz, then the inductive reactance equals $6.28 \times 60 \times 0.1$, or just under 38 Ω. If the same circuit is used, but the applied frequency is increased to 500 Hz, the inductive reactance becomes $6.28 \times 500 \times 0.1$, or 314 Ω.

Raising the ac frequency still further, to 2000 Hz, brings the inductive reactance up to $6.28 \times 2000 \times 0.1$, or 1256 Ω.

If the frequency remains constant, but the inductance is increased, then the reactance will also be increased. For example, you've already found that 100 mH in a 60 Hz circuit results in an inductive reactance of about 38 Ω. If you increase the inductance to 500 mH (0.5 H), and keep the frequency at 60 Hz, the inductive reactance comes out to $6.28 \times 60 \times 0.5$, or approximately 188 Ω.

You'll notice the relationship between the frequency and the inductance to the inductive capacitance is just the opposite of that of the frequency and the capacitance to the capacitive reactance.

Determine the inductive reactance of a circuit consisting of 25 mH (0.025 H) of inductance at 50 Hz, 300 Hz, and 4000 Hz. Then change the inductance to 300 mH (0.3 H) and solve for the same three frequencies.

Coils

In electronics, the component called a *coil* (or *inductor*) is just that—a coil of insulated wire wound around some core. This core might be made of powdered iron or some other magnetic material, or it might simply be air or a small cardboard tube.

The inductance of a coil is determined by a number of factors: the width of the core, the diameter of the wire, the number of turns of the wire around the core, and the spacing between the turns of the coil, to name just a few of the more important factors.

The material the core is made of is also important. A core with low magnetic reluctance can increase the strength of the magnetic field, thereby increasing the strength of the induced voltage. Figure 8-1 shows the construction of a typical coil.

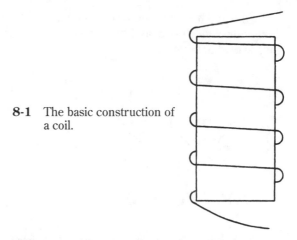

8-1 The basic construction of a coil.

Some coils are adjustable. Usually the core is constructed so it can be moved slightly in and out of the center of the coil with a screw called a *slug*. This arrangement makes the core appear to be partially made of air and partially of (usually) powdered iron, thereby altering the reluctance of the core and thus the inductance of the coil.

Many coils have the wires visibly exposed (but insulated, of course). But some are sealed in metal cans to avoid interaction with other components. Without this *shielding*, the magnetic field could induce a voltage in other nearby components. Obviously inducing a voltage where it's not intended can be detrimental to circuit operation.

The wires in a coil must usually be insulated, because they are generally wound quite closely together. If separate turns of uninsulated wire shift position and touch, allowing current to pass between them, a *short circuit* exists, making the coil appear to have fewer turns, as far as the current is concerned. See Fig. 8-2.

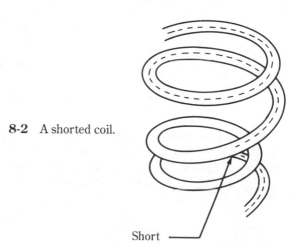

8-2 A shorted coil.

Short

Figure 8-3 shows the most common schematic symbols for coils. A and B are fixed inductance coils, and the arrows through symbols C and D indicate that these coils are adjustable.

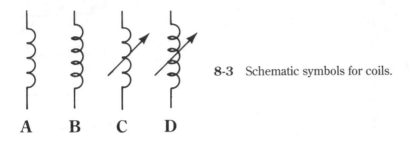

8-3   Schematic symbols for coils.

A   B   C   D

Sometimes special schematic symbols are used to indicate the core material of the coil. In this system, all of the symbols in Fig. 8-3 are for *air-core* inductors. If two dotted lines are drawn beside the symbol, as in Fig. 8-4A, it means the core is made of powdered iron. Two solid lines, as in Fig. 8-4B, indicate that the core is made of stacks of thin sheet iron.

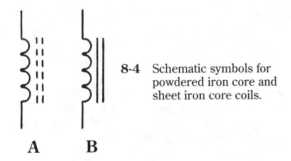

8-4   Schematic symbols for powdered iron core and sheet iron core coils.

A   B

Most schematic diagrams, however, simply use the standard symbols shown in Fig. 8-3, and the type of coil is specified in the parts list. Coils (and inductance values) are usually represented by the letter *L*, because *I* is commonly used for current.

Winding coils

A number of factors determine the inductance of a coil. These include the core material and diameter, the number of turns in the coil, and how closely they are spaced. For a single-layer coil (no overlapping windings) on a nonmagnetic core, the formula for determining inductance is as follows:

$$L = \frac{0.2d^2N^2}{3d + 91}$$

where L is the inductance of the coil in millihenries (mH), d is the diameter of the coil winding in inches, 1 is the length of the coil winding in inches, and N is the number of turns in the coil.

For example, assume you have a coil on a 0.75 inch diameter core, consisting of 150 closely wound turns of #32 enameled wire. The length of the coil is 1.2 inches. The inductance works out to:

$$L = \frac{0.2 \times 0.75^2 \times 150^2}{3 \times 0.75 + 9 \times 1.2}$$
$$= \frac{0.2 \times 0.5625 \times 22500}{2.25 + 10.8}$$
$$= \frac{2531.25}{13.05}$$
$$= 193.96552 \approx 194 \text{ mH}$$

If the number of turns (N) is increased, and the diameter (d) is held constant, the inductance (L) will be increased. The amount of increase will depend on whether the length is increased by the added windings, or if it is held constant to the original value of 1, by squeezing the turns more tightly together.

Return to the earlier example and double the number of turns ($2 \times N$). All of the other values, including length (l) will remain the same:

$$d = 0.75$$
$$l = 1.2$$
$$N = 150 \times 2 = 300$$

How does this affect the inductance of the coil? Use the formula to find out:

$$L = \frac{0.2 \times 0.75^2 \times 300^2}{3 \times 0.75 + 9 \times 1.2}$$
$$= \frac{0.2 \times 0.5625 \times 900000}{2.25 + 10.8}$$
$$= \frac{10125}{13.05}$$
$$= 775.86207 \approx 776 \text{ mH}$$

The change in inductance works out to:

$$\frac{776}{194} = 4$$

or, the square of the increase in the number of turns. That is, $2^2 = 4$.

Now, try the same problem, but assume doubling the original number of windings also doubles the length of the coil. That is, the spacing between the windings is not changed:

$$d = 0.75$$
$$l = 1.2 \times 2 \times = 2.4$$
$$N = 150 \times 2 \times = 300$$

$$L = \frac{0.2 \times 0.75^2 \times 300^2}{3 \times 0.75 + 9 \times 2.4}$$

$$= \frac{0.2 \times 0.5625 \times 90000}{2.25 + 21.6}$$

$$= \frac{10125}{23.85}$$

$$= 424.5283 \approx 424 \text{ mH}$$

When increasing the number of turns (N) increases the coil length (l), while the diameter (d) remains constant, the original inductance value (L) will be multiplied by a factor slightly greater than the multiple of 1 and N. In the example, l and N are multiplied by 2, and the inductance increases by a factor of:

$$\frac{424}{194} = 2.185567$$

In practical applications, you will probably know the desired inductance value (L), and will need to determine how to achieve that value. Start out by selecting a likely core form of a given diameter (d), and pick a reasonable length. Now the basic inductance formula can be algebraically rearranged to solve for the necessary number of turns (N):

$$N = \sqrt{\frac{L(3d + 91)}{0.2d^2}}$$

As an example, assume you need a 100 mH coil. You have a 1.2 inch diameter coil form. You'll set the coil length (l) at 1 inch. This means the number of turns in our coil should work out to:

$$N = \sqrt{\frac{100 \, (3 \times 1.2 + 9 \times 1)}{0.2 \times 1.2^2}}$$

$$= \sqrt{\frac{100 \, (3.6 + 9)}{0.2 \times 1.44}}$$

$$= \sqrt{\frac{100 \times 12.6}{0.288}}$$

$$= \sqrt{\frac{1260}{0.288}}$$

$$= \sqrt{4375}$$

$$= 66.14 \approx 66 \text{ turns}$$

Use the same coil form and length (d and l constant), and change the desired inductance to 250 mH. This time the required number of turns works out to:

$$N = \sqrt{\frac{250\ (3 \times 1.2 + 9 \times 1)}{0.2 \times 1.2^2}}$$

$$= \sqrt{\frac{250 \times 12.6}{0.288}}$$

$$= \sqrt{\frac{3150}{0.288}}$$

$$= \sqrt{10937.5}$$

$$= 104.58 \approx 104\text{-}\tfrac{1}{2} \text{ turns}$$

Several magnetic materials are often used for coil cores. These include iron, powdered iron, ferrite, and nickel alloy. When a magnetic core is used, a new factor enters into the equation for determining the inductance of the coil. This new factor is the permeability of the core material, and is represented by the Greek letter mu, μ.

The equation for determining the inductance of a coil with a magnetic core is as follows:

$$L = \frac{4.06N^2\ \mu\ A}{0.27_c \times 10^8 \times 1}$$

Once again, N stands for the number of turns, l is the coil length in inches. A is the cross-sectional area of the coil in square inches. Try a typical example. Set N equal to 1000 turns, μ to 5000, A to 0.4 inches, and l to 3 inches. This means the inductance of the coil will work out to:

$$L = \frac{4.06 \times 1000^2 \times 5000 \times 0.4}{0.27 \times (10^8) \times 3}$$

$$= \frac{4.06 \times 1000000 \times 5000 \times 0.4}{27000000 \times 3}$$

$$= \frac{8120000000}{81000000}$$

$$= 100.24691 \approx 100 \ \mu\text{H}$$

If you don't change the core, the values of μ, l, and A will remain constant. This means you can change the inductance value (L) simply by changing the number of turns (N).

Return to the previous example and increase the number of turns by a factor of 3:

$$N' = N \times 3 = 1000 \times 3 = 3000$$

Now see what this does to the inductance (L):

$$L = \frac{4.06 \times 3000^2 \times 5000 \times 0.4}{81000000}$$

$$= \frac{9000000 \times 8120}{81000000}$$

$$= \frac{73080000000}{81000000}$$

$$= 902.2222 \approx 900 \ \mu\text{H}$$

Increasing N by a factor of 3 increased the value of *L* by a factor of 9, or 3^2. The inductance varies as the square of the number of turns.

To design a coil for a specific inductance value (*L*), you can solve for the value of *N* (number of turns) by rearranging the basic equation like this:

$$N = \sqrt{\frac{0.27 \ (10^8) \ L1}{4.06 \ \mu A}}$$
$$= \sqrt{\frac{6650246.3054L1}{\mu A}}$$

Try an example by using the same coil from the earlier problems:

$$A = 0.4$$
$$l = 3$$
$$\mu = 5000$$

Assume you need a 500 μH coil, and solve for the number of turns required (*N*):

$$N = \sqrt{\frac{0.27 \ (10^8) \times 500 \times 3}{4.06 \times 5000 \times 0.4}}$$
$$= \sqrt{\frac{40500000000}{8120}}$$
$$= \sqrt{4987684.7}$$
$$= 2233.3125 \approx 2233\text{-}\tfrac{1}{4} \text{ turns}$$

Another type of coil uses a *toroidal core*. This coil is a toroid of magnetic material, as shown in Fig. 8-5. A *toroid* is a ring, or doughnut shaped object.

8-5 A toroidal core is a doughnut-shaped piece of magnetic material.

This type of coil core offers a number of important advantages, which include small size and compactness, high *Q*, and, perhaps most important of all, self-shielding. Toroidal cores can be operated at extremely high frequencies.

The formula for determining the inductance (*L*) of a toroidal coil is as follows:

$$L = 0.011684N^2 \times \mu \times h \times \log_{10} (OD/ID)$$

where *N* is the number of turns, μ is the permeability of the core material, *h* is the height of the core in inches, *OD* is the outside diameter of the core in inches, and *ID* is the inside diameter of the core in inches.

As a typical example, assume the following values:

$$N = 60$$
$$\mu = 400$$
$$h = 0.15$$
$$OD = 0.75$$
$$ID = 0.25$$

In this case, the inductance (L) will work out to:

$$
\begin{aligned}
L &= 0.011684 \times 60^2 \times 400 \times 0.15 \times \log_{10}(0.75/0.25) \\
&= 0.011684 \times 3600 \times 60 \times \log_{10}(3) \\
&= 2523.744 \times 0.4771213 = 1204.1319 \approx 1200 \ \mu\text{H}
\end{aligned}
$$

Adding or removing turns in a coil on a given toroidal core (μ, h, OD, and ID) increases or decreases the inductance (L) by the square of the change in the number of turns. For instance, if the number of turns is increased by a factor of 2.5:

$$N' = N \times 2.5$$

the inductance value will become

$$L' = L \times 2.5^2 = L \times 6.25$$

Similarly, if the number of turns is reduced by a factor of 1.75:

$$N' = \frac{N}{1.75}$$

the inductance value will be changed to:

$$L' = \frac{L}{1.75^2} = \frac{L}{3.0625}$$

The basic formula can be rearranged algebraically to determine the number of turns required for a specified inductance, using a specific toroidal core:

$$N = \sqrt{\frac{L}{0.011684 \ \mu h \ \log_{10}(OD/ID)}}$$

As an example, solve for N, using the following values:

$$L = 250 \ \mu\text{H}$$
$$\mu = 400$$
$$h = 0.15 \text{ inch}$$
$$OD = 0.75 \text{ inch}$$
$$ID = 0.25 \text{ inch}$$

The value of N should be equal to:

$$N = \frac{250}{0.011684 \times 400 \times 0.15 \times \log_{10} (0.75/0.25)}$$

$$= \frac{250}{0.70104 \times \log_{10} (3)}$$

$$= \frac{250}{0.70104 \times 0.477}$$

$$= \frac{250}{0.33448}$$

$$= 747.42642 \approx 747\text{-}\tfrac{1}{2} \text{ turns}$$

If you want to double the inductance (L), the number of turns will have to be increased by a factor of $\sqrt{2}$ or approximately 1.41.

RL circuits and time constants

A circuit consisting of a coil and a resistor, like the one shown in Fig. 8-6 also has a definite associated time constant, just as resistive-capacitive (RC) circuits do (see chapter 6). For *RL* (resistive-inductive circuits, the time constant (in seconds) is found by dividing the inductance (in henries) by the resistance in ohms:

$$T = L/R \qquad\qquad \textbf{Equation 8-2}$$

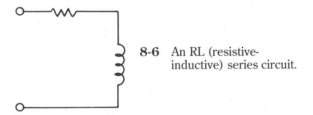

8-6 An RL (resistive-
inductive) series circuit.

In an RL circuit, the time constant is the time required for the induced current to reach 63% of its full value. For example, suppose the coil in Fig. 8-6 has an inductance of 100 mH (0.1 H), and the resistor has a resistance of 1000 Ω. $T = L/R = 0.1/1000 = 0.0001$ second.

Increasing the inductance to 200 mH (0.2 H), changes the time constant to 0.2/1000, or 0.0002 second. Increasing the inductance increases the time constant.

On the other hand, if the inductance is kept at 100 mH (0.1 H), and the resistance is increased to 2000 Ω, the time constant becomes 0.1/2000, or 0.00005 second. Increasing the resistance decreases the time constant.

What would the time constant be for an RL circuit consisting of a 50 mH (0.05 H) coil, and a 470 Ω resistor? What if the inductance is changed to 25 mH (0.025 H)?

Self-test

1. Which of the following characterizes inductance?

A Tends to oppose changes in voltage
B Tends to oppose changes in current
C Tends to oppose dc
D Opposes all frequencies equally
E None of the above

2. The reactance of a 25 mH coil at 5000 Hz is which of the following?

A 0.0013 Ω
B 785,000 Ω
C 785 Ω
D 13 Ω
E None of the above

3. What is the reactance of a 25 mH coil at 600 Hz?

A 785 Ω
B 94 Ω
C 0.011 Ω
D 94,000 Ω
E None of the above

4. What is the inductance of a single-layer coil on a 0.8 inch diameter nonmagnetic core with a length of 1.25-inch, and 320 turns of wire?

A 960 mH
B 3.8 mH
C 3.8 H
D 1200 mH
E None of the above

5. Assume you need a 150 mH coil on a 0.75 inch nonmagnetic diameter, 1 inch long. How many turns of wire will be required?

A 15,000
B 15
C 122-½
D 507-½
E None of the above

6. If you have a coil consisting of 500 turns on a magnetic core with a cross-sectional area of 0.35 inch, and a permeability rating of 750, and the coil is 1.5 inches long, what is the inductance?

A 6580 mH
B 6.6 mH
C 13 mH
D 100 mH
E None of the above

7. Assume you have a 50 turn toroidal coil 0.15 inch high, with an outside diameter of 0.8 inch, an inside diameter of 0.3 inch and a permeability of 400. What is the inductance of this coil? (Rounding off is permissible.)

A 4,224 mH
B 185 mH
C 73 mH
D 725 mH
E None of the above

8. How many turns are required to double the inductance of the coil described in question 7?

A 71
B 100
C 200
D 25
E None of the above

9. What is the time constant of a 500 mH coil and a 3,300 ohm resistor in series?

A 1650 seconds
B 0.0015 second
C 0.00015 second
D 6.6 seconds
E None of the above

10. In a RL circuit, the time constant is the time required for the induced current to reach what percentage of its full value?

A 100%
B 0%
C 63%
D 37%
E None of the above

9
Transformers

If a number of shielded coils are placed in a circuit in series (see Fig. 9-1), their inductance values simply add, the same as with multiple resistances in series. That is:

$$L_T = L_1 + L_2 + \ldots L_n \qquad \textbf{Equation 9-1}$$

n, of course, represents the total number of inductors in the series circuit.

9-1 Coils in series.

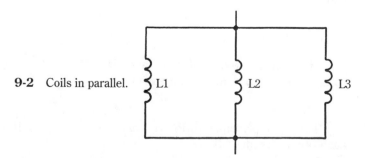

If the shielded coils are placed in a parallel circuit (see Fig. 9-2) the formula for finding the total effective inductance is similar to the resistance formula. The reciprocal of the total effective inductance equals the sum of the reciprocals of each of the parallel inductances. That is:

$$\frac{1}{L_T} = \frac{1}{L_1} \frac{1}{L_2} + \ldots \frac{1}{L_n} \qquad \textbf{Equation 9-2}$$

9-2 Coils in parallel.

You'll notice that shielded coils are specified in these equations. Shielded coils are needed because unshielded coils can interact. If the inductances are allowed to interact, the equations are complicated by a factor called the *coefficient of coupling*.

Coefficient of coupling

If two unshielded coils are brought within fairly close proximity of each other, their magnetic fields will interact. The magnetic lines of force from one coil will cut across the turns of the other so the coils will each induce a voltage in each other, as well as the voltage they induce in themselves. This inductance is called *mutual inductance*. Previously, you have considered only *self inductance*. The strength of the mutual inductance between coils is described as the coefficient of coupling.

If the coils are placed so that only a few of their magnetic lines of force interact, they are said to be *loosely coupled*. If, on the other hand, they are placed very close to each other so most of the lines of force from each coil cuts across the other, they are *closely coupled*.

This mutual inductance can either aid or oppose the self inductance of the coils, depending on the respective polarity of their magnetic fields. If both north poles or both south poles are facing each other, that is, each coil is wound in the opposite direction (see Fig. 9-3), the mutual inductance will be in opposition to the self inductance.

L1   L2

9-3 Coils wound in opposite directions produce negative mutual inductance.

If, however, a north pole is facing a south pole the coils are wound in the same direction (see Fig. 9-4) the mutual inductance will aid, or add to the self inductance.

L1   L2

9-4 Coils wound in the same direction produce positive mutual inductance.

Mutual inductance is represented algebraically by the letter M, and is measured in henries or millihenries, the same as self inductance (L). If two coils are in series and wound so that their mutual inductance aids their self inductance, the total effective inductance can be found with the following formula:

$$L_T = L_1 + L_2 + 2M$$ **Equation 9-3**

Similarly, if the mutual inductance is such that the magnetic fields of the two coils oppose each other, the formula is:

$$L_T = L_1 + L_2 - 2M$$ **Equation 9-4**

The same principle works with two coils in parallel. If their magnetic fields aid each other, the formula is:

$$\frac{1}{L_T} = \frac{1}{(L1 + M)} + \frac{1}{(L2 + M)}$$ **Equation 9-5**

If the magnetic fields of two coils in parallel oppose each other, the formula becomes:

$$\frac{1}{L_T} = \frac{1}{(L1 - M)} + \frac{1}{(L2 - M)}$$ **Equation 9-6**

The theoretical maximum amount of coupling between two coils is 100%. Obviously, this is when all of the magnetic lines of force of one coil cut across all of the turns of the other, and vice versa. In practice, 100% coupling can never be totally achieved. However, you can come quite close if both coils are wound upon a single, shared core. See Fig. 9-5.

9-5 Two coils wound on a single core.

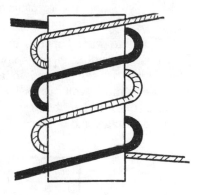

Transformer action

Interacting coils don't need to be in the same circuit. The magnetic field from the coil in one circuit can induce a voltage in the coil of a second, otherwise unconnected circuit. This process is called *transformer action*. A *transformer* is essentially two (or sometimes more) coils wound upon a single core, and arranged so their mutual inductance is at a maximum.

The typical construction of a transformer is shown in Fig. 9-6. The schematic

9-6 Construction of a typical transformer.

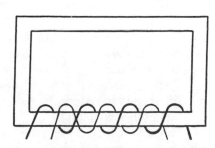

symbols for transformers are shown in Fig. 9-7. The symbol in Fig. 9-7A can be used as any type of transformer, or as an air-core transformer. Figure 9-7B represents a transformer with a powdered iron core, and Fig. 9-7C indicates the core consists of a stack of sheet iron wafers.

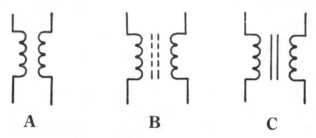

A   **B**   **C**

9-7   Schematic symbols. Air-core transformer (A); powdered-iron core transformer (B); sheet-iron core transformer (C).

Examine what happens in an ideal, transformer without losses transformer. You'll use the simple circuit shown in Fig. 9-8. Assuming that the ac source puts out 100 V rms, the self-induced voltage in the first (or *primary*) coil will also be 100 V rms. Remember, you are assuming there are no losses in the system. If the coil consists of 100 turns of wire, 1 V will be induced in each turn, so the total self-induced voltage is 100 V rms.

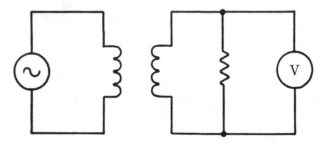

9-8   Circuit demonstrating an ideal, lossless transformer.

Assuming the coils have 100% coupling, the same amount of magnetomotive force will cut across the turns of the other (secondary) coil. That is, 1 V will be induced in each turn of the secondary. If the secondary also consists of 100 turns, 100 V rms will be provided to the load circuit. This kind of transformer is called an *isolation transformer* or a *1:1 transformer*, because the ratio between the two windings is one to one.

Such a device simply electrically isolates the load circuit from the power source. It is often used to reduce the risk of electrical shocks, or to prevent feedback (discussed in other chapters).

But what happens when the turns ratio is not exactly one to one? For example, what if the secondary has only 25 turns? The magnetic field of the primary will still be the same, of course, so 1 V rms will still be induced in each turn of the secondary. But, because there are only 25 turns in the coil, the total voltage available to the load circuit is only 25 V rms. This kind of transformer is called a *step-down transformer*, because the output voltage is stepped down or reduced from the input voltage.

However, the same amount of current is induced in the secondary coils, regardless of the number of turns. Remember, current is the same in all parts of a series circuit, so the current induced in one turn will flow through the entire coil. In other words, the power consumed by the secondary, is the same as the power consumed by the primary. When the voltage is stepped down, the current is stepped up.

Similarly, if the secondary has more turns than the primary, the voltage will be stepped up (increased) and the current will be stepped down (decreased). This device is called a *step-up transformer.*

For example, if the secondary of the imaginary transformer had 250 turns and the primary had 100 turns, the output voltage across the secondary coil would be 250 V. The 100 V rms source and 100-turn primary were selected for these examples, for simplicity. Not all transformers induce 1 V per turn. If the source voltage was increased to 230 V, 2.3 V would be induced in each turn. If the primary consisted of 200 turns, a 230 V input would induce 1.15 volts per turn.

Changing the input voltage, changes the output voltage too. Assume you have a transformer that steps down a 100 V input to a 25 V output. The turns ratio is 4:1. If 230 V were applied to the input, the output would be 57.5 V (all of these voltages are rms).

Suppose you have a transformer with 350 turns in the primary winding and 70 turns in the secondary. What is the output voltage with a 100 V rms input? What is the output if the input is increased to 700 V rms?

Center taps

Many transformers have an extra connection, or *tap* in the center of their secondary winding. The schematic symbol for a *center-tapped transformer* is shown in Fig. 9-9.

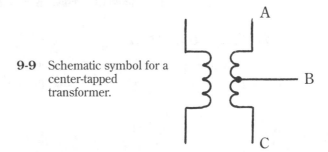

9-9 Schematic symbol for a center-tapped transformer.

The secondary can be considered two coils connected at the center tap. Assume the secondary consists of 100 turns of wire. If the center tap is in the exact center of this coil, there will be 50 turns on either side of the tap. If the primary induces 1 volt per turn in the secondary, the full output from A to C is 100 V rms (100 turns × 1 volt per turn).

From A to B there is only 50 turns. 50 turns $\times$ 1 V per turn = 50 V rms from point A to point B. Similarly, there will be 50 V across the 50 turn coil from B to C. Notice that AB + BC = AC

If the center tap is grounded, as in Fig. 9-10, the voltage across AB and BC will look like the graphs shown in Fig. 9-11. Notice that the two voltages are equal—50 V rms, with respect to ground, but there are 180 degrees out of phase. When voltage AB goes up, voltage BC goes down, and vice versa.

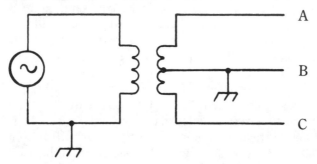

9-10   A typical transformer circuit with a grounded center tap.

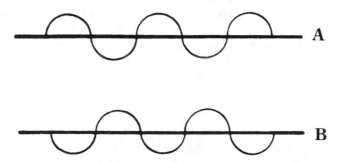

9-11   Voltages in Fig. 9-10 are voltage AB (shown in view A) and voltage BC (shown in view B).

Autotransformer

An autotransformer is a unique form of transformer that uses a single coil winding as both the primary and the secondary, using the center tap principle. See Fig. 9-12.

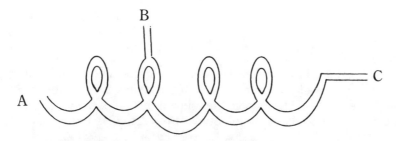

9-12   Construction of an autotransformer.

·Assume there are 100 turns in the entire coil, and 100 V rms is applied across AC. Because the entire voltage is distributed equally across the entire coil, the self induction will equal 1 volt rms per turn.

Now, suppose the center tap (B) is placed so that there are 70 turns of wire between A and B, and only 30 turns from B to C. If you took the voltage across BC, you'd have a 30-turn coil with one volt rms induced across each turn. The output voltage would be 30 V rms. The single coil acts like two very closely placed, but separate coils. An autotransformer can only be used in step-down applications, or step-up applications (the input voltage is applied across BC and the output is taken across AC), but not for isolation. Because the primary and the secondary are the same coil, there is no electrical isolation between them. Usually even in step-down or step-up applications some degree of isolation is required. For this reason, standard two coil transformers are more commonly used. The schematic symbol for an autotransformer is shown in Fig. 9-13. The center tap is often movable.

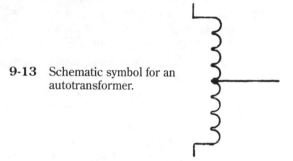

9-13 Schematic symbol for an autotransformer.

Some transformers have more than one secondary. See the schematic symbol in Fig. 9-14. The primary coil, AB, induces a voltage in both CD and EF. The cir-

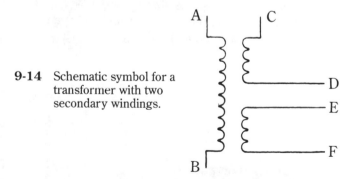

9-14 Schematic symbol for a transformer with two secondary windings.

cuits connected to these two secondaries can be completely isolated from each other electrically.

In addition to voltage changing (step up and step down), and isolation, transformers are often used for impedance matching. This concept is discussed in the next chapter.

Losses in a transformer

So far you have been dealing with ideal transformers without losses that waste no energy. In actual practice, of course, such perfection is never possible. You'll always have to put more power into a transformer (or any other electronic component, for that matter) because the transformer itself will use up, or lose some of the electrical energy.

Most of these losses can be made quite small, so they can usually be ignored if you're only interested in rough, ballpark figures. However, they can be quite significant in certain cases. For instance, a transformer with 100 turns in each winding and with 100 V rms applied should have a nominal output of 100 V rms. In an actual circuit, it might only put out about 98.5 V rms. Another transformer with the same basic specifications, but greater internal losses might only put out 79 V rms or so.

Generally, commercially available transformers are marked with their intended input and output voltages. These figures usually take most internal losses into account. You will rarely have to be concerned with the actual turns ratio of a transformer, just the voltage ratio. You should, however, understand some of the major causes of energy losses in practical transformers.

One important limitation is that 100% mutual inductance can never be achieved between two coils. Some of the magnetic lines of force will be lost into the surrounding air. However, it is possible to achieve a very high coefficient of coupling. Manufacturers can also largely compensate for the difference by adding a few extra turns to the secondary winding.

Because 100% coupling cannot be achieved, some of the magnetic field is leaked off, out of the circuit. These lost magnetic lines of force are called *leakage flux*, and the effect is called *leakage reactance*. Leakage flux can be greatly reduced by winding the coils on an iron core or some other low reluctance substance.

The actual shape of the core can also have a significant effect on the leakage reactance. Figure 9-15 shows some of the most common core shapes. The *shell core* type (Fig. 9-15C) typically has the lowest leakage reactance. It is also the most expensive core to manufacture.

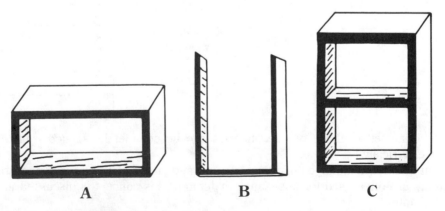

9-15 Typical transformer core shapes.

Another source of energy loss in a transformer is the dc resistance. Any practical conductor has some degree of resistance, which increases as the length of the conductor increases. A coil is essentially just a very long wire, wound into a small space. It acts like a very low value resistor. This dc resistance is usually very small—almost never as much as 100 Ω, but it does represent a small voltage loss.

The dc resistance between the separate windings of a transformer (from primary to secondary) should be virtually infinite, because there is no direct electrical connection between them. They are only connected magnetically. Of course, this isolation is not true of an autotransformer.

There is also the ac inductive reactance to be considered. Resistances and reactances cannot simply be added together. The next chapter explains how these two similar, but different effects can be combined into a single specification called *impedance*. Remember that inductive reactance increases as the applied frequency increases.

Because the coils are conductors, and they are separated by insulation, unwanted capacitances can be set up to add to the power loss.

In a transformer, dc resistance is sometimes referred to as copper loss, because the coils themselves are usually made of copper wire.

Another important type of energy loss in a transformer is *iron loss*. There are actually two forms of iron loss. The first of these is called *hysteresis loss*. Because the iron core is with a strong magnetic field (generated by the surrounding coil) it becomes magnetized itself. As the alternating current flows through the coil, the magnetic charge on the coil is forced to continually alternate its polarity. This process, naturally, uses up a certain amount of energy, because the iron core will oppose any change in its condition. This energy must be stolen from the electrical current, because it is the only energy source within the system. In other words, power is lost due to this effect.

Hysteresis loss can be reduced by constructing the core of some material that is very easy to magnetize and demagnetize. Some materials of this type are silicon steel, and certain other alloys.

The other form of iron loss is due to the electrical current that is induced in the conductive core by the fluctuating magnetic fields of the coils. This current (called an *eddy current*) is more wasted power that is not delivered to the desired load circuit. To reduce eddy currents, transformer cores are often made of a pile of very thin metallic sheets (called *laminations*) rather than a solid mass of iron or whatever alloy is being used. These laminations are individually insulated from each other by a layer of varnish or oxide, so eddy currents cannot flow through the entire core and pick up strength.

Another inevitable power loss in transformers is due to an effect called *reflected impedance*. As current flows through the primary winding, a magnetic field is created. This magnetic field induces a voltage in the secondary winding. So far, this action is simply the desired transformer action. But because there is now current flowing through the secondary, it also generates a magnetic field of its own, which induces an interfering current back into the primary. This energy is taken from the secondary winding's circuit (i.e., the load).

Despite all these various losses, commercially available transformers are generally surprisingly efficient. And, as you'll see in other chapters, these devices are used in a great many types of electronic circuits.

Self-test

1. If three shielded 100 mH coils are connected in series, what is the total effective inductance?

A 33.333 mH
B 300 mH
C 100 mH
D 67.777 mH
E None of the above

2. If three shielded 100 mH coils are connected in parallel, what is the total effective inductance?

A 33.333 mH
B 300 mH
C 100 mH
D 67.777 mH
E None of the above

3. A transformer consists of which of the following?

A A capacitor and an inductor
B An inductance and a resistance
C A parallel resonant circuit
D Two coils wound on a common core
E None of the above

4. A transformer with 100 turns in the primary winding and 25 turns in the secondary winding is which of the following?

A An isolation transformer
B A step-down transformer
C A step-up transformer
D An autotransformer
E None of the above

5. If 250 Vac is fed to the primary coil of the transformer described in question 4, what will be the voltage induced across the secondary winding (assuming 100% coupling)?

A 125 V
B 500 V
C 2.5 V
D 250 V
E None of the above

6. In a step-up transformer, what is the effect on the current?

A It is the same in both windings
B The current in the secondary is greater than the current in the primary
C The current is blocked at low frequencies
D The current in the secondary is less than the current in the primary
E None of the above

7. If the center tap of a transformer is grounded, what will be the phase relationship between the two ends of the secondary winding?

A In phase
B 360 degrees out of phase
C 180 degrees out of phase
D 90 degrees out of phase
E None of the above

8. An autotransformer contains how many coils?

A None
B One
C Two
D Three
E None of the above

9. What is the name of the effect of some of the magnetic field leaking off due to less that 100% coupling?

A Electromagnetic effect
B Eddy currents
C Leakage resistance
D Self-inductance
E None of the above

10. Which of the following does not contribute to losses in a transformer?

A Self-inductance
B Iron losses
C Stray capacitance
D Leakage resistance
E None of the above

10
Impedance and resonant circuits

In the previous chapter, you read about *impedance*, and learned that it was a combination of reactance (inductive or capacitive) and dc resistance. In other words, impedance is the total effective ac resistance of a component or circuit.

In this chapter, you will explore the concept of impedance, and discover how it can be determined mathematically. You'll also look at what happens when a circuit contains both inductive and capacitive components.

Impedance in inductive circuits

Any practical inductor has some dc resistance, as well as its inductive reactance. It also has some capacitance, but usually this value is small enough to be ignored in practical circuits, unless very high frequencies are involved. For now, you'll ignore this factor. So a circuit containing a resistor, a coil, and an ac voltage source can be redrawn as in Fig. 10-1. The resistor marked R_L does not exist as a separate component. It is the internal dc resistance of the coil itself. It is shown only to make the following discussion clearer.

For the following example, assign these values to the components. L is 0.1 H (100 mH), R_L is 22 Ω, and R_X is 1000 Ω. The ac voltage is 100 V rms. You'll examine what happens as different frequencies are applied.

If the frequency of the ac source is 60 Hz, the inductive reactance is about 38 Ω. It might seem that you could find the total effective resistance of the circuit, simply by adding $X_L + R_L + R_X$, giving a total of 1060 Ω. Unfortunately, this simple approach will not give an accurate result. The reason is that dc resistance is 90 degrees out of phase with inductive reactance.

An inductance offers more resistance to current than to voltage. You'll recall that ordinarily, the current and voltage from an ac source are in phase. After passing through a pure inductance, however, the voltage will lead the current by 90 degrees. See Fig. 10-2.

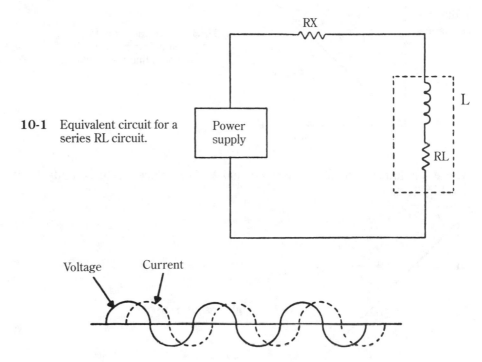

10-1   Equivalent circuit for a series RL circuit.

10-2   Voltage leads current after passing through an inductance.

When dc resistance is also in the circuit, the final phase difference between the voltage and the current will be proportionately less than 90 degrees. In a purely resistive circuit, of course, the voltage and the current remain in phase.

Another reason why resistance and reactance cannot simply be added together is that reactance is not a true resistance at all—it is only an apparent resistance. In a true dc resistance, power is consumed by the resistor (usually the energy is converted to heat), but no power is actually consumed by a reactance (either inductive, or capacitive).

You could find the combined value of reactance and resistance (that is, the impedance of the circuit) with a vector diagram, as in Fig. 10-3. But because the phase relationship between a resistance and inductive reactance is always 90 degrees, you can solve the problem mathematically.

Notice that the vector diagram of the resistance and inductive reactance is always in the form of a right triangle. If you know the length of the sides connected at the right angle, you can solve for the third side with the algebraic formula, $c = \sqrt{a^2 + b^2}$. Or you can rewrite the formula, using the appropriate electrical terms:

$$Z = \sqrt{R^2 + X^2}$$ **Equation 10-1**

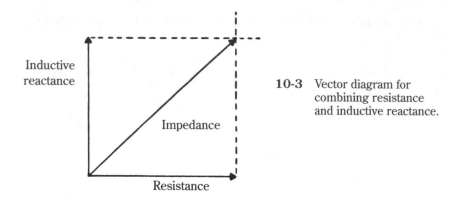

10-3 Vector diagram for combining resistance and inductive reactance.

where Z is impedance, R is resistance, and X is reactance. All three units are in Ω. See Fig. 10-4.

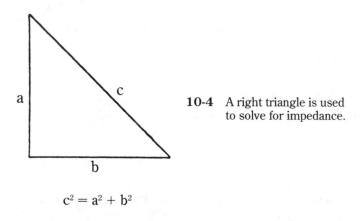

10-4 A right triangle is used to solve for impedance.

$$c^2 = a^2 + b^2$$

Returning to your example circuit, you find that the impedance equals $\sqrt{1022^2 + 38^2} = \sqrt{1044484 + 1444} = \sqrt{1045928} =$ about 1022.7 Ω. For a second example, you'll keep the same components, but increase the ac frequency to 200 Hz. The dc resistance is still 1022 Ω, but the inductive reactance is now approximately 126 Ω. Therefore, $Z = \sqrt{1022^2 + 126^2} = \sqrt{1044484 + 15876} + \sqrt{1060360} =$ just under 1,030 Ω.

If the frequency of the ac source is now increased to 5000 Hz, the inductive reactance goes up to 3,142 Ω, so the impedance equals $\sqrt{1022^2 + 314^2} = \sqrt{1044484 + 9872164} = \sqrt{10916648} =$ just slightly more than 3304 Ω.

You can see that as the frequency increases, both the inductive reactance and the total impedance of an RL circuit are also increased. The value of the dc resistance, of course, remains constant.

Now, change the value of R_L to 15 Ω, R_X to 2200 Ω, and L to 0.25 H (250 mH) and find the circuit impedance at 60 Hz, 200 Hz, and 5000 Hz.

If you draw vector diagrams for each of these examples, you find that the impedance phase angle in a circuit consisting of only resistance and inductance is always 45 degrees.

When you have a parallel RL circuit, such as the simple one shown in Fig. 10-5, the situation is a bit more complicated.

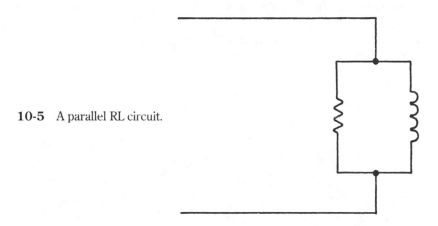

10-5 A parallel RL circuit.

The formula for a circuit of this type is:

$$Z = \frac{RX}{\sqrt{R^2 + X^2}}$$

Equation 10-2

Again, all three variables are in ohms. Assume you have 1000 Ω of resistance in parallel with an inductive reactance of 52 Ω. The impedance would be equal to (1000 × 52)/$\sqrt{100^2 + 52^2}$ = 52000/$\sqrt{1000000 + 2704}$ = 52000/$\sqrt{1002704}$ = 52000/1001 = just under 52 Ω.

If the inductive reactance is increased to 890 Ω, the impedance becomes equal to (1000 × 890)/$\sqrt{1000^2 + 890^2}$ = 890000/$\sqrt{1000000 + 792100}$ = 890000/$\sqrt{1792100}$ = 890000/1339 = about 665 Ω. The total impedance will always be less than either the dc resistance, or the inductive reactance.

You'll notice that you have been ignoring the dc resistance of the coil in these examples. To take this additional factor under consideration, the circuit becomes like the one in Fig. 10-6.

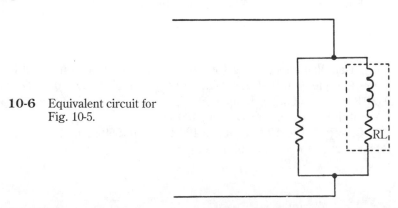

10-6 Equivalent circuit for Fig. 10-5.

To solve the total impedance of this circuit, use vector diagrams. Usually the dc resistance of the coil can simply be ignored.

A circuit with both an inductance and a resistance in each leg of the parallel paths, can be solved mathematically, because the impedances are similar (i.e., in phase). Such a circuit is illustrated in Fig. 10-7.

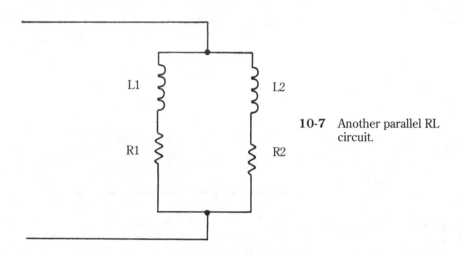

10-7   Another parallel RL circuit.

To solve for the impedance of this kind of circuit, simply solve for each parallel path as a separate series RL circuit. Suppose R1 equals 220 Ω, R2 equals 470 Ω, and L1 has an inductive reactance of 97 Ω, and L2 has an inductive reactance of 550 Ω.

The impedance in leg A equals $\sqrt{220^2 + 97^2} = \sqrt{48400 + 9409} = \sqrt{57809} = $ approximately 240 Ω. Z_B, on the other hand, equals $\sqrt{470^2 + 550^2} = \sqrt{220900 + 302500} = \sqrt{523400} = $ about 723 Ω.

You now have Z_A (240 Ω) in parallel with Z_B (723 Ω). Because impedances are essentially ac resistances, they can be combined in the same manner as dc resistances, as long as their phase angles are identical. In other words, impedances in series add:

$$Z_T = Z_1 + Z_2 + \ldots Z_n \qquad \text{\textbf{Equation 10-3}}$$

Impedances in parallel are solved by the formula:

$$\frac{1}{Z_T} = \frac{1}{Z_1} + \frac{1}{Z_2} + \ldots \frac{1}{Z_n} \qquad \text{\textbf{Equation 10-4}}$$

Dissimilar impedances (out of phase) cannot be combined in this manner. In your example, you find that $1/Z_T = 1/Z_A + 1/Z_B = 1/240 + 1/723 = 0.004167 + 0.001383 = 0.00555$. Taking the reciprocal, you find that the total effective impedance of the entire circuit is about 180 Ω.

Change R1 to 1000 Ω, R2 to 100 Ω, X_{L1} to 550 Ω, and X_{L2} to 85 Ω, and find the total effective impedance of the circuit.

Impedance in capacitive circuits

The situation with an RC circuit is basically similar. Capacitive reactance also consumes no real power. Ac flows in and out of a capacitor, but not through it. In a capacitor the current leads the voltage by 90 degrees, so the vector diagram of resistance and pure capacitive reactance would resemble Fig. 10-8. Notice that it is a mirror image of a resistance/inductive reactance combination. The resulting phase angle will always be 45 degrees (or, 315 degrees if the resistance is used as the 0-degree line).

10-8 Vector diagram for combining resistance and capacitive reactance.

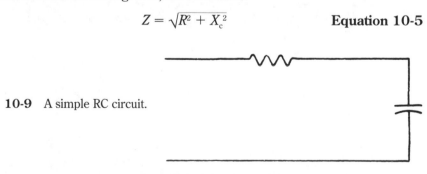

Like an inductor, practical capacitors have some degree of dc resistance and a small (usually negligible) inductance. The dc resistance through a capacitor is in parallel with its reactance, and is generally extremely high, so the effective impedance is essentially equal to the capacitive reactance. For this reason, the dc resistance is ordinarily ignored in all but the most critical calculations.

Combining dc resistance with capacitive reactance works in the same way as combining inductive reactance with dc resistance. For example, for a simple RC series circuit like the one in Fig. 10-9, the formula is:

$$Z = \sqrt{R^2 + X_c^2}$$

Equation 10-5

10-9 A simple RC circuit.

This formula is the same formula used to solve for the impedance in RL series circuits. However, it is important to remember that capacitive reactance is 180 degrees out of phase with inductive reactance. Also, because capacitive reactance decreases as the applied frequency increases, the circuit impedance of an RC series circuit also decreases with an increase in the applied ac frequency.

Impedance in inductive-capacitive circuits

Most practical electronic circuits have all three of the basic electrical elements: resistance, inductance, and capacitance. How would you go about finding the impedance of a circuit with a resistor, a capacitor, and a coil in series, like the one in Fig. 10-10?

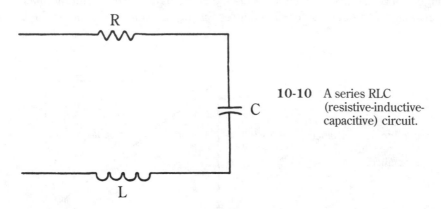

10-10 A series RLC (resistive-inductive-capacitive) circuit.

Because the inductive and capacitive reactances are 180 degrees out of phase you obviously can't simply add them together. Take a look at the graphs in Fig. 10-11. Signal B is smaller than signal A and 180 degrees out of phase. Signal C is the result of combining these two waveforms. Notice that the output signal resembles (and is in phase with) the larger input signal (A), but has a lower amplitude. The difference between signal A and signal B is signal C.

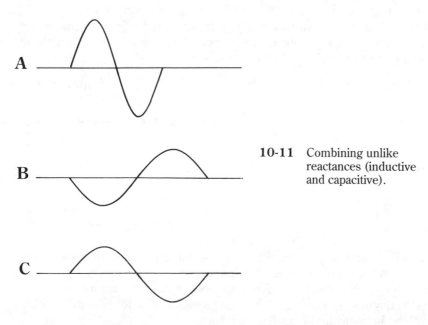

10-11 Combining unlike reactances (inductive and capacitive).

In other words, when two similar signals are 180 degrees out of phase, their combined value can be found by subtracting the smaller from the larger signal. You can obviously apply this same approach with combining inductive and capacitive reactances because they are, by definition, 180 degrees out of phase with each other. The formula is:

$$X_T = X_L - X_C \qquad \textbf{Equation 10-6}$$

X_L is the inductive reactance, X_C is the capacitive reactance, and X_T is the total effective reactance. If X_T is positive (as it will be whenever X_L is larger than X_C, then the resulting reactance behaves like an inductive reactance (increasing as the frequency increases). On the other hand, if X_T works out to a negative value (that is, X_C is larger than X_L), then the total reactance appears to be capacitive (decreasing as the frequency is increased). A circuit including both capacitance and inductance will act like an inductive reactance at some frequencies, and like a capacitive reactance at other frequencies.

X_T can then be combined with the dc resistance in the usual way to find the circuit impedance.

$$Z = \sqrt{R^2 + X_T^2} \qquad \textbf{Equation 10-7}$$

or

$$Z = \sqrt{R^2 + (X_L - X_C)^2} \qquad \textbf{Equation 10-8}$$

Of course, because the reactance is squared, the impedance will always have a positive value.

If an inductance and a capacitance are connected in parallel, as in Fig. 10-12, then the effective reactance of the combination is:

$$X_T = \frac{(X_L)(X_C)}{(X_L - X_C)} \qquad \textbf{Equation 10-9}$$

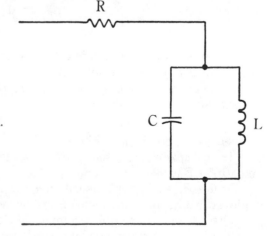

10-12   Inductance and capacitance in parallel.

Again, if X_T is positive the effective reactance is inductance, but if X_T is negative, the reactance is effectively capacitive.

Resonance

Capacitive reactance decreases as the applied frequency increases, and inductive reactance increases along with the applied frequency. At some specific frequency for every combination of a capacitor and an inductor, the capacitive and the inductive reactances will be equal. This frequency is the *resonant frequency*, and it results in a unique circuit condition called *resonance*.

In a series-connected LC (inductive-capacitive) circuit, the impedance equals $\sqrt{R^2 + (X_L - X_C)^2}$ if X_L equals X_C, these factors will cancel out leaving $Z = \sqrt{R^2 + 0^2} = \sqrt{R^2}$, or simply $Z = R$. At resonance, the impedance in a series LC circuit will be determined solely by the dc resistance. The impedance is at its minimum value at the resonant frequency.

Now, look at what happens at resonance in a parallel LC circuit? In this case:

$$X_T = \frac{(X_L)(X_C)}{(X_L - X_C)}.$$

If $X_L = X_C$ then $X_L - X_C$ must be zero. Because this is the denominator of the equation, the value of X_T becomes infinite by definition, because any number divided by zero equals infinity. Clearly a parallel connected LC circuit has its maximum impedance at resonance.

Resonant circuits are extremely important in the study of electronics, and you can learn how they are used in other chapters.

Any combination of an inductance and a capacitance will have a resonant frequency. Any given combination will be resonant only at a single frequency.

The formula for finding the resonant frequency for a specific inductance-capacitance combination (either series, or parallel connected) is:

$$F = \frac{1}{2 \pi \sqrt{LC}} \qquad \textbf{Equation 10-10}$$

F is the frequency in hertz, L is the inductance in henries, C is the capacitance in farads, and 2π is approximately 6.28.

As an example, suppose you have a 0.1 H (100 mH) coil, and a 10 μF (0.00001 F) capacitor. The resonant frequency for this particular combination equals $1/(2\pi \sqrt{LC})$, or $1/(6.28 \times \sqrt{0.1 \times 0.00001})$. This works out to 1/0.00628, or a resonant frequency of approximately 159 H.

Try calculating the resonant frequency for a circuit with a 0.25 H (250 mH) coil and a 100 μF (0.0001 F) capacitor.

In design work, you'll usually need to approach this equation from another angle. Probably, you'll know what frequency you want the circuit to resonate at, and you'll have to determine which components to use.

In this type of situation, you can arbitrarily select an inductance value and rewrite the resonance equation to solve for the capacitance as follows:

$$C = \frac{1}{4 \pi^2 F^2 L}$$ **Equation 10-11**

$4\pi^2$ is approximately 39.5. Assume you need a circuit that is resonant at 1000 Hz. You can arbitrarily decide to use a 0.1 H (100 mH) coil. This component will make the required capacitance equal to $1/(39.5 \times 1000^2 \times 0.1)$, or $1/(39.5 \times 1000000 \times 0.1)$, or 1/395000. Taking the reciprocal, you find that you need a capacitance of about 0.0000003 F, or 0.3 µF.

Alternatively, you can choose a capacitance value, and solve for the inductance. The formula for this method is quite similar:

$$L = \frac{1}{4 \pi^2 F^2 C}$$ **Equation 10-12**

Again, you'll assume you need resonance at 1000 Hz, and you'll select a 1 µF (0.000001 F) capacitor. $L = 1 (39.5 \times 1000^2 \times 0.000001) = 1/(39.5 \times 1,000,000 \times 0.000001) = 1/39.5 = 0.025$, or about 25 mH.

You can see that different combinations of components can produce the same resonant frequency.

Go back and check that last problem, by recalculating the resonant frequency from your component values, 25 mH (0.025 H) and 1 µF (0.000001 F). $F = 1/(2 \pi\sqrt{LC} = 1/(6.28 \times \sqrt{0.025 \times 0.000001}) = 1/(6.28 \times \sqrt{0.00000025}) = 1/(6.28 \times 0.0001581) = 1/0.000993 =$ about 1007 H. The 7 Hz error came from rounding off the inductance value. This degree of accuracy is sufficient for most practical purposes.

Another thing to consider is what happens when one of the component values is increased. Will the resonant frequency increase, or decrease?

Using the last example, increase the inductance to 50 mH (0.05 H). The resonant frequency now would equal $1/(6.28 \times \sqrt{0.05 \times 0.000001}) = 1/(6.28 \times \sqrt{0.0000005}) = 1/(6.28 \times 0.0002236) = 1/0.0014043 =$ slightly over 712 Hz. When the inductance was increased, the resonant frequency was decreased.

Similarly, if you now increase the capacitance to 25 µF (0.000025 farad), the resonant frequency becomes $1/(6.28 \times \sqrt{0.05 \times 0.000025} = 1/(6.28 \times \sqrt{0.0000013}) = 1/(6.28 \times 0.001118) = 1/0.0070213 =$ approximately 142 Hz. Increasing the capacitance also decreases the resonant frequency.

To practice using these formulas, design a circuit that resonates at 250 Hz, and one that resonates at 5000 Hz.

Impedance matching and power transfer

Any dc power source, such as a battery, has some amount of internal resistance. Similarly, any ac power source has a certain degree of internal impedance. This impedance will be constant for any given frequency, and the impedance of the load circuit can be any impedance that the designer makes it.

In Fig. 10-13, the internal impedance of the power source and the load imped-

ance of the circuit are shown as resistances for simplicity. Keep in mind that these
are actually ac impedances.

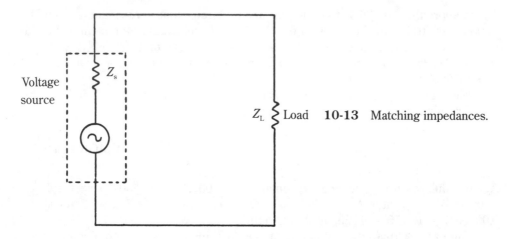

10-13   Matching impedances.

If you assume the ac source has an internal impedance of 50 Ω, Table 10-1
shows the current and the power apparently consumed by the load at various values
of the load impedance. Notice that the power available to the load circuit increases
until it reaches a maximum level; then the power transfer starts to drop back down
again. This point of maximum power transference will always occur when the source
impedance is equal to the load impedance. When these impedances are unequal,
the source will not put out the maximum amount of energy it is capable of. This is
called an *impedance mismatch*.

Table 10-1. Matching impedances.

E	Z_S	Z_L	Z_T	I	P_L
10	50	10	60	0.167	0.278
10	50	20	70	0.143	0.408
10	50	30	80	0.125	0.469
10	50	40	90	0.111	0.494
10	50	50	100	0.100	0.500
10	50	60	110	0.091	0.496
10	50	70	120	0.083	0.486
10	50	80	130	0.077	0.473
10	50	90	140	0.071	0.459
10	50	100	150	0.067	0.444

Often it isn't possible or practical to design the load circuit so that its impedance
directly matches that of the signal source. In this case, some form of *impedance
matching* is needed. One of the simplest methods for matching is with an *impedance
matching transformer* (as mentioned in the last chapter). The inductance of the coils

in such a transformer is carefully chosen to give the proper impedance ratio between the primary and secondary windings. Impedance can be either increased or decreased with such a transformer.

In practical circuits, the impedances will never match exactly, but a certain amount of leeway is usually acceptable. For instance, a 500 Ω load can generally be used with a 600 Ω single source with no problems. However, the circuit won't work very efficiently if it is driven by a 10,000 Ω source impedance. For best results, the source impedance and load impedance must be approximately equal.

Self-test

1. What is impedance?

A The effect of dc resistance at ac frequencies
B An imaginary factor used to simplify circuit design
C The combination of capacitive reactance, inductive reactance, and dc resistance
D The combination of capacitive reactance and inductive reactance only
E None of the above

2. Ignoring capacitive effects, what is the impedance of a 250 mH coil with an internal resistance of 55 Ω at 60 Hz?

A 94.2 Ω
B 10,900 Ω
C 149.2 Ω
D 109 Ω
E None of the above

3. Ignoring capacitive effects, what is the impedance of a 100 mH coil (with an internal resistance of 45 Ω) in parallel with a 4700 Ω resistor at a frequency of 500 Hz?

A 314 Ω
B 317 Ω
C 237 Ω
D 5014 Ω
E None of the above

4. Ignoring any inductive effects, what is the impedance of a RC series capacitor made up of a 56 kΩ resistor and a 0.033 μF capacitor at a signal frequency of 450 Hz?

A 10,730 Ω
B 57,019 Ω
C 66,730 Ω
D 45,270 Ω
E None of the above

5. Ignoring any effects of dc resistance, what is the total reactance of a 250 mH coil in series with a 4.7 μF capacitor at a signal frequency of 1000 Hz?

A 35 Ω
B 1570 Ω
C 1604 Ω
D 1536 Ω
E None of the above

6. Ignoring any effects of dc resistance, what is the total reactance of a 250 mH coil in series with a 4.7 μF capacitor at a signal frequency of 450 Hz?

A 97 Ω
B 84 Ω
C 706 Ω
D 781 Ω
E None of the above

7. Ignoring any effects of dc resistance, what is the total reactance of a 250 mH coil in series with a 4.7 μF capacitor at a signal frequency of 60 Hz?

A 111 Ω
B 659 Ω
C −471 Ω
D −113 Ω
E None of the above

8. In a series resonant LC circuit, what is the impedance at the resonant frequency?

A Determined solely by the dc resistance
B The maximum impedance value
C Zero
D Infinity
E None of the above

9. In a parallel resonant LC circuit, what is the impedance at the resonant frequency?

A Determined solely by the dc resistance
B The maximum impedance value
C Zero
D Infinity
E None of the above

10. If you need an LC circuit to be resonant at 2500 Hz, and use a 150 mH coil, what should the capacitance value be?

A 27 μF
B 0.027 μF
C 0.015 μF
D 0.15 μF
E None of the above

11
Crystals

In the last chapter you learned how a resonant circuit can be built from a separate capacitor and coil. Resonant circuits can also be made with a single component called a *crystal* (often abbreviated as *XTAL*).

Construction of a crystal

Figure 11-1 shows the basic structure of a crystal. A thin slice of quartz crystal is sandwiched between two metal plates. These plates are held in tight contact with

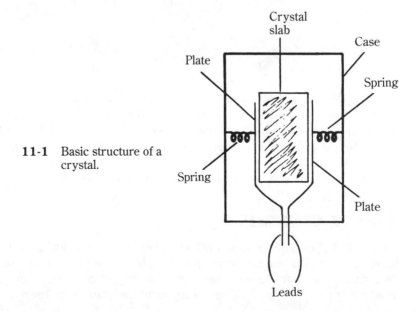

11-1 Basic structure of a crystal.

143

the crystal element by small springs. This entire assembly is enclosed in a metal case that is *hermetically sealed* to keep out moisture and dust. Leads connected to the metal plates are brought out from the case for connection to an external circuit. The schematic symbol for a crystal is shown in Fig. 11-2.

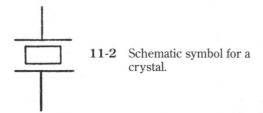

11-2   Schematic symbol for a crystal.

The piezoelectric effect

A crystal works because of a phenomenon called the *piezoelectric effect.* Two sets of axes pass through a crystal. One set, called the *X axis,* passes through the corners of the crystal. The other set, called the *Y axis,* is perpendicular to the X axis, but in the same plane. Figure 11-3 shows two axes in a crystal, looking down through the top of the crystal.

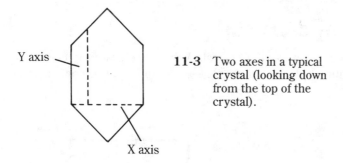

Y axis

X axis

11-3   Two axes in a typical crystal (looking down from the top of the crystal).

If a mechanical stress is placed across a Y axis, an electrical voltage will be produced along the X axis. The reverse also is true—if an electrical voltage is applied across an X axis, a mechanical stress will be created along the Y axis.

This piezoelectric effect can cause the crystal to *ring* (vibrate), or resonate at a specific frequency under certain conditions. In practice, only a thin slice of crystal is used. This slice can be cut across either an X or a Y axis.

Using crystals

Figure 11-4 shows the equivalent electrical circuit for a typical crystal. Depending on how the crystal is manufactured, it can replace either a series resonant (minimum impedance) or a parallel resonant (maximum impedance) capacitor-coil circuit. Generally, a crystal designed for series resonant use, cannot be used in a parallel resonant circuit. The resonant frequency of a crystal is determined primarily by the thickness and size of the crystal slice.

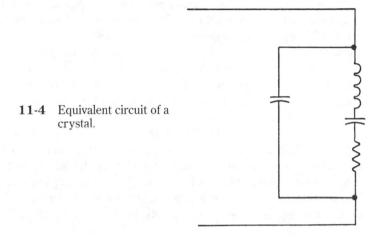

11-4   Equivalent circuit of a crystal.

Crystals can also be made to resonate at integer multiples of its main resonant frequency. These multiples are called *overtones*, or *harmonics*. The primary frequency is called the *fundamental*. For example, a crystal designed to resonate at 15,000 Hz (the fundamental) will also resonate (but to a lesser degree) at 30,000 Hz (second harmonic), 45,000 Hz (third harmonic), 60,000 Hz (fourth harmonic), and so on. Notice that no first harmonic is mentioned, because one times the fundamental equals the fundamental frequency. The resonance effect will become steadily less pronounced at higher harmonics.

Crystals are usually more expensive than separate capacitors and coils, but their resonant frequency is much more precise and stable. Capacitor-coil resonant circuits often drift off frequency (that is, the components change their values slightly), particularly under changing temperature conditions.

Crystals are also somewhat temperature sensitive (although not as much so as capacitors and coils) so when very high accuracy is required (as in broadcasting applications), the circuit is enclosed in a *crystal oven*, which maintains a constant temperature.

Reliability is another advantage of crystals. Their failure rate tends to be somewhat lower than capacitors and coils. However, they can be damaged by high overvoltages, or extremely high temperatures. Also, a severe mechanical shock (such as being dropped onto a hard surface) could crack the delicate crystal element. If cracked, the entire crystal package must be replaced. There is no way to replace just the crystal slice.

Occasionally the hermetic seal develops a small leak, allowing contaminants (like dust, or water) to get inside the metal case. This contamination can interfere with the electrical contact between the metal plates and the crystal element. All of these defects are relatively rare, but they do occur frequently enough to warrant mentioning them here.

The most important disadvantage of crystals is that the resonant frequency generally cannot be changed. In a capacitor-coil circuit, one or both components can be made variable, but there is no such thing as a variable crystal. Sometimes special external circuitry can be added for a limited degree of fine tuning, but generally a crystal resonant circuit is fixed. The only way the frequency can be changed is by physically replacing the crystal. For this reason, crystals are usually inserted into sockets, rather than actually being permanently soldered into the circuit. The leads of the crystal plug into holes in the socket that make electrical contact with the external circuit. Crystals can easily and quickly be removed and replaced. Using sockets also sidesteps the potential problem of thermal damage resulting from soldering the crystal leads directly.

Crystals are available, cut for many different frequencies. The most common type of crystal is tuned for a frequency of approximately 3.58 MHz. This frequency is used for the standard color burst signal in a color television transmission, so there is a very large market for such crystals. Many other circuits, which often have nothing at all to do with television, are designed to use the same frequency, simply to take advantage of the wide availability and low cost of 3.58 MHz crystals.

Crystal lore

For centuries, humanity has been fascinated with the intricate and mysterious structures of crystals. Many legends and myths have grown up around these unusual stones. In many ages they have been assumed to possess wondrous magical or mystical powers.

Today there is a lot of talk about the New Age movement, and crystals play an important part in many (though certainly not all) New Age theories. People involved in the New Age movement often speak of *Crystal Power*. Different New Age sources make somewhat different claims about crystals, but there are two basic themes that reappear frequently—that crystals are alive, and that crystals can store and emit energy that can somehow interact with the human mind. A number of New Age authors try to claim that these ideas are supported by modern science and technology (especially electronics), but this simply is not true. Either they don't understand scientific principles, or they are deliberately distorting the facts to suit their beliefs. Notice that this in no way proves that any of the New Age theories about crystals are wrong, just that these particular ideas are not supported by current science and technology. Of course, there is much modern science doesn't know, so it could be that such theories have some possibility, even if they don't seem very likely.

The idea that crystals are alive is a little slippery. A lot depends on just how you define *alive*. Many metaphysicians argue that ultimately everything is alive in some spiritual sense. In that case, of course crystals would be included along with everything else. But there is nothing inherently special about crystals then. Such metaphysical concepts can be very useful for the spiritual seeker, but they have little importance or even meaning from a purely scientific point of view. If everything is alive, then the word *alive* doesn't have much value. The word is useful only if it can be used to distinguish things that are alive from things that are not alive. If everything is alive, and nothing is not alive, what is the point of mentioning whether or not things are alive?

(This argument is not on a metaphysical level at all. In a metaphysical sense, this concept might be totally true, but that is irrelevant to what is being discussed here—pragmatic, material science.)

But many new agers go beyond this general metaphysical concept and argue that crystals are alive in some sort of special, unique way. It is unclear just how they are defining alive. In the strictly scientific sense, living things comprise primarily organic biochemicals, and this fact is certainly not the case for crystals. A common argument is that crystals grow like other living things. Yes, crystals do grow, but not at all like biological entities. New crystalline material is accepted from external sources, and is layered over the existing crystal, resulting in a larger crystal. In a biological entity, food or other nourishment is taken in from some external source. The nature of this external matter is changed through the process of digestion. Unusable portions are excreted as waste in some fashion. The usable nourishment is internally burned off to create biological energy, or to aid in growth of the entity, which grows from inside out, quite unlike the crystal.

Experiments that supposedly demonstrate that crystals move have been made from time to time, but often the procedures are questionable, and the movement involved is so slight that it is for all intents and purposes negligible. Curiously, many who most firmly believe the controversial results of these experiments, also strongly believe in psychic mental powers, such as telekinesis. Well, if they believe *telekinesis* (the ability to move objects by thought energy alone), why don't they assume that is the cause of the tiny movement exhibited by the crystals in these experiments?

Ultimately, when such movement actually occurs at all, it probably has some unanticipated physical explanation, such as a slight vibration in the surface the crystal is resting on.

The second major New Age idea about crystals is even more intriguing, and just as scientifically unsound. Many new agers believe that crystals emit and store some form of energy waves that can interact with mentally projected psychic or thought energy. Crystals supposedly can affect moods, aid psychic powers and physical healings, and protect against evil or negative spiritual energies. Some or all of these claims might be partially or totally true, but none of them have been scientifically demonstrated or documented, despite some published claims. Some experiments have indicated that something is happened, but it might have as much to do with suggestibility as anything else. No measurable energy or anything else has been detected when crystals are used in these ways. If the crystal is putting out, or modifying energy in any way, it must be some totally undiscovered type of energy completely undetectable by all existing technology. Sure, it's possible, but it scarcely seems likely.

The crystal believers very frequently point to the use of crystals in electronic circuits, as if this use supported their beliefs in some way. It does not. At least as far as electronics is concerned, a crystal neither generates nor stores energy. Yes, the piezoelectric effect does allow a mechanical stress to be converted into an electrical voltage, or vice versa. No energy is being created (or stored). It is merely changed from one type to another.

A crystal most certainly does not store any known type of energy. Mechanical energy or electrical energy can be put into it, but this energy will immediately be put right back out as the opposite type of energy (mechanical changes to electrical, or vice versa). Some of the energy is consumed in the conversion process (which is true any time energy is changed in form), but none at all remains stored in the crystal itself.

In most electronic circuits using crystals, the crystal is employed as a resonant tank. When it is electrically excited (stressed) it vibrates at a very specific and predictable frequency. It acts much like a coil and a capacitor in a resonant circuit. There is nothing mystical about it. No detectable energy waves are emitted from the crystal, except perhaps a very weak ultra-audible tone. Even if it was at an audible frequency, you would have to get very close to the crystal to ever hear it. This little bit of audio energy will never be very strong, and it will be completely dissipated a foot or two away from the crystal.

Some apparently uninformed New Age authors make some sort of connection between the resonant vibrations of a crystal and brain wave activity. Sorry, the frequencies are totally wrong. Crystals resonate at very high frequencies, usually well in the RF range. Brain waves, on the other hand, are very low in frequency. There is no logical connection between the two.

It is vaguely possible that there is some sort of energy phenomenon around crystals that modern science is totally unaware of, but this argument begs the question of just how the New Age sources know so much about it, when no one else can even detect it.

This discussion is not intended to sound harsh or closed minded. The New Age movement is filled with some very intriguing and exciting ideas, but this sort of garbled pseudoscientific nonsense must be weeded out so you can focus your time and research on the more plausible ideas.

There might possibly be something to some of the New Age concepts about crystals, but, at the present time, there is absolutely no scientific evidence to support any such claims. New Age authors who claim such links to current technology are either misinformed or dishonest. The use of crystals in electronics has nothing at all to do with any psychic claims.

Self-test

1. What material are crystals made of?

A Nickel alloy
B Quartz
C Iron
D Glass
E None of the above

2. What is the purpose of the hermetically sealed housing used with crystals?

A To stabilize the temperature
B To protect the fragile crystal from being cracked

C To shield the crystal from external magnetic fields
D To keep out moisture and dust
E None of the above

3. What is the principle behind the operation of a crystal?

A The piezoelectric effect
B The photoelectric effect
C The electromagnetic effect
D The ionization effect
E None of the above

4. What is the result of placing a mechanical stress along the Y axis of a crystal?

A An electrical voltage will appear across the X axis
B The crystal might be damaged
C An electrical voltage will appear across the Y axis
D The mechanical motion will be amplified across the X axis
E None of the above

5. A crystal can be used in place of what type of circuit?

A An RC circuit
B A series resonant circuit
C A parallel resonant circuit
D Either a series or a parallel resonant circuit
E None of the above

6. What determines the resonant frequency of a crystal?

A External components
B The hermetic seal
C The size and thickness of the crystal material
D The temperature of the crystal
E None of the above

7. If a crystal has a resonant frequency of 47,000 Hz, what is the third harmonic?

A 50,000 Hz
B 141,000 Hz
C 188,000 Hz
D 300,000 Hz
E None of the above

8. When should a crystal oven be used?

A In very cold environments
B Always
C When extremely high accuracy is required
D When the crystal is being operated at a high harmonic of its resonant frequency
E None of the above

9. What is the greatest disadvantage of using crystals in resonant circuits?

A Size
B Cost
C The resonant frequency cannot be easily changed
D Poor accuracy
E None of the above

10. What is the biggest advantage of using crystal in resonant circuits?

A Size
B Greater accuracy and stability
C Less fragile
D Cost
E None of the above

12
Meters

In practical electronics work, it is very often necessary to measure the basic parameters in a circuit (that is, voltage, current, and resistance) You can calculate the nominal values, but component tolerances can throw the actual values off quite a bit. For example, a circuit containing three resistors, each with a tolerance of ±20% can be off value by as much as ±60% (actually, it's very rare for the values to be that far off, because some of the component errors will probably cancel each other out; but it is possible.) Also, the calculations for complex circuits can be extremely tedious, time-consuming, and difficult to keep track of.

Various defects can cause components to change their values drastically, so for troubleshooting a method of actually measuring the circuit values is absolutely essential. It is the only way to pinpoint the defective component.

Fortunately, there is a relatively simple way to measure voltage, current, and resistance. These measurements are made with devices called *meters*.

The D'Arsonval meter

By far, the most common type of meter is the *D'Arsonval meter*. The construction of this kind of meter is shown in Fig. 12-1.

A permanent magnet in the shape of a horseshoe is positioned around a coil of wire wound on a piece of soft iron. This coil—which is called the *armature*, or the *movement* of the meter—is on a pivot that allows it to move freely. When an electrical current flows through the coil with the polarity shown, the movement develops a magnetic field. Its north pole will be facing the north pole of the permanent magnet, and its south pole will be facing the south pole of the permanent magnet. Because like poles repel, the movement will turn on its pivot so its poles will no longer face the poles of the permanent magnet. Just how far the movement will turn will depend on the relative strength of the magnetic fields involved.

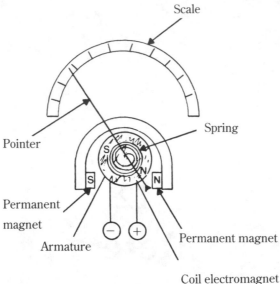

12-1   Construction of a basic D'Arsonval meter.

The magnetic field of the permanent magnet is constant, of course, but the magnetic field of the armature will depend on the strength of the electrical current flowing through it. Therefore, the motion of the armature will be directly proportional to the amplitude of the applied electrical current.

A *pointer* can be attached to the center of the armature, so that the amount of applied current is indicated on a *calibrated scale*. This scale is marked in equal units spaced so that the current value can be read directly.

A small spring opposes the rotation of the armature. This small amount of physical opposition is easily overcome by the applied current and its resulting magnetic field, but when current stops flowing through the coil, the spring forces the movement (and thus, the pointer) to return to the zero position.

The current applied to this type of meter must have the correct polarity. If the polarity of the current is reversed, the south pole of the electromagnet (armature) will be facing the north pole of the permanent magnet, and vice versa. Because unlike poles attract, the armature will not move.

Some meters can be adjusted so that the zero position is in the center of the scale. This zero position will allow the pointer to move in either direction, thus indicating current of either polarity. Most meters, however, do not have this capability.

A meter movement is quite delicate and fragile. Meters are usually enclosed in protective cases (made of transparent plastic), but they can still be permanently damaged if dropped or otherwise mishandled. Also, the coil in the armature is made of very fine wire so it will be light and move easily. This means the wire cannot carry much current. If too much current is applied directly, the armature can be quickly ruined. Fortunately, there are methods of decreasing the current actually applied to the meter, and still allowing the meter to give an accurate reading. This subject is discussed in this chapter. With a reasonable amount of care, a D'Arsonval meter movement is sturdy enough for practical use and can last for years.

This type of meter (which is sometimes also called a *moving-coil movement meter*) is quite popular for a number of reasons. It is relatively inexpensive, highly accurate and *sensitive* (that is, you can readily measure very small currents). The scale is uniform and easy to read, because the movement of the pointer is directly proportional to the applied current, resulting in evenly spaced calibration markings. A D'Arsonval meter drains very little current from the circuit being tested, so it is quite efficient, and doesn't significantly affect the values being measured. Finally, the D'Arsonval meter can easily be adapted to read current, voltage, or resistance. This type of meter is so commonly used, that whenever a meter is mentioned in electronics work, it can generally be assumed to be a D'Arsonval type unit unless otherwise specified.

Ammeters

When a meter is used to measure electric current, it is referred to as an *ammeter* (from ampere meter). The standard schematic symbol for a meter is shown in Fig. 12-2A. Letters are generally placed within the circle to indicate exactly what kind of meter is being used. For example, the letter *A* in the symbol in Fig. 12-2 B means the unit is an ammeter.

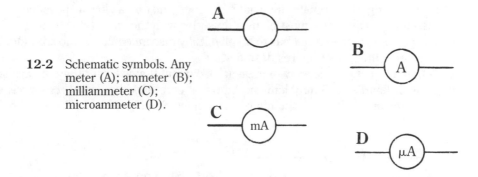

12-2   Schematic symbols. Any meter (A); ammeter (B); milliammeter (C); microammeter (D).

Because the ampere is a rather large unit for an electronic circuit, *milliammeters* (Fig. 12-2C), or *microammeters* (Fig. 12-2D) are frequently used instead. All three types of ammeters work in exactly the same way—the only difference is the range of values they are capable of measuring.

To measure current, the meter must actually be inserted into the circuit itself. In other words, the meter is placed in series with the circuit. This arrangement is shown in Fig. 12-3. Remember that in a series circuit the current is equal at all points, so if an ammeter is placed in series with the rest of the circuit, the same current that flows through the circuit will flow through the meter.

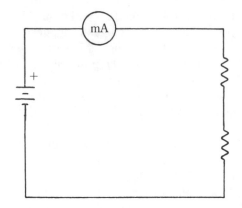

12-3 An ammeter in series
with the circuit to be
measured.

Because the meter is in series with the circuit, its internal resistance must be as low as possible, to avoid upsetting the normal current flow. For example, suppose you have a circuit with a total resistance of 10,000 Ω, and is powered by a 10 V source. The nominal current will be equal to $E/R = 10/10{,}000 = 0.001$ A (1 mA).

When the meter is inserted in the circuit, its internal resistance will add to the resistance of the circuit. For instance, if the meter internal resistance is 5000 Ω, the total resistance in the circuit becomes 15,000 Ω (10,000 + 5,000), so the current changes to 10/15,000 or 0.00067 A, (0.67 mA). The current value is dropped 30%. Obviously, that is a significant difference.

On the other hand, if the meter internal resistance is only 50 Ω, the total circuit resistance will be 10,050 Ω, and the current will be equal to 10/10,050, or 0.000995 A (0.995 mA). This is within 0.05% of the correct nominal value. Clearly, for an accurate current reading, the ammeter internal resistance must be as low as possible.

Because an ammeter must be used in series with the circuit it is testing, one of the connections in the circuit has to be physically disconnected, so the ammeter can be inserted. Often this requires desoldering.

Quite often it is necessary to measure a current that is larger than the available meter with handle. This problem can be taken care of with a *shunt resistance*, that is, a resistor in parallel with the meter movement. See Fig. 12-4.

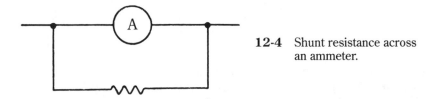

12-4 Shunt resistance across
an ammeter.

By carefully selecting the proper ratio between the shunt resistance and the internal resistance of the meter itself, you can measure virtually any amount of current flow. Most ammeters are actually milliammeters or microammeters with an appropriate shunt resistance, because the tiny coil in a meter movement usually can't carry a full ampere without damage.

The shunt resistance is generally quite small. As an example, suppose you have a meter with an internal resistance of 50 Ω, and which has a *full-scale reading* of 1 mA (or 0.001 A). This means that a current of 1 mA will cause maximum deflection of the armature and pointer. A greater current might damage the meter.

Now, suppose you need to use this meter to read currents up to 100 mA (0.1 A). The meter itself can handle only 1% of the desired full-scale reading, so, obviously, the shunt resistance will have to carry the other 99%.

Using Ohm's law you can find the full-scale voltage dropped through the meter. $E = IR$, or, in this example, 0.001 A × 50 Ω, or 0.05 Vt. For a full-scale reading of 0.1 A, the shunt will have to carry a current of 0.099 A (0.001 + 0.099 = 0.1). Because the shunt resistance is in parallel with the meter, the voltage drop will be the same. Therefore, $R = E/I$ = 0.05/0.099, or approximately 0.5 Ω. Notice that this drops the apparent resistance of the meter to 1/50 + 1/0.5, or 0.495 Ωm.

From the above example, note that the full-scale reading of an ammeter cannot be increased by too large a factor, or the shunt resistance will become impracticably small. In fact, this particular example would probably be rather impractical, because resistance values below 10 Ω are fairly rare.

Of course, when the range of a meter is changed, the calibration markings on the scale face will no longer be accurate. However, when the increased full scale reading is an exact multiple of 10 (as in the example) the same calibration markings can be used, and the appropriate number of zeros can be added mentally. For instance, if the meter in the example gave a reading of 0.5 mA on its old scale, it would mean that the actual current was 0.5 × 100, or 50 mA (0.05 A).

Of course, all shunt resistors should have the tightest tolerance possible. Resistors with a 1% tolerance are generally essential, but 5% resistors can be adequate in some uncritical applications. The tolerance should never be more than 5%.

Voltmeters

To measure the current flowing in a circuit, the meter is placed in series with the circuit. If, on the other hand, a meter is connected in parallel across a resistance within a circuit (such as in Fig. 12-5), the deflection of the pointer will be proportional to the voltage drop across the resistance.

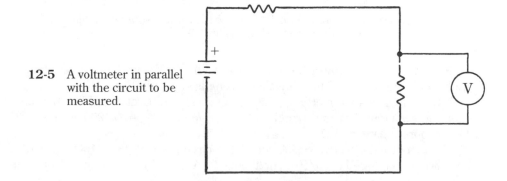

12-5 A voltmeter in parallel with the circuit to be measured.

Actually, the meter is still measuring the current flowing through it, but because current is determined by voltage and resistance (I = E/R), making the resistance constant, will cause the current through the meter to vary in step with the voltage. The meter's scale can easily be calibrated so that the reading is given directly in volts. Such a device is called a *voltmeter.*

In an ammeter, the resistance should always be kept as low as possible, because it is in series with the circuit being tested, because a large resistance would produce a large voltage drop that would not be present if the meter was not in the circuit.

In a voltmeter, on the other hand, the resistance should be as large as possible. Generally, a rather large fixed resistor called a *multiplier* is connected in series with the meter, as shown in Fig. 12-6. This multiplier resistance serves two important functions—it protects the meter movement from excessive current, and it helps prevent *circuit loading.*

12-6   Adding a multiplier resistance to a voltmeter.

Suppose you want to take measurements in the simple resistor/voltage source circuit shown in Fig. 12-7A. Assume each of the two resistors has a value of 1000 Ω, and the voltage source is 10 V. The total resistance in the circuit is 1000 + 1000 or 2000 Ω. The current in the circuit equals 10/2000, or 0.005 A (5 mA), which means the voltage drop across each resistor equals 1000 × 0.005, or 5 V.

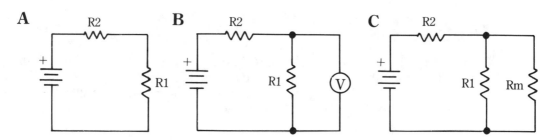

12-7   A simple circuit. The basic circuit (A); the circuit with a voltmeter added (B); equivalent circuit of the circuit and the meter (C).

Now, when you connect the meter, as in Fig. 12-7B, it places a resistance in parallel with the resistor. The equivalent circuit is shown in Fig. 12-7C.

If the meter has an internal resistance of 50 Ω, the parallel combination of the meter and R_1 is found by the formula, $1/R_T = 1/R_1 + 1/R_m = 1/1000 + 1/50 = 21/1000$ = approximately 48 Ω.

Now, the total effective resistance in the circuit is just 48 + 1000, or 1048 Ω. The current is changed to 10/1048, or about 0.01 A (10 mA). The voltage drop across the R1–Rm combination is 0.01 × 48, or a mere 0.48 V. This reading will be the reading shown on the meter. Quite obviously, this is extremely inaccurate.

Now, suppose you include an extra series resistor (that is, a multiplier) inside the meter, so that the total resistance of the meter is 20,000 Ω. In this case, the parallel combination of R_1 and R_M can be calculated as $1/R_T = 1/1000 + 1/20,000 = 21/20,000$ or about 952 Ω. This is much closer to the original value of 1000 Ω.

If you increase the meter resistance further to 1,000,000 Ω (1 megohm), the parallel combination value comes even closer to the nominal value of R_1 alone—just slightly over 999 Ω.

For the most accurate readings a meter should be essentially non-existent as far as the circuit's operation is concerned. For a voltmeter, increasing the resistance of the meter will increase the accuracy of the measurement.

However, if you're comparing voltages between similar circuits (as in servicing, when you compare a defective unit to a working standard), the meters used in each circuit should have the same internal resistance, or identical voltages will not produce the same readings. If the difference is large, you might not be able to compare the readings in any meaningful way.

Ohmmeters

The third type of common meter circuit measures dc resistance and is called an *ohmmeter*. Of course, the meter itself is actually responding to the current flowing through it, but by passing a known voltage through an unknown resistance, you can measure the current and use Ohm's law to calculate the resistance. You don't have to actually do any calculations when you use an ohmmeter—the scale is calibrated to read the resistance directly in Ω.

The basic ohmmeter circuit is shown in Fig. 12-8. The resistor labeled Rm is included to calibrate the meter. For example, if the battery is 3 V, and the total

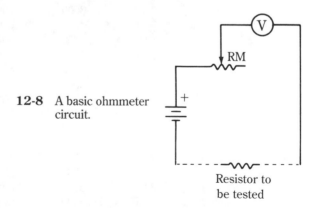

12-8 A basic ohmmeter circuit.

Resistor to
be tested

resistance of the meter circuit is 3000 Ω, and you short the test leads together (0 Ω external resistance), the current flowing through the meter will equal 3/(3000 + 0), or 0.001 A (1 mA). If the meter has a full-scale reading of 1 mA, the pointer will move all the way over to the high end of the scale, which is marked 0 for an ohmmeter. Rm is used to aim the pointer directly at the zero mark when the leads are shorted. As the battery ages, the required resistance to do this will change.

On the other hand, if the test leads are disconnected (infinite external resistance) no current at all will flow through the meter, because there isn't a complete circuit path available. Of course, the pointer will remain at its rest position (low end of the scale). On an ohmmeter, this position is labeled ∞, or infinity.

Now, if you connect a 3000 Ω resistor between the test leads, the total circuit resistance will equal 3000 (internal resistance) + 3000 (external resistance), or 6000 Ω. Therefore, the current flowing through the meter must equal 3/6000, or 0.0005 A (0.5 mA). The pointer will move the center of the scale.

You'll notice that the lower the resistance between the test leads, the farther the pointer moves up the scale. This, of course, is the exact opposite of what happens with voltmeters and ammeters, where the pointer moves further for higher values.

There is another major difference in the way the scale of an ohmmeter must be calibrated. This difference is indicated in Table 12-1, which compares the current flow with the test resistance. Unfortunately, the current through the meter does not

Table 12-1.
Why an ohmmeter scale is not linear.

E	R_M	Test resistance	R_T	I(mA)
3	3000	0	3000	1.00
3	3000	500	3500	0.86
3	3000	1000	4000	0.75
3	3000	1500	4500	0.67
3	3000	2000	5000	0.60
3	3000	2500	5500	0.55
3	3000	3000	6000	0.50
3	3000	3500	6500	0.46
3	3000	4000	7000	0.43
3	3000	4500	7500	0.40
3	3000	5000	8000	0.38
3	3000	5500	8500	0.35
3	3000	6000	9000	0.33
3	3000	6500	9500	0.32
3	3000	7000	10000	0.30
3	3000	7500	10500	0.29
3	3000	8000	11000	0.27
3	000	8500	11500	0.26
3	3000	9000	12000	0.25
3	3000	9500	12500	0.24
3	3000	10000	13000	0.23
3	3000	10500	13500	0.22
3	3000	11000	14000	0.21
3	3000	∞	∞	0.00

change in a direct linear fashion with the external resistance. This nonlinearity is present because the internal resistance is, of necessity, a constant value.

The internal resistance can be altered to change the total range of the meter (if the internal resistance is 10,000 Ω, a mid-scale reading would indicate an external resistance of 10,000 Ω), but within a given range, the resistance is fixed as far as the actual testing is concerned.

ac meters

An ohmmeter can only be used to test dc resistance—not reactance or impedance. It might seem that an ac ohmmeter could be built with an ac power source, using a circuit something like the one shown in Fig. 12-9. However, any measurement made with such a device would be meaningful only at the specific frequency of the source voltage used in the test. The voltage source could be made so that the frequency is variable, but such a circuit would add greatly to instrument cost and complexity. Testing would be a long, drawn-out process. Besides, such a device would give no indication of phase relationships.

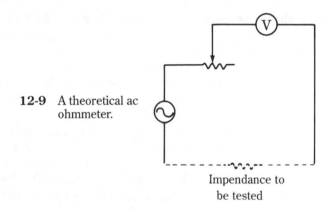

12-9 A theoretical ac ohmmeter.

Impendance to be tested

Unfortunately, there is no easy way to directly measure impedance. Usually it must be calculated from voltage and current measurements. Fortunately, meters can be constructed to read ac voltages and currents, with the help of a device called a *diode*, which only lets current pass through it if the polarity is correct. If the polarity is reversed, the diode blocks the current. Fig. 12-10 shows what happens to an ac signal as it passes through a diode. Only half of the actual ac waveform is applied to the meter itself. Because the two halves of an ac waveform are mirror images of each other, the total value can easily be derived. Diodes are discussed in detail in another chapter. Meters for ac generally measure rms values of sine waves. If the waveform is something other than a sine wave, the reading will not be equal to the true rms value.

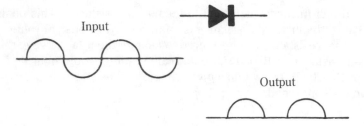

Input

Output

12-10   What happens to an ac signal when it passes through a diode.

VOMs

Probably the most commonly used piece of equipment in electronics work is the *VOM. VOM* stands for *volt-ohm-milliammeter.* Various resistors are switched in and out of series or parallel to set up the meter for each type of measurement. An internal battery is included for resistance measurements. Some VOMs do not have the capability to measure current directly, and many are designed for dc use only.

Usually any of a number of different value resistances can be switched into the circuit, so measurements can be made within different ranges. Whichever range is easiest to read for a specific value can be easily selected, 10 mV would probably be very difficult to detect on a meter scale that went up to 100 V.

The scale face of a VOM has a number of sets of calibration markings, so each of the metering functions can be read directly. Usually the different ranges are multiples of 10 of the basic range, so converting the reading to the appropriate range is simply a matter of mentally adding the correct number of zeros. For example, a reading of 1.5 on an ×100 range would indicate a value of 150.

Closely related to the VOM is the VTVM, or *vacuum-tube voltmeter.* This device is operated by an ac power supply and has an extremely high input impedance, and therefore it has a very high degree of accuracy. The disadvantages of the VTVM are its greater cost and complexity, and the fact that it must be plugged into an ac wall socket, which limits its portability.

A VTVM's sensitivity and high impedance can be simulated by a special type of VOM that is built around a device called a *field-effect transistor,* or *FET.* This component is dealt with in another chapter.

In the last decade or so, meter-type VOMs have been largely (but not completely) replaced by *digital multimeters,* or *DMMs.* These devices are not built around a mechanical meter. Instead, the value of the measured signal is displayed directly in numerical form using LEDs (light-emitting diodes) or LCDs (liquid-crystal displays). These display devices will be covered in another chapter of this book.

For most applications, DMMs and VOMs or VTVMs are pretty much interchangeable. DMMs usually have very high input impedances, so they are quite accurate. In addition, the numerical display is easier to read than a meter pointer. There is no need to worry about any error from looking at the meter face at an angle. So why haven't DMMs replaced mechanical meters altogether? For some applications they really don't work very well. This problem is especially noticeable when you are measuring values that change over time (for example, the charge on a capacitor). Although the movement of a meter pointer is easy to follow either up or down (or even back and forth), a DMM will just display an unreadable and meaningless blur of rapidly changing numbers. A well-stocked modern electronics workbench will have both a VOM (or VTVM) and a DMM.

Self-test

1. Which of the following cannot be easily measured with a simple meter circuit?

A Resistance
B Current
C Impedance
D Voltage
E None of the above

2. What is the most common type of meter movement?

A D'Arsonval
B Fixed coil
C Digital
D Farad
E None of the above

3. What type of meter is used to measure current?

A Ohmmeter
B Wattmeter
C Moving-coil meter
D Ammeter
E None of the above

4. How much internal resistance should an ammeter have?

A As much as possible
B A variable amount determined by Ohm's law
C As little as possible
D It doesn't matter

5. To increase the capacity of an ammeter, what should be added to the circuit?

A A series resistance
B A shunt resistance in parallel with the meter
C A shunt capacitance in parallel with the meter
D A series inductance
E None of the above

6. How is a voltmeter used?

A It is placed in parallel across the component being measured
B It is placed in series with the component being measured
C It is placed within the magnetic field of the component being measured
D None of the above

7. Which type of meter requires its own power source?

A A voltmeter
B An ohmmeter
C A wattmeter
D An ammeter
E None of the above

8. What is a VOM?

A A combination voltmeter and ohmmeter
B A voltage only meter
C A combination ohmmeter, milliammeter, and voltmeter
D A measurement of the movement of a meter's pointer
E None of the above

9. For the greatest accuracy, what should the input impedance of a VOM be?

A 50,000 Ω/V
B 1000 Ω/V
C As large as possible
D As small as possible
E 1,000,000 Ω/V

10. What do ac voltmeters measure?

A The peak voltage of a sine wave
B The rms voltage of any waveform
C The average voltage of a sine wave
D The rms voltage of a sine wave
E None of the above

13
Other
simple components

To this point in the book, you have read about *passive* components. In other chapters, you can look at *active* components, such as tubes and transistors. An active component is one that is capable of amplification. A passive component does not amplify and does not need a power source other than the signal passing through it.

The most important types of passive components have already been covered. These components are resistors (chapter 3), capacitors (chapter 6), and inductors (chapters 8 and 9). This chapter will discuss a few additional passive devices that are used in many electronic circuits. You will mostly be working with switching devices. The last portion of this chapter will turn to motors and lamps.

What is a switch?

In the circuits described so far, the only way to stop the current from flowing is to physically disconnect one of the components. Obviously, this would be impractical in most types of equipment. Yet, a continuous current flow is generally undesirable—power is wasted, continuous operation might be annoying or even dangerous, and it might interfere with the operation of other circuits. Also, it is often necessary to selectively allow current to pass through various subcircuits or components at different times. The solution to these problems is a simple device called a *switch*.

Types of switches

A switch is simply a device that makes and breaks an electrical connection, thereby completing or opening a current path. The simplest type of switch is the *knife switch*, which is shown in Fig. 13-1. When the handle of the switch is in the position shown in Fig. 13-1A, no current can flow through the attached circuit, because there is not

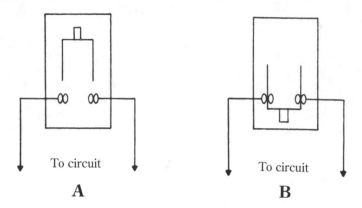

13-1   A knife switch. Switch opened (A); switch closed (B).

a complete path for the current to follow. If the handle is moved into the position shown in Fig. 13-1B, the metal handle completes the circuit and current can flow.

A switch like this is quite simple to make and use, but it is rarely used in modern circuits because it is quite bulky, and the electrical connections are exposed, which could mean a serious shock hazard.

A more common type of switch is shown in Fig. 13-2. This kind is called a *slide switch*. When the *slider* (movable portion) is in the position shown in Fig. 13-2A, the circuit is open and no current can flow. But, when the slider is moved into the position shown in Fig. 13-2B, the metal strip on the bottom of the slider touches the two connections to the external circuit, thus completing the current path.

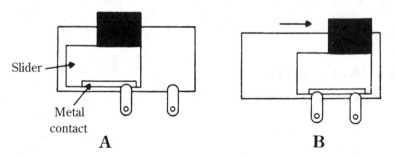

13-2   Construction of an SPST (single-pole, single-throw) slide switch.
Switch opened (A); switch closed (B).

Another popular type of switch is the *toggle switch*, shown in Fig. 13-3. It operates in a manner similar to the slide switch, but the slider is in the shape of a ball that rolls into and out of position.

Still another commonly used switch type is the *pushbutton*. This device is discussed elsewhere in this chapter.

13-3 A toggle switch.

SPST switches

The switches you have been reading about so far have only two connections. Because the slider, or *pole*, makes contact in only one position, this form of switch is said to be a *single-pole, single-throw switch*. This name is generally abbreviated as *SPST*. The schematic symbol for a SPST switch is shown in Fig. 13-4. This switch is usually shown in its open position for clarity. The same symbol is used for most kinds of switches (that is, slide, toggle, or whatever).

13-4 Schematic symbol for an
SPST switch.

When the switch is open, there is a break in the circuit. No current flows. When the switch is closed it acts like a simple piece of wire, and current can flow through it. See Fig. 13-5.

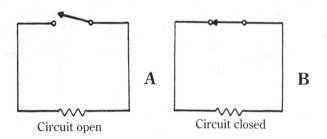

A   B

Circuit open   Circuit closed

13-5 Equivalent circuits for an SPST switch. Switch
opened (A); switch closed (B).

Another way an SPST switch can be used is shown in Fig. 13-6. This circuit has two resistors in parallel, but if the switch is open, current can flow through R1 only. Electrically, R2 does not exist. However, when the switch is closed, current can flow through both resistors. R1 and R2 act like any parallel resistance combination. Referring again to Fig. 13-6, if the battery generates 6 V, and each resistor has a value of 100 Ω, the milliammeter will indicate 60 mA when the switch is open. However, when

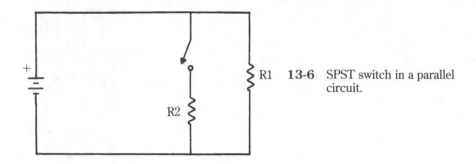

R1   **13-6**   SPST switch in a parallel circuit.

R2

the switch is closed, connection R2 is in parallel with R1, the total effective resistance in the circuit drops to 50 Ω, and the current increases to 120 mA.

SPDT switches

A single switch can be used to control two separate circuits if a third connection is added, as in Fig. 13-7. When this switch is in the position shown in Fig. 13-7A, the current can flow between connector 1 and connector 2. Connector 3 will be open (disconnected). Moving the slider into the position shown in Fig. 13-7B will allow current to flow between connector 2 and connector 3, but connector 1 will now be left open.

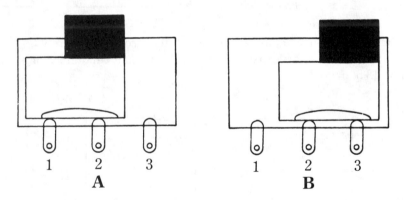

1   2   3   1   2   3
A   **B**

13-7   An SPDT (single-pole, double-throw) slide switch. Switch opened (A); switch closed (B).

Because this type of switch has a single pole and two possible positions, it is called a *single-pole, /double-throw* or *SPDT* switch. The schematic symbol for this kind of switch is shown in Fig. 13-8. The slider can be shown in either position. Notice that connector 2 (Fig. 13-7) is common to both circuits, regardless of the slider position. The connector numbers are not shown in actual schematic diagrams.

A simple circuit using an SPDT switch is shown in Fig. 13-9. R2 and R3 are in parallel with R1, but current will flow through only one resistor at any given time, depending on the position of the slider. In Fig. 13-9A, current can flow through R1 and R2, but not R3. In Fig. 13-9B, the situation is reversed. Current can flow through R1 and R3, but not R2.

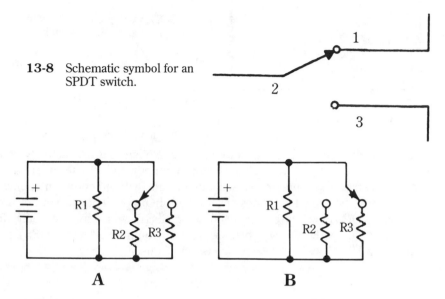

13-8 Schematic symbol for an SPDT switch.

13-9 An SPDT switch in a parallel circuit.

If you assume the battery puts out 6 V, R_1 is 100 Ω, R_2 is 220 Ω, and R_3 is 4,700 Ω, the milliammeter will read 87 mA (0.087 ampere) in Fig. 13-9A, and in Fig. 13-9B it will read 98 mA (0.98 A).

Some SPDT switches have a *center off* position. In this case, the slider would be positioned as shown in Fig. 13-10. Neither circuit 1-2, nor circuit 2-3 is complete, so no current flows through the switch at all.

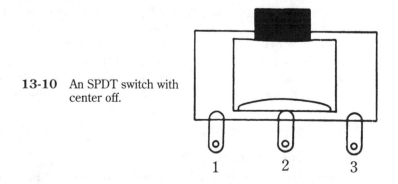

13-10 An SPDT switch with center off.

An SPDT switch could be used as an SPST unit by just leaving one of the end connectors unused. See Fig. 13-11.

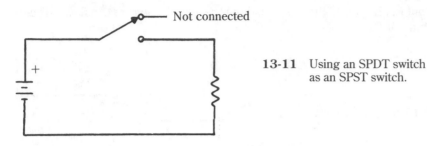

13-11   Using an SPDT switch
as an SPST switch.

DPST switches

A similar type of switch is the *double-pole, single-throw* switch (*DPST*). The schematic symbol for this device is shown in Fig. 13-12. Notice that it is actually two entirely separate (electrically) SPST switches with a common slider. The slider has two separate metal strips to complete two separate circuits with no common connections. Of course, these two SPST switches always work in union. They are either both off, or both on. A DPST switch can be used as an SPST switch by just using one set of contacts. Actually, DPST switches are fairly uncommon. When a DPST function is required, a DPDT switch (see below) is usually used.

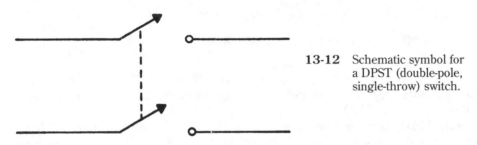

13-12   Schematic symbol for
a DPST (double-pole,
single-throw) switch.

DPDT switches

Double-pole/double-throw (*DPDT*) switches have six connection terminals (see Fig. 13-13) and are essentially two SPDT switches with a common slider, just as a DPST switch is essentially two SPST switches with a common slider. Obviously a DPDT switch can be substituted for any of the other switch types (SPST, SPDT, DPST, or a combination of an SPST and an SPDT) simply by using only the appropriate terminals. By using all six contacts, you can achieve two electrically separate SPDT actions with the throw of a single switch. The schematic symbol for a DPDT switch is shown in Fig. 13-14. Like the SPDT switch, DPDT switches are often equipped with a center-off position that leaves all six terminals open.

Multiposition switches

The four basic types of switches discussed above (SPST, SPDT, DPST, and DPDT) are the most commonly used configurations, but other configurations are sometimes required in specific circuits. Generally, slide switches are available only up to DP4T (two poles with four positions each). If more poles, or positions are needed, a

Bottom view

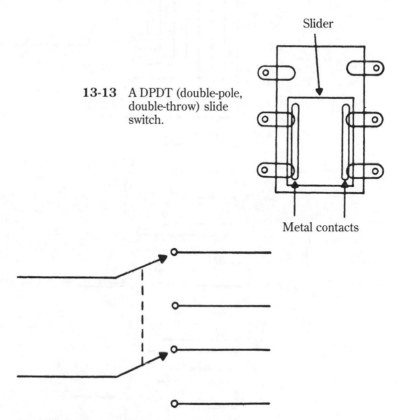

Slider

13-13 A DPDT (double-pole, double-throw) slide switch.

Metal contacts

13-14 Schematic symbol for a DPDT switch.

rotary switch is used. This type of switch gets its name from the fact that a knob is rotated to control the position of the slider(s).

The schematic symbol for a rotary switch varies somewhat, of course, depending on the number of poles and positions in the specific switch. Figure 13-15A shows the schematic symbol for an SP12T rotary switch, and Fig. 13-15B is the symbol for a 3P6T unit. Many other combinations are also possible.

There are two basic varieties of rotary switches. The *nonshorting type* disconnects the circuit at one position completely before the connection of the next position is made. The other kind of rotary switch is the *shorting*, or *make-before-break* type. With this type of switch, in switching from position A to position B, the switch makes contact with both position A and position B for a brief instant, before the connection at position A is broken. In most circuits it doesn't really matter which type is used, but some specialized circuits require one type or the other.

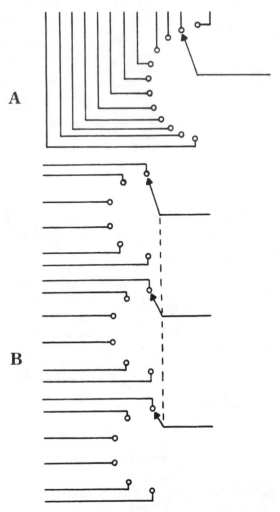

13-15   Schematic symbols. SP12T rotary switch
(A); 3P6T rotary switch (B).

Momentary-contact switches

Sometimes it is necessary to open or close a circuit connection only briefly, then return it to its original condition. In many such cases, manually moving the switch back and forth is inconvenient or impractical. For this sort of situation, a *momentary-action* or *momentary-contact* switch is used. This kind of switch is loaded with a spring that always returns the slider to a specific rest position, unless it is manually held in the other position. The rest position can be either *normally open* (*NO*) or *normally closed* (*NO*), depending upon the requirements of the specific circuit.

Although momentary slide switches and toggle switches are available, this type of switch action is usually in the form of a *pushbutton switch*. Pushbutton switches can be SPST, SPDT, or DPDT (or, occasionally DPST), but the most common configuration for momentary action switches is SPST. If it is a momentary action switch, this information will be indicated in the parts list or in a note on the schematic. Notice that a double-throw, momentary-contact switch has one set of contacts that are normally open, and another set of contacts that are normally closed.

Most pushbutton switches are of the momentary action type, but some work as regular switches that can be left in either position. A pushbutton switch that does not have momentary contacts is called a *push-on, push-off switch.*

Potentiometer switches

Another type of commonly used switch fits onto the back of a potentiometer. This kind of switch is usually an SPST switch. The switch is controlled by the potentiometer knob. When the potentiometer is in its maximum resistance position, the switch is open (off). But as soon as the control knob is advanced from this extreme position, the switch is clicked shut, turning the controlled circuit on. From then on the potentiometer operates normally.

The potentiometer and the switch will usually be used in the same part of the circuit buy not always. The potentiometer and the switch are mechanically tied together, but electrically distinct. They could be used in two entirely separate circuits, although this arrangement could cause some confusion in operation.

Potentiometer switches are most often used to turn the main power supply of a circuit on and off. For example, an amplifier might have a switch connected to its volume control-potentiometer, so that when the volume is turned all the way down to its minimum setting, the entire circuit is switched off.

Relays

In many applications, it might not be practical to use any of the basic manual switches discussed so far. For instance, you might need a circuit that must be switched on when the voltage in another circuit rises above some specified level. Of course, you could watch a voltmeter connected to the second circuit and manually flip a switch in the first circuit at the appropriate moment, but that is obviously a highly impractical and inconvenient approach to the application. Some sort of automatic switching would clearly be more efficient.

An automatic switch of some kind would also be necessary in a remotely controlled circuit. Often the circuit being switched is not readily accessible. It would be inconvenient at best to run a pair of wires carrying the full power-supply voltage or electrical data over a long distance to a convenient control point. It would be far better to just send a small control voltage over light-duty connecting wires to a remote-controlled electrical switch.

The simplest form of automatic switching in an electrical circuit is through a device called a *relay.* A relay basically consists of two parts—a coil and a magnetic switch. When an electrical current flows through a coil, a magnetic field is created around it. This magnetic field is proportional to the amount of current flow through the coil. At some specific point, the magnetic field will be strong enough to puff the switch slider from its rest, or *de-energized* position to its momentary, or *energized* position. If the electrical power through the coil drops, the strength of the magnetic field will drop off to zero, releasing the switch slider and allowing it to spring back to the original de-energized position.

The switch section of a relay can be any of the basic switching types discussed in this chapter (SPST, SPDT, DPST, or DPDT). For single-throw units the switch contacts might be either normally open or normally closed. Of course, with an SPDT or a DPDT switching arrangement, you have both a normally open and a normally closed contact simultaneously. The schematic symbols for an SPST and an SPDT relay are shown in Fig. 13-16. Relays are usually identified in schematic diagrams and parts lists by the letter *K*.

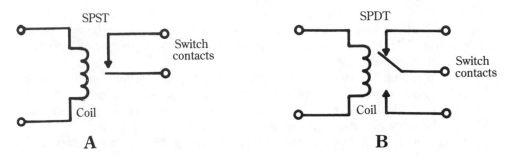

13-16   Schematic symbols. SPST relay (A); SPDT relay (B).

A relay coil and switching contacts are virtually always used in electrically isolated circuits. That is, the current through one circuit controls the switching of another circuit. Relays vary greatly in size, depending primarily on the amount of power that they can safely carry. Separate ratings are usually given for the coil and the switch contacts, because they are generally used in separate circuits. Relays range from tiny units intended for 0.5 W transistorized equipment to huge megawatt (millions of watts) devices used in industrial power-generating plants. At either extreme, the principle of operation is precisely the same.

Of course, the most important rating for a relay is the voltage required to make the switch contacts move to their energized position. This action is called *tripping* the relay. Typical trip voltages for relays used in electronic circuits are 6, 12, 24, 48, 117, and 240 V. Both ac and dc types are available.

The controlling voltage through a relay coil should be kept within about ±25% of the rated value. Too large a voltage could burn out the delicate coil windings. On the other hand, too small a control voltage could result in erratic operation of the relay.

Sometimes it might be necessary to drive a relatively high-power circuit with a rather low-power control signal. This application can be handled the type of circuitry shown in Fig. 13-17. High voltage supply B is operated only when the relay is energized, minimizing power consumption.

Occasionally the available control signal will not be sufficient to drive a large

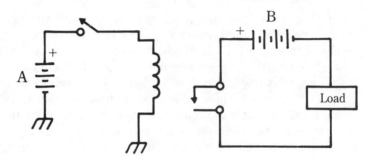

13-17   A relay is used so that a low voltage can control a
higher-voltage load.

enough relay for the circuit to be controlled. The solution might be to add an additional medium-power relay to act as an intermediate stage, as shown in Fig. 13-18.

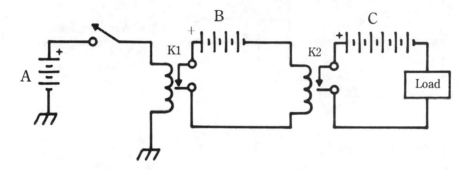

13-18   A small relay can be used to drive a heavier relay.

The coil winding could self-destruct if the current through it changes suddenly, perhaps due to opening a series switch as shown in Fig. 13-19A. The voltage drops

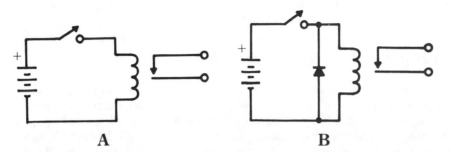

13-19   A parallel diode is often used to protect a relay coil from high-voltage
spikes, which can occur during switching.

from a positive voltage to 0 in a fraction of a second. This fast drop causes the magnetic field around the coil to collapse rapidly. This abrupt change in the magnetic field will induce a brief high-voltage spike in the relay. This spike could be high enough to damage or eventually destroy the switch contacts.

A diode is often placed in parallel with the relay coil to suppress such high-voltage transients. This arrangement is shown in Fig. 13-19B. With this arrangement, the diode limits the voltage through the relay coil to the power supply voltage (unless, of course, the spike is large enough to damage the diode itself).

A transistor amplifier is often used to drive a high-current relay from the low-current source, such as a small battery. The low-current source limits undue power drain. A typical circuit of this type is shown in Fig. 13-20.

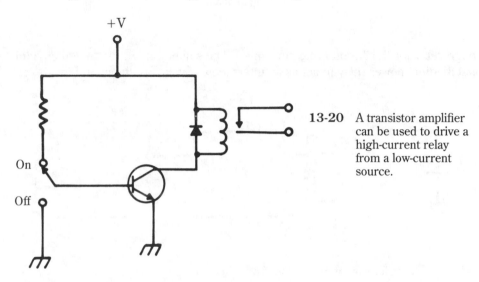

13-20 A transistor amplifier can be used to drive a high-current relay from a low-current source.

Some relays are *latching* relays. One control pulse closes the switch contacts, which will remain closed, even if the control pulse is removed. A separate control pulse opens the switch contacts. Each time the relay is triggered, it latches into the appropriate position until the next trigger signal is received. Not surprisingly, this type of device is called a *latching relay*.

Fuses and circuit breakers

Another kind of automatic switching is used specifically for circuit protection. The voltage through a circuit can be controlled by the design of the power supply, except for relatively rare transients. But, the current drawn through a circuit depends on the resistance and impedance factors within the load circuit. If the resistance drops because of a short circuit, or some other defect, the current could rapidly rise to a level that can damage or destroy some of the components. What is needed is a way to disconnect the power supply from the load circuit before the current reaches a dangerous level. This is most often done with a special device called a *fuse*. A fuse is basically just a thin wire that is carefully manufactured so that it will melt when the current passing through it exceeds a specific value.

The schematic symbol for a fuse is shown in Fig. 13-21, and Fig. 13-22 illustrates a simplified circuit using a fuse. If the current drawn by the load circuit exceeds the current rating of the fuse for any reason (such as changing the value of R1), the fuse will blow, opening the circuit. No further current will flow through the circuit.

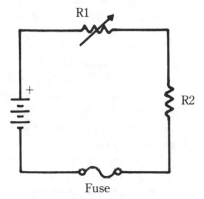

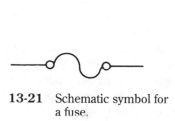

13-21 Schematic symbol for a fuse.

13-22 A simple circuit using a fuse.

Fuses are usually enclosed in glass (or sometimes metal) tubes for protection. The fuse wire is very thin, and could easily be damaged. Figure 13-23 shows a typical fuse.

13-23 Basic structure of a fuse.

Sometimes fuses are soldered directly into a circuit, but because once a fuse element has been melted, it must be replaced before the circuit can be reused, this installation method is rather impractical in most applications. For more convenient fuse replacement, some kind of socket is generally used for fuses. Most commonly the fuse is held between a set of spring clips. Another frequently used method is to fit the fuse into a special receptacle with a screw cap.

Frequently replacing fuses can be a nuisance, so sometimes a component called a *circuit breaker* is used. This switch is a special switch that will automatically open if the current through it exceeds some specific amount. To close the switch again, you just manually push a *reset* button.

Occasionally *transients* (brief irregular signals) can cause a fuse to blow, or a circuit breaker to open even if there is no defect in the load circuit at all. But if a new fuse immediately blows, or if the circuit breaker repeatedly opens when it's reset, it indicates that something is wrong and repairs are needed. *Never* replace a fuse with a higher rated unit. You could end up blowing some expensive electrical components to protect the fuse, and that certainly doesn't make much sense.

Motors

Motors aren't switching devices, but they warrant a brief discussion here. Rather than including a separate short chapter on motors, this chapter is something of a potpourri. A motor is a *transducer*, a device that converts one form of energy into another form of energy. Additional transducers will be discussed in chapters 19, 22, and 30. In the case of a motor, electrical energy is converted into mechanical energy. That is, an electrical signal can cause something to physically move.

A motor is a practical application of the electromagnetic fields discussed in chapter 7. There are many different types of motors; some are extremely tiny, and others are huge. Some can only move very small weights, and larger motors can move tons; some run on dc and others run on ac. Regardless of these differences, all motors are basically the same, at least in their operating principles.

An electric current is fed through a set of coils, setting up a strong magnetic field. The attraction of opposite magnetic poles and the repulsion of like magnetic poles results in the mechanical motion of the motor.

A simplified cutaway diagram of a typical motor is shown in Fig. 13-24. Notice that there are two sets of coils. One is stationary, and is known as the field coil. The other coil, which is known as the armature coil, can freely rotate within the magnetic field of the field coil. The motor shaft is connected directly to the movable armature coil. As the armature coil moves, the motor shaft rotates.

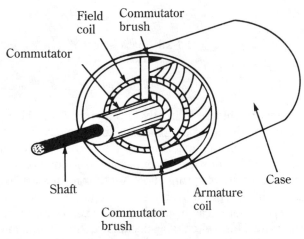

13-24   Simplified cutaway diagram of a typical motor.

The commutator reverses the polarity of the current with each half rotation of the armature and shaft. This reversal keeps the armature coil constantly in motion. In Fig. 13-25A, the armature coil is positioned so that its magnetic poles are lined up with the like poles of the field coil. The like magnetic poles repel each other, forcing the armature coil to rotate, as shown in Fig. 13-25B. At some point, the attraction of unlike poles will take over, pulling the armature into the position shown in Fig. 13-25C.

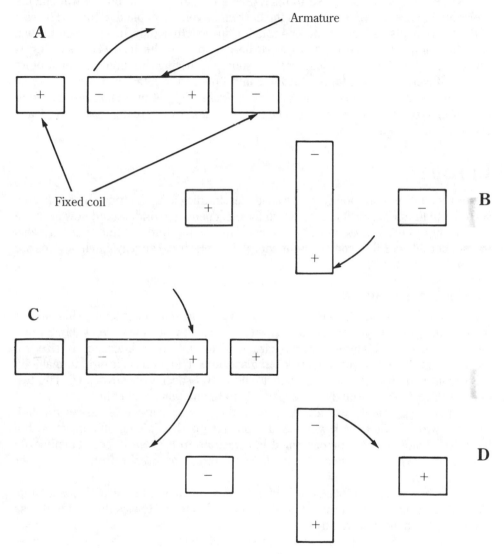

13-25 The commutator reverses the polarity of the current with each half-roation to keep the armature in constant motion.

The commutator reverses the polarity of the current, so once again the like poles of the armature coil and the field coil are lined up. The whole process repeats for the second half rotation (Fig. 13-25D), bringing you back to the position shown in Fig. 13-25A, and a new cycle begins.

Increasing the current through the coils (assuming everything else remains equal) will increase the *torque* of the motor. That is, it can turn a larger load if a larger current is supplied. Using a given motor to move a load, the heavier the load is, the more current the motor will be forced to draw from the power supply.

Some motors are designed to operate at a constant speed; others will change their rotation speed with changes in the current or voltage applied to them. The size of the load can also affect the motor's rotation speed. Obviously, heavier weights will slow the motor, because it has to work harder to turn the load. This slowing is especially noticeable if a constant current source is driving the motor. Some motors are specifically designed for the load to control the actual rotation speed.

Direct-current motors and ac motors are generally not interchangeable. Using the wrong type of power source could damage or destroy the motor. Excessive loads can also damage small motors.

Lamps

A lamp is another common type of transducer. In this case, electrical energy is converted into light energy. It isn't too hard for most people to understand how a simple lamp or light bulb works. Most people are surprised to find out that there are also transducer devices that convert light energy into electrical energy. Such devices are covered in chapter 22.

Incandescent lamps

The most common and familiar type of lamp is the incandescent lamp. This lamp is the type of light bulb you're acquainted with and use in your home every day. Similar lamps are used in many electronics circuits. The only real difference is size. An incandescent lamp is a vacuum enclosed glass bulb. The bulb is air tight to maintain the vacuum. Within the bulb is a short length of a special resistive wire. The two ends of this wire are brought out separately to the metallic socket in the bulb.

When a sufficient electrical current at the correct voltage is passed through this thin wire, or *filament*, it will heat up and start to glow, giving off a great deal of light. The filament must be contained in a vacuum to prevent it from burning out too quickly. When the filament wire burns through or breaks from some other cause, the bulb must be replaced.

Contrary to popular belief, Edison did not really invent the incandescent lamp. He just came up with the first practical device of this type. He was the first to devise a suitable filament that would last.

Incandescent lamps vary widely in size and shape, as well as power requirements. You can find incandescent lamp bulbs designed to operate at almost any voltage ranging from less than one volt up to several hundred volts.

In electronics work, you generally use only small, low-voltage bulbs. These lamps are often called flashlight bulbs, because they are commonly used in flashlights. Small incandescent lamps are used in many electronics circuits (especially older designs) as indicator devices. Often, the bulb will be painted with translucent paint, or a translucent plastic cap will be placed over the bulb to give the light a specific color, such as red or green.

At one time, small incandescent lamps were just about the only practical choice for low-power indicator devices. Today they are largely being replaced by LEDs (see chapter 19) because the lamps are bulky and fragile. They are also quite inefficient in terms of power. They generate more heat than light. Because many modern semiconductor components are heat sensitive, this is not just wasteful; it could actually cause harm.

Some other types of lamps are filled with a gas, rather than being evacuated. A fluorescent tube is a common example.

Neon lamps

A number of electronic circuits (again, mostly older designs) use neon lamps as indicating devices. The construction of a neon lamp is shown in Fig. 13-26. In a neon lamp, there is air-tight glass bulb, just like in the lamp discussed above, but this bulb does not contain a vacuum; it contains neon gas. Instead of a filament, there are two separated electrodes. If a voltage is connected across these electrodes, the neon gas between them will become ionized and start to glow. Neon glows with a characteristic orange color. A high voltage is normally required to light a neon lamp; typically, close to 100 V is required.

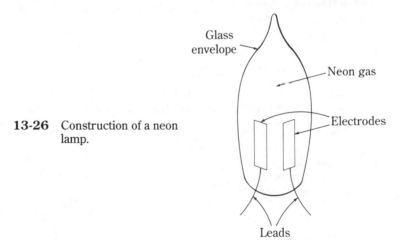

13-26 Construction of a neon lamp.

Because of the way the neon lamp suddenly switches on or *fires* when its threshold level (minimum operating voltage) is exceeded, this device is often used for triggering and voltage regulation applications, as well as an output indicator.

Self-test

1. What is the simplest type of switch?

A SPDT switch
B Toggle switch
C Knife switch

D Relay
E None of the above

2. How many circuits can a DPDT switch simultaneously control?

A One
B Two
C Three
D Four
E None of the above

3. Which of the following is not a standard switch type?

A SPST
B SPDT
C DPST
D DPDT
E None of the above

4. If R_1 in the circuit shown in Fig. 13-6 is 2200 Ω, and R2 is 3900 Ω, what is the circuit resistance when the switch is closed?

A 2200 Ω
B 3900 Ω
C 6100 Ω
D 1407 Ω
E None of the above

5. What is the term for a switch that disconnects the circuit completely at one position before the connection of the next position is made?

A Rotary
B Shorting
C Nonshorting
D Make-before-break
E None of the above

6. What happens when a momentary action switch is released?

A Nothing
B The switch contacts latch into the new position
C The switch contacts return to their normal rest position

7. What is contained in a relay?

A An RC circuit
B A coil and a fuse
C A coil and a diode
D A coil and a set of switch contacts
E None of the above

8. Why is a diode placed across the coil portion of a relay?

A To increase the current flow
B To speed up the switching

C To protect the relay from high-voltage transients when the magnetic field collapses

D To protect the relay against incorrect polarity

E None of the above

9. What is the name of a device that protects a load circuit from excessive current flow?

A Relay

B Suppression diode

C SPDT switch

D Fuse

E None of the above

10. When should a fuse be replaced with a higher rated unit?

A If it blows

B Never

C When the original value is not available

D When fuses of the original value blow as soon as they are replaced

E None of the above

14

Reading
circuit diagrams

As the old cliché says, a picture is worth a thousand words. This expression is as true in electronics as it is anywhere, maybe even more so. It's impossible to imagine working in the electronics field without using diagrams and drawings. There are several basic types of diagrams commonly used in the electronics field. For the most part, these various diagram types can be grouped into three broad categories:

Pictorial diagrams
Block diagrams
Schematic diagrams

Each of these will be discussed in this chapter.

Pictorial diagrams

The most basic diagram you will encounter in electronics work is the *pictorial diagram*. Though widely used, the pictorial diagram is probably the least useful, although it can be helpful in certain cases. A pictorial diagram is simply a drawing of the way a circuit or piece of equipment should look. This drawing can be useful for hobbyists building a project, or in repairing a piece of equipment that has been modified from its original design. A typical pictorial diagram is shown in Fig. 14-1. It tells you nothing more than a photograph of the circuit would tell you.

A slightly more sophisticated form of the pictorial diagram is the *exploded diagram*. An example is shown in Fig. 14-2. In an exploded diagram, the various parts are shown in their relative positions to one another, but are moved apart so you can see them easier. Lines show how the parts are interconnected. Generally, exploded diagrams are not used for circuits. They are used to show how circuit boards and bulky components (such as heavy power transformers) are mounted within a case and how the case is assembled.

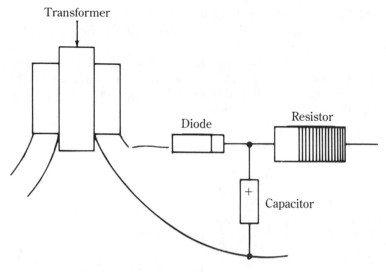

14-1   Pictorial diagram.

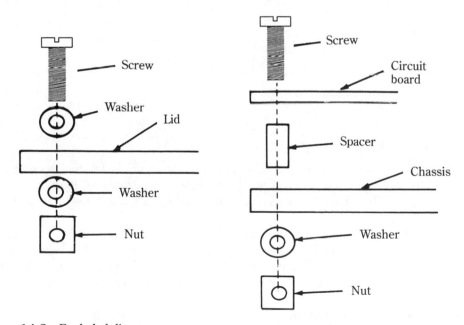

14-2   Exploded diagram.

Component placement diagrams are often used with projects that are built on printed circuit boards. If you are not familiar with printed circuits, don't worry about it; they are covered in chapter 15. A component placement diagram shows the correct position for each component mounted on the printed circuit board. This feature is important, because if a component is misplaced (with one or more leads in the wrong hole), the circuit will not function properly. Off-board components, or connections to other boards, are also indicated. Usually just a labelled lead line coming off the board at the appropriate point is shown. To keep things simple and to avoid wasted effort, the actual off-board device(s) is generally not drawn, although it might be shown in some diagrams, if the technician drawing the diagram thinks this will offer more information to someone using the diagram later. A typical component placement diagram is shown in Fig. 14-3.

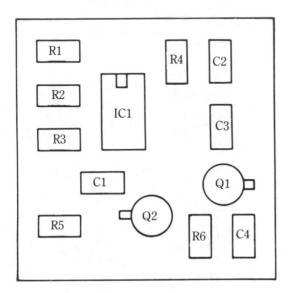

14-3 Component placement diagram.

Block diagrams

Pictorial diagrams show how a circuit is physically constructed, but tell you nothing about how the circuit works, or even what its intended purpose is. *Block diagrams* are more useful for understanding circuit operation. Some technicians refer to block diagrams as *functional diagrams*.

Except for very simple circuits, most electronic systems are made up of multiple subcircuits. Each subcircuit performs a specific function or set of related functions. The various subcircuit functions work together to achieve more powerful and versatile results. In a block diagram, each subcircuit is shown simply as a block. The actual components used to make up the subcircuit are ignored. When you are reading a block diagram, you are only concerned with what the circuit does, not how it does it. Each subcircuit is considered a *black box*. You know what the black box is supposed to do, but you use it as if you have no way of looking inside. If the input is A, then the output is B. The function or name of each subcircuit is written in the appropriate block.

Occasionally a few components are shown separately in a block diagram. The separate diagram usually is drawn if the component is not part of any specific subcircuit and in essence functions as a subcircuit by itself. *Feedback components* that connect several blocks, or stages, are usually drawn separately, as are interstage switches, plugs, and jacks. A typical block diagram is shown in Fig. 14-4.

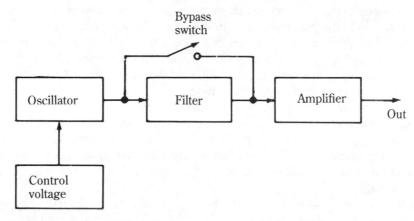

14-4   Block diagram.

In some block diagrams, special shapes are used to indicate certain functions. For example, circles are normally used to indicate oscillators or signal sources, and triangles are used to represent amplifiers. The diagram in Fig. 14-4 is redrawn using this system in Fig. 14-5. There is no functional difference in the two types of block diagrams. Some people find the varied shapes of the second version easier to read at a glance. Generally, which system you use will be strictly a matter of personal preference.

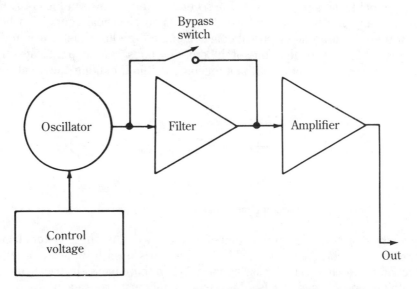

14-5   A different way of drawing the block diagram of Fig. 14-4.

Schematic diagrams

The most important and most frequently used type of diagram in electronics work is the *schematic diagram*. A schematic diagram (sometimes known as a *wiring diagram*) shows all of the components in the circuit and how they are electrically interconnected. The position of any given component in a schematic diagram doesn't necessarily correspond to its actual position in the physical circuit. The arrangement of the components in a schematic diagram is influenced more by the clarity of the diagram than by any specific construction details.

Straight (usually) lines are used to represent interconnecting wires and leads between components. As a rule, the schematic diagram should be drawn so that as few lines as possible cross each other where there is no electrical connection. Unfortunately, in most circuits of any complexity, a few line crossings are inevitable. Certain conventions are followed to prevent confusion. There are three commonly used standards.

In the first of these systems, a dot is used to indicate an electrical connection between two crossing leads. If there is no dot, there is no electrical connection. This system is shown in Fig. 14-6.

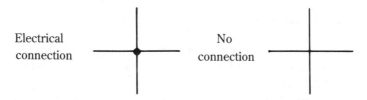

14-6   First system used to indicate crossed wires.

Some technicians prefer to use the second system, shown in Fig. 14-7. Here, a crossing of any two lines assumes that there is an electrical connection between them. If there is to be no electrical connection, and the lines just cross in the diagram, a small loop is made in one of the crossing lines. The loop indicates that the one wire *jumps* over the other without touching (without making electrical contact).

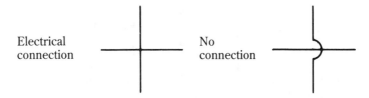

14-7   Second system used to indicate crossed wires.

The third system is a combination of the first two. This third system leaves the least room for error. As shown in Fig. 14-8, a dot is used to indicate an electrical connection between a pair of crossing lines. When there is no electrical connection, the jumping loop is used. No two lines ever simply just cross each other, because this is potentially confusing. This convention is used in all the schematic diagrams in this book.

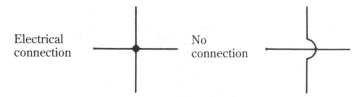

Electrical connection   No connection

14-8   Third (and least ambiguous) system used to indicate crossed wires.

Standardized schematic symbols are used to indicate various component types. These symbols are introduced in the appropriate chapter for each type of component. Some of the most important (and most commonly used) component symbols for the components you have learned about so far are shown again in Fig. 14-9. Those components are: resistor (A), capacitor (B), coil or inductor (C), transformer (D), SPST switch (E), and pushbutton switch (F).

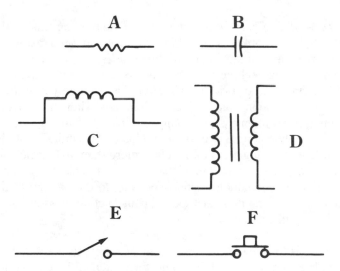

14-9   Schematic symbols. Resistor (A); capacitor (B); coil (C); transformer (D); SPST switch (E); push button switch (F).

Many schematic diagrams also indicate certain electrical parameters that can be measured at specific test points within the circuit. The indicated parameters are most commonly voltages, currents, or waveforms. If the circuit is functioning properly, you should be able to measure the same values as those indicated in the schematic diagram. It is usually important to use test equipment with the same specifications as the equipment used to determine the original values shown in the schematic, or you might not get the same results.

15

Construction techniques

It is probably reasonable to assume that most people reading this book will want to build some of their own projects, whether from plans in a magazine or book, from their own design, or from a kit. In any of these cases, you will need to know something about construction techniques. This knowledge is even needed for many commercial kits that have rather brief instructions that assume you know what you're doing. (A notable exception was the Heath Company. Their manuals are generally excellent, leading the first-time project builder step by step through the construction process without really talking down to the old-hand electronics hobbyist.)

Obviously, constructing a circuit involves more than just gathering the components together and tossing them into a box. Some sort of electrical connection must be made between the component leads, according to the pattern of the schematic diagram. There are several popular types of construction in wide use today. These methods are briefly discussed in this chapter.

Start with the construction technique known as *breadboarding*, which is the one best-suited methods for the experiments in the next chapter (and in chapters 23 and 31). It is strongly recommended that you perform each of the experiments yourself. You will learn more this way than if you just read about them.

Breadboarding

Breadboarding is a method of temporarily hooking up electronic circuits for testing and experimenting. Of course, you could *hard wire* (connect components with wire and solder) each test circuit, as if it were intended for permanent use. But hard wiring can rapidly become extremely expensive.

You could cut costs somewhat by desoldering each circuit when you're through with it and reuse the components. But desoldering tends to be very tedious, time consuming, and quite inconvenient. Besides, the repeated heating and reheating of component leads can damage some components.

Fortunately, there is a much more convenient device that allows you to set up temporary circuits—the breadboard. In its simplest form, this is merely a solderless socket that the various component leads and wires can quickly be plugged into or pulled out of. A typical solderless socket is shown in Fig. 15-1. The various holes in this type of socket are electrically connected to each other. The most frequently used pattern is shown in Fig. 15-2.

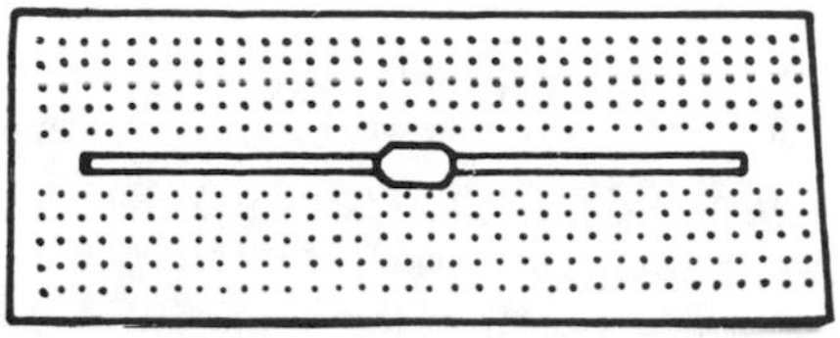

15-1 A typical solderless socket.

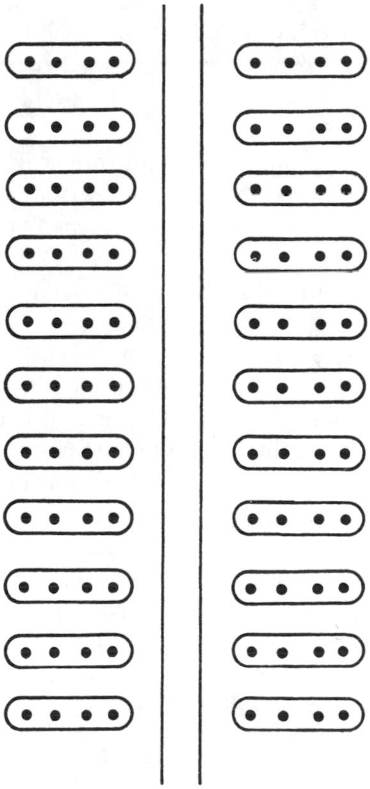

15-2 The most commonly used interconnection pattern for a solderless socket.

These sockets can make experimentation and circuit design much easier, but they are even more useful as part of a complete breadboarding system. These systems consist of a solderless socket and various commonly used subcircuits, such as *power supplies* (which produce a desired dc voltage from the standard ac house current, thus saving battery costs) and *oscillators* (which produce an ac signal, usually with a variable frequency). These subcircuits can be separate, self-standing units used along with a simple solderless socket, but it's generally more convenient to have them grouped together within a single, compact unit. At any rate, these subcircuits will be needed far too often to make breadboarding them each time they're needed reasonable. The way these (and other) basic circuits work is described in detail in other chapters.

If you don't have a power supply available, you can perform the experiments in the next chapter with dry-cell batteries. There will be no difference in circuit operation.

Unfortunately, there is really no substitute for the variable-frequency oscillator. If you do not such a device, you won't be able to perform Experiments 7 through 9. Just read these experiments carefully for now. Later in this book, you'll learn how to build an oscillator. Once you have learned this, you can return to chapter 16 and complete these experiments.

You should have no problem performing the rest of the experiments in chapter 16.

Finally, most breadboarding systems have one or more potentiometers and switches handily available for use in experimental circuits.

Once you have designed your circuit, breadboarded the project, and gotten all the *bugs* out (bugs are errors that prevent your project from working properly), you will probably want to rebuild some circuits in a more permanent way. Solderless breadboarding sockets are great for testing and experimenting with prototype circuits, but they really aren't much good when it comes to putting the circuit to practical use. Breadboarded circuits, by definition, have nonpermanent connections. In actual use, some component leads will easily bend and touch each other, creating potentially harmful short circuits. Components can even fall out of the socket altogether when the device is moved about. Interference signals can easily be generated or picked up by the exposed wiring.

Generally, packaging a circuit built on a solderless socket will be tricky at best. They tend not to fit well in standard circuit housings and boxes. Also, a solderless socket is relatively expensive. It is certainly worth the price if it is repeatedly reused for many different circuits. But if you tie it up with a single permanent circuit, you are only cheating yourself. Less expensive construction methods are available that are more reliable, more compact, and offer better overall performance.

Soldering

Most permanent circuit construction methods involve *soldering*. When soldering, you melt a special metal (called *solder*) over the connection point of two or more leads, binding the leads together and creating a strong, reliable mechanical and electrical connection.

There are different types of solder available for various purposes. Some of these are summarized in Table 15-1. The most common type used in modern electronics work is 60-40 solder. This type of solder is composed of 60% tin and 40% lead. It has a rosin core. For electronics work, you must use rosin-core solder only. Never use acid core solder on any electronic circuit. The acid is highly corrosive and will eat through many of the components. Acid-core solder is used only for metal bonding applications.

Table 15-1. Typical types of solder.

Metal used	Core	Melting point °F	°C	Applications
TL* 50–50	Rosin	430	220	electronic
TL* 60–40	Rosin	370	190	electronic low heat
TL* 63–37	Rosin	360	180	electronic low heat
TL* 50–50	Acid	430	220	metal bonding NOT FOR ELECTRONIC USE
Silver	—	600	320	high heat high current

* TL indicates tin-lead alloy. Percent of each metal in the alloy is shown by the numbers.

Some solder is sold with no core at all. It is used with a separately applied rosin paste. The rosin helps make a good electrical connection between the leads being soldered together.

Before soldering, clean the leads. A few quick rubs with some fine sandpaper will be sufficient in most cases. Arrange the leads in a mechanically solid connection before soldering. If the mechanical connection is weak, you will most certainly end up with a poor solder joint.

For most modern electronics work, you should use a low-power soldering iron. Generally a 20 to 30 W unit will be best. Many electronic components are very sensitive to heat. Semiconductors (discussed in other chapters) are especially sensitive. Even with a low-power soldering iron, do not apply heat near sensitive components for too long a time, or you will damage them.

Do not use a high-power soldering iron or soldering gun. Most low-power soldering irons are shaped like the one shown in Fig. 15-3. This type of iron is sometimes called a soldering pencil. You can also find some low-power soldering guns with an easy-to-hold, pistol-like handle.

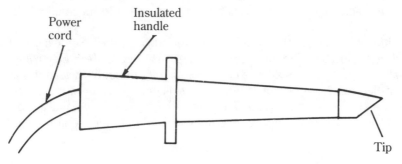

Power cord

Insulated handle

Tip

15-3   Most low-power soldering irons use the pencil shape.

Better soldering irons are grounded. Grounding reduces problems with static electricity, which can damage some electronic components, especially CMOS (complementary metal-oxide semiconductor) ICs (integrated circuits). (See chapter 28 for more information). A grounded soldering iron usually isn't absolutely necessary, but it is desirable.

Warm the soldering iron fully before you begin soldering. For most soldering irons, this means you need to plug it in for about 5 to 10 minutes before you actually start soldering. For maximum safety, place the hot soldering iron in a soldering iron stand when not actually being used. A typical soldering iron stand is shown in Fig. 15-4. You can buy such stands for just a few dollars, and they can reduce considerably the potential hazards of accidental burns to your body or fire. They are very cheap insurance. Don't scrimp on safety.

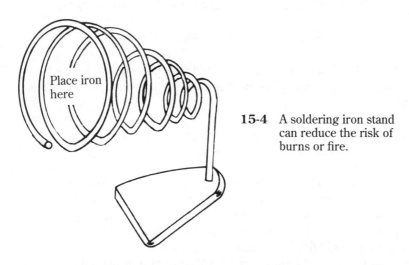

Place iron here

15-4   A soldering iron stand can reduce the risk of burns or fire.

Clean the soldering iron tip periodically on a damp sponge. If too much rosin and miscellaneous build-up accumulates on the tip of the soldering iron, the heat transfer will be significantly reduced. The odds of producing cold solder joints (explained below) will increase enormously.

Before actually soldering, *tin* the tip of the soldering iron; melt a little bit of solder (not too much) over the tip. Use just enough to coat the tip with solder.

Good soldering is normally a two-handed process. (Some specialized soldering aids will let you solder one-handed, but this is by far the exception, rather than the rule.) Hold the handle of the soldering iron in one hand and the solder in the other. Apply both to the joint being soldered as shown in Fig. 15-5. Do not apply the iron directly to the solder as shown in Fig. 15-6. You do not want to melt the solder and let it drip over the joint. The idea is to heat up the joint so that the solder will flow evenly over it.

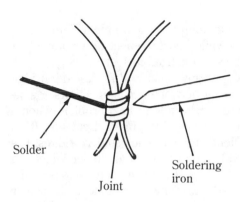

15-5 Apply heat and solder separately to the joint.

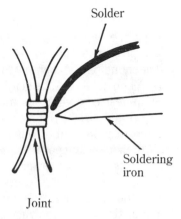

15-6 Do not apply the soldering iron directly to the solder.

Make the solder joint as quickly as possible, remove the unused solder and the soldering iron from the joint without moving or jarring the joint before the melted solder has a chance to cool and set. It takes a couple seconds for solder to set. If you move the joint before the solder hardens, you will likely wind up with a cold solder joint.

Cold solder joints are one of the most common causes of problems in newly built electronic circuits, especially those built by beginners. Examine all solder joints very carefully. A small, high intensity lamp and a magnifying glass can be very helpful. A good solder joint will look smooth and shiny. If any solder joints look rough or grainy, they are probably cold solder joints. The cure is simple enough—just reheat the connection to remelt the solder and let if flow more smoothly over the joint. In a cold solder joint, there is not a good electrical connection between the soldered leads. In some cases, the mechanical connection is pretty weak too.

The term *cold solder joint* really refers to a number of possible problems. One of the easiest to visualize is a bubble within the solder joint. This kind of problem is shown in Fig. 15-7. From the outside, the leads appear to be soldered together, but they aren't.

Cold solder joints are not always apparent when you examine the joint. Even an experienced expert will not be able to identify all cold solder joints. If the completed

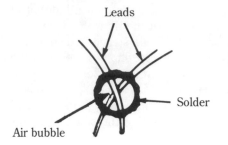

15-7 A cold solder joint could be the result of an air bubble.

project does not work properly, a cold solder joint might be the problem. Sometimes, if the project is not too complex, it is worthwhile just to reheat all of the solder connections rather than try to track down the specific trouble spot.

When soldering, be very careful not to create any *solder bridges*. A solder bridge is an undesired connection made by excess solder between two connections or adjacent printed circuit traces. Solder bridges are most likely to be a problem when a printed circuit board (discussed in this chapter) is used. Integrated circuits (or ICs), with their closely spaced pins are also frequent candidates for solder bridges. ICs will be discussed in other chapters. A careful examination of all solder joints can help save you a lot of grief. Check for solder bridges after each connection is made. Then when all soldering is complete, go back and carefully re-examine all the solder joints again before applying power to the circuit. Remember, a solder bridge is a form of short circuit. If a power line or a signal gets into the wrong part of the circuit, the circuit will not work properly, and, in some cases, some components (usually the more expensive ones) can be damaged or destroyed.

Do not use too much solder. Use enough to cover the joint thoroughly, but don't fill your project with great, ugly lumps of solder. In addition to looking unattractive and unprofessional, too much solder can be the source of many problems. The more solder you use, the greater the chances of solder bridges, and the greater the likelihood that it all won't melt thoroughly, and cold solder joints will result. If you use too much solder, you will have to apply more heat longer to each joint, increasing the chances of damaging delicate semiconductors. Use what you need, but don't resort to overkill.

Desoldering

If you work in electronics (whether professionally or as a hobby), sooner or later you're going to have to do some desoldering. Desoldering is not the fun part of electronics. It's not awful, but it tends to be a tedious and obnoxious task at best. There are probably no electronics technicians or hobbyists who say they enjoy desoldering. It is sometimes necessary, and it can save you a lot of money. *Desoldering* is just the opposite of soldering, as the name suggests. The joint is heated and the solder is removed. This action permits you to remove and replace a component in the circuit.

Suppose you make a mistake and solder the wrong two leads together, or perhaps you happened to get a bad component. Perhaps, a project or other circuit has been in use for a while, and you'd now like to modify the circuit in some way. In any of these instances, desoldering is the only alternative to junking the whole thing and starting over from scratch, which not only an extremely inelegant solution, it tends to be ridiculously expensive.

The tricky part of desoldering is removing the old solder. Don't be deceived by advertising claims, no device is going to make desoldering truly easy or convenient. Some devices make the job a little less of a chore, but it's still something you'll want to avoid as much as possible. Aside from the sheer nuisance value, desoldering can overheat temperature-sensitive components (especially semiconductors). Desoldering almost always takes longer than soldering. Basically, there are two main approaches to removing old solder in a desoldering operation:

Capillary action
Suction

Each of these methods are discussed below.

Capillary action

The *capillary action* approach to desoldering uses a braided cable. You press this cable against the joint as the solder is reheated. Thanks to some principles of physics (which you don't need to go into here), the melted solder will be drawn up into the braid. This effect is known as capillary action.

Desoldering braid is available from many electronics dealers, including the ever-present Radio Shack chain. Unfortunately, some people seem to have a lot of difficulty with this method. They can never seem to get enough solder sucked up by the braid. Others find the capillary-action method such a snap, they use ordinary stranded wire in place of the special desoldering braid.

If you can get capillary action to work for you, it tends to be the most efficient method. But, if after numerous tries you can't get it to work, don't get too upset. You're not alone. It seems to be a specialized skill that some people have a natural knack for, and others don't.

Suction

Melted solder can also be removed with suction. Some sort of vacuum device pulls the molten solder up into some type of container. Almost anybody can use these suction devices, but generally, they are not quite as efficient as desoldering braid. A good suction device can be fairly expensive. They often tend to spit out tiny globules of solder that can cause short circuits if you are not careful. They are also prone to clogging when portions of sucked-up solder harden in the intake nozzle. The simplest and least expensive suction based desoldering devices are simple rubber bulbs with a nonstick nozzle. Such a device is shown in Fig. 15-8.

To use the bulb, squeeze and hold it, forcing most of the air out of it. Bring the nozzle into position over the melted solder, and release the bulb. There is now a minor vacuum within the bulb, because of the expelled air. Nature abhors a vacuum,

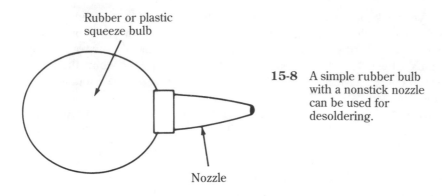

Rubber or plastic
squeeze bulb

Nozzle

15-8   A simple rubber bulb
with a nonstick nozzle
can be used for
desoldering.

so a strong suction force will appear at the nozzle until the pressure outside and inside the bulb is equalized. (This just takes a fraction of a second.) If the nozzle is properly positioned over molten solder, the liquid solder will be pulled up into the bulb. You will usually need several repetitions of this process to remove a sufficient amount of solder from the joint to allow removal of the component.

The next step up in suction-based desoldering devices uses a spring loaded piston. In the set-up position, the spring is compressed, and a plastic plug blocks the nozzle. Some sort of triggering mechanism is included on the device. When the trigger is activated, the spring is released, and the plug/piston is pulled quickly back out of the way, creating a momentary vacuum and suction force. Again, the liquefied solder is pulled up into the nozzle. There are many variations on this type of device. Some work a lot better than others, of course.

If you frequently have to do a lot of desoldering, you might want to purchase an electric desoldering pump. These rather expensive devices work in a manner very similar to the common vacuum cleaner.

Point-to-point wiring

Now, read about some actual permanent construction techniques for electronic circuits. For fairly simple circuits using just a handful of components, you could use point-to-point wiring. No real base for the components is used. The leads of the components are simply soldered together. Often solder terminals are used. A solder terminal is a strip of Bakelite or other plastic with a screw-down foot for mounting onto the case of the project. One or more metal loops are provided as connection points for the leads to be soldered. Some typical solder terminals are shown in Fig. 15-9. The leads to be soldered are mechanically connected to the metal loop, or terminal, as shown in Fig. 15-10. The connection is then soldered.

This construction method is only suitable for very simple circuits. For a circuit of any complexity, you can very easily run into problems with "rats nest" wiring. *Rats nest* is a fairly self-explanatory name for jumbled wiring that goes every which way, full of tangles. Such jumbled wiring is next to impossible to trace if any error is made or if the circuit needs to be serviced or modified at a later date.

Loose, hanging wires can create their own problems, such as stray capacitances

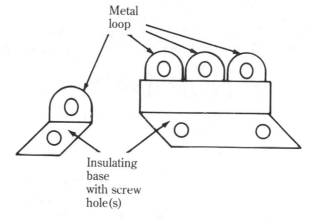

15-9 Solder terminals are often used for point-to-point wiring.

15-10 The component leads are mechanically connected to the loop of the solder terminal.

and inductances between them, which allows signals to get into the wrong portions of the circuit. The result is erratic or incorrect operation. Rats nest wiring is just begging for internal breaks within the connecting wires, and short circuits between them. Quite a bit of mechanical stress can be placed on some of the connecting wires. Momentary, intermittent shorts might not cause permanent damage in all cases, but they can result in some strange circuit performance that can be maddeningly frustrating to diagnose and service.

Perf boards

Simple to moderately complex circuits can be mounted on *perf board* (perforated board). A perf board is a nonconductive board with a regular pattern of holes or perforations drilled through it. Component leads are mounted directly through these holes. Alternately, special clips can be mounted in the holes and the component leads connected to the clips. The clips are commonly known as *flea clips*.

Perf board construction is essentially point-to-point wiring on a fixed base. The potential problems of rats-nest wiring can also show up in this construction method.

In any minimally complex circuit, one or more *jumper wires* can be required. A jumper wire is used to make a connection from one part of the board to another. If a jumper wire crosses any other wire or component, it must be insulated.

Take your time and experiment with component placement before you start soldering. Try to find an arrangement that will minimize the number of jumper wires and crossings used. Use straight line paths for jumpers whenever possible.

Another reason to experiment with the component placement before you start soldering is to make sure all of the components will actually fit on the board. If you start soldering without checking this out first, you could find yourself facing some unpleasant surprises, like ending up with no place to mount that big filter capacitor.

Printed circuits

For moderate to complex circuits, or for circuits from which a number of duplicates will be built, a printed circuit (PC) board gives very good results. A nonconductive board is used as a base. Copper traces on one side (or, in very complex circuits, on both sides) of the board act as connecting wires between the components. Very steady, stable, and sturdy connections can be made, because the component leads are soldered directly to the supporting board itself.

Take great care in laying out a PC board to eliminate wire crossings as much as possible. Obviously, two copper traces cannot cross over each other (unless they are on opposite sides of the board). If a crossing is absolutely essential, an external wire jumper must be used.

Normally, components are mounted on the opposite side of the board from the copper traces. The component leads are fitted through holes drilled in the board and soldered directly to the copper pad on the opposite side of the board. Snip off the excess lead to reduce the chances of a short circuit and to create a more professional appearance. A typical PC board solder joint is shown in Fig. 15-11.

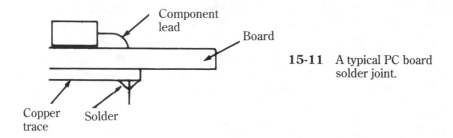

15-11 A typical PC board solder joint.

Printed circuit board construction results in strong mechanical connections and very short leads. Lengthy leads actually get in the way. Short leads can help minimize interference and stray capacitance problems.

Stray capacitances between adjacent traces can adversely affect circuit performance. In some critical circuits, a guard band between traces can help reduce this potential problem. Especially in circuits using ICs (integrated circuits—see chapters 24 through 29), the copper traces are often placed very close to one another. You must be extremely careful to avoid solder bridges. Use only small amounts of solder. If you use too much solder, it will flow and bridge across adjacent traces. Also watch out for short circuits from other causes. A small speck of loose solder, or a piece of a excess lead from a component could easily bridge across two (or more) adjacent traces, creating a short.

When soldering to a printed circuit (or PC) board, be very careful not to use too much heat or to apply heat for too long a time. Excessive heat can cause the copper foil trace to lift off the board. The unsupported trace is very fragile and will break.

Tiny, nearly invisible hairline cracks in the copper traces can also be a problem if you're not careful. Generally, fairly wide traces that are widely spaced are the easiest to work with and the most reliable. However, this method isn't always practical with all circuits—especially where ICs are used.

Blank boards for use as printed circuits are widely available. These are nonconductive boards with one side (sometimes both sides) completely covered with copper foil. The desired pattern is put onto the board using a special resist ink. This process can be done either by photographic methods, or it can be drawn on directly with a special resist pen. The board is then soaked in a special acid solution, which eats away the exposed copper. The resist protects the portions of the copper foil it covers. The board is removed from the acid solution, and washed. The resist ink is removed, and the desired copper traces are left on the board.

Universal PC boards

There is a fairly recent form of printed circuit construction. Designing and etching a customized printed circuit board is a time-consuming and somewhat tricky job. Now you can buy various universal PC boards. These boards have a generalized pattern of copper traces, and can be used for many different circuits.

Wire wrapping

There is one type of permanent circuit construction that does not require soldering. This method is the *wire-wrap* method. It is used primarily in circuits using large numbers of integrated circuits.

In a wire-wrapped circuit, a thin wire (typically 30 gauge) is wrapped tightly around a square post. The edges of the post bite into the wire, making a good electrical and mechanical connection without soldering. Components are fitted into special sockets that connect their leads to the square wrapping posts.

If you have just a few discrete components (resistors, capacitors, etc.), you can fit them into special sockets or solder them directly, and then wire wrap the connections to the ICs. If you use both soldering and wire wrapping, you have used what is called *hybrid construction*. In circuits involving many discrete components, the wire-wrapping method of construction tends to be rather impractical.

Wire-wrapped connections can be made (or unmade) quickly and easily, without risking potential heat damage to delicate semiconductor components. Moreover, it usually is not difficult to make changes or modifications in the circuit.

Manual and electrical wire-wrapping tools are available. The tool is needed to wrap the wire tightly enough around the square post. Most wire-wrapping tools can also be used for unwrapping. Many of these tools can also cut the wire and strip off the insulation.

There are some disadvantages to this type of construction. Wire wrapping is awkward for use with discrete components. The thin wire-wrapping wire is very fragile and easily broken. It can carry only very low-power signals. In complex circuits, the wiring can be difficult to trace.

When many integrated circuits are involved (some advanced circuits require several dozen), wire wrapping can be a very convenient construction method.

16

Experiments 1

This chapter is a collection of simple experiments intended to illustrate some of the concepts discussed in the previous chapters and to give you some practical experience with them. Performing these experiments is of course optional, but it is highly recommended because actually participating and seeing the principles in action is usually a more effective learning experience than simply reading about them. Table 16-1 lists all of the equipment and parts you will need to perform the experiments in this chapter. All of the items in this list have been described in previous chapters.

Table 16-1. Equipment and components
needed for the experiments in this chapter.

1	100 Ω resistor
2	1000 Ω resistors
1	10,000 Ω resistor
1	0.01μF disc capacitor
1	0.1μF disc capacitor
1	10μF electrolytic capacitor
1	100μF electrolytic capacitor
1	0.1 mH (millihenry) coil
1	power transformer—primary = 120 Vac: secondary = 6.3 Vac with center tap
1	6 V SPDT relay
1	10,000 Ω potentiometer (preferably with a linear taper)
VOM	(including dc voltmeter, dc milliameter, ohmmeter, and ac voltmeter)
Breadboarding system	(including solderless socket, power supply and oscillator)

Experiment 1   Ohm's law

For this experiment, build the circuit shown in Fig. 16-1 . As you can see, this the circuit is about as simple as a circuit can get—it consists only of a resistor and a dc voltage source with a 3 V output.

With some breadboarding systems, only fixed dc voltages are available. You might have to use a 5 V supply. This will alter your results, of course, but the principles are the same. And, of course, if you don't have a dc power supply, you can make a 3 V battery from two 1.5 V dry cells in series.

In any case, carefully measure the dc voltage with a voltmeter before starting the experiment. Be sure the voltmeter range can handle more than the maximum voltage you expect to measure. If you're using a 3 V supply, the voltmeter should be able to handle at least 4 V full scale. For a 5 V supply, a 6 V full scale meter is about the minimum.

Measuring the voltage in advance is especially important if you are using batteries, because the voltage can vary a great deal, depending on the age and condition of the cells. Two 1.5 V cells in series will generally have a combined voltage somewhere between 2.5 and 3.25 V. Enter the measured voltage of the source in the position labeled *Source voltage* in Table 16-2. Now use a 100 Ω resistor (marked brown-black-brown) to complete the circuit shown in Fig. 16-1. This resistor can be either a ½ or a ¼ W unit. The tolerance (fourth band) does not matter.

Table 16-2. Worksheet for Experiment 1—Ohm's law.

	Source voltage: _____ V		
R	*E*	*I* (measured)	*I* (calculated)
100 Ω	_____	_____	_____
1000 Ω	_____	_____	_____
10,000 Ω	_____	_____	_____

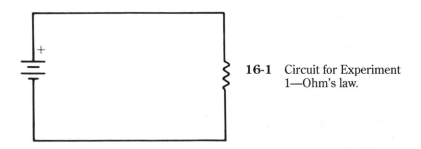

16-1   Circuit for Experiment 1—Ohm's law.

Measure the voltage drop across the resistor, as shown by the dotted lines in the schematic. Enter this value in the column labeled *E*, beside the heading *100 Ω*.

Remove the 100 Ω resistor and replace it with a 1000 Ω unit (brown-black-red). Again measure the voltage drop across the resistor and enter the result in the appropriate space in Table 16-2.

Next, repeat the procedure with a 10,000 Ω resistor (brown-black-orange) and enter the measured value.

The voltage dropped by each of the three resistors should be identical to the source voltage. This relationship will always be true of any circuit, regardless of the resistance value. The voltage dropped by the total resistance of a circuit will always be equal to the source voltage.

Now, adapt the circuit by inserting a milliammeter between the voltage source and the resistor, as shown in Fig. 16-2. This meter should be able to measure up to 5 milliamps (0.005 A, or 5000 μA). If the full-scale value is too large, however, you won't be able to get accurate readings of very small currents. A 5 mA or a 10 mA meter would be ideal.

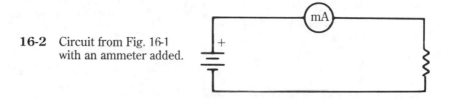

16-2   Circuit from Fig. 16-1
with an ammeter added.

Return the 100 Ω resistor (brown-black-brown) to the circuit, and record the current read on the meter in the column labeled *I (measured)* in Table 16-2.

Now repeat the procedure with the 100 Ω resistor (brown-black-red); then with the 10,000 Ω resistor (brown-black-orange). Be sure to enter each value as accurately as possible. Disconnect the circuit and calculate the current for each resistor using Ohm's law (*I = E/R1*. Enter the results in the column marked *I (calculated)*. Try to ignore the measured values while you do this.

When you have finished, compare the *I* (measured) values with the *I1* (calculated) values. They should be quite close, although there will probably be some minor variations due to measurement errors (no meter is ever 100% accurate) and the probability that one or more of the resistors is not precisely its nominal value.

If you like, you can calculate the exact resistance values with the formula *R = E/I* (be sure to use the measured values for *I*). Or, you could measure each of the resistors with an ohmmeter to find their true value.

Table 16-3 shows the results that the author obtained from performing this experiment.

Table 16-3. Author's results for Experiment 1—Ohm's Law.

	Source voltage (E) 3.0 V		
R	*E*	*I* (measured)	*I* (calculated)
100 Ω	3.0 V	2.7 mA	3 mA
1000 Ω	3.0 V	0.31 mA	0.3 mA
10,000 Ω	3.0 V	0.03 mA	0.03 mA

Experiment 2 Resistors in series

The next circuit you'll be working with is shown in Fig. 16-3. This circuit is quite similar to the circuit used in Experiment 1, except this time there are two resistors instead of just one. The resistors are connected in series.

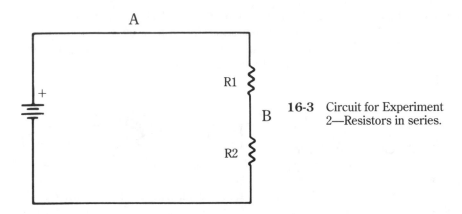

A

R1

B

R2

16-3   Circuit for Experiment 2—Resistors in series.

Use the same voltage source you used in the last experiment. R1 will be a 1000 Ω resistor (brown-black-red) for all the steps of this experiment.

In this first step, use a 100 Ω resistor (brown-black-brown) for R2. Measure the voltage dropped across R1, then measure the voltage dropped across R2, and finally, the voltage across both resistors together. See Fig. 16-4. Enter all three measurements in the appropriate spaces in Table 16-4.

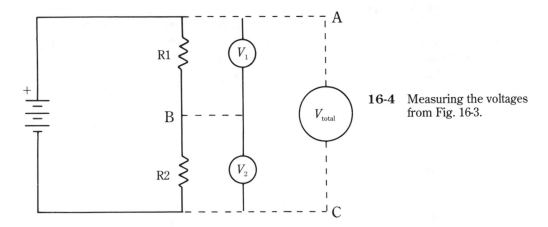

R1

V_1

A

B

V_{total}

R2

V_2

C

16-4   Measuring the voltages from Fig. 16-3.

Now install your milliammeter at point A and record the current. Then do the same for points B and C.

Finally, disconnect the voltage source, and use an ohmmeter to measure the actual value of the two resistors together. Enter this figure in the row labeled *R (total)*.

Table 16-4.
Worksheet for Experiment 2—Series resistances.

Source voltage (*E*) _____

	1000 Ω	1000 Ω	1000 Ω
R₁			
E (R₁)	_____	_____	_____
R₂	100 Ω	1000 Ω	10,000 Ω
E (R₂)	_____	_____	_____
E (total)	_____	_____	_____
I (A)	_____	_____	_____
I (B)	_____	_____	_____
I (C)	_____	_____	_____
R (total)	_____	_____	_____

Repeat the previous steps with a 1000 ohm resistor (brown-black-red) for R2. Then substitute a 10,000 Ω resistor for R2 and repeat all of the above measurements.

There are several things you should notice in the chart of your results. First, in each circuit, *E (total)* should equal the source voltage. Also, in any given circuit *I(A)*, *I(B)*, and *I(C)* should all be exactly equal. The same amount of current flows through all portions of a series circuit. Finally, *R (total)* should be approximately equal to $R_1 + R_2$. There will probably be some variation because of individual component tolerances.

If you use the nominal value of $R_1 + R_2$ as *R*, you can use Ohm's law to calculate the nominal current flowing through the circuit ($I = E/R$). You should come up with a figure that is close to the measured current value.

If you used the measured value of *R (total)* in the equation, you should come very close to the measured current value. Any minor error is due to imprecision in the measurement.

Experiment 3 Resistors in parallel

Now you'll experiment with the parallel resistance circuit shown in Fig. 16-5. R1 again will be a constant 1000 Ω (brown-black-red). The source voltage will also remain the same as for the previous experiments.

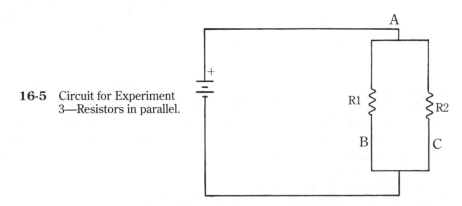

16-5   Circuit for Experiment 3—Resistors in parallel.

Beginning with a 100 Ω resistor (brown-black-brown) for R2, measure the voltage drop across each resistor. Then measure the current at points A, B, and C. Finally, disconnect the voltage source and measure the combined value of the paralleled resistors with an ohmmeter. Enter all of these values in Table 16-5.

Table 16-5. Worksheet for Experiment 3
—Parallel resistances.

	Source voltage (E)		
R_1	1000 Ω	1000 Ω	1000 Ω
$E(R_1)$	————	————	————
R_2	100 Ω	1000 Ω	10,000 Ω
$E(R_2)$	————	————	————
$R(total)$	————	————	————
$I(A)$	————	————	————
$I(B)$	————	————	————
$I(C)$	————	————	————

Now replace R2 with a second 1000 Ω resistor (brown-black-red) and repeat all of the measurements. Then, using a 10,000 Ω resistor (brown-black-orange) repeat the experiment one more time.

Notice that the voltage across each resistor is equal to the source voltage, but the currents flowing through each resistor (measured at points B and C) are different. The current measured at point A (total circuit current) should equal $I_{(B)} + I_{(C)}$.

Also notice that $I_{(B)}$ (the current through R1) should be a constant value. The current should be constant because neither the voltage drop across the resistor, nor the resistance are changed. Because $I = E/R$, the value of I should also remain stable.

When $R_1 = R_2$, I(B) should be equal to I(C). There might be some slight variation because of the tolerances in the resistor values. R1 might, for instance, be 4% above its nominal value, and R2 might be 6% below.

The component tolerances also explain why the measured value of R *(total)* probably won't be precisely equal to the calculated value (that is, $1/R_t = 1/R1 + 1/R2$).

According to the formula, the nominal total resistance should be about 91 Ω when R2 is 100 Ω, 500 Ω when R2 is 1000 Ω, and 909 Ω when R2 is 10,000 Ω. Your measured values should be reasonably close to these figures.

Experiment 4 A series-parallel circuit

The circuit for this experiment is shown in Fig. 16-6. R1 is in series with the parallel combination of R2 and R3. R_1 is 100 Ω, R_2 is 1000 Ω, and R_3 is 10,000 Ω. R(P) is the combined resistance of R2 and R3 in parallel. R *(total)* is the combined resistance of all three resistors.

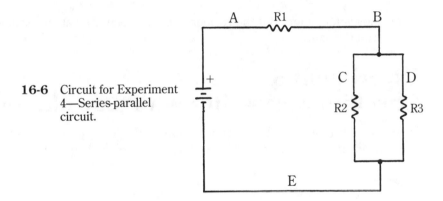

16-6 Circuit for Experiment 4—Series-parallel circuit.

Calculate the nominal values of *R(P)* and *R (total)* and enter your results in the appropriate spaces in Table 16-6. Then, with the voltage source disconnected, measure the actual resistance of *R (P)* and *R (total)*. The measured and the calculated results should be similar.

**Table 16-6. Worksheet for Experiment 4
—Series-parallel circuit.**

	Calculated	Measured
R_1	*100 Ω*	
R_2	*1000 Ω*	
R_3	*10,000 Ω*	
R(P)	_____	_____
R(total)	_____	_____
E(source)		_____
E(R₁)		_____
E(RP)		_____
I(A)		_____
I(B)		_____
I(C)		_____
I(D)		_____
I(E)		_____

Connect the voltage source and measure the voltage drop across R1 and across R(P). Then measure the current at points A through E. Enter all of these readings in the table.

Notice that $E_{(R1)} + E_{(RP)}$ = the source voltage. Also currents $E_{(A)}$, $I_{(B)}$, and $I_{(E)}$ should all be equal. They should be equal because all of these currents are in series.

Finally, the currents in the parallel section of the circuit should add up to equal the total circuit current. That is, $I_{(C)} + I_{(D)} = I_{(A)}$.

You might want to repeat this experiment with a different source voltage. For example, if you are using a battery, you could add a third cell for a source voltage of about 4.5 V. All of the resistances will remain the same, but the voltage and current values will change. But the relationships between the values discussed above will remain the same.

Experiment 5
Checking a capacitor with a dc ohmmeter

Take a 0.1 µF (microfarad) capacitor and short the leads together, as in Fig. 16-7. This will discharge the capacitor if it contains any residual charge.

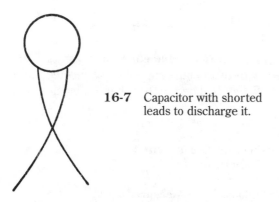

16-7   Capacitor with shorted leads to discharge it.

Set your ohmmeter to its highest range and connect it across the leads of the capacitor. See Fig. 16-8. Watch the meter pointer very carefully while you're doing this. The pointer should jump down the scale, indicating some finite dc resistance, then it will slowly creep back up the scale towards the infinity (open circuit) mark. Discharge the capacitor by shorting the leads and repeat the procedure several times, until you get a good feel for what is happening.

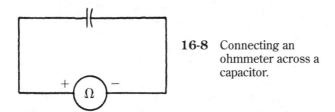

16-8   Connecting an ohmmeter across a capacitor.

The equivalent circuit for this experiment is shown in Fig. 16-9. Because an ohmmeter consists of a voltage source, a meter, and a range resistor, the capacitor is charged by the voltage source through the resistor. That is, it is an RC circuit.

When the capacitor has no charge (that is, when the ohmmeter leads are first applied) quite a bit of current flows through the circuit. As the capacitor reaches its charged condition, less and less current can flow. Finally, the capacitor is fully charged, and no further current can flow in the circuit at all.

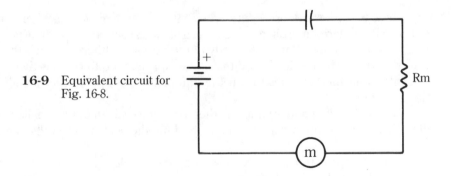

16-9   Equivalent circuit for
Fig. 16-8.

If you draw a graph of the resistance shown on the meter against the elapsed time, it would look like Fig. 16-10. The same graph shows the charge on the capacitor plates against time.

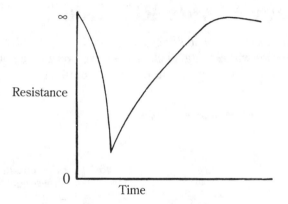

16-10   Graphing resistance reading versus
charging time—Experiment 5.

If the pointer doesn't move at all, the capacitor could be open, or the meter range might be too low for a readable indication. A ×1 MΩ range should give a readable movement when charging a 0.1 μF capacitor. If the pointer jumps to 0, or a very low resistance value, and stays there, it means the capacitor is shorted.

With many capacitors, the pointer won't return completely to the infinity position because dc leakage through the dielectric is providing a limited current path. As long as the final value is very high (50 to 100 MΩ or more), you can consider that the capacitor is in good shape.

Now, perform the experiment again and carefully time the motion of the meter from its minimum to its maximum resistance position.

Next, try the same procedure with a 0.01 μF capacitor. Notice that the pointer doesn't move as far down the scale as with the 0.1 μF unit, and it returns to the infinity position much faster. This indicates that the capacitance is less. Capacitors smaller than about 0.01 μF usually won't give a readable indication on most ohmmeters.

Now, try the same procedure with a 10 μF electrolytic capacitor. Be sure to observe the polarity markings. Some electrolytic capacitors have their positive lead identified by a + or a dot on the body of their casing. Others indicate the negative lead with a −. The red lead from the ohmmeter must go to the positive lead of the capacitor, and the black lead from the ohmmeter must go to the negative lead of the capacitor.

Notice that the pointer jumps further down the scale than the smaller capacitors did. Also, it takes much longer to move back up the scale to the fully charged condition.

Finally, repeat the experiment with a 100 μF electrolytic capacitor (observe polarity). It should take close to a second for the pointer to return to the infinity position. Actually, with a capacitor this size you will probably see some leakage resistance in the meter indication, but the pointer should be fairly close to the infinity mark when it stops moving.

Experiment 6 A dc RC circuit

Place a resistor and a capacitor in series, as shown in Fig. 16-11. Short the leads of the capacitor together with a metal-blade screwdriver, or a piece of bare wire. Then connect the ohmmeter leads across the RC series combination.

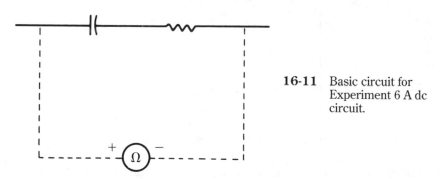

16-11 Basic circuit for Experiment 6 A dc circuit.

Start with a 100 Ω resistor (brown-black-brown) and a 0.1 μF capacitor. Notice that this combination takes longer to charge than just the capacitor alone. Repeat this procedure several times, until you can determine exactly how long it takes the capacitor to charge. Enter this value in Table 16-7.

**Table 16-7. Worksheet for Experiment 6
—A dc RC circuit.**

	100 Ω	1000 Ω	10,000 Ω
0.01 μF	_____	_____	_____
0.1 μ	_____	_____	_____
10 μ	_____	_____	_____
100 μ	_____	_____	_____

Do this experiment with each of the resistor-capacitor combinations indicated in the table. Don't forget to observe the polarity markings on the electrolytic capacitors.

Notice that increasing either the resistance or capacitor proportionately increases the time required for charging the capacitor (that is, the time constant of the combination). The 10,000 Ω resistor (brown-black-orange) and 100 μF capacitor combination should take the longest time (nominally a full second—disregarding the internal resistance of the meter).

Experiment 7 An ac RC circuit

Again you'll use a resistor in series with a capacitor, but in this experiment you will examine what happens when the combination is powered by an ac voltage source. You will need a variable frequency oscillator for this experiment.

Set your variable frequency oscillator to a fairly low frequency and measure the ac voltage coming from the oscillator itself (inexpensive oscillator circuits don't put out the same voltage at all frequencies).

Connect the oscillator across the resistor/capacitor combination, as shown in Fig. 16-12, and measure the voltage drop across the resistor.

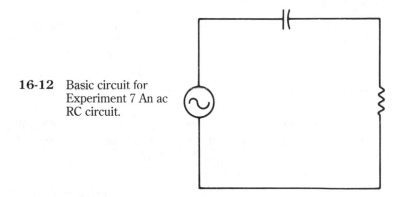

16-12   Basic circuit for Experiment 7 An ac RC circuit.

Perform this experiment—once with the 0.1 μF capacitor, and once with the 0.01 μF unit. Use the 1000 Ω resistor (brown-black-red) in each case. *Caution: do not try to perform this experiment with the electrolytic capacitors!* They cannot hold up under an ac voltage.

Enter your voltage readings in Table 16-8. Disconnect the circuit and set the oscillator to a medium frequency and measure the ac voltage. Connect the oscillator to the RC combination again and repeat the experiment. Finally, repeat the experiment once more with the oscillator set to generate a high frequency.

With the author's oscillator, the source voltage decreased as the frequency increased (this isn't always true). Even so, the voltage passed by the capacitor increased noticeably with each increase in frequency because a capacitor has a lower reactance at higher frequencies.

Table 16-8. Worksheet for Experiment 7—An ac RC circuit.

Frequency	Source voltage	0.1 μF	0.01 μF
Low	_____	_____	_____
Medium	_____	_____	_____
High	_____	_____	_____

Also notice that the 0.1 μF capacitor passed more voltage than the smaller, 0.01 μF unit because the larger capacitor has a lower impedance.

Experiment 8   Coils

Measure your source voltage carefully, then build the circuit shown in Fig. 16-13 with a 100 Ω resistor (brown-black-brown) and a 0.1 mH coil. Measure the voltage drop across the resistor. It should be just about equal to the source voltage.

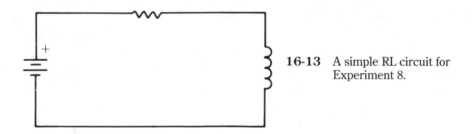

16-13   A simple RL circuit for Experiment 8.

Now disconnect the dc power source and measure the voltage drop across the resistor at three different ac frequencies. Be sure to measure the source voltage at each frequency. Enter your results in Table 16-9.

Table 16-9. Worksheet for Experiment 8—Coils.

Frequency	Source voltage	Voltage drop
Low	_____	_____
Medium	_____	_____
High	_____	_____

Notice that the results of this experiment should display the exact opposite of the pattern found in the experiment with capacitors. As the source frequency increases, the voltage passed by the coil decreases. In other words, the voltage dropped across the coil increases with frequency.

Remember, an inductance exhibits more reactance to high frequencies, at very low frequencies, and at dc a coil acts like a simple length of wire.

Experiment 9 Resonance

Connect the series circuit shown in Fig. 16-14. Vary the frequency of the oscillator while carefully watching the pointer of the ammeter. As the frequency is increased, the current drawn through the circuit should also increase, until some maximum value is reached. Increasing the frequency beyond this point should cause the current to fall off again. The maximum current occurs at the resonant frequency of the coil-capacitor combination.

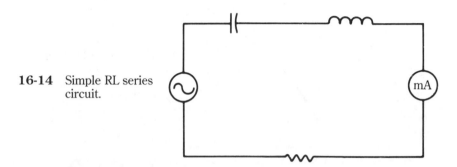

16-14 Simple RL series circuit.

If you graphed the results of this experiment, it would look something like the chart shown in Fig. 16-15.

Next, set up the circuit shown in Fig. 16-16.

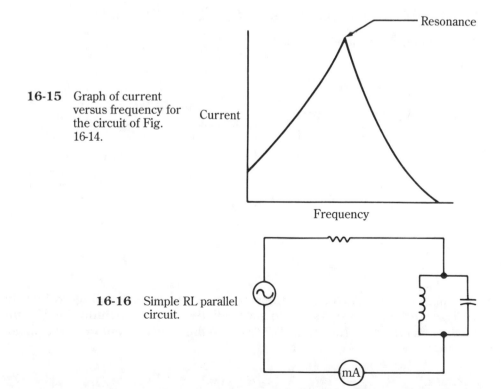

16-15 Graph of current versus frequency for the circuit of Fig. 16-14.

16-16 Simple RL parallel circuit.

Again, set the oscillator at its minimum position, and slowly increase the frequency, while watching the ammeter. The current should start out at a fairly high value, and decrease to a minimum value at the same frequency it reached a peak in the last step. From that point on, the current should increase with the applied frequency. In this case you have passed through parallel resonance. The graph for this circuit would resemble the one in Fig. 16-17.

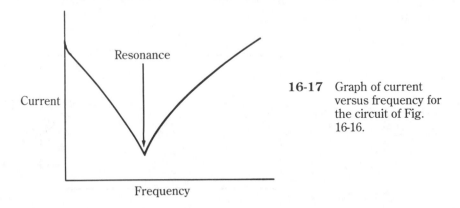

16-17   Graph of current versus frequency for the circuit of Fig. 16-16.

Replace the 0.1 μF capacitor with a 0.01 μF unit, and repeat the experiment. Is the resonant frequency higher or lower than before?

Experiment 10 Transformer action

Very carefully construct the circuit shown in Fig. 16-18.

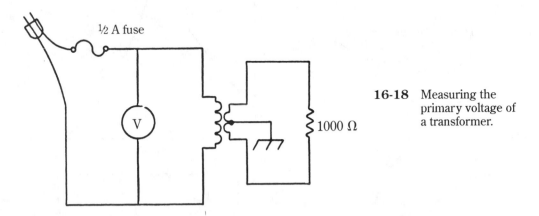

16-18   Measuring the primary voltage of a transformer.

Warning: Do not touch any of the wires once the circuit is plugged into the wall socket! Carelessness could be fatal! Be sure to include the ½ amp fuse—it is for your protection. The fuse will limit the current in case of an accident.

Wrap all connections and bare wires with several layers of electrical tape for insulation!

Be sure the voltmeter is connected to the circuit before ac power is applied!

Plug in the line cord, read the voltage on the meter and immediately unplug the cord. (It probably wouldn't hurt anything to leave it plugged in for a few minutes, but why take chances?) You should have gotten somewhere between 100 to 120 Vac.

Now, with the circuit unplugged, move the voltmeter to the position shown in Fig. 16-19. Again, carefully plug in the line cord, read the meter and unplug the circuit. This time you should have gotten about 6.3 V. This transformer is a step-down transformer. It transforms one ac voltage to a lower ac voltage.

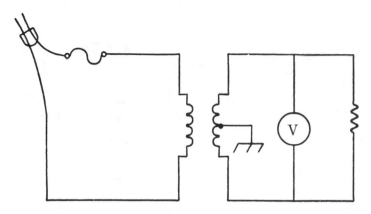

16-19   Measuring the secondary voltage of a transformer.

This same transformer could be reversed and used as a step-up transformer, but **do not try it!** The very high voltage produced could be extremely dangerous in an unshielded circuit like this.

Now, connect the voltmeter between one of the secondary end leads and the center tap as shown in Fig. 16-20A. Of course, the circuit should be unplugged while you are working on it. Plug the circuit in, read the voltage, and unplug the circuit. Move the voltmeter so it is connected between the center tap and the other end lead of the secondary, as illustrated in Fig. 16-20B. Plug in the line cord, read the voltage, and unplug the cord.

In these last two steps you should have read exactly one half of the total secondary voltage. Since this transformer has a 6.3 Vac secondary, the voltage between the center tap and either end of the secondary will be about 3.15 Vac.

Experiment 11 Relay action

Hook up the circuit illustrated in Fig. 16-21. Watch the metal contacts inside the relay as you slowly rotate the shaft of the potentiometer. At some point along the potentiometer rotation, you should see them move and hear a faint but distinct click.

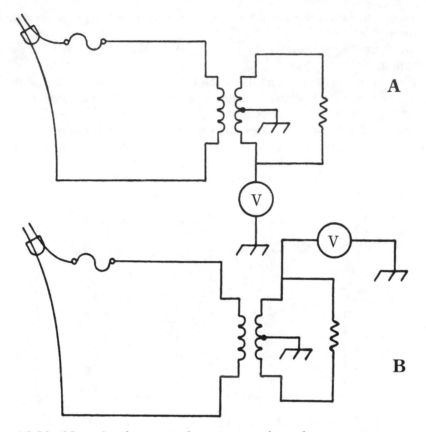

16-20   Measuring the output of a center-tapped transformer.

The ohmmeter will now show that this set of contacts are now shorted together. This condition will remain stable as long as the potentiometer is not turned in the opposite direction.

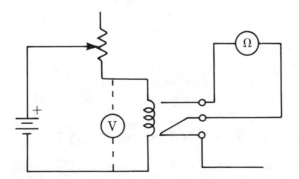

16-21   Simple SPDT relay circuit for Experiment 11.

Slowly turning the potentiometer back in the opposite direction will produce the opposite result, of course. You will reach a point where the relay contacts will click back to their original position, and the ohmmeter will show infinite resistance, or an open circuit.

If you connected the voltmeter as shown in the dotted lines, you'd find the relay contacts remain open as long as the voltage shown on the meter is below 6 V. Any voltage over 6 V will energize the relay and cause the contacts to close.

Reconnect the ohmmeter as shown in Fig. 16-22 and notice that the relay works in exactly the same way, but backwards. That is, increasing the voltage above 6 V will cause the switch contacts to open, or vice versa.

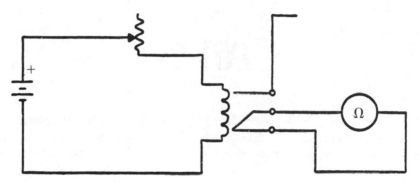

16-22   Figure 16-21 with an ohmmeter across the normally closed contacts of the relay.

If you don't understand how the voltage is being varied in this experiment, re-member that the relay coil has a certain amount of dc resistance, so you essentially have two resistors in series (as far as dc voltage is concerned). Changing the value of one series resistor (that is, turning the shaft of the potentiometer), will alter the amount of voltage dropped by the other resistor. Refer back to Experiment 2 for further clarification.

17
Tubes

By themselves, the circuits described so far are of relatively little practical value because all of the components discussed in the previous chapters have been *passive* devices. They can reduce (or *attenuate*) a signal, but they cannot increase (or *amplify*) it.

For practical electronic circuits you also need *active* devices. Active devices are components that can amplify or in some other way actively alter a signal. The first practical active device, and probably the simplest, is the *triode vacuum tube*. But before you can examine how this device works, you need to look at a couple of related passive devices.

Light bulbs

The vacuum tube is closely related to the common light bulb. In fact, a light bulb could be called a single element vacuum tube.

Figure 17-1 shows the construction of a typical light bulb. A thin, specially prepared wire is enclosed in a glass bulb and all of the air is pumped out, creating a vacuum within the bulb. Electrical connections to the wire (called the *filament*) can be made from outside the bulb via a metal base.

When an electric current passes through the filament, its resistance causes it to heat up. The special type of wire used for the filament will glow when heated, producing light. Some of the filament material is inevitably destroyed by this process, which is why light bulbs eventually burn out. If you look inside a burned-out light bulb, you'll see that the filament wire is broken.

The resistance of the filament determines the wattage consumed by the bulb (and thus the brightness of the emitted light). For example, if the filament is 144 Ω, and works off of standard house current (nominally 120 V), the current drawn by the bulb will be equal to the voltage divided by the resistance. (Ohm's law—$I = E/$

218

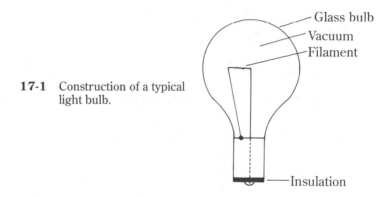

17-1   Construction of a typical
light bulb.

R). In this example, $I = 120/144$, or about 0.83 A. This means the power consumed
by this particular light bulb ($P = EI$) is approximately 100 W.

It takes more energy to heat up the filament to the glowing point, than to maintain its temperature once it is heated. In other words, the resistance of the filament
is higher when it is cold. This means when power is first applied to a light bulb, the
current drawn will flow in a large surge before settling down to its nominal value.
This surge current can be several times larger than the nominal current flow. For
this reason, no power is saved by turning out a light if it will be turned back on
within a few minutes.

The diode

In addition to emitting light, the heated filament in a light bulb also emits a stream
of electrons. If a second element is placed within the vacuum-tube envelope, and
given a positive charge, it will attract these electrons. That is, a current can be made
to flow between the elements within the bulb, or tube. Because this type of tube has
two elements, it is called a *diode*. Actually, most practical diodes have three elements,
as shown in Fig. 17-2.

The positively charged element is called the *plate*, or *anode*. The stream of electrons is emitted from the *cathode*, which is given a negative charge by the external
circuit. The filament, or *heater* is generally not considered an active element in the
tube. It simply heats up the cathode so it can emit electrons easily. Heating the
cathode directly would result in less efficient operation and a tube with a shorter
life expectancy.

Usually the heater circuit is electrically isolated from the main circuit. In most
tube equipment, there is a separate power source (or transformer winding) just for
powering the filaments of the tubes. For the longest possible life, the filaments
should be heated with an ac voltage, rather than dc.

The most common schematic symbols for diodes are shown in Fig. 17-3. Sometimes the filament is not shown in the schematic diagram at all, as in Fig. 17-3B and
17-3C. The symbol in Fig. 17-3C isn't often used for vacuum tube diodes (see the
next chapter), but it occasionally shows up in certain schematics.

Figure 17-4 shows a simple circuit for testing the action of a diode. When the

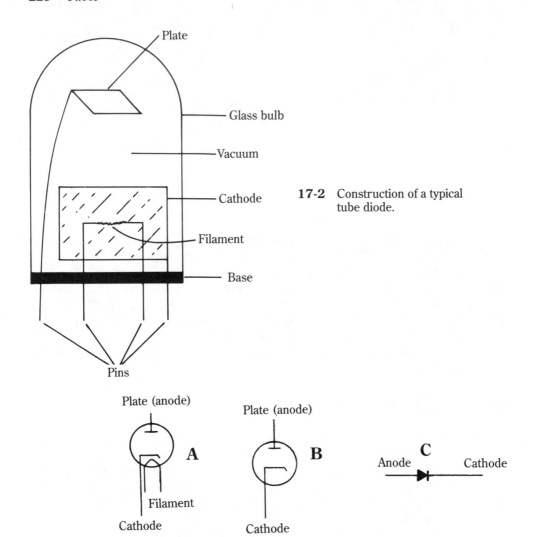

17-2   Construction of a typical tube diode.

17-3   Schematic symbols for diodes.

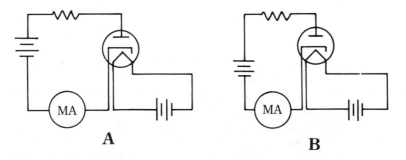

17-4   Test circuit for a diode. Forward bias (A); reverse bias (B).

power source is connected as shown in Fig. 17-4A, a current flows through the ammeter. The value of this current will be determined primarily by the resistor. The diode electrically looks like a very small resistance—almost a short circuit. The diode is *forward biased*.

However, if the polarity of the dc voltage source is reversed, as in Fig. 17-4B, no current will flow (or very, very little), because the plate cannot emit electrons. The diode is now *reverse biased*, and its resistance is extremely high.

The basic principle of a diode is that current can flow through it in one direction but not in the other. An ideal diode has zero resistance if measured from cathode to anode but infinite resistance from anode to cathode. Practical diodes have some resistance when forward biased, but the value will be very low. Similarly, some current will flow through a diode when it is reverse biased, but the resistance will be so high the current will be of a negligible value.

Considering the way a diode behaves in a dc circuit, what would happen if it were placed in an ac signal path? In an ac circuit, only that portion of the applied signal with the correct polarity can pass through the tube, and the rest of the signal will be blocked. Figure 17-5 shows the effect of a basic diode circuit on a simple sine wave.

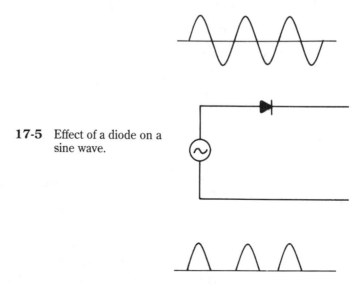

17-5 Effect of a diode on a sine wave.

If you add a capacitor, as shown in Fig. 17-6, its charging and discharging times will tend to smooth out the waveform, producing a more or less dc voltage from an ac source. See the chapter on power supplies for additional information.

Triodes

As useful as a diode is, it still can't amplify. To amplify, you need to add a third active element to our tube (ignoring the heater). This new element is called the *grid* (sometimes the *control grid*), and such a three element tube is called a *triode*.

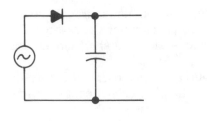

17-6   Effect of a diode and a filter capacitor on a sine wave.

Figure 17-7 shows the construction of a typical triode, and Fig. 17-8 shows the most common schematic symbols for the device. As with the diode, the heater is sometimes omitted from the schematic diagram, because it is not a part of the actual circuit. The heater connections are always assumed.

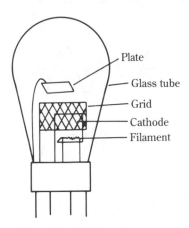

17-7   Construction of a typical triode.

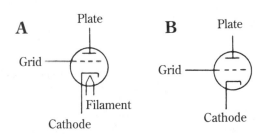

17-8   Schematic symbols for a triode.

Figure 17-9 shows a simple demonstration circuit for a triode tube. For simplicity and convenience, the heater circuit is not shown in the diagram—the heater circuit is identical to the one in the diode circuit discussed previously. As a matter of fact, all tube circuits use the heaters in essentially the same way—this is why they can be omitted from the diagrams. When you see a tube in a circuit, you automatically know you need to apply a voltage across the heater. The level of this voltage varies from type to type, and will be specified by the manufacturer.

The grid in a triode tube is a metallic mesh. That is, it has holes in it that allow electrons from the cathode to pass through it on their way to the plate. Just how many electrons can pass through the grid (that is, the current) depends on its electrical charge. If the grid is made very negative with respect to the cathode, it will repel all of the electrons (which are also negatively charged), and let none of them

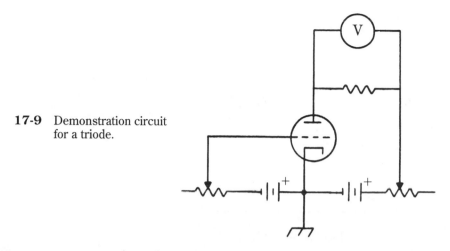

17-9 Demonstration circuit for a triode.

pass through to the plate. The voltage at which all current through the tube is blocked is called the *cutoff point* of the tube.

As the voltage on the grid is made more positive (or less negative) with respect to the cathode, more and more electrons can pass through the mesh and get to the plate. At some point, all of the electrons emitted by the cathode will reach the plate. This point is called the *saturation point* of the tube.

If the grid is made even more positive past the saturation point, it will start to attract the electrons itself, once again preventing them from reaching the plate. Usually the grid is slightly negative with respect to the cathode in practical circuits.

Using a hypothetical tube, examine some of the effects that take place in this kind of circuit. For this example, use the cathode as your reference point. That is, the cathode is grounded, which means its voltage is, by definition 0 V.

Assume the grid voltage (E_g) is 0 V. If the plate voltage (E_p) is also 0, obviously no current will flow through the tube. If you increase the plate voltage to 25 V, about 1.8 mA (0.0018 A) of current will flow through the plate circuit. If the load resistor (R_l) is 5000 Ω, the voltage drop across it will be 0.0018 × 5000, or 9 V. The rest of the plate voltage is used up (dropped) by the tube itself.

Increasing the plate voltage further, to 50 V, will cause a 4 mA (0.004 A) current to flow. The voltage drop across R_l is now equal to 0.004 × 5000, or 20 V, and 30 V is dropped by the tube itself.

Increasing the plate voltage to 75 V increases the current flow to 7.25 mA (0.00725 A). The voltage drop across the load resistor is now equal to 0.00725 × 5000, or 36.25 V.

Finally, increasing E_p to 100 V will increase the current flow to 11 mA (0.011 A). R_l drops 0.011 × 5000, or 55 V under these conditions.

100 V in the plate circuit is the saturation point of this particular tube with a 0 V grid voltage. The current drawn through the tube cannot be increased further without risking damage to the tube.

Figure 17-10 shows a graph of this plate voltage to current ratio. Notice that it is not a straight line, but a curve. Figures 17-11 and 17-12 show similar graphs for the same tube, but with the grid voltage at −2 V and −4 V, respectively. These

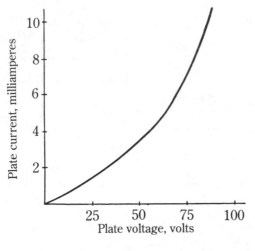

17-10 Characteristic plate current curve for a typical tube—$E^g = 0$ V.

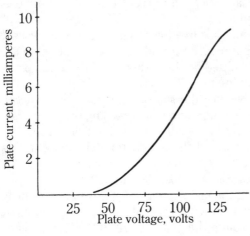

17-11 Characteristic plate current curve for a typical tube—$E^g = -2$ V.

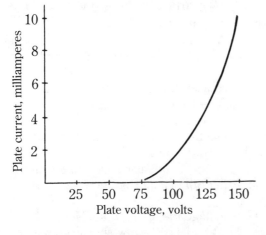

17-12 Characteristic plate current curve for a typical tube—$E^g = -4$ V.

graphs are collectively called a *family of plate characteristic curves* for this specific tube. Other tubes will have somewhat different curves, but they will always exhibit basically the same shape.

Obviously you could eliminate the tube altogether and just vary the resistance through R_L directly. In actual practice, the plate voltage is usually held at a constant level, and the grid voltage (E_g) is varied.

Assume a plate voltage of 100 V. You already know that if E_g equals 0 and E_p equals 100, then the current will be 11 mA, and the voltage drop across the 5000 Ω load resistor will be 55 V.

If the grid voltage is changed to –1 V (remember, the grid should be negative with respect to the cathode), the current through the plate circuit will be 8 mA (0.008 A). The negative charge on the grid is repelling some of the electrons from the cathode. The voltage drop across the load resistor will be equal to 0.008 × 5000 or 40 V.

At a grid voltage of –2 V only 5 mA (0.005 A) will flow through the plate circuit. The load resistor will drop 0.005 × 5000 or 25 V.

The entire graph for a constant E_p of 100 V, and a variable E_g is shown in Fig. 17-13. Notice that when E_g is −6 V or less, no current will flow through the plate circuit at all. This point is the cutoff point. Compare the graph in Fig. 17-13 to the one in Fig. 17-12. Notice that although it takes a 100 V range in the plate voltage to produce an 11 mA range of plate current, it takes only a 6 V range of grid voltage to produce the same plate current range. A relatively small change in grid voltage produces a relatively large change in the plate current, and this produces a fairly large voltage drop change across the load resistor.

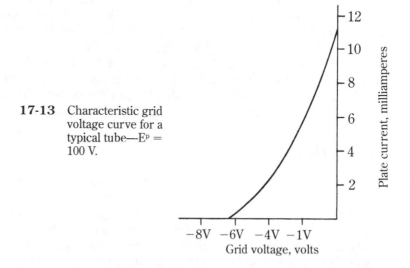

17-13 Characteristic grid voltage curve for a typical tube—Eᵖ = 100 V.

If an ac signal is applied to the grid, the signal across the load resistor will be a larger replica of the input signal. This process is called amplification (see Fig. 17-14).

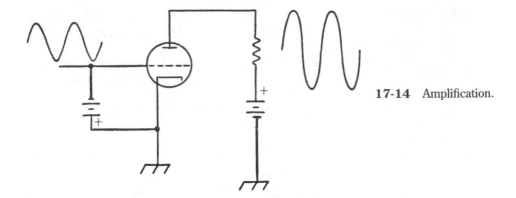

17-14   Amplification.

Of course, the energy across the load has to be provided by the plate voltage source—you can't get something for nothing. The voltage drop of the load resistor will always be less than the voltage applied to the plate circuit.

The amount of amplification in any given circuit is called the *gain*. How much gain a specific tube is capable of is called the *amplification factor*, which is usually represented by the Greek letter μ (mu). The amplification factor is determined by the ratio of the change in grid voltage needed to produce a given change of current and the change in plate voltage required for the same amount of current change. That is:

$$\mu = \frac{\Delta\, Ep}{\Delta\, Eg}$$ **Equation 17-1**

where μ is the amplification factor, $\Delta\ E_p$ is the change in plate voltage (Δ is *delta*, and is used to represent a changing value). E_g is the change in grid voltage. In the sample tube, increasing the plate voltage 20 V will increase the output current about 2.5 mA, and a change of about 1 V in the grid will produce the same change in current. Therefore, μ equals 20/1, or an amplification factor of 20. As in the diode, current can flow through a triode in only one direction—from cathode (–) to plate (+). Reversing the polarity of the plate voltage will automatically result in zero current flow, regardless of the value of either E_p or E_g, which is true of all tubes.

Tetrodes

A major problem with triodes is due to *interelectrode capacitance*. That is, the electrodes within the tube act like the plates of a capacitor. See Fig. 17-15.

The capacitance between the plate and the grid is particularly significant, because it can allow ac current from the plate circuit to leak back into the grid circuit, putting a severe limitation on how much gain the tube can put out. This effect can be greatly reduced by adding a second meshed element called a screen grid, which is placed between the original control grid and the plate. Figure 17-16 shows the schematic symbol for this type of four element tube, which is called a *tetrode*.

The screen grid is connected so that it is positive with respect to the cathode, but somewhat negative with respect to the plate. A capacitor is usually connected

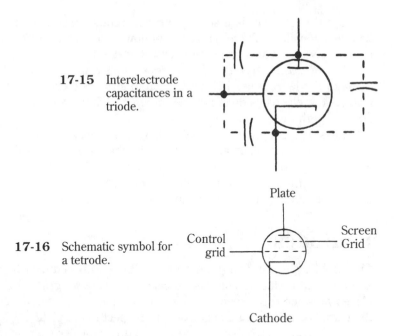

17-15 Interelectrode capacitances in a triode.

17-16 Schematic symbol for a tetrode.

Plate

Control grid

Screen Grid

Cathode

from the screen grid to the cathode. This will have no effect on the dc voltage levels, but any ac signal that manages to get into the screen grid circuit will be shorted to the cathode, which is generally at ground potential (0 V). The diagram for the basic tetrode circuit is shown in Fig. 17-17.

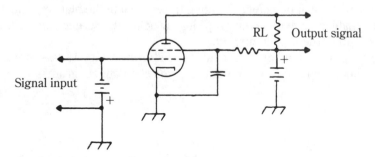

RL   Output signal

Signal input

17-17 Basic tetrode circuit.

Because the screen grid is physically closer to the cathode than the plate is, its positive charge has a greater effect on pulling the electrons through the holes in the control grid than does the plate. This means the plate voltage has very little effect on the current flow through the tube. A very large change in plate voltage would be needed to equal a very small change in the control grid voltage. Of course, this means the amplification factor of such a tube is quite high. A typical triode might have an amplification factor of 20 to 25, but a tetrode amplification figure is often more than 600.

Of course, changing the voltage on the screen grid could alter the current flow through the tube, but in practical tetrode circuits the screen grid is virtually always held at a constant voltage. The current flow through a tetrode is determined almost exclusively by the voltage on the control grid.

Because the screen grid is an open mesh, most of the electrons pass right through the large holes in it and go on to strike the even more positively charged plate. A few electrons do strike the screen grid, however, causing a small current to flow through the screen grid circuit.

Passing through the positively charged screen grid tends to speed up the electrons in their path, causing them to strike the plate with considerable force. If this force is large enough, many of the electrons can ricochet off the plate and return to the screen grid. Obviously this is undesirable, because it represents a loss of current flow through the plate circuit. This problem is called *secondary emission*.

Pentodes

The problem of secondary emission can be greatly reduced by the addition of yet another grid element. This one is called a *suppressor grid*. The suppressor grid is placed between the screen grid and the plate, and it is usually connected directly to the cathode, so it is quite negative with respect to the plate.

The main electron stream is speeded up by the screen grid. The electrons pass through the holes in the suppressor grid so fast the negative charge doesn't have a chance to repel them, but it does slow them down a bit. Any secondary electrons that bounce off of the plate are repelled by the suppressor grid negative charge, so they are forced to return to the positively charged plate.

The plate voltage in a *pentode* (five-element tube) can vary over an extremely large range without appreciably changing the current in the plate circuit. As a matter of fact, the plate voltage can even drop slightly below the screen grid voltage without a serious drop in the output current.

The schematic symbol for a pentode is shown in Fig. 17-18. As with all other tubes, the heater circuit is often omitted from schematic diagrams.

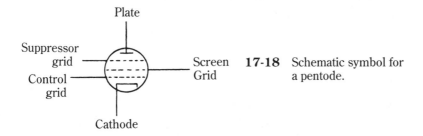

17-18 Schematic symbol for a pentode.

In most pentodes, the suppressor grid is brought out to its own terminal pin and is connected to the cathode via the external circuit. In some pentodes, however, the suppressor grid is internally connected to the cathode. This type of tube is usually shown schematically as in Fig. 17-19.

The amplification factor of a pentode can be extremely high. Some tubes have

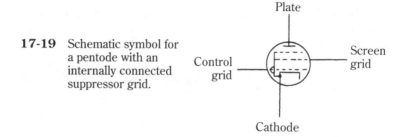

17-19 Schematic symbol for a pentode with an internally connected suppressor grid.

an amplification factor of 1500 or higher. Compare this value to the amplification factor of a simple triode!

Of course, because pentodes have more elements, and are more complicated to manufacture, they are more expensive. Triodes and tetrodes are usually used whenever possible to achieve the desired results.

Multiunit tubes

Some tubes actually contain more than one set of electrodes in a single bulb. In other words, more than one tube is contained in a single glass envelope. The most common combinations are dual diodes, dual triodes, and diode-triode combinations. Tetrodes and pentodes are rarely found in multiunit tubes.

Some dual tubes have a common cathode, and many share a common heater filament. This means the element is used in both tubes.

Figure 17-20 shows the schematic symbol for a dual triode. In many circuits the two sections of the tube can be used in entirely different circuits.

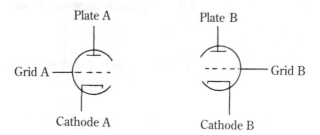

17-20 Schematic symbol for a dual triode.

Cathode-ray tubes

There are a number of special tube types available for specific, unique applications. One that merits special discussion here is the *cathode-ray tube*, or *CRT*. The key principle in a cathode-ray tube is that certain special materials, called *phosphors*, will glow when struck by an electron beam.

The basic structure of a cathode-ray tube is shown in Fig. 17-21. The elements that make up the section called the *electron gun* are shown in more detail in Fig. 17-22.

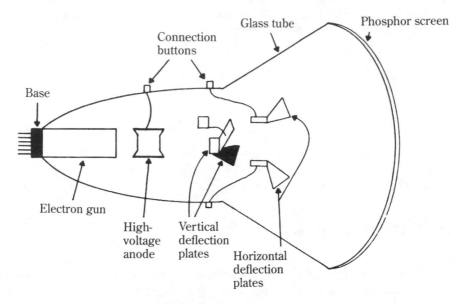

17-21   Basic structure of a CRT (cathode ray tube).

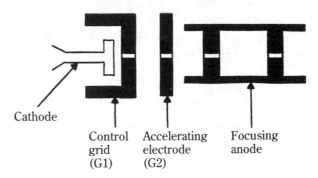

17-22   Basic structure of an electron gun from a CRT.

The cathode is indirectly heated (there is a separate heater filament) and emits a stream of electrons, as in any tube. There is one difference, however. In ordinary tubes, the cathode generally emits electrons from its sides, but the cathode in a CRT is designed so that it emits electrons primarily from the end facing the *phosphor screen*.

The cathode is enclosed in a metal cylinder that acts as the control grid. There is a minute opening at the end of this grid, facing the screen. This hole is for the electrons to pass through. Because it is so small, it forces the electrons to travel in a narrow beam.

By making the control grid negative with respect to the cathode, some of the electrons are repelled, and thus, aren't allowed to pass through the opening. If the control grid is made negative enough, it will cut off the electron beam to the rest of the tube entirely.

In other words, changing the voltage to the control grid with respect to the cathode controls (or *modulates*) the intensity of the electron beam. Holding the voltage on the control grid constant and varying the cathode voltage would have exactly the same effect. The intensity of the electron beam is determined by the difference between these two voltages. Both methods are commonly used in practical circuits.

The more intense the beam (that is, the greater the number of electrons) striking the phosphors, the brighter they will glow. So, modulating the cathode-control grid voltages will control the amount of light emitted.

Once the electron beam has passed through the control grid, it moves through a second grid element, called the *accelerating electrode*, or *grid 2*. This electrode is a metal cylinder or disk with a small opening for the electron stream to pass through. A high positive voltage is applied to the accelerating electrode. This voltage is held constant—that is, it is not modulated.

As the name implies, the purpose of this element is to accelerate, or speed up, the electrons as they pass through. In this respect, it is somewhat similar to the screen grid in a regular tetrode.

Because the accelerating electrode is highly positive it drains off some of the electrons from the passing stream. But the electrons are moving too fast for the positive voltage to deflect them from the narrow beam created by the narrow opening in the end of the control grid.

Next, the electron beam passes through the *focusing anode*. Again, the name suggests the function—this element focuses, or tightens the stream of electrons into a still finer beam.

The focusing anode is a metal cylinder that is open at both ends. Inside the cylinder are two metal plates with tiny holes in the center. The element acts similarly to a glass focusing lens in an optical system.

In addition to focusing the electron beam, this electrode also speeds it up still further. A rather large, constant positive voltage is applied to the focusing anode.

These four elements (the cathode, the control grid, the accelerating electrode, and the focusing anode) comprise the electron gun. The electron gun is so named because it "shoots" a narrow beam of electrons at the phosphor screen. Electrical connections to these electrodes are brought out through metal pins in the base of the tube, just as with ordinary tubes. Once the electron beam leaves the electron gun, it passes through a second anode. Because an extremely high (several thousand volts) positive potential is applied to this element, it is called the *high voltage anode*. The electrical connection for this element is brought out to a metallic button on the body of the tube.

Within the electron gun, the accelerating electrode and focusing anode (sometimes called *anode #1*) are both held at a positive voltage, and might tend to attract a large number of electrons out of the beam if the higher positive voltage of the high voltage anode (*anode #2*) didn't have such a strong attraction that it pulls the electrons on through. Despite this high attraction, even the high voltage anode doesn't drain many electrons out of the beam. Because the electron beam is very tightly focused, and moving at an extremely high speed, and because the high voltage anode is an open cylinder, almost all of the electrons pass through it to strike the phosphor screen.

If the tube consisted only of the elements described so far, the electron beam would always strike the exact center of the screen. Obviously, this wouldn't be particularly useful. You need a way to deflect the beam so that it can strike any portion of the screen you choose. There are two basic ways of accomplishing this—*electrostatic deflection* and *electromagnetic deflection*.

The cathode-ray tube shown in Fig. 17-21 is of the electrostatic deflection type. In this kind of tube there are four *deflection plates*, with electrical connections made to metal knobs on the outside of the glass envelope.

The plates at the top and bottom of the tube are called the *vertical deflection plates*. The other set, at the sides, are called the *horizontal deflection plates*. The electron beam passes between all four plates. For simplicity, ignore the horizontal deflection plates (the ones on the sides) for the time being. If both vertical deflection plates have the same voltage applied to them, they will have no effect on the path of the electron beam, and it will strike the center of the screen. See Fig. 17-23.

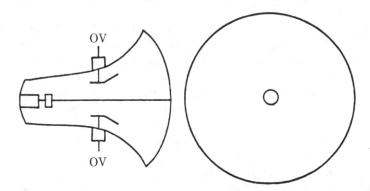

17-23   A CRT with equal voltages applied to the vertical plates.

Now, if the lower plate is made more negative than the upper plate, the lower plate will repel the stream of electrons, and the upper plate will attract it. This means the electron beam will move at an upward angle. It will strike the screen near the top—see Fig. 17-24.

The exact location of the lighted spot on the screen will depend on the voltage difference between the deflection plates. The greater the difference between the plate voltages, the further the spot will be displaced from the center of the screen. It is very important to realize that the displacement is dependent on the difference of voltage on the plates—not necessarily their absolute values. When you say the lower plate is negative, you are speaking of its relation to its partner—not necessarily with respect to ground. For instance, if the lower plate has an applied voltage of –25 V (with respect to ground), and the upper plate is at +25 V, the voltage difference is 50 V. The exact same effect on the electron beam can be achieved if the lower plate is at +100 V over ground and the upper plate is at +150 V.

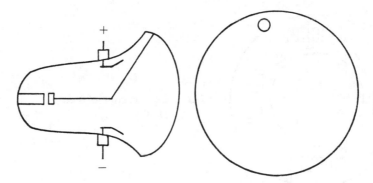

17-24 A CRT with the lower vertical plate negative with
respect to the upper vertical plate.

Of course, if the relative polarities of the deflection plates are reversed, as in Fig. 17-25, the effect on the electron beam will also be reversed. A negative upper plate and a positive lower plate will cause the electron beam to move down the screen. The horizontal deflection plates work in the same way, moving the electron beam from side to side. By combining the effects of the horizontal deflection plates and the vertical deflection plates, the electron beam can be aimed so that any desired spot on the phosphor screen can be illuminated.

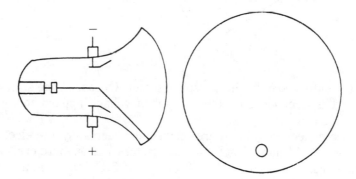

17-25 A CRT with the lower vertical plate positive with respect
to the upper vertical plate.

The electromagnetic deflection system works basically in a similar manner, but instead of internal deflection plates, electromagnets are placed around the neck of the tube in an assembly called a *yoke*. See Fig. 17-26. The yoke is positioned on the neck of the tube so the electromagnets are placed in the places shown in Fig. 17-27. Notice that these positions correspond directly to the positions of the deflection plates in an electrostatic deflection cathode-ray tube.

Because an electron can be attracted or repelled by a magnetic field (it can be considered as a microscopic magnet itself), the relative strength of the electro-magnet magnetic fields can control the angle of the electron beam, and thus, the

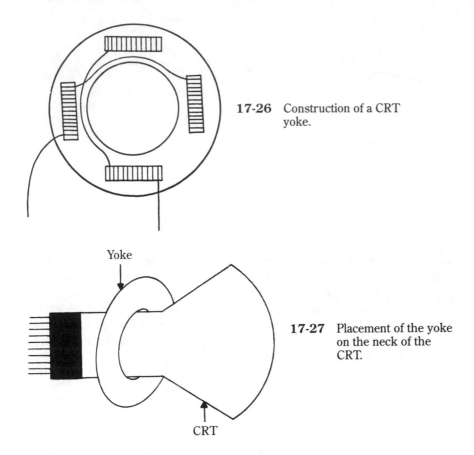

17-26 Construction of a CRT yoke.

17-27 Placement of the yoke on the neck of the CRT.

position of the lighted spot on the phosphor screen. Of course, the strength of each magnetic field is dependent on the amount of voltage applied to the appropriate electromagnet.

These two deflection systems are very similar. Generally, the electromagnetic deflection type CRT is more complex to manufacture, and is, therefore, more expensive, as a rule, than the electrostatic deflection type CRT. However, the electromagnetic deflection system allows for more precise control of the electron beam's angle. This means the image formed on the screen is sharper, or has higher *resolution* (finer detail).

In *oscilloscopes* and *radar monitors*, high resolution isn't particularly critical, so the less expensive electrostatic deflection type CRTs are usually used. A *television picture tube*, on the other hand, demands a very high degree of resolution, so an electromagnetic-deflection CRT is usually employed for that application.

If you apply a repeating ac wave shape to the horizontal deflection plates (or electromagnets) the electron beam will move back and forth across the screen in step with the ac frequency. The same voltage is applied to each of a pair of deflection plates (magnets), but one is inverted 180 degrees, so as one voltage increases, the other decreases, so the difference between the two plate voltages will vary in the same manner as the applied signal. See Fig. 17-28.

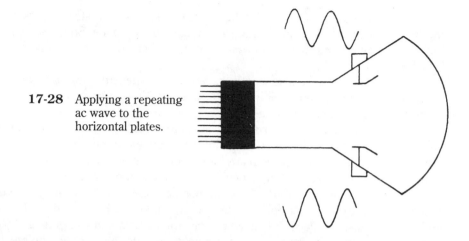

17-28 Applying a repeating ac wave to the horizontal plates.

Usually the best waveform for moving the lighted dot across the screen is the *sawtooth*, or *ramp wave*. This wave shape is shown in Fig. 17-29. Notice that the voltage starts at some specific minimum value and gradually builds up to a maximum level. Then it quickly drops back to the original minimum value, and the entire cycle is repeated.

17-29 A ramp wave.

At the minimum voltage point of the cycle, the electron beam is angled to strike the far left edge of the screen (facing the screen from the front of the tube). The left deflection plate is exhibiting maximum attraction, and the right deflection plate is exhibiting maximum deflection. As the voltage increases, the left deflection plate gradually loses some of its attraction, and the right deflection plate loses some of its repulsion. The lighted dot moves across the screen from left to right. When it is in the center of the screen, both deflection plates are at an equal voltage. From this point on, the right deflection plate starts to attract the electron beam, and the left deflection plate starts to repel it. The lighted dot continues to move across the screen, until, at the maximum applied voltage, it is at the far right edge of the screen. This part of the cycle is called the *sweep*. The line drawn by the electron beam across the screen is called the *trace*.

During the next part of the cycle, the applied voltage drops quickly back to the original minimum level, causing the electron beam to snap back to its original far left position. This process is called the *retrace*, or *flyback*.

In most practical circuits, the electron gun is cut off (no electron beam at all) during the flyback time. It is impossible to produce a sawtooth wave with an instantaneous flyback. It takes a certain finite amount of time to go from the maximum voltage to the minimum voltage. If the beam were allowed to strike the screen during the retrace time, it could produce a confusing trace image. So the screen is only illuminated by the left to right movement of the electron beam. During the retrace it is dark.

The frequency of this sawtooth waveform is called the sweep frequency, because it determines how rapidly (and how many times per second) the electron beam will sweep across the screen.

If, at the same time the horizontal plates are being fed by the sweep signal, you apply another waveform to the vertical deflection plates, something quite interesting (and useful) takes place. Between any two given instants, the electron beam will be moved a small amount, so each instantaneous value of the vertical deflection voltage will be displayed in a different horizontal position on the screen. In other words, if a sine wave of the same frequency as the sweep signal is applied to the vertical deflection plates, the electron beam will draw the pattern shown in Fig. 17-30 on the phosphor screen. If the frequency of the sine wave is doubled, two complete vertical cycles will take place in the time required for a single horizontal cycle, so two complete waveforms will be displayed on the screen, as in Fig. 17-31.

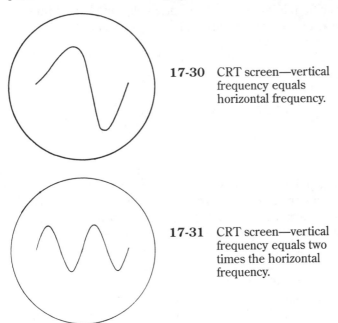

17-30 CRT screen—vertical frequency equals horizontal frequency.

17-31 CRT screen—vertical frequency equals two times the horizontal frequency.

The sweep frequency is selected to be fast enough so that the trace will appear to be a solid, continuous line. Actually, at any given instant, the electron beam is striking only one tiny spot on the screen. The phosphors glow due to a property called *fluorescence*. Another property of these materials, which is known as *phosphorescence* allows them to continue glowing for a brief time after the electron beam stops striking the spot. This property, coupled with the persistence of vision (the eye continues to see a light source for a brief moment after it is removed) gives the illusion of a solid image.

The exact chemical properties of the phosphors used determine the phosphorescence time. Different applications require different amounts of afterglow. A typical oscilloscope generally uses a phosphor that produces a green trace with a moderate afterglow time. If the oscilloscope is intended to display noncyclic voltage patterns of very short duration, a greater degree of phosphorescence is necessary. For television pictures, on the other hand, a relatively short afterglow time is preferable. In a black and white picture tube, the phosphors glow white. In a color picture tube, three types of phosphors are used together. These phosphors glow red, green and blue. This will be explained in the chapter on television.

The screen of a cathode-ray tube can either be round (as in most oscilloscopes and radar monitors), or rectangular (as in most television picture tubes). With the round type, the size is specified by the diameter of the screen, and with the rectangular shape, the size is defined by the diagonal. See Fig. 17-32.

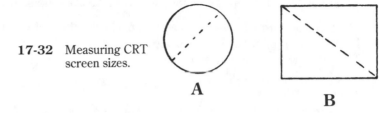

17-32 Measuring CRT screen sizes.

A   **B**

Because the electron beam strikes the phosphor screen at an extremely high speed, secondary emission could be a problem, producing reflections at undesired portions of the screen. This problem is generally prevented by lining the interior surface of the glass tube with a conductive graphite coating called the *Aquadag*. This Aquadag is tied electrically to the high-voltage anode. Because it has a high positive potential, any electrons bouncing off of the screen will be attracted to the Aquadag coating, rather than striking the screen a second time.

Self-test

1. What are the active parts of a diode tube?

A Anode, grid, and cathode
B Anode and plate
C Anode and cathode
D Cathode and grid
E None of the above

2. What is the simplest type of tube capable of amplification?

A Triode
B Anode
C Diode
D Tetrode
E None of the above

3. Under what conditions will a diode conduct?

A At all times
B When it is forward biased
C When it is reverse biased
D When it is amplifying
E None of the above

4. Which of the following best describes the grid in a tube?

A A large flat plate
B A metallic mesh
C A cone-like shape
D A filament
E None of the above

5. What happens if the grid is made more positive than the saturation point?

A Electrons are drawn to the grid and do not reach the plate
B No further amplification takes place
C The tube elements might be damaged
D The tube stops conducting
E None of the above

6. *What is the term specifying the maximum gain a tube is capable of?*

A μ-characteristic curve
B Ω—amplification factor
C β—attenuation factor
D μ—amplification factor
E None of the above

7. What is the purpose of the screen grid?

A To allow greater amplification
B To reduce the effect of interelectrode capacitances
C To reduce impedance of the tube
D To make the tube more durable
E None of the above

8. How many electrodes does a pentode have?

A Two
B Three
C Four
D Five
E Six

9. What is the name of an electrode found in a pentode but not in a tetrode?

A Control grid
B Screen grid
C Suppressor grid
D Signal grid
E None of the above

10. What type of tube is used to display signals on an oscilloscope?
    A Tetrode
    B Cathode-ray tube
    C Filament tube
    D Pentode
    E None of the above

18
Semiconductors

In the first chapter, you learned that certain substances allow electrons to flow through them fairly easily. Such substances are called conductors. Other substances, called insulators, tend to oppose the flow of current.

There is a third important class of substances with properties somewhere between conductors and insulators. These substances are called semiconductors. As you will soon see, semiconductors are extremely important to modern electronics.

Semiconductor properties

All substances, whether they are conductors, insulators, or some substance between will offer some resistance to current flow. Conductors present a very small resistance, and insulators present a very large resistance. As might be expected, a semiconductor offers a moderate amount of resistance to the flow of electrons through it.

Copper is an excellent conductor. A cubic centimeter of this substance has a resistance of about 1.7×10^{-6} (0.0000017) Ω, clearly a very minute amount of resistance. On the other hand, a cubic centimeter of slate (a good insulator) has a resistance of about 100 MΩ (100,000,000 Ω). Compared to copper, virtually no current can flow through slate. Now compare both of these substances to germanium. A cubic centimeter of this material has a resistance of approximately 60 Ω. Germanium is a semiconductor. Another common semiconductor is silicon.

A germanium atom has four electrons in its outermost ring. The electrons in the outermost ring of any atom are called *valence electrons*. The valence electrons in germanium pair up with the electrons of other germanium atoms in a crystalline structure. The pattern of these interlinked atoms is shown in Fig. 18-1. The atoms within the crystal are held together by a force called the *covalent bond*. As this term suggests, the atoms share their valence electrons.

Pure germanium has no particularly unique electrical properties, beyond being

240

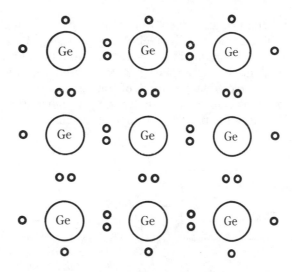

18-1   Germanium crystal structure.

a fair material for making small resistors. But if selected impurities are added to a germanium crystal, a number of interesting effects can be achieved. The process of adding impurities to a piece of semiconductor material is called *doping*. First, look at what happens if a pure germanium crystal is doped with a small amount of arsenic. Assume a single arsenic atom has been added.

The arsenic atom will try to act like a germanium atom, but, because arsenic has five valence electrons, there will be an extra electron left over. See Fig. 18-2.

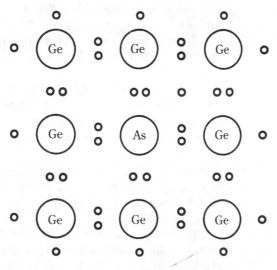

18-2   Germanium crystal doped with arsenic.

This extra electron can drift freely from atom to atom throughout the crystal. The crystal as a whole is electrically neutral, because the total protons equal the total electrons. If the germanium crystal is doped with a number of arsenic atoms, there will be an equal number of surplus electrons drifting through the crystal. The crystal itself will still be electrically neutral, of course.

If a voltage source is connected to the crystal, as shown in Fig. 18-3, the extra electrons will be drawn to the positive terminal of the voltage supply, and removed from the crystal.

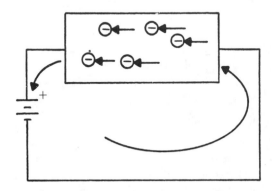

18-3   Voltage source applied to a Germanium crystal doped with arsenic.

Because the crystal now has fewer electrons than it has protons, it possesses a positive electrical charge and draws electrons out of the negative terminal of the voltage source. These electrons will move through the crystal and out to the positive terminal of the voltage source. In other words, current will flow through the crystal. So far you still don't have anything special, but be patient.

There are two types of doped semiconductors. The type you have just discussed is called an *N-type semiconductor*, because negatively charged electrons move through it. The other kind of doped semiconductor is called a *P-type semiconductor*. It is quite similar to the N-type, except an impurity with just three valence electrons is used to dope the crystal. Indium is frequently used. In this case, some of the covalent bonds are incomplete. That is, there are holes where electrons belong. See Fig. 18-4.

The various covalent bonds will steal electrons from each other to fill their holes, causing the positions of the holes to apparently drift. You can say you have a flow of holes. Actually electrons are being moved about, as in any electric circuit, but in this situation it is simpler to think of the holes as moving. Remember, a hole is simply the absence of an electron. By thinking of the holes as positively charged particles (because subtracting an electron will leave a positive charge) you can greatly simplify discussion of semiconductor action. If the impurity adds extra electrons (like arsenic), it is called a *donor impurity*. If it adds extra holes (fewer electrons) (like indium) it is an *acceptor impurity*.

Electrical current will flow through either an N-type or a P-type semiconductor. In a P-type semiconductor, you assume a flow of holes from positive to negative, instead of the usual flow of electrons from negative to positive, but it really amounts to exactly the same thing. See Fig. 18-5.

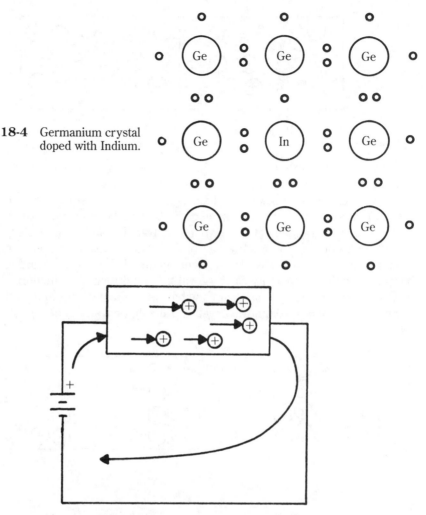

18-4 Germanium crystal doped with Indium.

18-5 Voltage source applied to a Germanium crystal doped wtih Indium.

Electrons and holes are referred to as *current carriers* or simply *carriers*. Both types of semiconductors contain both types of carriers, but one kind of carrier will be much more plentiful than the other. In an N-type semiconductor, electrons are the *majority carriers*, and holes are the *minority carriers*. That is, there are more electrons than holes. In a P-type semiconductor, the situation is reversed. Holes are the majority carriers and electrons are the minority carriers.

Neither type of semiconductor exhibits any special electrical properties when used separately, but when the two types are welded together you find a very unique situation.

The point at which different types of semiconductors are joined is called a *junction*, or, more precisely, a *PN junction*. When no external voltage is applied to a PN junction, the carriers are randomly placed, as in Fig. 18-6. Remember that despite

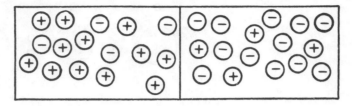

18-6   A PN junction with randomly placed carriers.

the extra electrons or holes, the net charge of each type of semiconductor is electrically neutral.

Now, suppose you hook up a voltage source with its positive terminal connected
to the N-type semiconductor, and its negative terminal connected to the P-type semiconductor. This arrangement is shown in Fig. 18-7. The holes in the P-type semiconductor will be drawn towards the end of the crystal with the negative charge, and
the excess electrons in the N-type semiconductor will be drawn towards the positive
charge. Virtually no majority carriers will be found near the junction. This means
virtually no electrons can cross from one type of semiconductor to the other. Almost
no current will flow through the crystal. You call a crystal under these conditions
reverse biased.

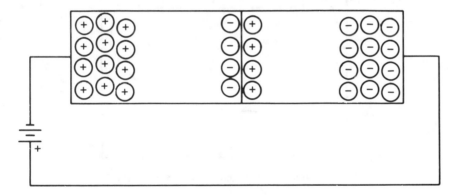

18-7   A reverse-biased PN junction.

If you now reverse the polarity of the voltage source, as in Fig. 18-8, you will
find a completely different situation. The positive charge on the P-type material will
attract its minority carriers (that is, electrons). Some of these electrons will leave
the semiconductor and flow towards the voltage positive terminal of the source.
Because some electrons have been removed from the P-type material, and it still has
the same number of protons, it now has an overall positive charge.

Meanwhile, the negative terminal of the voltage source is connected to the
slab of N-type material, repelling its majority carriers (electrons) towards the
junction. Because there are many electrons being pushed towards the junction,
and a positive charge pulling them from the other side, they are forced through
the narrow junction area to neutralize the positively charged P-type side. Mean-

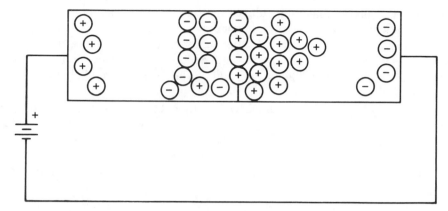

18-8   A forward-biased PN junction.

while, the voltage source is drawing away more electrons from the P-type material, so it retains its positive charge. Current flows through the semiconductor under these conditions.

You could also look at the whole procedure from the point of view of a flow of holes. The voltage supply's negative terminal on the N-type material adds electrons to it. These added electrons fill the holes in the N-type material (minority carriers). Because there are now more electrons than in the neutral state, this material acquires a negative charge that pulls holes from the P-type side. These holes (majority carriers in the P-type section) are also forced towards the junction by the positive terminal of the voltage source. In both cases, you are describing exactly the same phenomenon, only your point of reference has changed.

Semiconductor diodes

No doubt you've recognized the similarity between a PN semiconductor junction and the diode tube discussed in the last chapter. Both devices will pass current in only one direction, and block current if the polarity is reversed. In fact, the device you have been describing in the last section is a *semiconductor diode*. The schematic symbol for a semiconductor diode is shown in Fig. 18-9. Notice that this is identical to one of the symbols used to represent tube diodes.

18-9   Schematic symbol and two typical cases for semiconductor diodes.

Current flows in the direction indicated by the small arrow. This arrow is not part of the schematic symbol. Semiconductor diodes can be used for most of the same applications as tube diodes. There are a number of advantages to using semiconductor devices instead of vacuum tubes. Semiconductor diodes tend to be less expensive and smaller than their tube counterparts. Their operation produces less heat, and no separate filament circuit is needed.

The only major disadvantage is that tubes, as a rule, can operate at higher power levels without damage than can semiconductors. However, modern semiconductors will comfortably handle virtually all power levels you're likely to encounter in electronic circuits.

Note that tube diodes generally have a higher resistance when reverse biased, because the minority carriers in a semiconductor diode will allow a small amount of current to flow. This rarely is of any significance in practical circuits.

As with tube diodes, semiconductor diodes are most commonly used in rectifying (see the chapter on power supplies), and demodulation (see the chapter on radio).

The most important specification for a semiconductor diode is the *PIV*, or *peak inverse voltage*. Sometimes this specification is referred to as *PRV*, or *peak reverse voltage*. This title is pretty much self-explanatory. It is the maximum voltage that can be applied to a diode with a reverse bias without the diode breaking down. With ordinary diodes this voltage must never be exceeded. You will learn more about how diodes work in the next set of experiments.

Zener diodes

A specialized variation of the semiconductor diode is the *zener diode*. This type of diode responds to a reverse polarity voltage in a unique way. The schematic symbol for a zener diode is shown in Fig. 18-10.

18-10   Schematic symbol for a zener diode.

In the circuit shown in Fig. 18-11, the zener diode is reverse biased, and the input voltage is variable via the potentiometer. The zener diode is a 6.8 V unit. The meaning of this specification will soon be clear.

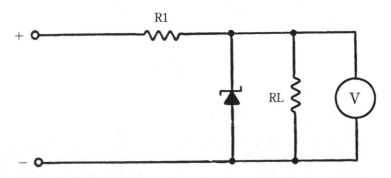

18-11   Basic zener diode circuit.

When the applied voltage is zero, the voltmeter will read 0 V, of course. When the input voltage is increased to 1 V, the meter reads just under 1 V. Resistor R1, which is included to limit the current through the diode, drops a small amount of the source voltage. With 1 V reverse bias, the diode does not conduct—the circuit acts essentially as if the diode wasn't there at all.

This will hold true up until the point when the source voltage exceeds the voltage rating of the zener diode (6.8 V in this example). This voltage is the level at which the diode begins to conduct when reverse biased. It is often called the *avalanche point* of the diode, because the current through the diode abruptly rises from practically zero to a very high value, limited only by the low internal resistance of the diode. The low internal resistance is why it is necessary to include R1 in the circuit. Resistor R1 increases the series resistance and therefore lowers the circuit current.

Because the zener diode sinks any voltage greater than its reverse bias avalanche point to ground, the voltmeter will read 6.8 V, even if the source voltage is raised to 7 V, 8 V, or even higher.

This basic zener diode circuit also serves to *regulate* the voltage to the load. That is, the voltage remains fairly constant, regardless of the amount of current drawn by the load. To understand regulating, see what happens when you vary the load resistance (R_l) and leave the zener diode out of the circuit. This experiment circuit is shown in Fig. 18-12. If R_1 is a constant 500 Ω, Table 18-1 shows the current and voltage drop across the load resistor for various values of R_L.

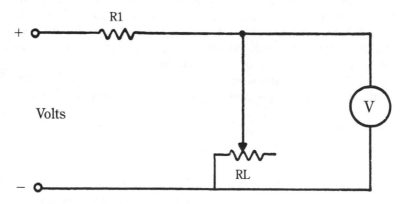

18-12   Varying the load resistance to a zener diode circuit.

Notice that the voltage across the load varies a great deal as the load resistance changes. The load resistance has to be close to 10,000 Ω before E_L gets close to the desired 6.8 V (with a 7 supply).

When the zener diode is in the circuit, however, it will hold the output voltage to a fairly constant 6.8 V. Because the load resistance is in parallel with the zener diode, E_L is also a constant 6.8 V. Remember, when two components are in parallel, the voltage dropped across them is equal. In other words, Fig. 18-11 is a simple *voltage regulation* circuit.

If the effective load resistance drops for any reason (that is, if the load circuit

**Table 18-1. Effects of varying
the load resistance in an unregulated circuit.**

R_L	E (source) = 7 V R_1 = 500 Ω R (total)	I (mA)	E (RL)
100	600	11.7	1.17
200	700	10.0	2.00
300	800	8.8	2.63
400	900	7.8	3.11
500	1000	6.4	3.82
600	1100	5.8	4.08
800	1200	5.4	4.31
900	1400	5.0	4.50
1000	1500	4.7	4.67
1100	1600	4.4	4.81
1200	1700	4.1	4.94
1300	1800	3.9	5.06
1400	1900	3.7	5.16
1500	2000	3.5	5.25
10,000	10500	0.7	6.67

starts to draw more current), this will cause an increase in the voltage drop across R1 (corresponding to the decreasing voltage drop across R_L. This decreases the voltage to the diode. As less voltage is applied to the diode, it draws less current. That means the voltage drop across R1 must decrease, forcing the output voltage to stabilize at the level determined by the zener diode.

Figure 18-13 is a graph showing the relationship of current and voltage through a zener diode. Zener diodes are available for voltages up to about 200 V.

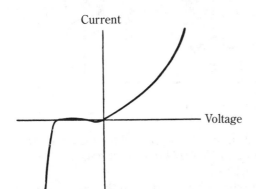

18-13   Current versus voltage graph through a zener diode.

Varactor diodes

The PN junction of any diode has a certain amount of internal capacitance, along with its semiconductor properties. The value of this internal capacitance is dependent on the width of the junction itself. A *varactor diode* is a special type of diode that takes advantage of this concept. Varying the reverse bias voltage to this kind of diode will vary the effective size of the PN junction. In other words, the diode acts like a voltage variable capacitor.

When a varactor diode is used as the capacitor in a resonant circuit (either series or parallel), the resonant frequency can be electrically controlled. Obviously, this device is ideal for automatic tuning systems. The specific applications will be discussed in later chapters. The schematic symbol for a varactor diode is shown in Fig. 18-14. Sometimes a varactor is called a *voltage controlled capacitor*.

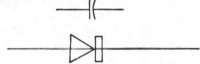

18-14   Schematic symbol for
a varactor diode.

Shockley diodes

Another special purpose diode is the *Shockley diode*. Unlike most other diodes, the Shockley diode is made up of more than a single PN junction. Its construction includes two of each type of semiconductor—NPNP. For this reason this component is also known as a *four-layer diode*. Like most other diodes, the Shockley diode has two terminals—an anode and a cathode. The schematic symbol for this device is shown in Fig. 18-15.

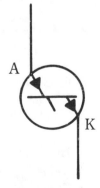

18-15   This is the schematic
symbol used to
represent a Shockley
diode.

The Shockley diode has switching properties similar to those of a neon glow lamp. It has an inherent *trigger voltage*. Below the trigger voltage, the device is in its "off" state, and exhibits a very high resistance. If the applied voltage exceeds the trigger value, the diode will be switched on and the resistance will drop to an extremely low value of just a few ohms. The trigger voltage is also known by other names, including *threshold voltage*, *firing voltage*, and *avalanche voltage*.

If a third electrode, called a *gate* is added to a Shockley diode, the result will be a *SCR* (silicon-controlled rectifier). This more complex device is discussed in chapter 21.

Fast-recovery diodes

Ordinary diodes are designed to operate on fairly low frequencies. One of the most common applications is to rectify 60 Hz sine waves. A certain finite amount of time is required for the diode to recover. This is the time it takes for the diode to turn back off when the polarity of the applied voltage is reversed. In low-frequency applications, the recovery time is not particularly critical.

In high-frequency applications, such as in a television flyback circuit, the recovery time becomes critical because the diode must respond to very short duration spikes. An ordinary diode could cause erratic or incorrect operation of the circuit. For high-frequency applications a special purpose diode called a *fast-recovery diode* is used.

Self-test

1. Which of the following is a semiconductor?

A Rubber
B Copper
C Germanium
D Carbon
E None of the above

2. A P-type semiconductor has a shortage of which of the following?

A Neutrons
B Holes
C Electrons
D Doping
E None of the above

3. How many junctions are there in a semiconductor diode?

A None
B One
C Two
D Four
E It varies

4. The arrow in the schematic symbol for a diode points which way?

A Toward the cathode
B Toward the anode
C In the direction of current flow
D Toward magnetic north
E None of the above

5. What is the most important specification for a semiconductor diode?

A Forward resistance
B Reverse resistance
C Peak inverse voltage
D Current capacity
E None of the above

6. What happens when the voltage applied to a zener diode exceeds its avalanche point?

A The current through the diode abruptly rises to a very high value
B The current through the diode abruptly drops to zero
C The diode is destroyed
D The capacitance of the diode increases
E None of the above

7. What type of circuit would a zener diode be most likely used in?

A Amplifier
B Rectifier
C Oscillator
D Voltage regulator
E None of the above

8. What is another name for a varactor diode?

A Zener diode
B Voltage-controlled capacitor
C Four-layer diode
D Fast-recovery diode
E None of the above

9. Which of the following is not another name for *trigger voltage*?

A Threshold voltage
B Avalanche voltage
C Firing voltage
D Peak inverse voltage
E None of the above

10. What is another name for a four-layer diode?

   A Varactor diode

   B Fast recovery diode

   C Shockley diode

   D Voltage controlled inductor

   E None of the above

19
LEDs

A special type of diode is the *LED*, or *light-emitting diode*. This device is shown in Fig. 19-1. Like any other diode, an LED will pass current in only one direction. But as the name implies, it will glow or emit light when forward biased. When reverse biased an LED will remain dark.

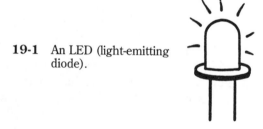

19-1 An LED (light-emitting diode).

Although some clear (white light) LEDs have been developed, most of these devices emit colored light. Red is by far the most common color for LEDs, but green and yellow are also frequently used. In addition to these visible color, some LEDs are designed to emit light in the infrared region, which is outside the visible spectrum.

Figure 19-2 shows the most commonly used schematic symbols for LEDs. As

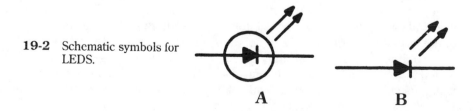

19-2 Schematic symbols for LEDS.

A   B

you can see, the symbol in Fig. 19-2B simply omits the circle shown in Fig. 19-2A; otherwise the two symbols are identical.

LEDs are used primarily as indicator devices. That is, an operator can tell whether or not a specific voltage is present in a circuit by whether the LED is lit up or dark. For example, in the circuit shown in Fig. 19-3, the LED lights whenever the circuit is activated (switch closed) and serves as a reminder to turn off the equipment when it is not in use. When this LED is lit it also indicates that the circuit is getting power and is presumably operating properly.

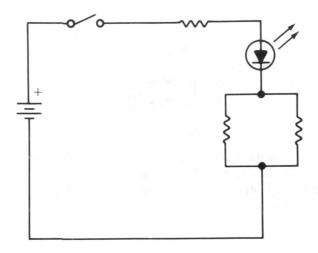

19-3   Simple circuit using an LED as a power-on indicator.

Within certain limits, the higher the voltage applied to an LED, the brighter it will glow. And, of course, lowering the applied voltage will dim the LED. This cannot be used to measure exact values, but it can be used for simple relative comparisons.

LEDs are relatively durable, but they are intended for use in low-power circuits only. Typically no more than 3 to 6 V should be applied to an LED. Excessively high voltage could burn out the semiconductor junction and render the LED useless. Some consideration should also be given to the amount of current flowing through an LED. Excessive current can damage or destroy the component very quickly.

The LED, being a diode, exhibits a very high resistance when reverse biased. According to Ohm's law ($I = E/R$) this means the diode will draw very little current. When forward biased, on the other hand, the LED resistance drops to a very low value, allowing the current to climb to a relatively high level. Depending on the rest of the circuit, the LED might attempt to pass more current than it can safely handle.

To limit the current through the LED to a safe value, a series resistor is often added to the circuit, as shown in Fig. 19-4. This resistor should have a relatively low value, usually between about 100 and 600 Ω. A 330 Ω resistor is commonly used.

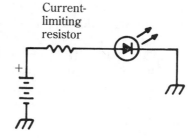

19-4 A series resistor is used
to reduce the current
flow through a LED.

Three-state LED

Because LEDs will glow only if they are forward biased and not when they are reverse biased, they can be used to test voltage polarity. Figure 19-5 illustrates the
circuit for a simple polarity checker. The resistor limits the current through both of
the LEDs, so only one is needed.

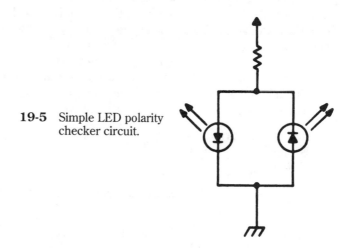

19-5 Simple LED polarity
checker circuit.

For the most convenient indication, use two LEDs that have contrasting colors.
For instance, LED 1 might be red, and LED 2 could be green.

If the lead is connected to an unknown voltage source (both the tester and the
circuit being tested should have a common ground) you can easily tell the polarity
of the unknown signal. If the voltage is positive with respect to ground, red LED 1
win light up, and green LED 2 will remain dark. If the polarity is reversed, green
LED 2 will glow instead of red LED 1. If neither LED lights up, the applied voltage
must be zero or very close to it.

Single unit dual LEDs are available. These are simply two differently colored
LEDs within a single package. These LEDs are internally connected like in the circuit of Fig. 19-5, without the current-limiting resistor.

The dual LED will glow one color when a voltage of one polarity is applied to it, and the other color will glow when the polarity is reversed. The device is known as a *three-state LED*, because there are three ways it can light up. You have already covered two. Before you get to the third, consider what happens when an ac voltage is applied to an LED.

By definition, the polarity of the ac signal keeps reversing itself. For half of each cycle the LED is forward biased and fit. For the other half of each cycle the polarity is reversed so the LED is reverse biased and dark. In other words, the LED blinks on and off.

If the applied frequency is low enough, you would actually be able to see the LED blink on and off in step with the applied signal. When the frequency is increased, the LED will still blink on and off, but it will do so too rapidly for the human eye to follow each separate blink. The LED will appear to be continuously lit, although it might seem somewhat dimmer than what a similar dc voltage would produce.

Now, return to the three-state LED and examine what happens when an ac voltage is applied. For this discussion, assume the two internal LEDs are red and green. For half of each ac cycle the red LED will be fit, and the green LED will be dark. For the other half of each cycle the green LED will be lit and the red LED will be dark. If the ac frequency is very low, you will be able to see the alternation between red and green. Because the two LEDs are so closely placed within a single package, the device will appear to be changing color.

When the applied frequency is raised, the two colors will both appear to be continuously on. They will tend to blend together producing, a yellow glow.

The three-state LED has three different color states:

Red—dc polarity A
Green—dc polarity B
Yellow—ac

The three-state LED is an extremely useful indicator device.

Multiple-segment displays

LEDs are useful indicators, but they can be even more useful if a number of them are used together to indicate a wider range of circuit conditions.

Figure 19-6 shows a simple circuit in which three LEDs indicate the position of a rotary switch. The switch could have additional poles that simultaneously perform other functions. If the switch is in position A, only LED 1 will be fit. LED 2 and LED 3 will remain dark. Advancing the switch to position B will extinguish LED 1 and light LED 2. LED 3 will remain dark. If the switch is moved to position C, both LED 2 and LED 3 will be lit, and LED 1 will be dark. If only LED 3 is lit and the other two LEDs are dark, you know the switch must be in position D. Position E lights all three LEDs, while in position F, all three LEDs are off.

Study this circuit diagram carefully to make sure that you understand how the three LEDs are being controlled in each of the switch positions. This kind of multiple indication system can be extremely helpful in operating complex circuits.

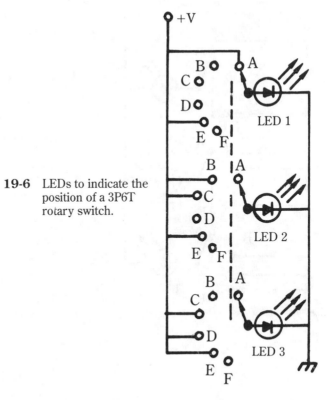

19-6 LEDs to indicate the position of a 3P6T rotary switch.

Notice that the cathodes of all three LEDs are electrically tied together. Effectively, all three LEDs share a single cathode. Such a system is called a *common-cathode display.*

Alternatively, the LEDs could be connected as a *common-anode display* as shown in Fig. 19-7.

Notice that in these cases the word *common* does not necessarily refer to the circuit common ground point. It simply refers to a shared element that is common to each of the component LEDs.

Bar graphs

Figure 19-8 shows another multiple LED display. In this circuit, the higher the voltage, the more LEDs will light. For an example, assume that each LED requires at least 1.5 V to light, and that each resistor has a value of 1000 Ω. For simplicity, ignore the internal resistances of the LEDs.

The resistors act as a voltage divider network. Because all four resistors are equal, one quarter of the applied voltage will be dropped across each resistor.

For instance, if 2 V is applied to the circuit, you would be able to read the full 2 V at point A. Resistor R1 would drop 0.5 V (one quarter of the source voltage), leaving 1.5 V at point B. This voltage is enough to illuminate LED 1. Another 0.5 V is dropped by R2 so only 1 V can be read at point C, so LED 2 is off. LED 3 also

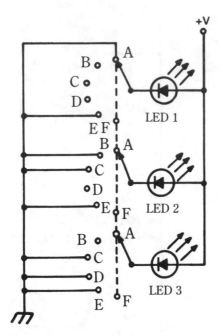

19-7   The same circuit as Fig. 19-6, but with the LEDs connected from common-anode operation.

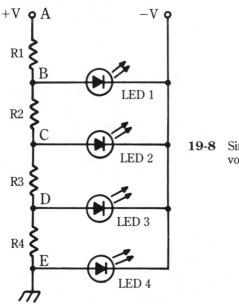

19-8   Simple four-LED voltmeter circuit.

remains dark because after the 0.5 V drop across R3, there is only 0.5 V at point D. Point E of course is grounded, so it is always at zero potential.

Now, what happens if you increase the source voltage to 4 V? One volt will be dropped across each resistor in this case, so you'll have 3 V at point B, 2 V at point

C, and 1 V at point D. Because points B and C are greater than the minimum turn-on voltage (1.5 V), LED 1 and LED 2 will light up, and LED 3 will still stay dark.

Raising the input voltage to 6 V will result in a 1.5 V drop across each resistor. Point B will be at 4.5 V, point C will be at 3 V, and point D will be at 1.5 V. Of course, this means all three LEDs will be lit.

Naturally, this type of circuitry can readily be expanded to include more than just three LEDs. But remember, there is an inherent limitation on how much voltage can safely be applied to an LED. To measure higher level signals, some sort of attenuation stage will be necessary. This kind of display is often called a *bar graph*, because the LEDs are usually arranged as a line or bar, as shown in Fig. 19-9. The longer the lighted portion of the bar, the greater the input voltage.

19-9 A bar graph lights up all of the indicator LEDs below the measured level.

A variation on the bar graph is the *dot graph*. In this case only the highest appropriate LED is lit. All lower LEDs stay dark. A dot graph is shown in Fig. 19-10.

19-10 A dot graph lights up only the single LED that represents the measured level.

Of course, this method of measurement is not as precise as a meter, but it is quite sufficient, and very convenient in certain pieces of equipment. For example, a dot or bar graph is ideal for a VU (volume-unit) meter in a tape recorder.

Dot and bar graphs are so useful that a number of manufacturers sell strips of LEDs in bar-graph form. Dot and bar graph driver circuits are also available in IC form. These circuits use active comparators, rather than the passive resistances of Fig. 19-8. For example, the TL490C and TL491C are ten-step analog level detectors designed for use in bar graphs in a 16-pin IC. The user merely needs to add an external reference voltage (to set the range), the LEDs themselves, and their current- limiting resistors. The rest of the circuitry is included within the IC chip (see chapter 24).

The LM3914, LM3915, and LM3916 take things a step further. The driver IC and a row of 10 LEDs are sold premounted on a small PC board (printed circuit). The board measures only 1.99 inches by 0.850 inch. The chip is protected by an opaque plastic cover. The row of LEDs are also under a plastic cover strip with a square window for each individual LED, making the bar graph look more like a bar. These devices might be used for either bar graphs or dot graphs.

Seven-segment displays

Perhaps the most widely useful multi-LED display arrangement is the seven-segment display. This consists of seven LEDs shaped like narrow rectangles and arranged in a figure-8 pattern, as shown in Fig. 19-11.

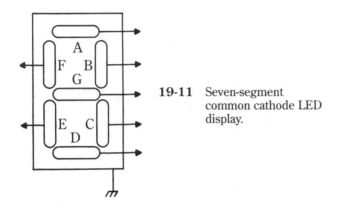

19-11   Seven-segment common cathode LED display.

A seven-segment display might be of either the common-cathode or the common-anode type. There is only a single lead for the common element, and the other leads are brought out individually. If all seven LEDs are lit, the display will, of course, look like the digit 8. But, by lighting only selected LEDs, or segments, a single digit from 0 to 9 can be formed. For example, if segments A, B, D, E, and G are lit, but segments C and F are dark, the number 2 will be formed.

Figures 19-12 through 19-21 show how each digit can be formed. If a number higher than 9 must be displayed, more than one seven-segment display is used.

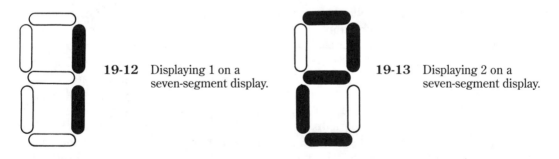

19-12   Displaying 1 on a seven-segment display.

19-13   Displaying 2 on a seven-segment display.

Certain letters of the alphabet can also be displayed on a seven-segment display, although some require a little imagination to read. Can you determine which segments must be lit to display the following letters: A, B, C, E, F, G, H, I, J, L, O, P, S, and U? Can you find a way to display any other letters?

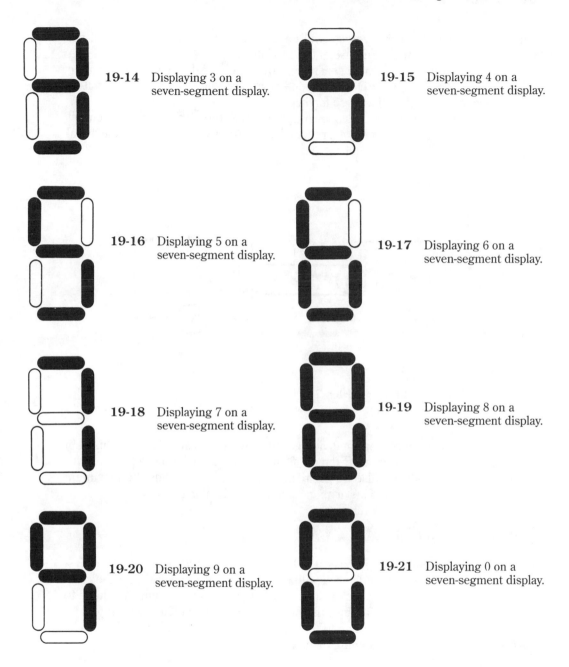

19-14 Displaying 3 on a seven-segment display.

19-15 Displaying 4 on a seven-segment display.

19-16 Displaying 5 on a seven-segment display.

19-17 Displaying 6 on a seven-segment display.

19-18 Displaying 7 on a seven-segment display.

19-19 Displaying 8 on a seven-segment display.

19-20 Displaying 9 on a seven-segment display.

19-21 Displaying 0 on a seven-segment display.

LEDs as light detectors

LEDs are ordinarily used to produce light, but they also can be used to detect the presence and approximate level of an external light source. If a light-emitting diode is exposed to a strong light source, a small voltage will be generated between its leads. The magnitude of this voltage will be determined by the brightness of the light source and the actual structure of the LED itself.

LED light detectors are most sensitive to the type of light they were designed to emit. For instance, a red LED will respond best to red light, a green LED will be more sensitive to green light, and so forth.

If two LEDs are connected as shown in Fig. 19-22, the signal applied to circuit A will be transferred to circuit B (at a reduced amplitude) without a direct electrical connection between the two circuits.

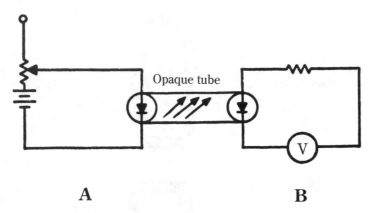

A   **B**

19-22   An LED optoisolator circuit.

To avoid interference from outside light, the two LEDs should be enclosed in some kind of light-tight opaque structure, such as a cardboard tube with its interior painted black. They should also be positioned so that LED 1 sheds the maximum amount of light onto LED 2. This arrangement is called an *optoisolator*.

Flasher LEDs

An interesting variation on the basic LED might be difficult to recognize at first glance. If you look very carefully, you will see a small black speck within the clear epoxy case of this LED. The speck is a tiny oscillator IC. Just applying a voltage to the two leads of this device will cause the LED to blink on and off at a 3-Hz rate. The device is known as a *flasher IC*, and the schematic symbol is shown in Fig. 19-23. Flasher ICs make very eye-catching displays and indicator devices.

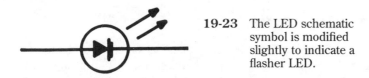

19-23   The LED schematic symbol is modified slightly to indicate a flasher LED.

The flasher LED includes its own internal current dropping resistor. It uses a +5 V power supply and draws about 200 mA. Higher voltages might be applied if a series dropping resistor is used, as shown in Fig. 19-24. For a 9 V power supply, the dropping resistor should have a value of 1000 Ω, ½ W.

19-24 This simple flasher LED circuit requires only a voltage supply and a current limiting resistor.

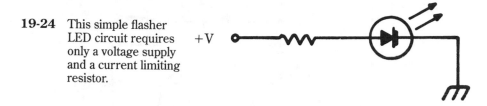

The flasher LED can even use an ac power source by placing a diode in parallel across the LED. This circuit is shown in Fig. 19-25. You can speed up the flash rate by adding a capacitor in parallel across the dropping resistor, as shown in Fig. 19-26.

19-25 A flasher LED can be powered by ac if a diode with reverse polarity is placed in parallel with the flasher LED.

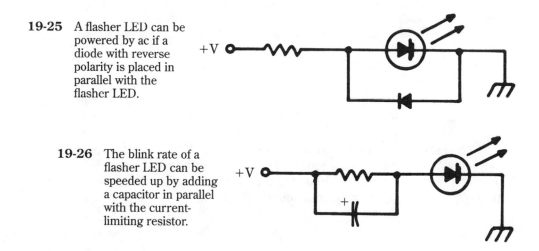

19-26 The blink rate of a flasher LED can be speeded up by adding a capacitor in parallel with the current-limiting resistor.

Experiment with various component values to obtain the desired flash rate. Capacitance values should be kept in the 500 to 3000 μF range. Flash rates above about 10 to 12 Hz tend to blend together to the human eye. The LED will appear to be continuously lit.

Figure 19-27 shows a dual-flasher circuit. LED 1 is a flasher LED, and LED 2 is a standard LED, perhaps of a contrasting color. When the components are wired as shown here, the second LED will be under the control of the flasher IC. When LED 1 is lit, LED 2 will be dark, and vice versa. This be used as an eye-catching display. Typical component values for this circuit are:

R1 = 680 Ω
R2 = 680 Ω

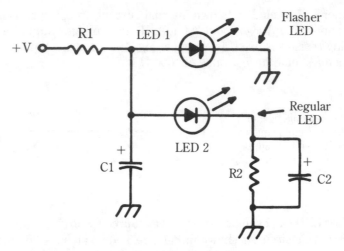

19-27   Here is an alternate flasher circuit. When the flasher LED goes off, the other (standard type) LED is lit up.

C1 = 47 μF
C2 = 250 μF

The flash rate is determined by the supply voltage. The useful range runs from about +3 to +6 V. For a +6 V supply, the LEDs will flash at the nominal flasher rate of 3 Hz. Raising the voltage above +6 V will cause the flasher LED to stop flashing and stay on. If this condition lasts too long, the IC can be damaged.

Decreasing the supply voltage will cause the flash rate to increase. Around +3 V the flash rate will exceed the 10 to 12 Hz limit of perception. Below this voltage, both the LEDs will appear to be continuously lit, but dim.

LCDs

LEDs are handy for many indicator and display applications. They are small, inexpensive, sturdy, and easy to use. However, they have disadvantages. They are difficult to see when the ambient light is bright. Also, they tend to eat up a lot of current, especially in circuits with a number of LEDs. This current drain is often a problem in battery-operated circuits where multiple-digit seven-segment displays are used. A more recently developed alternative is the *LCD*, or *liquid crystal display.*

An LCD display panel is an optically transparent sandwich, usually with an opaque backing. The inner faces of the two panels that make up the sandwich contain a thin metallic film. On one of the panels, the film has been deposited in the form of the desired display, such as the standard seven-segment display, shown in Figs. 19-11 through 19-21. Figure 19-28 shows a 3-½ digit display with colon for use in digital clocks.

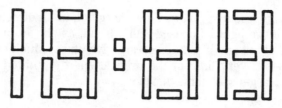

19-28 A multiple-digit display is an expansion of the basic seven-segment display system.

Between the two panels is a special fluid called a *nematic liquid*. Ordinarily this fluid is transparent. But when an electrical field is passed between the back metalized panel and one of the metalized segments on the front panel, the liquid between these portions of the panels, will darken and become opaque. The segment will appear as a black mark. When the electrical field is removed, the liquid will become transparent again.

A very small amount of current is required to darken an LCD segment. The segments are easily visible in most ambient lighting conditions, except in dim lighting conditions. LCDs can be designed to display almost anything.

In modern commercial electronic products, LCDs are increasingly replacing LEDs. There are even miniature TV sets using LCD screens to form the images. Early LCD devices were relatively slow, and moving pictures would be extremely blurred. Recent improvements in faster LCDs have made applications such as LCD screen TV sets possible.

Laser diodes

A very special type of LED is a *laser diode*. An LED and a laser might seem like very, very different things at first glance. But an LED is a light-emitting diode, and a laser beam is nothing but concentrated, tightly focused and coherent light. The word *laser* is actually an acronym standing for light amplification by simulated emission of radiation.

Light waves, like sound waves, can occur at many different frequencies. Each specific frequency corresponds to a particular color. When you see light with one frequency, you will see red, and another light frequency will be perceived as green. Red light has the lowest visible frequency, and violet light has the highest visible frequency for human beings. Light frequencies too low to be seen by the human eye are said to be in the infrared region. Similarly, light frequencies too high to be visible are in the ultraviolet region.

Ordinary white light is analogous to white noise used in audio work. It is a random combination of all possible frequencies over the entire visible spectrum. Even when there are multiple components at a specific, single frequency, they are almost certainly out of phase with one another, sometimes reinforcing each other, and other times partially canceling each other out.

An ordinary light source throws light energy off in all directions, as shown in Fig. 19-29. Even if the light is focused into a beam, as shown in Fig. 19-30, it will very quickly spread out and become increasingly diffuse. Much of the light energy is essentially wasted. Very little of the original energy emitted from the source reaches a specific target spot some distance away.

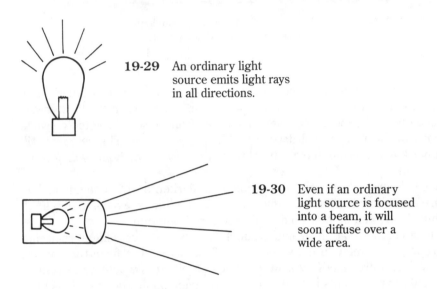

19-29 An ordinary light source emits light rays in all directions.

19-30 Even if an ordinary light source is focused into a beam, it will soon diffuse over a wide area.

A laser beam is made up of the same electromagnetic energy waves as ordinary light, but it is put together in a markedly different way, resulting in some striking differences in its effects.

A laser beam is highly *monochromatic*. The prefix *mono* means one, and *chromatic* refers to color, so *monochromatic* means one color. All of the frequency components in a laser beam are of one single frequency. Where ordinary white light is like audio white noise, a laser beam is like a sine wave, consisting of just the fundamental frequency and nothing else.

A laser beam is also very *coherent*. This means that all of the waves in the beam are working together. All the wave components are in phase with one another, so they are always reinforcing each other; they never cancel each other out, or weaken the effect of the others. Finally, a laser beam is very tightly focused, and it can hold that focus with very little diffusion over very large distances as shown in Fig. 19-31.

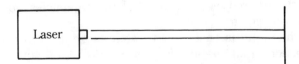

19-31 A laser beam is very tightly focused, and exhibits very little diffusion with distance.

These characteristics of the laser beam mean that very little of the source energy is wasted. Most of the energy from the source actually reaches a specific target spot some distance away. An ordinary 100 W light bulb is too hot to touch, but just a few inches away, your hand will feel very little, if any, heat from the bulb. A 100 W laser, on the other hand, can easily burn a hole through a metal plate a few feet away. Notice that both of these light sources started out with the same amount of energy—100 W. The difference is in how the laser concentrates and directs that energy.

One method is not better than the other. It depends on the intended purpose. For example, it would be very difficult to read under light from a laser. The beam would have to be focused specifically on the words you are currently reading—it would not cover the entire page. Also, unless it was a very low-power laser, it would probably tend to burn through the page before you finish reading it. So "wasteful" and uncoordinated ordinary light is still more desirable for most day to day activities.

But for many specialized purposes, where power must be focused into a very small and specific area, a laser can do tricks that would be impractical or impossible for an ordinary light source. For instance, lasers are very useful for delicate surgery, often without an actual incision in the patient's skin. A laser beam is also used to read the millions of tiny pits and islands on a compact disc. These pits and islands reflect the laser light back to a sensor at different angles in a pattern representing the encoded musical data.

Laser functions

Now, take a brief look at how a laser functions. Because of space limitations, this discussion is greatly simplified. If you are particularly interested in the subject of lasers, you might want to read a book specifically on that topic. *Laser Experimenter's Handbook, 2nd Edition* by Delton T. Horn (TAB Books #3115) might interest you.

Light is made of packets of energy called *photons*. A photon acts like both a particle and a wave, but you don't need to delve into that curious paradox here. In an ordinary, incoherent light source, photons are emitted in a manner called *spontaneous emission*. The individual photons are emitted randomly in time and direction and have a wide range of wavelengths, or colors.

A laser uses a process of stimulated emission of photons to create a coherent beam. The emitted photons are perfectly coordinated in terms of time (phase), direction, and wavelength (frequency or color). Stimulated emission can arise from the action of electrical or light energy on any of a variety of materials, which might be in the form of a gas, a solid, or (more rarely) a liquid. The individual material used is contained in a suitable confining enclosure called a *resonator.*

When an electromagnetic wave (including a light beam) of a particular wavelength passes through a suitable highly excited (energized) material, it stimulates the emission of more electromagnetic waves of the same frequency. The resulting radiation, or light, is in phase with, and in the same direction at the stimulating energy source. The end product of laser action is a high-intensity (amplified) and coherent light beam—the laser beam.

There are three basic ways a photon can interact with matter. These are shown in Fig. 19-32. Atoms (or groups of atoms) can contain only certain discrete amounts of internal energy. They can exist only at certain discrete energy levels. In-between energy levels are not possible.

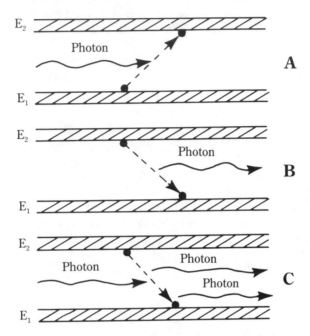

19-32   A photon can interact with matter in three
ways. Absorption of a photon (A); radiation
of a photon (B); stimulated emission of a
photon (C).

Normally, an atom stays in the lowest possible energy state, called the ground level, unless it is nudged into a higher energy state by some external means, or is stimulated. One way an atom can gain energy (or become excited) is by the absorption of a photon, as in Fig. 19-32A. But this absorption can only happen when the incoming photon has an energy that is exactly equal to the energy level separation E in that atom. In the figure, for instance, $E = E_1 - E_2$ If left to itself, the excited atom can lose energy by spontaneous emission, radiating a photon of energy E in any direction, as shown in Fig. 19-32B.

An excited atom also can be stimulated to emit a photon of energy E if another photon (also of energy E strikes that atom, as shown in Fig. 19-32C. As a result, two photons will leave the atom: the original photon striking the atom and the new photon emitted by the atom. Significantly for your purposes, these two photons will have the exact same wavelength (frequency), the same phase, and the same direction. In other words, they will be coherent. This process is the stimulated emission process that is the basis of laser operation. It functions as a coherent amplifier on an atomic scale.

For a laser to function, stimulated emission must predominate over absorption throughout the laser medium. This means more atoms must be forced into the excited (high-energy) state than are left in the normal (nonexcited) lower energy state. The distribution is called population inversion, or the reverse of the normal ratio of excited to nonexcited atoms in the mass population of atoms in the material as a whole.

An inverted population can be achieved by energy pumping. There are several different types of pumping, or mass excitation, including optical pumping, electronic pumping, collision pumping, chemical pumping, and gas expansion pumping. In semiconductor junction lasers (laser diodes) electronic pumping is used. The mass excitation and population inversion is brought about by the use of electronic currents.

The ruby laser

Before discussing the specifics of the laser diode, take a quick look at the somewhat simpler ruby laser. The earliest practical lasers were of this type. A simplified diagram of a pulsed ruby laser is shown in Fig. 19-33. The basic components of this device include:

- Light-emitting medium, or substance
- Ruby crystal
- Two reflective mirrors facing one another, providing the confining barriers for the medium
- The resonating cavity
- A controllable source of energy to excite the medium

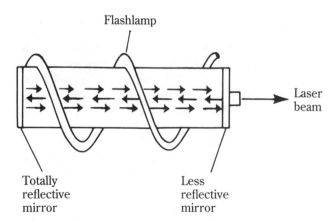

Flashlamp

Laser beam

Totally reflective mirror

Less reflective mirror

19-33 The earliest practical lasers were ruby lasers.

In this example, the controllable energy source is a xenon-filled flash lamp (like those used in strobe lights). It is in spiral form, wrapped about the diameter of the ruby crystal, as shown in the diagram.

To provide amplification, the optical path (the path travelled by the emitted photons) through the laser material is made long in one direction. Highly reflective mirrors are mounted at each end, arranged so that they can repeatedly send or reflect the photons back and forth along the laser length many times before the light is released from this resonating medium or enclosure cavity. The mirrors provide the necessary feedback to turn the optical amplifier into an optical oscillator. The principle is pretty much the same as the use of feedback in ordinary electronic oscillator circuits (see chapter 34). Oscillation in the laser takes place only in the enclosed medium and is bounded by the mirrors.

Of course, as long as all the laser energy is contained within the laser medium itself, it isn't very useful. It can't do any external work unless you can get all that amplified energy out of the enclosure. To allow the eventual release of the trapped light energy, one of the end mirrors is deliberately made slightly less reflective than the other, or one mirror is made somewhat transparent to permit the laser beam to emerge from the resonator once it has been sufficiently amplified.

The pulsed ruby laser operates as follows. First an electrical pulse or charge is set through the flash lamp, causing it to emit a burst of white light. A portion of this light energy is absorbed by the chromium atoms in the crystals. These atoms then emit red light in all directions. Whenever any of these internal red light photons hits one of the end mirrors it is reflected back and forth many times. During reflection, the red light is being amplified. Some of this trapped red light escapes through the one slightly less mirrored end of the resonator, coming out as highly monochromatic, directional, and coherent beam of powerful, concentrated light energy.

This description is very elementary and nontechnical, but it is sufficient to give you an idea of the basic principles of how the pulsed ruby laser works.

The injection laser

The ruby laser is an example of a broader class of crystal lasers. Many different types of crystal materials can be used, including semiconductor lasers, such as gallium arsenide (GaAs) and lead telluride. Because these materials can carry an electric current, electronic pumping of semiconductor lasers is possible. A PN junction, like those used in any semiconductor diode, is formed in the semiconductor crystal, as shown in Fig. 19-34. The junction is put into forward bias with the positive voltage on the P side and the negative voltage on the N side. Electrons flow through the valence band into the junction from the N side. Simultaneously, holes flow through the valence band into the junction from the P side. The conduction band is the upper energy level for the laser, and the valence band is the lower energy level. Thus, an inverted population is established between the upper and lower energy levels, permitting laser action to occur. Because the electron and hole flow are referred to as injection, these laser diodes are often called injection lasers.

The injection laser is very closely related to the common LED. When a voltage above a certain threshold level is applied to a LED, it glows, as discussed in this chapter.

An injection laser will work just like an LED with a low applied voltage. But in injection lasers there is a second critical threshold point, called Jth or Ith. When the

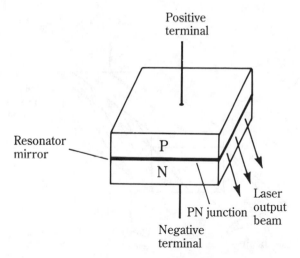

Positive
terminal

Resonator
mirror

P

N

PN junction

Negative
terminal

Laser
output
beam

19-34 A laser diode has a PN junction much like
an ordinary semiconductor diode.

applied current is below this threshold, the device functions exactly like an ordinary
LED. That is, the device glows with a relatively broad spectrum of incoherent wavelengths, and the light is emitted in a wide pattern of radiation, as shown in Fig. 19-35.

19-35 An LED emits
wideband, unfocused,
and incoherent light.

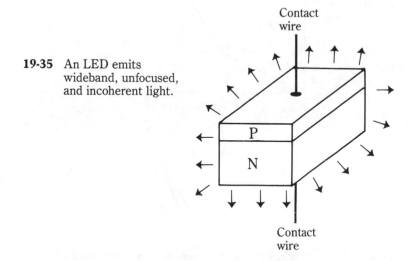

Contact
wire

P

N

Contact
wire

But if the applied current to an injection laser exceeds the threshold level, the
emitted light thins down to a very narrow beam. The emitted laser beam escapes
from both of the laser end facets, unless one is coated with a reflective (usually gold)
film, as shown in Fig. 19-36. The operation of an ordinary LED and an injection laser
are compared in Fig. 19-37.

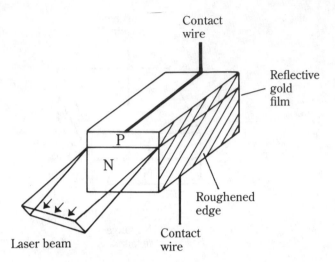

19-36   A laser diode emits a very narrow beam of laser light.

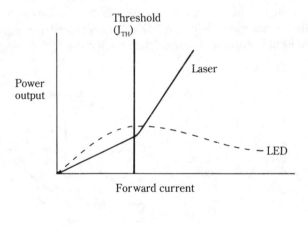

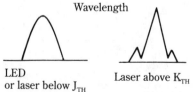

19-37   At low voltages, a laser diode operates much like an ordinary LED.

Injection lasers are very efficient light sources. They are generally no larger than 1 millimeter (0.04 inch) in any dimension. For the most efficient operation, they should be cooled far below room temperature. For instance, 50% efficiency with a continuous output of several watts has been achieved by cooling a GaAs laser to −253°C (−423°F).

A laser diode will work at room temperature but with less efficiency. A typical GaAs laser can emit infrared light continuously at room temperature with an output of 0.02 W and an efficiency of 7%. More powerful laser diodes have been developed in recent years.

Most laser diodes emit infrared light. This means the laser beam is invisible to the human eye. In any event, never look directly into a laser beam, even momentarily. Looking into a laser beam is potentially very dangerous and can cause permanent damage, blindness, or in some cases, death. Lasers are not toys, and they should always be used with great care and caution. Don't ever cut any corners on safety precautions.

Until fairly recently, the injection laser diode was an exotic and expensive component. Now, it is reasonably inexpensive (some surplus laser diodes can be bought for well under ten dollars), and is they are reasonably easy to find. The widespread popularity of CD (compact-disc) players has significantly increased the market for laser diodes. In the near future, you can expect to find such components in an increasing number of consumer products.

Self-test

1. What is the correct operating voltage for a typical LED?

A +12 to +18 V
B −6 to +5 V
C +3 to +6 V
D 0 to +120 V
E none of the above

2. When will an LED glow?

A When power is applied
B When it is forward biased
C When it is reverse biased
D When the PIV rating is exceeded
E None of the above

3. How is current through an LED limited to a safe value?

A A small-value series resistor
B A parallel diode
C A large-value parallel resistor
D A capacitor to ground
E None of the above

4. How many segments are needed to display any digit?

A Six
B Eight
C Seven

D Ten
E None of the above

5 What is the purpose of an optoisolator?

A Display indication
B Measurement
C To provide isolation between a controlling and a controlled circuit
D To reduce electrical interference
E None of the above

6. What is the normal flash rate of a flasher LED?

A Once per second
B 10 to 12 times per second
C 60 times a second
D Three times a second
E None of the above

7. What does *LCD* stand for?

A Liquid-crystal display
B Liquid-crystal diode
C Light-crystal display
D Light-crystal diode
E None of the above

8. Which of the following is not an advantage of LEDs?

A Small size
B Visible in dark environments
C Sturdy
D Visibility is not affected by bright ambient light
E None of the above

9. What is another name for a laser diode?

A Injection laser
B LCD
C Excitation laser
D Ruby laser
E None of the above

10. Which of the following is not a characteristic of laser light?

A Monochromatic
B Diffuse
C Coherent
D Focused
E None of the above

20

Transistors

Because there is a semiconductor equivalent to the vacuum-tube diode, it would be reasonable to ask, is there a semiconductor equivalent to the vacuum tube triode? The answer is yes and no. A *transistor* is a three-terminal semiconductor device that can perform most of the functions of a triode tube, but has some very different properties of its own.

There are a number of different types of transistor. The simplest, and most common is the *bipolar transistor*. As the name implies, this is a device with two PN junctions.

NPN transistors

Figure 20-1 shows the basic structure of one kind of bipolar transistor. You can see that this device consists of a thin slice of P-type semiconductor material sandwiched between two thicker slabs of N-type semiconductor material. Leads are brought out from each of these semiconductor sections.

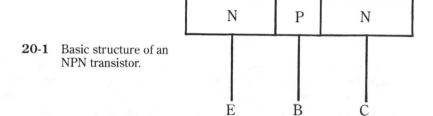

20-1 Basic structure of an NPN transistor.

One of the N-type sections is identified as the *emitter,* and the other N-type section is called the *collector.* The center P-type section is called the *base.* These terms are explained below. For obvious reasons, a transistor built according to this model is called a *NPN transistor.* Later in this chapter you will read about its mirror image, the *PNP transistor.*

The entire semiconductor sandwich is enclosed in a protective plastic or metal case. When the case is metal, one of the leads is often electrically connected to the case. Most frequently, this is the corrector, but there are exceptions. When in doubt, check the manufacturer's data sheet or use an ohmmeter to test for continuity (zero resistance) between each of the leads and the case.

Figures 20-2 and 20-3 show some typical transistors. The device shown in Fig. 20-2 is intended for use under fairly low-wattage conditions. Figure 20-3 shows a *power transistor,* which is designed to safely handle a moderately large wattage. Notice that one of the power transistors has only two leads. The third connection (the collector) is made directly to the case.

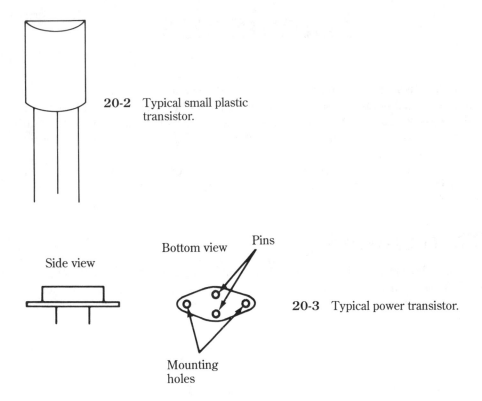

20-2 Typical small plastic transistor.

Side view

Bottom view   Pins

20-3 Typical power transistor.

Mounting holes

Power transistors can get very hot in operation, and this self-generated heat can damage the semiconductor material. To prevent this, power transistors are usually mounted on *heatsinks.* A heatsink is simply a piece of metal that conducts heat away from the component and dissipates it into the air. Many heatsinks are finned for maximum surface to air contact area. If the transistor is in a metal case, it is usually necessary to insulate it with a sheet of mica to prevent a short circuit. For maximum heat transfer, the transistor is often smeared with *silicon grease.*

The schematic symbol for a npn transistor is shown in Fig. 20-4. The lead marked *E* is the emitter, *B* is the base, and *C* is the collector. (The emitter and the collector are usually doped somewhat differently, so they are rarely electrically interchangeable.) Some schematics will have the leads marked in this manner, but usually it is assumed that you can tell from the symbol which lead is which. The lead with the arrow is always the emitter.

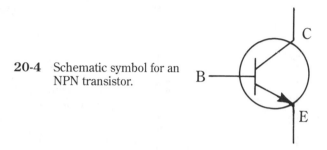

20-4   Schematic symbol for an NPN transistor.

Unfortunately, there isn't much standardization of lead positions on the actual transistors. See Fig. 20-5. For this reason, every technician and hobbyist should have a good transistor specification book that identifies the leads on various transistors.

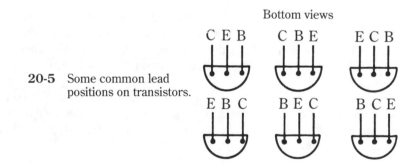

20-5   Some common lead positions on transistors.

Such a specification book is also usually a substitution guide for transistors. Literally thousands of different transistors have been manufactured all over the world over the years, and many are very difficult, if not impossible, to locate. Fortunately, many transistor types are interchangeable (at least in most circuits), so you can often substitute one type for another. However, you should keep in mind that some circuits are extremely fussy and will work with only one specific type of transistor. Such special requirements will usually be noted on the schematic or parts list. A good transistor substitution guide is an absolute necessity for anyone working in electronics, whether professionally, or as a hobby.

Return to the schematic symbol of the NPN transistor in Fig. 20-4. Notice that the arrow on the emitter points outward. Arrow direction identifies the transistor as a NPN transistor. To help you remember this, you can think, "NPN *never points in.*"

How an NPN transistor works

Figure 20-6 shows the basic electrical connections for normal operation of a npn transistor. Notice that there are two voltage sources. The two sources are for convenience in the discussion. Later, you'll learn how these two voltages can be obtained from a single power source.

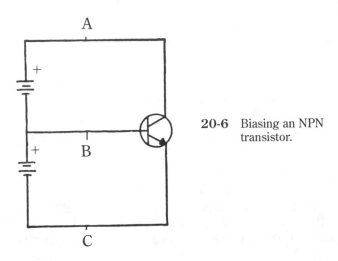

20-6   Biasing an NPN transistor.

Take careful notice of the polarities within this circuit. The base is more positive than the emitter but more negative than the collector. The actual voltage applied to the base (measured from the common ground) can be either positive or negative, but the polarity relationships between the transistor's leads always follow this pattern.

You'll recall that an N-type semiconductor has extra electrons, and a P-type semiconductor has extra holes (spaces for electrons). Because the P-type section in an NPN transistor is much smaller than either of the N-type sections, it has fewer holes than they have spare electrons.

The negative charge from the terminal connected to the emitter, forces the spare electrons in the emitter section towards the base region. The base-emitter PN junction is forward biased, so the electrons can cross into the base, filling the holes. But there are too many electrons and not enough holes. Because the base section now has more electrons than in its normal state, it acquires an overall negative charge that forces the extra electrons out of the base region.

Some electrons will leave through the base lead to the positive terminal of the base-emitter battery. The base lead is kept positive with respect to the emitter. But the collector lead is even more positive, drawing the extra electrons out of the collector section, leaving it with a strong positive charge. This will pull most of the electrons out of the base section, and into the collector section, where they are drained off into the positive terminal of the base-collector battery. In other words, the emitter emits electrons, and the collector collects them.

About 95% of the current flow will pass through the collector, and only about 5%

will leave the transistor via the base lead. Imagine that there are milliammeters at the points labeled A, B, and C in Fig. 20-6. If the current drawn by the emitter (meter C) is 10 mA (0.01 A), then meter A (collector current) will read 9.5 mA (0.0095 A), and meter B (base current) will show only 0.5 mA (0.0005 A) passing through it. Just how much current is drawn by the emitter is determined by the characteristics of the specific transistor being used, and the level of the voltage applied to the base terminal.

You can adapt the basic circuit of Fig. 20-6 to the circuit shown in Fig. 20-7. The setting of potentiometer R1 will determine the voltage to the base, which will, in turn, determine the current drawn by the transistor. Regardless of the amount of current drawn, only about 5% will flow through the base lead, and the remaining 95% will flow out of the collector lead, and through the load resistance, R2.

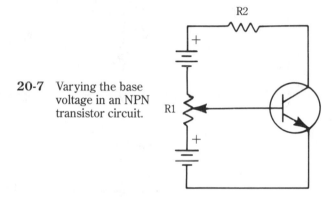

20-7 Varying the base voltage in an NPN transistor circuit.

Note that a very small change in the base current will result in a very large change in the collector current. For this reason, transistors are sometimes called *current amplifiers*. They amplify current rather than power. Of course, thanks to Ohm's law, the net effect is basically the same, because varying the current through the load resistor (R2) will vary the voltage dropped across it.

Figure 20-8 shows a more advanced version of this circuit. Notice that there is only one battery in this version. The combination of resistors R1, R2 and R3 is called a *voltage divider*. To see how this works, assume the battery generates 9 V, and all three resistors are of identical value. That is, each resistor drops one third of the voltage, or three V (ignoring, of course, the effects of the other components in the circuit, such as the internal resistance of the transistor itself).

Measuring between point A and ground, you naturally have the full 9 V. At point B (the collector lead), resistor R1 is within the section being measured, so the 3 V drop across it is subtracted. At point B, you would measure 6 V above the ground. This voltage is the voltage seen by the collector. The base, however, is connected to point C. There are two resistors between this point and the source voltage, so, subtracting the two 3 V drops, you find that the voltage applied to the base is only 3 V.

Resistor R3 subtracts another 3 V from the voltage allowed to reach the emitter. In other words, the emitter is at 0 V (ground), and the base is positive with respect to the emitter (+3 V). The collector, however, is even more positive: +6 V.

If you connected a voltmeter between point C (negative lead) and point B (positive lead), you'd get a reading of +3 V. This means the base, where it has a positive voltage with respect to ground, is negative with respect to the collector. This bias is the correct bias for a NPN transistor. Thus, you can see that you can obtain all of the required polarities in the transistor circuit with a single voltage source.

Now, if a very small ac voltage source is also applied to the base (as shown in Fig. 20-8) the voltage on the base will vary above and below its nominal dc value. This causes the collector current, and thus the voltage dropped across the load resistance (R_{l}) to vary in step with the varying voltage applied to the base. Because of the current gain through the transistor, the ac voltage across R_{L} will be much larger than the original ac voltage applied to the base. That is, the signal is amplified.

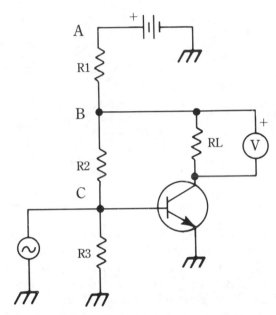

20-8   Common-emitter amplifier powered by a
single voltage source.

Transistor amplifier configurations

Because the emitter in Fig. 20-8 is at common ground potential, this type of amplifier circuit is called a *common-emitter amplifier*. The emitter is used as the common reference point for both the input and output signals.

In most practical common-emitter circuits, there will be a resistor between the emitter and the actual common ground point. The resistor is included to improve stability of the circuit. The emitter is still considered to be grounded.

The common emitter amplifier configuration exhibits a low input impedance and a high output impedance. Current, voltage, and power gain are all high. The output will always be 180 degrees out of phase with the input. That is, when the input signal goes positive (above the dc bias level), the output signal will go negative, and vice versa. The common-emitter amplifier is probably the most commonly used configuration, but there are others.

Figure 20-9 shows a *common-base amplifier* circuit. Notice that the polarity relationships between the transistor leads remain the same. The base is positive with respect to the emitter, but negative with respect to the collector. In other words, the emitter is at a negative voltage (below common ground), and the collector is at a positive voltage (above common ground). The base is grounded, so its nominal value is 0 V. If you are having trouble visualizing what is happening here, the voltage drop across R2 causes the voltage on the emitter to be below ground potential. That is negative.

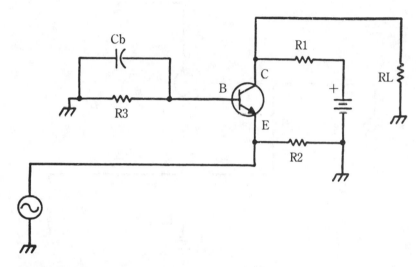

20-9   Common-base amplifier.

Resistor R3 and capacitor Cb which are between the base and the actual ground point, are for stability. Their values are quite small to keep the voltage drop across them negligible. For all intents and purposes, the voltage applied to the base is zero.

The power gain (current gain times voltage gain) of a common-emitter amplifier is slightly lower than that of a common-base amplifier using the same transistor, but its voltage gain is much higher. Another important difference between these circuit configurations is their input and output impedances. Remember, power is transferred between circuits most efficiently if their impedances match.

As already mentioned, the input impedance of a common-emitter amplifier is fairly low (typically between about 200 and 1000 Ω) and the output impedance is fairly high (typically between about 10,000 to 100,000 Ω). The impedances of a common-base amplifier are similar, but the difference between the input and the output are much more dramatic. The input impedance of a common-base amplifier is generally below 100 Ω, and the output impedance can be up to several hundred kΩ (1 kΩ is 1000 Ω).

Another difference between these circuit configurations is that the output signal of a common-base amplifier is in phase with its input signal. Remember, a common-emitter inverts the signal (throws it 180 degrees out of phase).

The third transistor amplifier configuration is rather unique. As you might have guessed, this is the *common-collector amplifier.* See Fig. 20-10. Notice that this circuit uses a positive ground point—the operating voltages within the circuit are all negative. The emitter is the most negative, and R1 drops some of the negative voltage so that the base is less negative (more positive) than the emitter, but it is still more negative than the collector. Thus, the relative polarity requirements are still met.

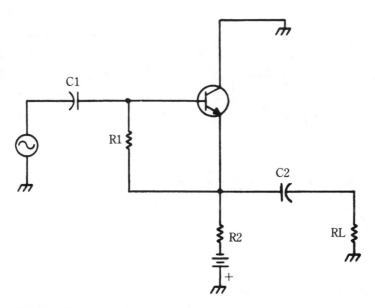

20-10   Common-collector amplifier.

One of the unique features of the common-collector amplifier configuration is that the voltage gain is always negative. That is, the output voltage is less than the input voltage. Another way of saying this is that the voltage gain is less than unity. There is, however, some positive power gain (voltage gain × current gain) in this type of circuit, but it is relatively small compared to the power gains of the common-base and common-emitter configurations.

For obvious reasons, the common-collector circuit doesn't make a very good amplifier, as such. But it is quite a useful circuit for impedance matching. With the other configurations, the input impedance is always lower than the output impedance. In the common-collector circuit, this is reversed. The output of a common-collector amplifier is in phase with its input signal.

The differences between these three basic circuit configurations are summarized in Table 20-1.

**Table 20-1. Comparing the
three basic transistor amplifier configurations.**

Common element	Base	Emitter	Collector
Input impedance	Very low	Low	Medium-high
Output impedance	Very high	High	Low
Current gain	Negative	High	High
Voltage gain	Medium	High	Negative
Power gain	High	High	Low
Output in phase with input?	Yes	No	Yes

Alpha and beta

There are literally thousands of different types of bipolar transistors available. They differ in a number of factors. For example, two transistors might differ in the maximum amount of power they can safely dissipate, their internal impedances, and most importantly, how much current gain they can produce. The current gain of any specific transistor is defined by two interrelated specifications. These are α (*alpha*), and β (*beta*).

Alpha is the current gain between the emitter and the collector. That is, for any given change in the emitter current (with the supply voltage held constant), the collector will change with a fixed relationship to the emitter. The basic equation for determining alpha is:

$$\alpha = \frac{\Delta I_c}{\Delta I_c}$$

Equation 20-1

The symbol Δ is read as *delta*. It is used to identify a changing value. I_c is the collector current, and I_e is the emitter current.

For a typical transistor a 2.6 mA (0.0026 A) change in the emitter current would result in 2.4 mA (0.0024 A) change in the collector current. Notice that the collector current changes less than the emitter current does. This is because there is always a negative current gain from the emitter to the collector in a bipolar transistor. Remember, only about 95% of the emitter current gets through to the collector. For the example, $\alpha = \Delta I_c / \Delta < I_e = 2.4/2.6 =$ approximately 0.92.

Alpha will always be less than unity (one). A small change in the base current, however, results in a large change in the collector current. 5% of the total current through a transistor flows through the base lead, and the remaining 95% flows through the collector.

For the sample transistor, the same 2.4 mA (0.0024 A) current change in the collector can be achieved with a mere 0.2 mA (0.0002 A) current change in the base circuit. The ratio between base current and collector current is β, or beta. The formula for beta is:

$$\beta = \frac{\Delta I_c}{\Delta I_b}$$ **Equation 20-2**

So beta for the sample transistor equals 2.4/0.2, or about 12. Beta is always greater than one. Alpha and beta are closely interrelated. If you know one, you can calculate the other with the following formulas:

$$\alpha = \frac{\beta}{1 + \beta}$$ **Equation 20-3**

$$\beta = \frac{\alpha}{1 - \alpha}$$ **Equation 20-4**

For example, you just found the beta for the sample transistor was 12, so you could calculate the alpha as 12/(12 + 1) = 12/13 = about 0.92. Of course, this is the same value you found when you figured the alpha directly from the emitter and collector currents.

Suppose you have another transistor with an alpha of 0.88. The beta would be equal to α/(1 − α) = 0.88/(1 − 0.88) = 0.88/0.12 or just slightly over 7. On the other hand, a transistor with an alpha of 0.97 would have a beta of 0.97/(1 − 0.97) = 0.97/0.03 = approximately 32. You can see that alpha and beta increase together. If one is increased, the other will also be increased.

PNP transistors

The other basic type of bipolar transistor is the *PNP transistor*. This device is a direct mirror image of the NPN type. The basic structure of a PNP transistor is shown in Fig. 20-11. Its schematic symbol is shown in Fig. 20-12. Notice that the arrow on the emitter points in for a PNP transistor. A handy memory aid is "*points in perpetually.*"

If you think in terms of the flow of holes rather than the flow of electrons, a

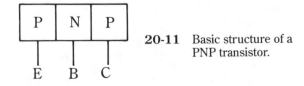

20-11   Basic structure of a PNP transistor.

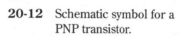 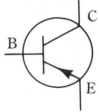

20-12 Schematic symbol for a PNP transistor.

PNP transistor works exactly like a NPN transistor, except all of the polarities are reversed. That is, the base must be negative with respect to the emitter, but positive with respect to the collector. Holes are forced out of the emitter, through the base, and into the collector.

PNP transistors can be used in any of the basic amplifier configurations (common-base, common-emitter, or common-collector), and alpha and beta are calculated in the same manner as with NPN transistors. The input/output phase relationships also remain the same.

Self-test

1. What is the most basic type of transistor?

A NPN
B Bipolar
C Triode
D Emitter

2. What are the three leads to a standard transistor called?

A Emitter, base, and collector
B Emitter, anode, and cathode
C Emitter, anode, and base
D Base, collector, and source
E None of the above

3. What can be done to ensure maximum heat transfer from a transistor?

A Ground the transistor case
B Ground the collector
C Keep the transistor moist
D Smear the transistor surface with silicon grease
E None of the above

4. Which of the following is not a standard transistor circuit type?

A Common-emitter
B Common-cathode
C Common-collector
D Common-base
E None of the above

5. How much of the current through a transistor will flow through the corrector?

A 5%
B 63%
C 95%
D 100%
E None of the above

6. What is the phase relationship between the input and output of a common-emitter amplifier?

A In-phase
B 90 degrees out of phase
C 180 degrees out of phase
D 360 degrees out of phase
E None of the above

7. What is the input impedance in a common-base amplifier?

A Very low
B Medium
C High
D Very high

8. Which is larger, alpha or beta?

A Alpha
B Beta
C Either
D Neither—the values are equal

9. Which of the following formulas is correct for determining the beta of a transistor from its alpha?

A $\beta = \dfrac{1}{\alpha}$

B $\beta = 1 - \alpha$

C $\beta = \dfrac{\alpha}{1 + \alpha}$

D $\beta = \dfrac{\alpha}{1 - \alpha}$

E None of the above

10. What is the biggest difference between a NPN transistor circuit and a PNP transistor circuit?

A The emitter and collector are reversed
B All polarities are reverse
C Input and output impedances are reversed
D Resistance values must be recalculated
E None of the above

21
Special-purpose transistors

Although the bipolar transistor is probably the most commonly used transistor, there are a number of other kinds of transistors that are designed for various special purposes.

Darlington transistors

Bipolar transistors (transistors with two complete diode junctions) can often become quite unstable if a high output current is required of them. A more stable, higher gain can be achieved by connecting two bipolar transistors in series as shown in Fig. 21-1. The emitter of transistor Q1 (transistors are often identified as Q in schematic diagrams) is connected to the base of transistor Q2. Both collectors are tied together. When transistors are connected in this manner, they are called a *Darlington pair.*

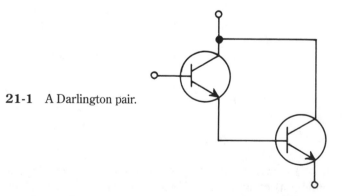

21-1   A Darlington pair.

A Darlington pair can be used in most circuits almost as if they were a single super transistor. The current at the emitter of Q2 is virtually the same as the current at the collectors. This arrangement allows for excellent balance. The transistors in a Darlington pair must be very closely matched for the best performance.

Often a specific *Darlington transistor* is used. A Darlington transistor is essentially two identical transistors within a single housing. Externally, such a device looks just like a regular transistor. To indicate that it is a single unit rather than two *discrete* (separate) transistors, a ring is usually drawn around the schematic symbol. See Fig. 21-2.

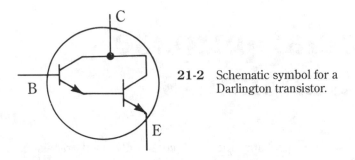

21-2   Schematic symbol for a Darlington transistor.

Unijunction transistors

So far you have read about bipolar transistors. That is, transistors with two complete diode junctions. A *unijunction transistor* (UJT), as the name implies, has only a single PN junction. The internal structure of this device is shown in Fig. 21-3. The schematic symbol is given in Fig. 21-4. Note that there are three leads—an emitter (the P-type section) and two base connections on either end of the length of N-type semiconductor material.

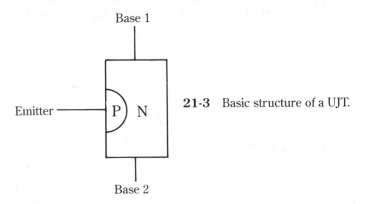

21-3   Basic structure of a UJT.

Electrically, the N-type section acts like a voltage divider, with a diode (the PN junction) connected to the common ends of the two resistances. Figure 21-5 shows a roughly equivalent circuit for a unijunction transistor. A voltage applied between

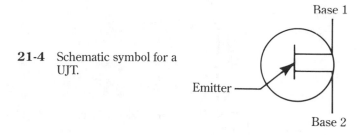

21-4 Schematic symbol for a UJT.

base 1 and base 2 reverse biases the diode. Of course, this means no current will flow from the emitter to either base.

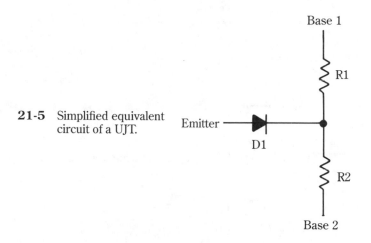

21-5 Simplified equivalent circuit of a UJT.

Now, suppose there is an additional variable voltage source connected between the emitter and base 2 (see Fig. 21-6). As this emitter-base 2 voltage is increased, a point will be reached when the diode becomes forward biased. Beyond this point current can flow between the emitter and the base.

Unijunction transistors are most often used in *oscillator* circuits. These circuits will be discussed in another chapter.

FETs

You read in the last chapter that the operation of a bipolar transistor doesn't quite directly correspond to that of a triode vacuum tube, so they are not suitable for some circuits. The *field-effect transistor* (FET) is a semiconductor device that more closely resembles a vacuum tube in operation.

The basic structure of an FET is shown in Fig. 21-7. Like the unijunction transistor, its body is a single, continuous length of N-type semiconductor material. But in an FET there is a small section of P-type material placed on either side of the N-type section. Both P-type sections are electrically tied together internally.

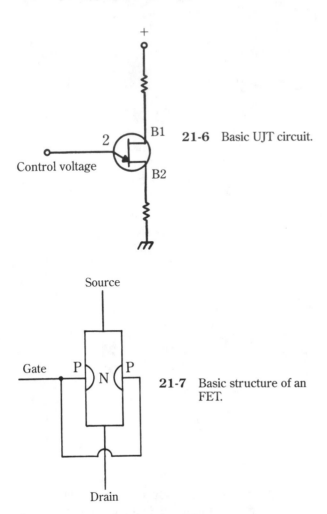

21-6   Basic UJT circuit.

21-7   Basic structure of an FET.

The lead connected to the two P-type sections is called the *gate*. Additional leads, called the *source* and the *drain*, are connected to either end of the N-type material. The schematic symbol for an FET is shown in Fig. 21-8.

To get a general idea of how an FET works, consider the mechanical system shown in Fig. 21-9. When the valve (gate) is opened, as in Fig. 21-9A, water can flow through the pipe (from source to drain). If, on the other hand, the valve is partially closed, as in Fig. 21-9B, the amount of water that can flow through the pipe is limited. Less water comes out of the drain.

Quite similarly, the gate terminal of a field-effect transistor controls the amount of electrical current that can flow from the source to the drain. See Fig. 21-10.

A negative voltage applied to the gate lead reverse biases the PN junction, producing an *electrostatic field* (electrically charged region) within the N-type material. This electrostatic field opposes the flow of electrons through the N-type section, acting somewhat like the partially closed valve in the mechanical model. The higher

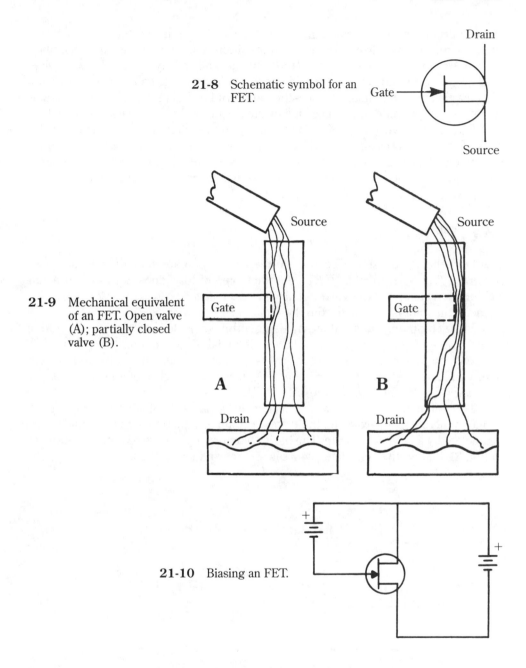

21-8   Schematic symbol for an FET.

21-9   Mechanical equivalent of an FET. Open valve (A); partially closed valve (B).

21-10   Biasing an FET.

the negative voltage applied to the gate, the less current that is allowed to pass through the device from source to drain.

Notice how this is directly analogous to the action of a vacuum tube. The gate, of course, corresponds to the grid, controlling the amount of current flow. The source is the equivalent to the cathode (it acts as the source of the electron stream) and the drain serves the same function as the plate (it drains off the electrons from the device). The path from the source to the drain is sometimes called the *channel*.

FETs can also be made with a P-type channel and an N-type gate. These devices work in the same way if you think of a positive voltage on the gate opposing the flow of holes through the channel. The schematic symbol for a P-type FET is the same as for an N-type unit, but the direction of the arrow on the gate lead is reversed.

FETs have a very high input impedance, so they draw very little current (because $I = E/Z$, increasing Z will decrease I). This means they can be used in highly sensitive measuring applications, and in circuits where it is important to avoid loading down (drawing heavy currents from) previous circuit stages. This is another way in which FETs resemble vacuum tubes.

MOSFETs

The field-effect transistors discussed in the last section are sometimes called *junction field-effect transistors*, or *JFETs*. Another type of FET does not have an actual PN junction. These devices are called *insulated-gate field-effect transistors* (*IGFETs*) because the gate is insulated from the channel.

Most commonly, this is done by using a thin slice of metal as the gate (rather than a piece of semiconductor crystal). This metal is oxidized on the side against the semiconductor channel. This insulates the gate because metal oxide is a very poor conductor. When a metal-oxide gate is used, the device is often called a *metal-oxide silicon field-effect transistor*, or *MOSFET.*

The semiconductor channel is backed by a *substrate* of the opposite type of semiconductor. If, for example, the device is built around an N-type channel, it will be backed by a P-type substrate. The basic structure of an N-type MOSFET is shown in Fig. 21-11, and the schematic symbols are shown in Fig. 21-12.

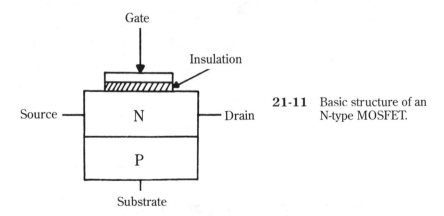

21-11 Basic structure of an N-type MOSFET.

Even though the gate is physically insulated from the channel, it can still induce an electrostatic field into the semiconductor to limit current flow. The substrate is generally kept at the same voltage as the source. The basic MOSFET biasing circuit is shown in Fig. 21-13.

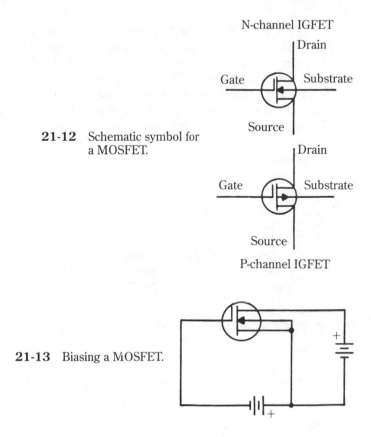

N-channel IGFET

21-12   Schematic symbol for
a MOSFET.

P-channel IGFET

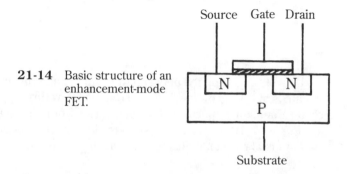

21-13   Biasing a MOSFET.

Enhancement-mode FETs

So far you have read about *depletion-mode FETs*. These devices reduce current flow
by increasing the negative voltage on the gate (assuming an N-type channel). Another type of FET operates in the *enhancement mode*. In this type of FET the source
and the drain are not parts of a continuous piece of semiconductor material. The
basic structure of this component is shown in Fig. 21-14.

21-14   Basic structure of an
enhancement-mode
FET.

A positive voltage is applied between the gate and the source. The higher this voltage is, the greater the number of holes drawn from the N-type source into the P-type substrate. These holes are then drawn into the drain region by the voltage applied between the drain and the source. In other words, increasing the voltage on the gate increases the current flow from source to drain.

The schematic symbols for enhancement mode FETs are shown in Fig. 21-15. Enhancement mode devices are always insulated gate field-effect transistors.

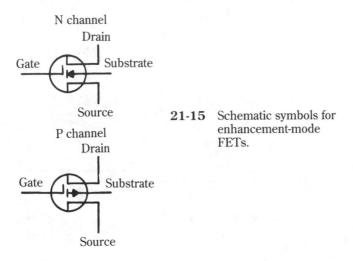

21-15 Schematic symbols for enhancement-mode FETs.

SCRs

Another special-purpose semiconductor device is the *silicon-controlled rectifier*, or *SCR*. Its schematic symbol is shown in Fig. 21-16. Notice that this is basically the same as the symbol for a diode, but there is a third lead, called the *gate*.

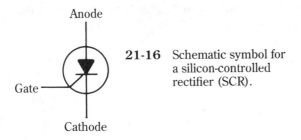

21-16 Schematic symbol for a silicon-controlled rectifier (SCR).

If a voltage is applied between the cathode and the anode (see Fig. 21-17A), but the gate is at 0 V (grounded), no current will flow through the SCR. Now, if a voltage greater than some specific value (that varies from unit to unit) is applied to the gate (see Fig. 21-17B), current will start to flow from cathode to anode against only a small internal resistance (as with an ordinary diode).

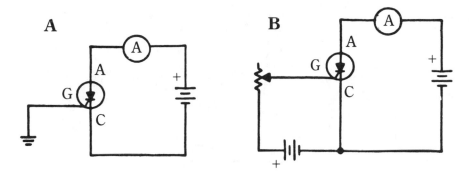

21-17 Basic SCR circuits.

This current will continue to flow, even if the voltage on the gate is removed. The only way to stop the current flow through an SCR once it has started to decrease the positive voltage on the anode (or remove it altogether). When the anode voltage drops below a predetermined level, current flow will be blocked, even if the anode voltage returns to its original value, unless the gate receives the required *triggering* voltage.

A special type of diode, called a *trigger diode* is usually connected in series with the gate (see Fig. 21-18). This diode is ordinarily reverse biased, and exhibits a sharp pulse when its trigger voltage is exceeded. This is shown graphically in Fig. 21-19. The trigger diode can provide cleaner, and more reliable triggering of the SCR.

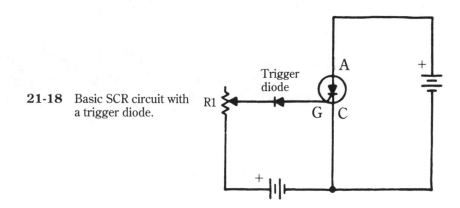

21-18 Basic SCR circuit with a trigger diode.

Silicon controlled rectifiers are typically used to control ac voltages. Figure 21-20 shows a simplified circuit of this type. When the voltage applied to the trigger diode reaches a specific point in the cycle, it turns the SCR on with a sharp voltage pulse. When the applied voltage to the anode drops below the holding level of the SCR, the current is cut off until the cycle is repeated.

Figure 21-21 illustrates the input signal and several possible output signals. Var-

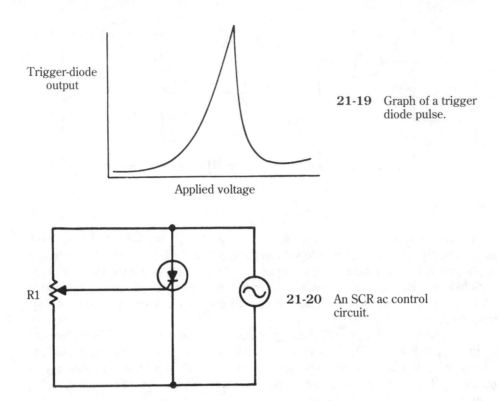

Trigger-diode output

Applied voltage

21-19 Graph of a trigger diode pulse.

R1

21-20 An SCR ac control circuit.

ying R1 in Fig. 21-20 will alter the voltage to the trigger diode, and thus, the point in the cycle when the SCR is turned on.

Because part of the ac cycle is cut out, and the voltage stays at zero during those times, the average value of the output signal must be lower than that of the input signal. The less time the current is allowed to flow through the SCR during each cycle, the lower the effective value of the output voltage. Figure 21-22 shows the basic structure of a silicon-controlled rectifier.

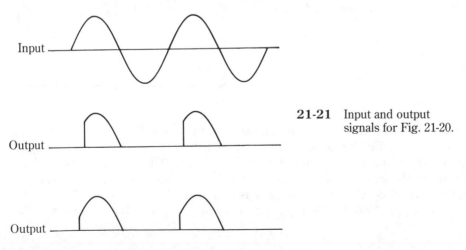

Input

Output

Output

21-21 Input and output signals for Fig. 21-20.

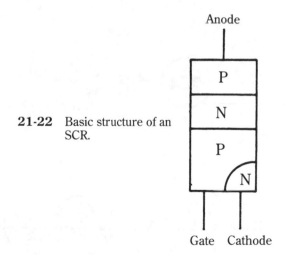

21-22 Basic structure of an SCR.

Diacs and triacs

Trigger diodes and SCRs are single-polarity devices. They can work only during one half of each ac cycle. Of course, this means the energy in the other half cycle is wasted. A *diac* is a dual-trigger diode that will produce an output pulse on each half cycle. It is effectively two separate diodes that are internally connected as shown in Fig. 21-23. The schematic symbol for a diac is shown in Fig. 21-24.

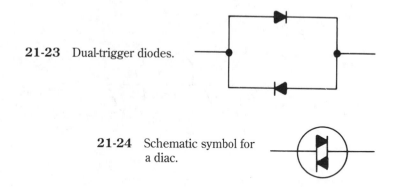

21-23 Dual-trigger diodes.

21-24 Schematic symbol for a diac.

Similarly, a *triac* is a dual SCR. It acts like two separate SCRs that are connected in parallel and in opposite directions, so current of either polarity can flow through the device. When one of the SCRs is conducting, the other one is cut off. The schematic symbol for a triac is shown in Fig. 21-25.

Figure 21-26 illustrates the operation of a typical diac/triac circuit. The circuit itself is shown in Fig. 21-27. Notice that this is essentially the same circuit that is used with the regular, single polarity trigger diode-SCR combination, except, of course, this circuit is operative during both half cycles. This means less potential power is wasted. This circuit is much more efficient.

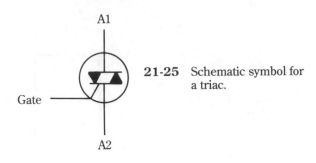

21-25   Schematic symbol for a triac.

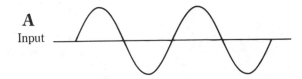

A
Input

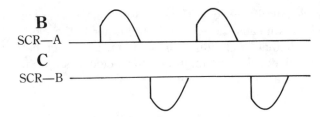

B
SCR—A

C
SCR—B

21-26   Triac or diac operation.

D
Output

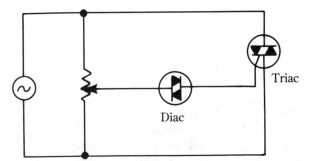

Triac

21-27   Triac or diac control circuit.

Diac

When point A is reached in the cycle, diode A conducts, triggering SCR A. At point B, SCR A is turned back off. See Fig. 21-26. During the second half cycle, diode A and SCR A are inactive, and diode B and SCR B are employed. At point C in the cycle, SCR B is turned on by the pulse from trigger diode B. Finally, at point D, SCR B is cut off again. Then the entire cycle is repeated. Figure 21-26D is a graph of the combined output of SCR A and SCR B. This is the signal that will be applied as the power source to the load.

VMOS transistors

A powerful new type of semiconductor was developed in the late 1970s. It still isn't quite as familiar to electronics hobbyists as it really should be, but it is widely used in industrial equipment and consumer products. These devices are known as *VMOS (vertical metal-oxide transistor)* transistors. They are a refinement of the MOSFET.

MOSFETS are low-voltage, low-power devices with a very high input impedance. In many applications it would be highly desirable to have the high impedances of MOSFETs in a moderate to high-power circuit. An ordinary MOSFET just can't handle the required power in many of these applications, but a VMOS device can.

The VMOSFET has a V shaped groove cut into it, as shown in Fig. 21-28. Notice that the V groove cuts through both the source and body regions of the component. Compare the construction of this device to the ordinary MOSFET shown in Fig. 21-11 and the enhancement mode MOSFET of Fig. 21-14.

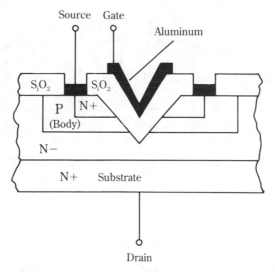

21-28   A VMOSFET has a V-shaped groove cut into it.

You could assume the *V* in *VMOS* represents the V-shaped groove, but actually, it stands for *vertical*. The groove is cut into the semiconductor material vertically, so the *V* in the name works either way. In effect, two channels are created, on either side of the V shaped gate. This permits a much higher drain-source breakdown voltage, and greater current density, so a VMOSFET can handle much larger power levels than a comparable MOSFET device.

Thanks to its high input impedance, a VMOSFET can be directly interfaced with a digital logic circuit, using either CMOS or TTL (transistor-transistor logic) gates. The high impedance means there will be negligible loading of previous circuit stages.

Another advantage of VMOS devices is the fact that they are capable of very fast switching operations, making these devices well suited for high frequency applications, well up into the megahertz region.

COMFETs

Which is better, a bipolar transistor or a FET? In terms of general specifications, a FET usually looks better, but this isn't necessarily true in practical circuits. The better choice is determined by the specific intended application. Bipolar transistors are well suited for certain jobs, and FETs do better at other tasks.

In some cases, it might be nice to have the best of both worlds. Another specialized transistor is the COMFET (*conductivity-modulated field-effect transistor*), which combines the characteristics of a bipolar transistor, a MOS power transistor, and even a thyristor. A simplified equivalent circuit for a COMFET is shown in Fig. 21-29.

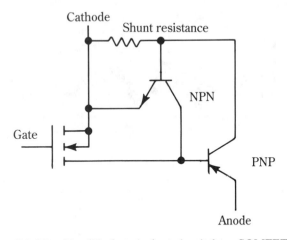

21-29   Simplified equivalent circuit for a COMFET.

The specifications for a COMFET are similar to those of bipolar power transistors, except with a much higher input impedance, like a power MOSFET. This high input impedance makes it convenient to drive the COMFET from a low-power, low-voltage source without excessive loading. A COMFET can be reliably driven directly by most digital logic gates, including TTL and CMOS devices. Bipolar transistors often require fairly expensive high-power drivers and complex drive in many multistage circuits.

When used as an electronic switch, the COMFET on-resistance ($Rds_{(on)}$) is significantly lower than that of an ordinary MOSFET with the same chip area. A typical on-resistance specification for a COMFET is less than 0.1 Ω, with a maximum drain current of 20 A flowing through the component.

COMFETs are suitable for use in high-power, high-voltage circuits. A typical device of this type can block up to 400 V in the forward direction, and up to 100 V in the reverse direction. A MOSFET can also handle fairly high voltage or power levels, but the relatively high on-resistance value can result in excessive voltage drops and power losses in the circuit. The COMFET neatly sidesteps such problems.

No practical component is ever perfect, of course. The COMFET has its disadvantages and drawbacks. For one thing, a COMFET will almost always be significantly more expensive than a comparable standard semiconductor device. The maximum switching speed for a COMFET is somewhat slower than for a conventional MOSFET device. Practical switching speeds for a COMFET are basically similar to those for ordinary bipolar transistors. Bear in mind that this is slow only in comparative terms. In many practical circuits, the switching speed limits won't cause any difficulty or practical disadvantages at all. The turn-on time for a typical COMFET is less than 100 nanoseconds. The turn-off time for such a device is typically between about 5 to 20 microseconds. Unless your specific intended application deals with very high frequencies, the speed is unlikely to be a problem. COMFETs are often an excellent choice for low to medium frequency applications.

The first COMFETs were all N-channel devices, but a few P-channel COMFETs are now also available. As is the case for most FETS, the N-channel units are still considerably more common than their P-channel counterparts. In most MOSFET devices a P- channel device requires a larger chip area than a comparable N-channel unit, but through modern technology , the P-channel COMFET can be made the same size as the N-channel version.

The on-resistance ($Rds_{(on)}$) for a P-channel COMFET is a little higher than for an N-channel COMFET. A typical value is about 0.35 Ω at 20 A, compared to the 0.1 Ω at 20 A rating mentioned earlier for a typical N-channel COMFET. P-channel MOSFETs usually have higher on resistances than comparable N-channel devices.

Microwave transistors

As electronics technology improves, the practical usable upper-frequency limit is continuously being pushed further upwards. In the early days of radio, around the time of World War I, any frequency above 1.5 MHz (1,000,000 Hz) was considered useless. In the late thirties, everything above 30 MHz was considered ultra high frequency. The term *UHF* is still used for the 30 MHz to 100 MHz frequency band. The prefix *ultra* suggests that this level was considered pretty much the ultimate in useful high frequencies. Higher frequencies were known to exist, and had been scientifically demonstrated, but they were not considered practical for use in actual circuits. The technology of the day just wasn't up to the task. Costs would have been unthinkably high for any practical commercial application.

That situation has definitely changed. Modern technology regularly uses far higher frequencies. The old UHF band is nowhere near the top of the range of frequencies being used. It would be too much trouble to rename the rf spectrum bands now, but if they were first being named today, the so-called UHF band would probably be dubbed something like *MHF* for *moderately High Frequency*.

Microwaves are now in widespread use for many applications. A microwave has a wavelength of less than a millionth of a meter. These higher frequencies not only help expand the crowded rf spectrum, they are also capable of some unique applications, all their own. An obvious, and familiar example is the popular microwave oven, which cooks food via the heat from extremely high frequency waves.

Today microwaves with frequencies in the gigahertz region are widely used. One gigahertz is equal to 1000 MHz, or 1,000,000,000 Hz. The usable frequency range today extends well past 100 GHz (gigahertz). If 30 MHz is an ultra-high frequency, then what can you call 100 GHz but a *ridiculously high* frequency? History tells you that you shouldn't be too surprised to find even higher frequencies in the future, although you are now reaching some true theoretical limits, as opposed to simple limits of "practicality." Even microwave circuits require some specialized components to handle the incredibly short wavelengths. Component leads are quite long compared to the wavelength(s) of interest. Even distances within some components themselves, such as the thickness of a semiconductor layer, can be of significance.

Ordinary bipolar transistors are limited in their high-frequency response by the transit time of charge carriers, that is, how long it takes electrons (or holes) to move across the base region. The obvious solution would be to reduce the width of the base region in the semiconductor sandwich. Within certain limits this will work, but it soon adds new problems of its own. The thinner the base region is, the higher the internal capacitance of the transistor, and the lower its tolerance to reverse biased voltages.

Even if you could get around these capacitance and reverse-bias problems, a thinner base region still would not entirely remove the high frequency limit of bipolar transistors. Semiconductor materials have a property known as the electron saturation velocity, which seems to be a fundamental and inescapable limit to the high frequency response of any bipolar transistor. The limitation is essentially built into the very concept of the component's design.

The Gunn diode

Now, obviously you wouldn't have begun this discussion at all if some practical solution hadn't been reached. The answer to the limited high-frequency response of ordinary transistors was to design some entirely new semiconductor components specifically for the job at hand—frequency response that extends through the microwave region.

One of the first researchers in this area to achieve some practical success was James Gunn of IBM. In 1963, Gunn was experimenting with the properties of N-type GaAs (gallium arsenide) semiconductors when he discovered that increasing the applied voltage above a certain threshold level would cause the current through the semiconductor material to become unstable. It was eventually discovered that the current in the GaAs crystal could be made to pulsate at microwave frequencies. Gunn explained this unusual effect in terms of negative resistance due to a decrease in electron mobility at the higher applied voltage. Negative resistance is an interesting phenomena in which Ohm's law partially breaks down. Ohm's law, you should recall, defines the relationships between the voltage (E), the current (I), and the resistance (R):

$$E = IR$$

Assuming the resistance (R) is held constant, increasing the voltage (E) will cause an increase in the current flow (I) and vice versa. In negative resistance, increasing the voltage decreases the current and vice versa. This description is a somewhat oversimplified description of the phenomenon, which is actually quite complex. A Shockley diode, discussed in chapter 18, can also exhibit negative resistance under certain conditions, which makes it suitable for use in microwave circuits too.

It seems that some materials, including GaAs, can allow electrons in either of two conduction bands, instead of just one. In the lower conduction band, the effective mass and energy of the electrons are low, so the electron mobility will be high. The material in this band acts like an ordinary ohmic resistance.

When the applied voltage is increased to about 3 kV to 3.5 kV per centimeter, the electrons will become more energetic, and will jump to the higher conduction band. You encountered this same two-band phenomena in the discussion of laser diodes in chapter 19.

In the higher energy band, the effective electron mass increases, and electron mobility decreases. This is logical enough—the electron is effectively more massive, so it can't be moved as easily as before.

Figure 21-30 illustrates this effect in graph form. When the applied voltage is less than the critical threshold voltage, labelled V_{th}, the electron velocity increase linearly with the applied voltage, as you'd expect. In other words, the device cooperates with the predictions of Ohm's Law. But now take a look at what happens when the applied voltage is increased past V_{th}. The net electron velocity starts to drop off with further increases in the applied voltage. The material is now acting like a negative resistance.

At some point, increasing the applied voltage further won't cause any further decrease in electron velocity. This is the saturation voltage, or V_s. From here on, the graph is a simple straight line until the applied voltage becomes large enough to damage or destroy the semiconductor material itself.

These principles are put into action in a component usually known as a *Gunn diode*, in honor of James Gunn. Because there is no actual PN junction, some technicians argue that this component is not properly a diode at all, and they call it a *Gunn device*. Frankly, it is getting a little too picky and it sounds rather awkward, so continue to call this component a Gunn diode. After all, strictly speaking a diode is a component with two (di-) electrodes—the anode and the cathode. The Gunn diode does fit this minimal description, even though it ultimately has few other similarities to the ordinary semiconductor diode.

The Gunn diode can be used to create a microwave frequency oscillator that

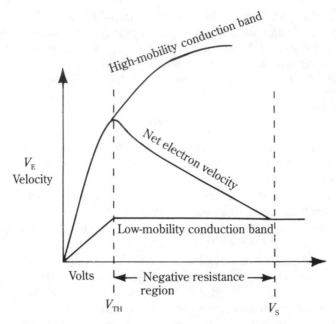

21-30 Graph of Gunn diode operation.

depends on the transfer of electrons between the high-mobility and low-mobility conduction bands. Such an oscillator is sometimes called a *TEO*, or *transferred electron oscillator.* The Gunn diode is capable of two different modes of oscillation.

A simplified diagram of the Gunn diode internal structure is shown in Fig. 21-31. Notice that the device is divided into three sections, or areas. The uppermost section offers a low (positive) resistance connection between the metal contact electrode, and the active center region. This section is very thin, typically of just 1 to 3 microns and has very low resistivity. A typical resistance value is 0.001 ohms per centimeter. The main functions of this section of the Gunn diode are to prevent migration of metallic ions from the electrode into the active region and yet to provide good electrical contact between the active section and the electrode, which is used, of course, to make connections to an external circuit.

The most important section of the body of the Gunn diode is the active region in the center, made of the N-type GaAs semiconductor material. The thickness of this section determines the oscillation frequency of the component. This can range from about 6 GHz at 18 micro-meters up to 18 GHz at 6 micro-meters. Not surprisingly, the thickness of this active section also affects the Gunn diode threshold voltage. A thinner active region (higher oscillation frequency) results in a lower threshold voltage and vice versa. For example, a 6 GHz Gunn diode will have a threshold voltage of about 5.85 V, and a 18 GHz Gunn diode will have a threshold voltage below 2 V.

Beneath the active region is the substrate layer. Low-resistance GaAs material is used here too, but it is metallized to allow bonding to the support structure of

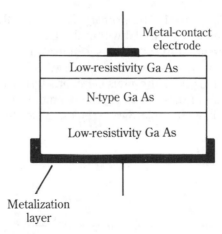

Metal-contact
electrode

| Low-resistivity Ga As |
| N-type Ga As |
| Low-resistivity Ga As |

Metalization
layer

21-31   Internal structure of a Gunn diode.

the component. This structure will usually be a heat-sinking package. The second component connection point is made to the substrate layer.

One disadvantage of Gunn diodes is that they are relatively inefficient, especially when oscillating at microwave frequencies. A typical Gunn diode will eat up about 20 to 50 times more dc power than it generates in the form of microwave RF output energy. But at least it can do the job.

The Gunn diode can function as a microwave oscillator in either of two operational modes—the transit-time mode and the delayed transit-time mode. The transit-time mode is sometimes called the *Gunn mode*, because it is based on Gunn's original theories in this area. The delayed transit-mode was developed somewhat later.

Look at the transit-time mode first. When a voltage below the threshold potential (V_{th}) is applied across the Gunn diode, there will be an uniform electric field through the semiconductor. You don't really have anything special here. The Gunn diode electrically looks like an ordinary positive resistance, and it complies with Ohm's law. Increasing the applied voltage will result in a proportional increase in the current flowing through the device.

When the applied voltage across the Gunn diode reaches the threshold potential (V_{th}), electrons are forced into the cathode end of the component faster than they can be collected from the anode. As a result, a region, or domain with an excess of electrons build up on the cathode side. At the same time, a shortage of electrons (or an excess of holes) appears on the anode side. The excessively negative (too many electrons) domain drifts as a unit through the body of the Gunn diode, until it is collected at the anode end. A new domain forms near the cathode as the old domain is collected at the anode.

Most of the time the current output from the Gunn diode will be fairly low, except when a domain is collected at the anode. At that point there will be a brief large current pulse, as shown in Fig. 21-32. The time period between output current bursts depends on the exact characteristics of the material used and the design of the particular Gunn diode used. The primary factors are the length and the drift velocity of the semiconductor material within the Gunn diode. This time period is called the *drift time*. The drift time is obviously equivalent to the operating frequency of the device. In effect, you have a sort of pulse wave. Because of the small dimensions inside a Gunn diode, and the physical characteristics of the GaAs material used in the construction of these devices, the operating frequency (the reciprocal of the drift time) is well into the microwave region, typically about 6 to 18 GHz.

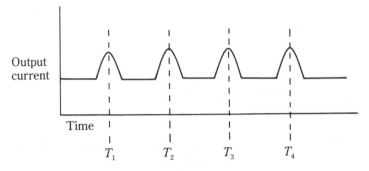

21-32   The output of a Gunn diode is a low, steady current punctuated by periodic high-current bursts.

The chief disadvantage of the transit-time mode is that it is quite inefficient. A lot of the input power is wasted. The delayed transit-time mode offers somewhat better efficiency (although it still isn't highly efficient, by any means). The transient-time mode also has a lower frequency limit, determined primarily by the thickness of the component active region.

In the delayed transit-time mode, the Gunn diode can adapt to the frequency of an external tank circuit, such as a high-Q resonant cavity. A typical equivalent circuit for a Gunn diode in the delayed transit-time mode is shown in Fig. 21-33. Notice that this is a greatly simplified equivalent circuit for illustrative purposes only. It is not a practical circuit. Notice also that the Gunn diode is shown here as a resistor with a negative resistance. You could think of it as sort of an antiresistor. The negative resistance Gunn diode is placed in parallel with a LC resonant tank circuit. The other resistor, marked *OL*, represents the ordinary resistance losses in the tank circuit. This value is typically very small. It is usually defined as a conductance rather than

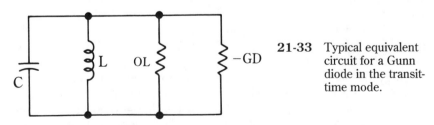

21-33   Typical equivalent circuit for a Gunn diode in the transit-time mode.

a resistance. A conductance, measured in *siemens* (*S*), is equal to the reciprocal of the comparable resistance, measured in ohms. Formerly, the unit of siemens was called a *mho*, which is *ohm* spelled in reverse. The relationship between ohms and siemens is:

Siemens = 1/ohms
Ohms = 1/siemens

For example, 0.0001055 Ω is an awkwardly small value. Expressed as a conductance value, it works out to:

$$r = 1/0.0001055$$
$$= 9479 \text{ siemens}$$

If the negative resistance of the Gunn diode ($-GD$) is much less than the conductance of OL, the circuit will oscillate. When the Gunn diode is biased by an applied voltage greater than the threshold potential (V_{th}), domains are created and drift through the device as the transit-time mode described above. The first few output pulses will excite the external circuit, causing it to *ring*, or break into oscillations. Soon a continuous RF sine-wave signal will be built up. The signal frequency is determined by the resonant frequency of the tank circuit.

On the negative peaks of the generated sine wave signal, the dc voltage applied across the Gunn diode will be forced below V_{th}. This action will tend to delay the next domain caused output burst, hence the name of this mode. Over time the output current pulse period will adjust itself to the tank circuit's resonant frequency, permitting much more efficient operation than in the simpler transit-time mode. In some technical literature, the delayed transit-time mode is referred to as the *limited space-charge accumulation* (or *LSA*) mode.

Because of the high microwave frequencies involved here, discrete tank networks comprised of physical coils and capacitors simply aren't practical. Instead, the Gunn diode is mounted in a special resonant cavity. This also helps improve the operating efficiency of the device.

IMPATT diodes

Another semiconductor component designed for use at microwave frequencies is the IMPATT diode. *IMPATT* is an abbreviation of impact avalanche transit time. Although the practical use of microwaves is a relatively recent phenomena, research into this area has been going on for some time. The IMPATT diode was first proposed in the early 1950s at Bell Laboratories.

Unlike the Gunn diode, the IMPATT diode is a true semiconductor diode, with a PN junction. Of course, this junction is used in a rather unusual way in this component. The IMPATT diode uses the phase delay between an applied RF voltage and an avalanching current in its PN junction as a negative resistance at microwave frequencies.

Structurally, the IMPATT diode is a little like a cross between an ordinary semiconductor diode and a Gunn diode, as shown in Fig. 21-34. Notice that the P-type section is much, much thinner than the N-type section. Like the semiconductor material in a Gunn diode, the N-type region in an IMPATT diode is divided into regions.

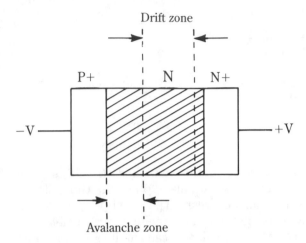

21-34 The structure of an IMPATT diode is like a cross between an ordinary diode and a Gunn diode.

In addition to the more or less standard PN junction, there is also a junction between a negative N-type section and a positive N-type section. The positive N-type section provides a low-resistance connection to the electrode and prevents the migration of metallic ions into the active section of the semiconductor. This structure is similar to the internal structure of a Gunn diode.

The active region in this case is the large center portion of (negative) N-type material. This section of semiconductor crystal is specially doped so that at the reverse-bias breakdown voltage for the PN junction, these region is fully depleted. Therefore a very small electrical field can force velocity saturation of electrons.

The critical breakdown voltage for an IMPATT diode is analogous to the avalanche breakdown voltage of an ordinary semiconductor diode. When reverse biased, the PN junction normally conducts little or no current. There is usually a small leakage current, but it is generally negligible for all practical intents and purposes. But if the reverse bias voltage across the diode is increased beyond a specific point, the PN junction breaks down, and starts conducting very heavily, as shown in the graph of Fig. 21-35. The PN junction goes into avalanche operation. The avalanche current comes from secondary emission. There are so many electrons forced into the area surrounding the PN junction, leakage current electrons have a much greater probability of colliding with other electrons. These increased electron collisions result in a very rapid rise in reverse current.

In a standard junction diode, this avalanche condition is to be avoided at all costs. The excessively high reverse current can quickly cause the PN junction to overheat and self-destruct, permanently damaging the component. The breakdown, or avalanche voltage is the same as the PRV (peak reverse voltage) or PIV (peak inverse voltage) rating given for all standard semiconductor diodes. This voltage

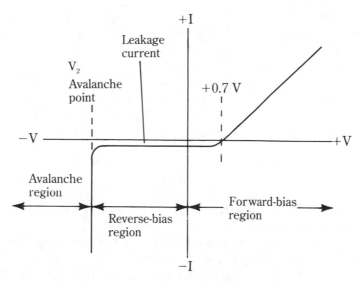

21-35   If the reverse bias to a diode exceeds the avalanche point, the PN junction breaks down, and starts conducting very heavily.

should never be exceeded when reverse biasing the diode. Some specialized types of semiconductor diodes, however, are designed to withstand and make positive use of avalanche breakdown. The zener diode, discussed in chapter 19 is a good example of this. The rise of the avalanche current is internally controlled within such devices, preventing self-destruction. The IMPATT diode is also designed for intentional avalanche breakdown conditions. In fact unless the PN junction is forced into avalanche, the IMPATT diode will not function at the intended microwave frequencies.

In operation, an IMPATT diode is connected in parallel with a high-Q resonant tank circuit like the Gunn diode. Again, because of the very high frequencies involved, discrete LC networks are not practical. The IMPATT diode is operated inside a resonant cavity. The PN junction is reverse biased to create a noise signal that causes enough shock to excite the tank circuit into oscillation. The rf output voltage is added to the dc bias voltage across the IMPATT diode. This combination forces the IMPATT diode to go into avalanche on positive peaks of the oscillation signal, creating energy bursts that re-excite the tank circuit, and allow the microwave frequency oscillations to continue.

TRAPATT diodes

Special microwave components like Gunn diodes and IMPATT diodes were specifically designed to operate at higher frequencies than ordinary semiconductor components can operate. Ironically, one of the chief limitations of the IMPATT diode is that some desirable microwave frequencies might be too low for it to handle. For the most part, IMPATT diodes can only operate at frequencies above 3 to 4 GHz. In some applications, it might be desirable, or even essential to use a frequency that is too high for an ordinary semiconductor component, but too low for an IMPATT diode. The solution is to find some method of extending the transit time through the device. But this is easier said than done.

In 1967, researchers at Bell Laboratories developed a modified IMPATT diode that could be excited into a different mode of operation that could operate at frequencies of about 0.9 GHz to 1.5 GHz with fair efficiency (about 25%). This new mode of operation was called the *anomalous mode*, reflecting the curious fact that no one at the time could quite explain just how it worked.

The new microwave component was named the *TRAPATT diode*, for trapped plasma avalanche transit time. Some technicians preferred the name *ARP diode* for avalanche-resonance pumping, based on an alternate theory of how the component functioned. Despite the fact that this label was shorter and more convenient, the *TRAPATT* name is the one that has stuck.

The basic internal structure of a TRAPATT diode is shown in simplified form in Fig. 21-36. Notice that this device is pretty much similar to the IMPATT diode in its constructions, although it is used in a somewhat different way. Many TRAPATT diodes can be used as if they were IMPATT diodes, although the reverse isn't as likely to be true. In fact many TRAPATT oscillator circuits actually begin operating in the IMPATT mode for a brief period (a few nanoseconds) when the circuit is first turned on, then they settle into a true TRAPATT mode. The switchover between modes is accomplished via a hard current pulse with a very short rise time.

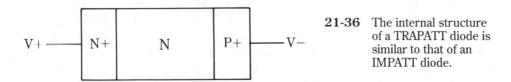

21-36   The internal structure of a TRAPATT diode is similar to that of an IMPATT diode.

Interestingly, many standard PN-junction silicon diodes can be made to oscillate in the TRAPATT mode. Unfortunately, this is very tricky to do reliably. In practical circuits, a device specifically designed for this purpose is used.

BARITT diodes

Yet another microwave device is the *BARITT* (barrier injection transient-time) diode. The internal structure of this device is shown in simplified form in Fig. 21-37. Notice that this component is made up of a pair of back-to-back PN junctions, somewhat like a bipolar PNP transistor, with no base connection. This similarity is more apparent than real, however.

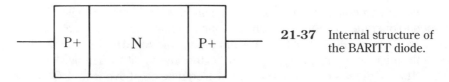

21-37   Internal structure of the BARITT diode.

In operation, one of the PN junctions in the BARITT diode is slightly forward biased, and the other is slightly reverse biased. If the applied bias voltage is less than a specific critical level (known as the *punch-through* voltage), the current flow through the BARITT diode will be no more than the ordinary leakage current of the reverse-biased junction. But once the applied bias voltage reaches or exceeds the critical punch-through voltage level, the device starts behaving quite differently. Under these conditions, the depletion region stretches across the entire N-type section, up to the forward-biased PN junction. All of the charge carriers (holes in this case) at the forward-biased junction will flow across the N-type section of the component. As a result, the current flow through the device will rise very rapidly, as shown in the graph of Fig. 21-38. For use in a practical microwave oscillator circuit, the bias voltage applied to a BARITT diode must be kept between the punch-through voltage (V_p) and the avalance threshold voltage (V_a).

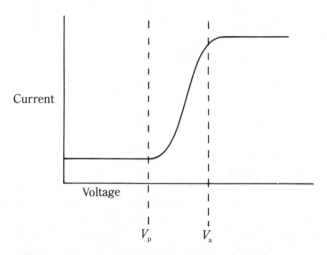

21-38 Above the threshold point, the current flow through a BARITT diode increases very rapidly.

PSIFETs

Ordinary FETS, like most other standard semiconductor components, are not well suited to microwave applications. This is especially true of IGFETs. These components are primarily used only in switching and audio frequency applications. But electronics is a constantly developing field, and recent breakthroughs in MOSFET technology has now provided specialized FET devices that can handle substantial power levels at microwave frequencies. Such microwave FETs offer numerous advantages over standard high-frequency silicon bipolar transistors.

FETs, as you should recall, offer much higher input and output impedances than bipolar transistors. The dc bias is not affected by fluctuations in temperature, and thermal runaway is not normally a problem. FETs are voltage-driven, rather than current-driven devices, so they can offer highly linear operation characteristics.

A power FET designed for microwave applications is called a *power-silicon FET,* or *PSIFET*. A simplified diagram of the internal structure of this device is shown in Fig. 21-39. Notice that this component is not really all that dissimilar to an ordinary JFET, as described earlier in this chapter. A short vertical channel is used. The gate and source connections are placed on the upper surface of the semiconductor slab, and the drain connection is at the bottom.

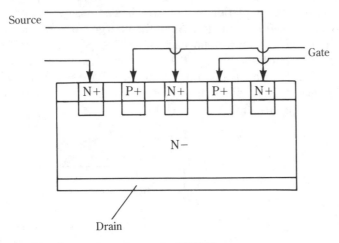

21-39   Internal structure of a PSIFET.

The gate voltage applied to a regular JFET controls the resistance, and therefore the current flow, of the lateral source-to-drain channel. In the PSIFET, this channel is constantly kept fully depleted. There are no free electrons or holes in the channel. The electric potential in the depletion area is controlled by the applied gate voltage. This electric potential acts as a barrier that tries to block any electron flow between the source and the drain. By applying a large enough voltage to the drain of the PSIFET, an electrostatic field will be created to counteract the depletion area's potential barrier. Under these circumstances, electrons can flow from the source to the drain of the component.

In a PSIFET, the charge carriers travel by drift, rather than by diffusion, so the carrier velocity is significantly higher than that of a standard bipolar transistor. PSIFETS therefore feature a high breakdown voltage, and a high gain-bandwidth product. Most significantly for your purposes here, the PSIFET can function at much higher frequencies, well into the microwave region of the rf spectrum. Because of their highly linear operation, PSIFETs are particularly well suited for use in linear amplifier circuits for communications systems.

Notice that unlike the other microwave components you have discussed in this section, the PSIFET is not a diode. It has three, instead of just two leads. This permits more sophisticated circuitry and applications.

Self-test

1. What is a Darlington pair?

A Two similar bipolar transistors in parallel
B Two similar bipolar transistors in series
C An NPN transistor and a PNP transistor in series
D A symmetrical transistor circuit
E None of the above

2. How are the three leads on a UJT labelled?

A Emitter, base 1, and base 2.
B Emitter, base, and collector
C Gate, drain, and source
D Emitter 1, emitter 2, and base
E None of the above

3. How many PN junctions are there in a FET?

A One
B Two
C Three
D It varies

4. Which type of transistor most closely resembles a vacuum tube in its operating characteristics?

A The bipolar transistor
B The UJT
C The Darlington transistor
D The FET
E None of the above

5. Which of the following is not a variation on the basic FET?

A JFET
B MOSFET
C IGFET
D EEFET
E None of the above

6. Which of the following best describes an SCR?

A A diode with an electrically operated switch.
B A modified FET
C Two triacs in a single housing
D An enhancement mode UJT
E None of the above

7. What is a diac?

A A dual SCR
B A dual-polarity trigger diode

C A simplified triac
D A Darlington pair
E None of the above

8. What are the three leads of an SCR called?

A The anode, the cathode, and the gate.
B The emitter, base 1, and base 2
C The drain, the gate, and the source
D Anode 1, anode 2, and gate
E None of the above

9. What is the primary advantage of a VMOSFET over a MOSFET?

A Less expensive
B Can operate at microwave frequencies
C Can operate at much higher power levels
D Never breaks into unwanted oscillations
E None of the above

10. Which of the following components can operate at microwave frequencies?

A VMOSFET
B Gunn diode
C UJT
D Darlington transistor
E None of the above

22

Light-sensitive devices

Semiconductors have an unusual property that has not been taken advantage of until recently. Semiconductor materials are light sensitive. That is, the amount of light striking them affects their electrical characteristics. Ordinarily, this photosensitivity is very undesirable in most applications, because the light level is generally uncontrolled. For this reason, most semiconductor components (transistors, diodes, ICs, etc.) are normally enclosed in a light-tight housing made of plastic or metal. This housing also protects the delicate semiconductor crystal from moisture, and stray particles (dust, etc.).

In certain applications, photosensitivity can be highly desirable, or even essential. An obvious example is a light meter. An electronic circuit cannot measure the light level unless it can "see" it.

A number of special semiconductor photosensitive devices (or light sensors) are now available. Some feature PN junctions (like ordinary diodes and transistors), and others are made from a single, junctionless slab of semiconductor crystal. The specific reaction to detected light depends on the construction of the photosensitive component and the specific semiconductor material used.

A junctionless semiconductor light sensor is generally known as a *photocell*. There are two basic types of photocells. They are the *photovoltaic cell* and the *photoresistor*.

Photovoltaic cells

A photovoltaic cell is a simple slab of semiconductor crystal, usually primarily made up of silicon. The semiconductor material is spread out onto a relatively large, thin plate for the greatest possible contact area. The more of the semiconductor material that is exposed to light, the stronger the response to the light will be.

In essence, a photovoltaic cell functions something like a dry cell (battery). This face is indicated in the schematic symbol for the device, shown in Fig. 22-1. Notice that this symbol is similar to the one used to represent an ordinary voltage cell or battery.

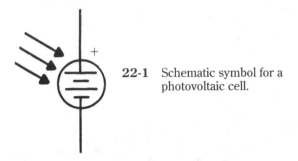

22-1   Schematic symbol for a photovoltaic cell.

When the silicon surface is shielded from light, no current will flow through the cell. But when it is exposed to a bright light a small voltage is generated, due to what is known as the *photoelectric effect*.

If an illuminated photovoltaic cell is hooked up to a load, a current will flow through the circuit. See Fig. 22-2. Just how much current will flow is dependent on the amount of light striking the surface of the cell. The brighter the light, the higher the current available.

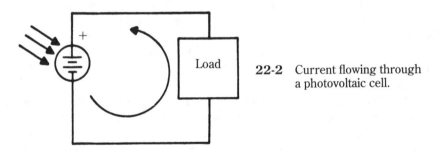

22-2   Current flowing through a photovoltaic cell.

The cell output voltage, on the other hand, is relatively independent of the light level. The voltage produced by most photovoltaic cells is in the range of about 0.5 V. The most obvious use for a photovoltaic cell is as a substitute for an ordinary dry cell. Of course, the 0.5 V output of a single cell is too low for most applications, so a number of photovoltaic cells can be added in series to form a battery, just as with ordinary dry cells. If the circuit you want to power from photovoltaic cells requires more power than your photocells can provide, more cells can be added as in a parallel battery.

You can power almost any dc circuit with a combination of series, parallel, or series-parallel connected photocells. This type of grouping of photocells is often called a solar battery, but it will work just as well under artificial light.

There's one factor that you should keep in mind: the more cells there are in a solar battery, the larger the total surface area will be, and the harder it will be to arrange the cells so they will all be lighted evenly. This means that usually solar batteries are best suited for fairly low power circuits.

A related use for photovoltaic cells is in a recharger for nickel-cadmium batteries. Figure 22-3 shows a typical arrangement. The diode can be almost any standard type, such as a 1N914. The diode is needed to protect the photocells. If the diode wasn't included in the circuit, and the cells were inadvertently darkened, the nickel-cadmium battery could start to discharge through them. This reverse polarity could quickly ruin the photocells.

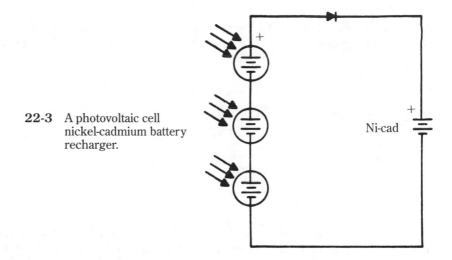

22-3   A photovoltaic cell nickel-cadmium battery recharger.

Ni-cad

Bear in mind that photovoltaic cells, like any other dc voltage source have definite polarity. That is, one lead is always positive and the other is always negative. Never reverse the polarity.

Another frequent application for photovoltaic cells is in light-metering circuits. A very simple light meter is shown in Fig. 22-4. Because the current flowing through a photocell is proportional to the amount of light striking its surface, measuring the current with a milliammeter will give a direct indication of the degree of illumination. The meter scale can be calibrated in whatever units are convenient.

The measurements made with this type of simple circuit won't be very precise, but the circuit is perfectly functional for comparative measurements.

The potentiometer adjusts the sensitivity of the meter. Generally this should be a trimpot that is set once for calibration, then left alone for the measurements. For best results the meter should be as sensitive a unit as possible, but care should be taken that the meter is not so sensitive that a strong light source can slam the pointer off the scale.

A photovoltaic cell can be used to trigger a relay. The basic setup for this is

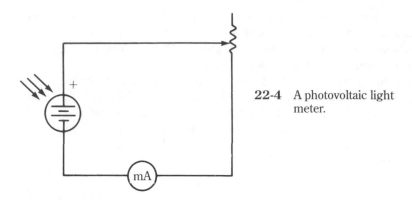

22-4   A photovoltaic light
meter.

shown in Fig. 22-5. Notice that the controlled circuit requires a separate power supply. The voltage generated by the photocell only opens or closes the relay contacts.

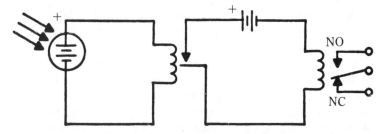

22-5   A light-operated relay.

Because the output of the photocell is fairly small, it can only drive a relatively light duty relay. If the circuit you want to control requires a heavier relay, there are two ways to solve the problem. One method is to use the light-duty relay to control the heavy duty unit as shown in Fig. 22-6. Alternatively, the output of the photocell can be amplified with a transistor, as shown in Fig. 22-7.

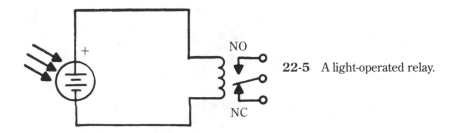

22-6   Using a light-duty relay to drive a heavy-duty relay.

The potentiometer is again used to adjust sensitivity. Notice that both of these methods require an extra voltage source in addition to the photocell and the controlled circuit power supply.

All of these relay circuits respond only to the presence or absence of light. They

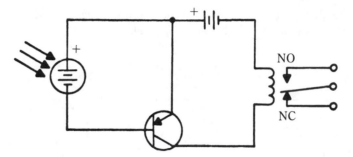

22-7   Using a photovoltaic cell to drive a heavy-duty relay
with the aid of a transistor amplifier.

are not adjustable to trigger on a specific amount of lighting, because a photocell
output voltage is essentially constant.

Photoresistors

Another popular light-sensitive device is the photoresistor, or *light-dependent resistor*.
As the name implies, a photoresistor changes its resistance value based on the level
of illumination on its surface. Photoresistors generate no voltage themselves. Photoresistors are usually made of cadmium sulfide, or cadmium selenide.

Functionally, a photoresistor is a light-controlled potentiometer. The light intensity corresponds to the position of the potentiometer shaft. These devices generally
cover quite a broad resistance range—often on the order of 10,000 to 1. Maximum
resistance—typically about 1 MΩ (1,000,000 Ω) is achieved when the cell is completely darkened. As the light level increases, the resistance decreases.

Because photoresistors are junctionless devices, like regular resistors, they
have no fixed polarity. In other words, they can be hooked up in either direction
without affecting circuit operation in any way. Figure 22-8 shows the schematic symbol for a photoresistor.

22-8   Schematic symbol for a
photoresistor.

Photoresistors are perfect for a wide range of electronic control applications.
They can easily be used to replace almost any variable resistor in virtually any
circuit.

Photoresistors can also be used in many of the same applications as photovoltaic cells, usually with just the addition of a battery and another resistor. They offer
the advantage of being sensitive to different light levels. For example, Fig. 22-9

shows a light-controlled relay circuit. In this circuit, R1 can be adjusted so that the relay switches at any desired light intensity level.

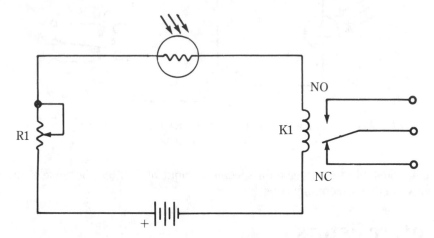

22-9   Light-activated relay using a photoresistor.

Figure 22-10 shows a simple light meter built around a photoresistor. Photoresistors can also be used in many applications where a photovoltaic cell would not be used. Because a photoresistor is a variable resistance, it can be used in place of almost any standard potentiometer in almost any circuit. If appropriate for the application at hand, a photoresistor can be substituted for a fixed resistor.

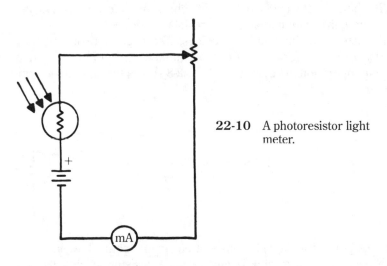

22-10   A photoresistor light meter.

Other photosensitive components

Any standard semiconductor device (with one or more PN junctions) can be made photosensitive, simply by placing a transparent lens in the component's housing so the semiconductor material is exposed to light. A great variety of light-sensitive components are now available. Most of these are light controlled versions of more familiar semiconductor devices.

Figure 22-11 shows the schematic symbol for a photodiode. A phototransistor is shown in Fig. 22-12. Figure 22-13 shows a *light-activated SCR* (*silicon-controlled rectifier*). This last component is commonly known by its acronym, *LASCR*.

22-11 Schematic symbol for a photodiode.

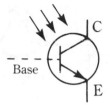

22-12 Schematic symbol for a phototransistor.

22-13 Schematic symbol for a LASCR.

Phototransistors are especially useful in a large number of applications because they can be used as amplifiers whose effective gain is controlled by light intensity. Usually (though not always) the base lead is left unconnected in the circuit, the base-collector current being internally generated by the photoelectric effect. Many phototransistors don't even have an external base lead. Many circuits that use standard bipolar transistors could be adapted and rebuilt with phototransistors. This could result in some very unique effects. Of course, the base lead could be used too, so the output would depend on both the signal on the base lead, and the light intensity.

An *optoisolator* is another extremely useful device. As the name implies, it isolates two interconnected circuits so that their only connection is optical.

In an earlier chapter, you read about a rudimentary optoisolator can be made from two LEDs. Most practical optoisolators, however, consist of a LED and a photocell, or phototransistor encapsulated in a single, light-tight package. The schematic symbols are shown in Fig. 22-14.

The LED is wired into the controlling circuit, and the photocell is wired into the circuit to be controlled. This provides a convenient means of control with virtually no undesirable crosstalk between the two circuits. Essentially the same effects can be achieved with a separate LED and photoresistor (or phototransistor), but they must be carefully shielded from all external light to prevent uncontrolled interference.

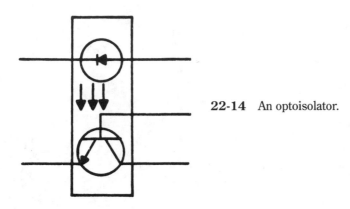

22-14   An optoisolator.

Self-test

1. What is the name for a component that changes its resistance in response to light intensity?

   A Photovoltaic cell
   B Photoresistor
   C Potentiometer
   D Photopotentiometer

2. What is the nominal voltage of a photovoltaic cell?

   A 0.5 V
   B 1.0 V
   C 1.5 V
   D Varies in proportion to the size of the cell
   E None of the above

3. What is the name for the property used in light sensitive devices?

   A Piezoelectric effect
   B Solar Energy
   C LASCR
   D Photoelectric effect
   E Photoresistance

4. Which of the following is not a light-sensitive transducer?

A Photodiode
B LASCR
C Photoresistor
D SCR
E None of the above

5. Which type of material is most likely to be photosensitive?

A Conductors
B Semiconductors
C Insulators

6. Which lead is often omitted in a phototransistor?

A Collector
B Emitter
C Base
D Gate
E None of the above

7. Why are ordinary transistors not sensitive to light?

A Different materials are used in their construction
B They only have two PN junctions
C They are enclosed in light-tight housings
D Different biasing is used
E None of the above

8. How many PN junctions does a photoresistor have?

A One
B Two
C Three
D None
E More than three

9. What semiconductor material is most often used in photovoltaic cells?

A Silicon
B Cadmium-sulfide
C Gallium-arsenic
D Cadmium-selenide
E None of the above

10. Which of the following performs a similar function to that of a photovoltaic cell?

A Capacitor
B dc battery
C Inductor
D LASCR
E None of the above

23

Experiments 2

Table 23-1 lists the equipment and parts you will need to perform the experiments in this chapter.

**Table 23-1. Equipment and
parts needed for the experiments.**

VOM
dc voltmeter, ac voltmeter, ohmmeter

Breadboarding system
Solderless socket, dc power supply (6V
and 9V, 6.3V ac power supply

Extra 3V Power Supply
2, 1.5-V dry cells connected as a series
battery

Components
1	100 Ω resistor (brown-black-brown)
2	1000 Ω resistors (brown-black-red)
2	10,000 Ω resistors (brown-black-orange)
1	100,000 Ω resistor (brown-black-yellow)
1	470,000 Ω resistor (yellow-violet-yellow)
1	10,000 Ω potentiometer
1	1N914 diode (or 1N4148)
1	3.6V zener diode (such as 1N747)
1	Red LED
1	Seven-segment, Common-cathode LED display
1	2N3904 npn transistor (or Radio Shack RS-2016)
1	2N306 pnp transistor (or Radio Shack RS-2034)
1	2N4350 FET (or Radio Shack RS-2035)

Experiment 1 dc resistance of a diode

Connect a 1N914 diode to an ohmmeter as shown in Fig. 23-1A. The negative ohm-meter lead (usually black) is attached to the end of the diode marked by a colored band. The ohmmeter positive lead (usually red) is connected to the other end of the diode.

Now read the resistance on the meter. You should get about 4500 Ω (4.5 kΩ). You might get a somewhat different resistance reading, but it should be fairly low (under 10,000 Ω). This indicates that the diode is conducting. It is forward biased.

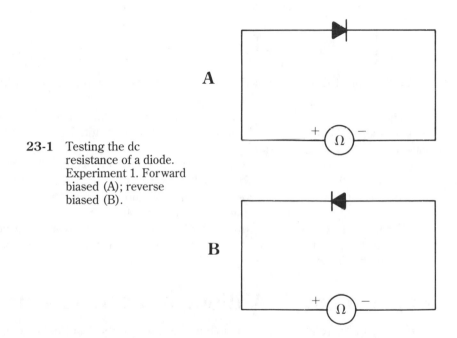

23-1 Testing the dc resistance of a diode. Experiment 1. Forward biased (A); reverse biased (B).

Now, reverse the ohmmeter leads on the diode, as shown in Fig. 23-1B. Now the banded end of the diode should be connected to the positive lead (red) of the ohmmeter, and the unbanded end is connected to the negative lead (black). In this case the meter pointer should scarcely budge from the full scale, infinity position. The diode is reverse biased.

Remember, a diode has a low internal resistance when it is forward biased and an extremely high internal resistance when it is reverse biased.

If you got a fairly low resistance reading in both directions, the diode is *shorted*. If a very large, or infinity reading is found in both directions, the diode is *open*. In either case, the component is defective, and therefore, useless. It should be discarded.

You can see that this method can be used to check diodes for defects with an ohmmeter. This test can also be used to determine the polarity of an unmarked diode.

Experiment 2 A diode in a dc circuit

Breadboard the circuit shown in Fig. 23-2A with a 1N914 diode and a 1000 Ω (1 kΩ) resistor (brown-black-red).

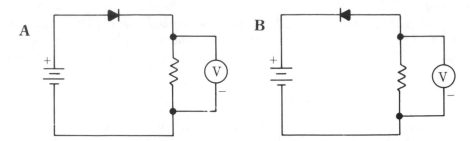

23-2   A dc diode circuit. Experiment 2. Forward biased (A); reverse biased (B).

The voltmeter should indicate a little over 5 V (about 0.7 V is dropped by a typical diode). Clearly, current is flowing through the circuit. The diode is forward biased.

The exact voltage reading you get is not terribly important. The key point is that current is flowing through the circuit. Now, reverse the polarity of the diode so that the circuit resembles Fig. 23-2B. Because the diode is now reverse biased, you should read very little, or no voltage across the resistor.

Virtually no current will flow through a circuit with a reverse biased diode. Some might leak through, giving you a small voltage drop across the resistor, but this value should be negligible.

Experiment 3 A diode in an ac circuit

Notice that the circuit in Fig. 23-3 is similar to the one used in the last experiment, except a 6.3 Vac source is used to power the circuit. You should read about a 3 V drop across the diode. Reversing the polarity of the diode should give you approximately the same voltage.

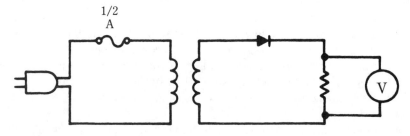

23-3   An ac diode circuit. Experiment 3.

No matter which way you connect the diode, current will flow through the circuit for half of each ac cycle. This fact is true because, regardless of the direction of the diode, the polarity is correct for one half cycle of each wave.

Now, repeat the experiment, and measure the *dc* voltage across the 1000 Ω (1 kΩ) resistor. When the diode is connected as shown in Fig. 23-3, you should read about +3 V dc. This is assuming the negative lead of the ohmmeter (black) is attached to the connection of the diode and the resistor. This means the diode conducts during the positive half of each ac cycle.

If the diode is reversed, however, current will flow only during the negative half cycles. The voltage should be about the same as before, but with reversed polarity. That is, there is about −3 V dropped across the resistor. With most voltmeters, you'll have to reverse your meter leads to read this voltage.

Experiment 4
Testing a zener diode with an ohmmeter

Repeat Experiment 1 with a 3.6 V zener diode, such as a 1N747. With the banded end of the diode connected to the negative lead (black) of the ohmmeter, you should get a fairly low resistance (about 4500 Ω—4.5 kΩ).

Reverse biasing the diode should result in a somewhat higher resistance reading, but the difference will probably not be as dramatic as with the 1N914 diode in Experiment 1. You should get a reverse-bias resistance of about 150,000 Ω (150 kΩ). The exact reading you get will depend on the voltage source of the ohmmeter you are using.

Experiment 5 Using a zener diode

Breadboard the circuit shown in Fig. 23-4. R2 and R3 should each be 1000 Ω (1 kΩ) (brown-brown-red). R1 is a 10,000 Ω (10 kΩ) potentiometer. Observe what happens to the voltage dropped across R3 as the source voltage is varied via the potentiometer.

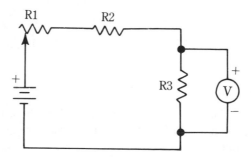

23-4   An unregulated circuit. Experiment 5.

Now, add your 3.6 V zener diode to the circuit, as shown in Fig. 23-5. Again vary the resistance of R1 and watch what happens to the voltage drop across R3. Notice that the voltage never goes beyond the 3.6 V rating of the zener diode. Any voltage above this level is routed to ground through the diode.

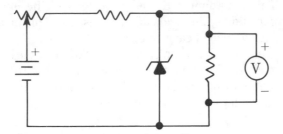

23-5   The circuit from Fig. 23-4 with zener diode regulation. Experiment 5.

Experiment 6   LEDs

Repeat Experiments 1 and 2 but use a light-emitting diode instead of the 1N914 diode. Notice that the resistance readings for an LED resemble those of a conventional diode, although the resistance of a forward-biased LED might be somewhat higher (perhaps about 200,000 Ω—200 kΩ).

Also, depending on the voltage of your ohmmeter, the LED might or might not light up when it is forward biased. It should not be lit when reverse biased.

For the in-circuit tests, use the circuit shown in Fig. 23-6. The resistor is essential to prevent excessive current from flowing through the LED, and quite possibly damaging it. Use a 1000 Ω (1 kΩ) resistor (brown-black-red) for this. The source voltage must be no more than 6 V.

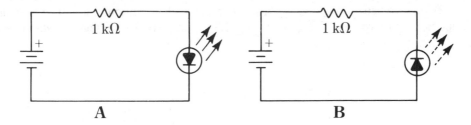

23-6   An LED test circuit. Experiment 6.

When the diode is forward biased, it should glow fairly brightly. When the polarity of the LED is reversed (that is, when it is reverse biased) as in Fig. 23-6B, the LED should remain dark.

Experiment 7
A seven-segment LED display

Breadboard the circuit shown in Fig. 23-7 with a common-cathode, seven-segment LED display. Connect each of the segment leads to point A, one at a time, and notice the effect.

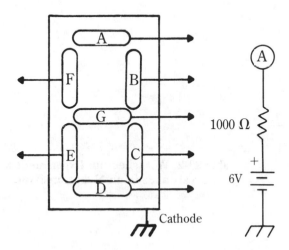

23-7   Seven-segment LED display test circuit. Experiment 7.

Connect both segments B and C to point A and see what number is displayed. Now connect segments B, C, F, and G to point A and observe the result.

Next connect segments B, C, E, F, and G (that is, all segments, except A and D) to point A and see what is fit on the display.

In these last three steps you should have seen (in this order) a 1, a 4, and an *H*.

Try connecting various other combinations of segments to point A and watch what happens to the display. See how many different numerals and letters you can form.

Experiment 8
Testing transistors with an ohmmeter

For purposes of testing, a transistor can be thought of as two back-to-back diodes, as shown in Fig. 23-8.

Measure the various internal resistances of a 2N3904 NPN transistor, and enter your results in Table 23-2. You will be making a total of six measurements; remember, measure each pair of leads with both polarities.

Now repeat the six measurements on a 2N3906 PNP transistor and enter your results in Table 23-3.

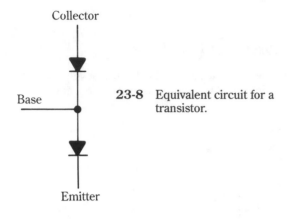

23-8   Equivalent circuit for a transistor.

+	E	B	C
−			
E	X		
B		X	
C			X

Table 23-2. Worksheet for Experiment 8— dc resistance in an NPN transistor.

+	E	B	C
−			
E	X		
B		X	
C			X

Table 23-3. Worksheet for Experiment 9— dc resistance in a PNP transistor.

Now compare Table 23-2 with Table 23-3. Notice how they are essentially mirror images of each other. The measurements should all be approximately the same, but with the polarities reversed.

Experiment 9 A common-base amplifier

Construct the circuit shown in Fig. 23-9. This is a common-base amplifier built around an NPN transistor.

Notice that one lead of the voltmeter is left unconnected in the schematic. If this lead is attached to point A, the meter will display the input voltage. If the lead is moved to point B, the output voltage will be indicated.

Connect the free voltmeter lead to point A and adjust the potentiometer until the input signal is exactly one V.

Now, move the positive voltmeter lead to point B and carefully measure the output voltage without touching the potentiometer. Enter this value in the appropriate space in the NPN column of Table 23-4.

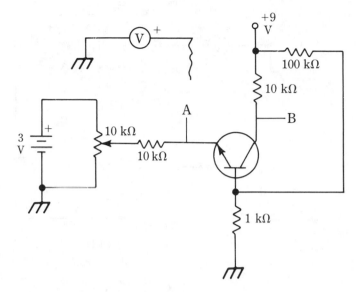

23-9 An NPN common-base amplifier. Experiment 9.

Table 23-4. Worksheet for Experiment 9—A common-base amplifier.

Input voltages	Output voltages NPN	PNP
1 V	_____	_____
1.5 V	_____	_____
2 V	_____	_____
2.5 V	_____	_____

Return the voltmeter lead to point A and repeat the above procedure for each input voltage given in Table 23-4.

Now examine the table carefully and notice how the output voltage varies in step with the input voltage.

Now, build the circuit shown in Fig. 23-10. This amplifier is essentially the same amplifier circuit as before, but it is now built around a PNP transistor. Notice the changes in polarity.

Repeat the entire experiment with this new circuit and enter the output voltages in the appropriate spaces in the PNP column of Table 23-4.

When you are finished compare the NPN column with the PNP column. The values listed in each column should be quite similar. There will perhaps be some minor variations due to component tolerances.

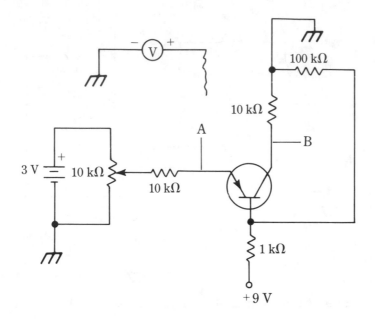

23-10   A PNP common-base amplifier. Experiment 9.

Experiment 10
Common-emitter amplifier

Repeat the procedure of Experiment 9 with the common-emitter amplifier circuit shown in Fig. 23-11. Enter your results into Table 23-5.

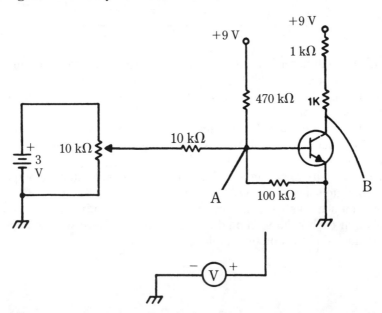

23-11   An NPN common-emitter amplifier. Experiment 10.

For this experiment you will just use the NPN transistor, because you know the PNP transistor would give essentially the same results once all the circuit polarities are reversed.

Table 23-5. Worksheet for Experiment 10—A common-emitter amplifier.

Input voltage	Output voltage
1 V	_____
1.5 V	_____
2 V	_____
2.5 V	_____

Experiment 11
Common-collector amplifier

Repeat the procedure of Experiment 9 again, this time using the circuit shown in Fig. 23-12. Enter your results in Table 23-6. Once again, you are using only the NPN version of the circuit. A PNP common-collector amplifier would produce the same sort of results.

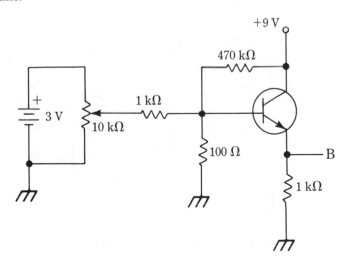

23-12 An NPN common-collector amplifier. Experiment 11.

Experiment 12 A field-effect transistor

Measure the resistance between the various leads of a 2N4350 FET (field-effect transistor) and enter your results into Table 23-7.

Construct the circuit given in Fig. 23-13 and repeat the procedure of Experiment 9 one more time, entering your results into Table 23-8.

Compare the output voltages of each of the four basic amplifier types: common-

**Table 23-6. Worksheet for
Experiment 11—A common-
collector amplifier.**

Input voltage	Output voltage
1 V	_____
1.5 V	_____
2 V	_____
2.5 V	_____

**Table 23-7. Worksheet for Experiment 12—
dc Resistance in a FET.**

+	S	D	G
− S	X		
D		X	
G			X

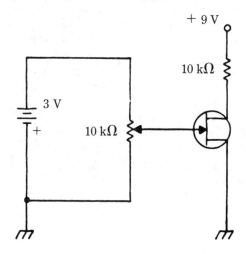

23-13 A FET (field-effect transistor)
amplifier. Experiment 12.

**Table 23-8. Worksheet for
Experiment 12—A FET amplifier.**

Input voltage	Output voltage
1 V	_____
1.5 V	_____
2 V	_____
2.5 V	_____

base (Table 23-4), a common-emitter (Table 23-5), common-collector (Table 23-6) and FET (Table 23-8).

Which basic amplifier has the greatest amount of gain? Which has the least? Can you explain why? If not, go back and re-read chapters 18, 19 and 20.

24

Linear integrated circuits and op amps

As electronics technology advances, physical circuit size decreases. Many modern devices would be impracticably expensive and physically unwieldy if not for recent advances in miniaturization. The switch from tubes to transistors was the first major step forward in this direction. But where miniaturization really came into its own was with the invention of *integrated circuits* (abbreviated as *ICs*).

Integrated circuits

Integrated circuits are made up of a series of silicon wafers that are specially treated to simulate a number of separate transistors, diodes, resistors and capacitors. A tiny package, that is smaller than a dime can take the place of a couple dozen or more (discrete) components.

Integrated circuits are called chips, or IC chips because the internal circuitry is etched onto a chip of semiconductor material (usually silicon). There are several levels of integration possible, depending on the complexity of the circuit simulated by the IC. The standard levels of integration are as follows, from lowest to highest:

SSI	Small-scale integration
MSI	Medium-scale integration
LSI	Large-scale integration
VLSI	Very large-scale integration

SSI (small-scale integration) devices are relatively simple circuits that could be built around a handful of discrete components. Still, a SSI IC represents a significant reduction in circuit size and usually in cost. SSI ICs are used as basic building blocks in more complex systems. They are the most commonly used type of integrated circuits. As a rough rule of thumb, an SSI chip has fewer than 100 on-chip components, including transistors and solid-state equivalents of passive components such as resistors and capacitors.

LSI (large-scale integration) devices include much more complicated and sophisticated internal circuitry. Sometimes a single integrated circuit can take the place of hundreds or even thousands of ordinary discrete components. LSI chips generally have between about 10,000 and 100,000 on-chip components.

LSI ICs are designed for specific, specialized functions and usually can't be used in many other applications. If, for example, you open up a typical pocket calculator, you will find a large IC, the display, the key switches, and very few (if any) external components. Virtually all of the electronic circuitry is contained within the IC. But it would probably be difficult, if not impossible to use this IC in anything but a calculator circuit. It is designed for one specific application, and that's all it can do. This is an example of an LSI device.

Between these two extremes (SSI and LSI) are MSI (medium-scale integration) devices. An MSI IC, as you might suspect, is more complex than an SSI unit, but not as complex as an LSI device. MSI ICs typically contain the solid-state equivalent of about 100 to 1000 discrete components. MSI ICs are usually designed to perform some specific function within a larger system. For example, an MSI chip might contain a preamplifier circuit or a voltage regulator circuit which can be used as stages in a much larger electronic system.

Recently, improved technology has permitted semiconductor manufacturers to develop VLSI (very large-scale integration) ICs. With more than 100,000 on-chip components. VLSI ICs are even more sophisticated and specialized than LSI chips. For example a CPU (central processing unit) is the thinking part of a computer (see chapter 29). Most personal computers use an LSI IC for the CPU. A VLSI chip could contain virtually the entire computer, including the memory and input/output ports, in addition to the CPU itself.

The first commercial IC appeared in 1961, and included four on-chip bipolar transistors. A LSI device, such as Motorola's 68000 CPU IC might contain up to 65,000 to 70,000 on-chip transistors. The VLSI chips now being designed will have as many as 250,000 transistors.

These advances are due to increasing capability for greater control, allowing finer detail. SSI ICs in the early 1970s had lines of about 20 micro-meters (millionths of a meter) wide. Current LSI devices have line widths as small as 3 or 4 micro-meters. Some experimental VLSI designs are shooting for line widths of 0.5 micro-meters.

Integrated circuits come in a number of standardized package types. Some are enclosed in round plastic, or metal cans. These devices look rather like slightly oversized transistors, except they often have ten or twelve leads, instead of just three.

Most modern integrated circuits, however, are housed in *dual-inline packages*, or *DIPs*. The integrated circuits shown in Fig. 24-1 are standard DIPs. These rectangular plastic packages have two parallel rows (or lines) of leads, hence the name. DIPs are most commonly found with 8, 14, or 16 pins. Figure 24-1 shows how these pins are numbered. There will be a notch or circle etched into one end of the casing to identify pin number one. Some MSI and LSI devices might be in DIPs with 24, 28, 40, or even more pins.

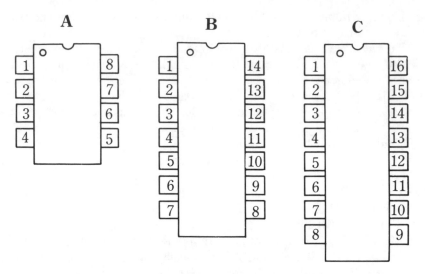

24-1   Typical IC (integrated-circuit) pin numberings 8-pin DIP (dual inline package) (A); 14-pin DIP (B); 16-pin DIP (C).

Most integrated circuits are shown in schematic diagrams simply as boxes. The leads are numbered for identification, but they don't necessarily have to be drawn in numerical order. In the schematic, the leads can be arranged in any convenient pattern for the greatest clarity. See Fig. 24-2.

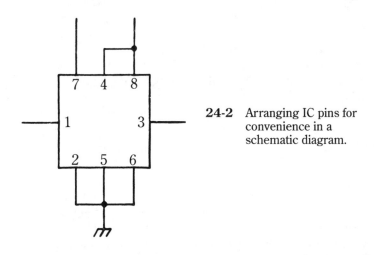

24-2   Arranging IC pins for convenience in a schematic diagram.

Certain types of ICs have special schematic symbols that indicate their function. For example, an amplifier is often drawn as a triangle (see Fig. 24-3). These special symbols are presented as needed.

Because of the many closely spaced pins, soldering and desoldering integrated circuits can be a problem. Like all semiconductors, too much heat can destroy an integrated circuit.

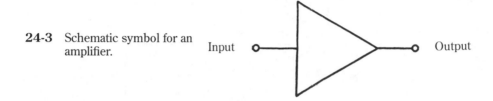

24-3 Schematic symbol for an amplifier.

Input ○   ○ Output

Use plug-in sockets whenever using integrated circuits. In some cases, the socket might actually cost more than the IC that is to be plugged into it. But the frustration and the time required to replace a defective integrated circuit makes the use of sockets good insurance.

New integrated circuits are being developed every day—especially MSI and LSI devices. So it would really be beyond the scope of this book to discuss them all. In this book you find discussions of only the most common types of basic building block devices. These, for the most part, will be SSI units.

There are two major categories of integrated circuits. In this chapter you will be considering *linear*, or *analog* circuits. In another chapter you can examine *digital* ICs. An analog circuit responds to an input signal with some output signal that varies in some (not necessarily direct) proportion to the input signal.

An amplifier is a perfect example of an analog circuit. The output signal is a larger version of the input signal.

If the relationship between the input and output signals is constant (for example, an amplifier might have an output that is 10 times the amplitude of the input signal, regardless of the frequency or wave shape of the input signal) then the circuit or device is said to be linear.

A filter is an example of an analog circuit that is not, strictly speaking, linear. Still virtually all analog integrated circuits are commonly called linear ICs.

Op amps

Probably the most common type of linear integrated circuit is the *operational amplifier*, or *op amp*. Figure 24-4 shows a simplified circuit diagram of a typical operational amplifier, built around discrete components. A practical operational amplifier circuit would require many more components than the basic ones shown here, so integrated circuit op amps are obviously vastly preferable to discrete devices.

There are a large number of IC op amps available. One of the most popular, and least expensive is the 741. Sometimes this number is preceded or followed by key letters that identify the manufacturer, the case style, the temperature range, and so forth. For most applications, these letters can be ignored.

This integrated circuit is available in several different package configurations, but 8-pin and 14-pin DIPs predominate. The *pin-out diagrams* for these packages are shown in Figs. 24-5 and 24-6. The basic specifications of the 741 are listed in Table 24-1.

The standard schematic symbol for an operational amplifier is shown in Fig. 24-7. Notice that there are two voltage supply inputs: +V and −V. These voltages

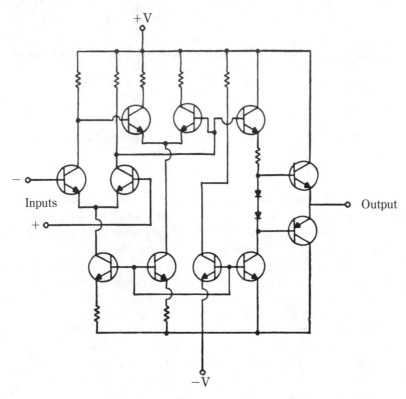

24-4 Simplified circuit diagram for an op amp (operational amplifier).

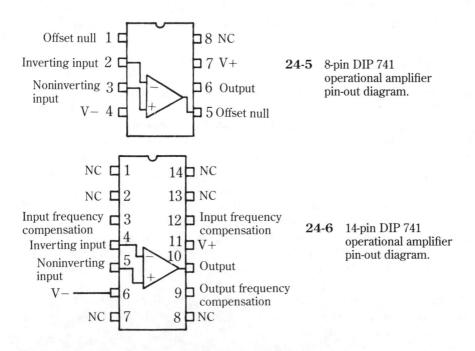

24-5 8-pin DIP 741 operational amplifier pin-out diagram.

24-6 14-pin DIP 741 operational amplifier pin-out diagram.

Table 24-1.
Specifications for a 741 op amp.

Maximum gain	200,000
Input offset voltage	1.0 mV
Input offset current	20 nA
Input bias current	80 nA
Output resistance	75 Ω
Common-mode rejection	90 dB
Slew rate	0.5 V/μS

All specifications are for a ± 15 V power supply.

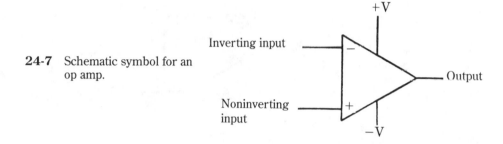

24-7   Schematic symbol for an op amp.

should be equal, but of opposite polarity with respect to ground. See Fig. 24-8. The power supply connections are often omitted from schematic diagrams, because like the filament circuit of a tube, these power supply connections are automatically assumed by the presence of the IC. Remember, each and every integrated circuit package must have the appropriate power supply connections to function.

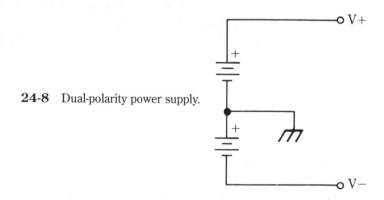

24-8   Dual-polarity power supply.

Returning to the schematic symbol (Fig. 24-7), notice that there are two inputs to an operational amplifier. One is marked with a plus (+). This is called the *noninverting input*. The output will be in phase with an input signal at this terminal. The other input is marked with a (–). This input is called the *inverting input*. The output signal will be 180 degrees out of phase with an input signal at this terminal. In other words, the signal is inverted.

There are almost countless applications for this basic device—certainly far too many to be dealt with in depth here. You can read about a few of the many applications for an operational amplifier. The two simplest applications, of course, are inverting and noninverting amplifiers.

Inverting amplifiers

Figure 24-9 shows the basic circuit of an inverting amplifier. The output is 180 degrees out of phase with the input. When the input signal increases (goes more positive), the output voltage decreases (goes more negative), and vice versa.

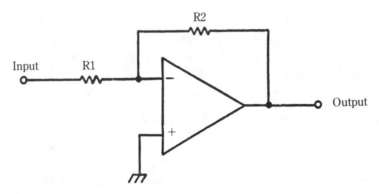

24-9   An inverting amplifier circuit.

Resistor R2 returns some of the output signal back to the input. Because the output is 180 degrees out of phase with the input, this returned signal will subtract from the input signal applied to the operational amplifier itself, reducing the effective gain of the amplifier. You'll understand why this is necessary shortly. The process of using the output to cancel out some of the input is called *negative feedback*.

The amount of gain (*G*) produced by the amplifier will be determined by the ratio of the resistance of R1 to that of R2. Specifically, the formula is:

$$G = -\frac{R_2}{R_1}$$ **Equation 24-1**

The negative sign is simply a mathematical indication that the signal is inverted by the circuit. Assume R1 is 10,000 Ω. If R2 has a value of 100,000 Ω, the gain will be equal to −100,000/10,000 or −10.

If R2 is increased to 1 MΩ (1,000,000 Ω), the gain increases to −1,000,000/10,000 or −100.

What happens if R2 is increased to infinity? (That is, if the feedback path is removed altogether.) The theoretical gain would be −∞/10,000 or simply infinite gain. (∞ is a symbol used to represent an infinite quantity.)

In actual circuits, infinite gain is impossible. The upper limit of gain is determined by the internal characteristics of the op amp itself. The maximum gain is given in the specification sheet for the IC. For the 741 the maximum gain can be up to 200,000, but you'll almost never be able to use that much gain. With no feedback circuit, an inverting operational amplifier will theoretically amplify a 1 mV (0.001 V) signal to −200,000 mV, or − 200 V at the output. But the output of an operational amplifier (or any amplifier, for that matter) is always limited to the power supply voltage. If the input signal exceeds the value that will produce an output equal to the source voltage, there will be no further change in the output. The amplifier is said to be *saturated*. Saturation is why the feedback circuit must be used in practical circuits.

Another unique condition occurs when R1 and R2 are of equal value. For example, if both are 10,000 Ω, the gain works out to −10,000/10,000 or simply −1. In other words, the output equals the input except, of course, for the phase inversion. This is called *unity gain*.

An amplifier circuit with unity gain is called a *voltage follower*, because the output signal follows, or duplicates the input signal. This might not seem very useful, but sometimes it can come in quite handy for impedance matching and for *buffering*.

A *buffer amplifier* prevents a later circuit from *loading* (putting an excessive current drain on) the signal source. In other words, various subcircuits within a system can be effectively isolated from each other via a buffer amplifier. Also, at times the phase inversion is necessary, but additional amplification is undesirable. In such a case a unity gain inverting amplifier is the obvious solution. All in all, a voltage follower is actually a useful (and frequently used) circuit.

Reducing the value of R2 to below that of R1 will result in less than unity gain. The output level will be less than the input amplitude. For instance, if R1 was left at 10,000 Ω, and R2 was reduced to 3,000 Ω, the gain would be −3000/10,000 or −0.3. This would rarely be very useful. In fact, by reducing R2 to zero, you can build an amplifier with no output at all. ($G = -0/10,000 = 0$. Output = input × G = input × 0 = 0.) It is doubtful that anybody could find a practical use for that!

Noninverting amplifier

Figure 24-10 shows a circuit similar to the one above, except the noninverting input is used rather than the inverting input. But the feedback is still applied to the inverting input. The output will be in phase with the input, so the circuit is called a *noninverting amplifier*.

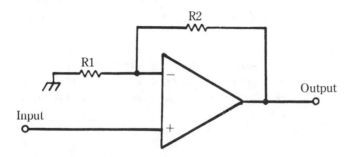

24-10   A noninverting amplifier circuit.

Why is the feedback applied to the inverting rather than the noninverting input? To find out why, assume you have a voltage gain of 200,000 with no feedback circuit. Ignoring the practical problems discussed in this last section, a 1 mVdc input would produce a 200 Vdc output.

Now, if you add a feedback resistor to return part of the output to the noninverting input for another pass through the amplifier (see Fig. 24-11), what will happen? Assume the resistance of the feedback resistor drops 195 V, so only 5 V is applied back to the input.

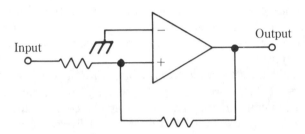

24-11   A noninverting amplifier with positive
feedback (a nonfunctional circuit).

Now, you have two voltages being applied to the input of the operational amplifier. The original 1 mV (0.001 V) signal, and the 5 V feedback signal. (Of course, this is quite a large feedback signal—this is to dramatize the effects.) In other words, the effective input signal is now 5.001 V. This signal is amplified by the full gain of the operational amplifier (that is, 200,000). The output jumps to 1,000,200 V. And some of this is fed back through for additional amplification. Obviously the output level would simply continue skyrocketing towards infinity.

The feedback signal must be out of phase with the original input signal, so that it will subtract some of the input and reduce the effective gain. This can be accomplished by routing the feedback signal to the inverting input of the op amp.

If, for example, the original signal by itself produces a 10 V output, and the feedback, by itself produces a 2 V output, these two values would subtract because they are 180 degrees out of phase with each other. The total effective voltage would be 8 V, in phase with the original input signal.

Of course, you can only measure the total effective output voltage. There is no way to measure the result of the input signal and the feedback signal separately.

Notice that most of the feedback voltage is dropped by R2 and shunted to ground via R1. Obviously in practical amplification circuits the negative feedback signal should be of a lower amplitude than the input signal you want to amplify.

In all other respects, the noninverting amplifier circuit operates in the same way as the inverting amplifier. Even the formula for determining the circuit gain remains almost the same:

$$G = 1 + \frac{R_2}{R_1}$$
                                               Equation 24-2

Integration

What if the feedback component was something other than a resistor? What if R2 in a basic inverting amplifier was replaced with a capacitor, as shown in Fig. 24-12. This circuit is called an *integrator*. Examine how it works.

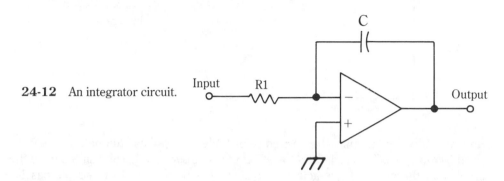

24-12 An integrator circuit.

Obviously, the gain would vary with frequency, because the capacitor's reactance is frequency dependent. For example, assume R1 is 10,000 Ω, and C is 0.01 µF (1 $\times$ 10⁸ F). At any specific frequency the gain would be determined by the formula:

$$G = \frac{-X_c}{R1}$$

Equation 24-3

Because you know $X_c = 1/(2\pi FC)$, you can rewrite the formula as:

$$G = \frac{-1}{2\pi FC} \times \frac{1}{R1}$$

Equation 24-4

This means that for the sample circuit, the gain equals $(-/(6.28 \times F \times 1 \times 10^{-8})) \times 1/10,000 = 1/(6.28 \times 10^{-8}) \times 1/F \times 0.001 = 1592.3567 \times 1/F$, or about $-1600/F$. Using this formula, you find that if a 60 Hz signal is applied to this circuit, the gain will be equal to $-1600/60$ or approximately -27.

Similarly, if the input frequency is 500 Hz, the gain becomes equal to $-1600/500$, or about -3.2.

At an applied frequency of 2000 Hz, the gain drops to $-1600/2000$ or approximately -0.8. As you can see, at this frequency (and at higher frequencies) the output is less than the input.

As the applied frequency increases, the reactance of the capacitor in the feedback circuit decreases, canceling out more and more of the input signal at higher frequencies.

Figure 24-13 is a chart showing the frequency response of this integrator circuit. Notice that it closely resembles the frequency response graph for a simple lowpass filter. In fact, an integrator is an *active* lowpass filter.

The term *active* in this context means that the circuit amplifies rather than simply attenuating. A simple RC filter will subtract some signal at all frequencies (loss).

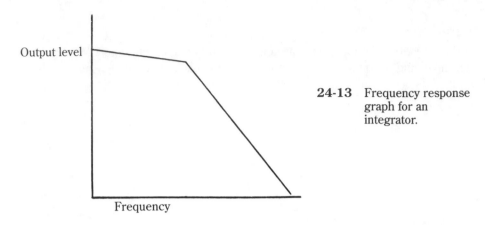

24-13 Frequency response graph for an integrator.

An active filter will amplify the desired frequencies, as well as attenuating the undesired frequencies. This produces the effect of a sharper cutoff. You can tell from the graph that the slope of this active filter is steeper than that of the simple passive filter discussed earlier. Lowpass filtering is often called *integration*. The terms are generally interchangeable.

Differentiator

The opposite of integration is *differentiation*. In a differentiator circuit, such as the one shown in Fig. 24-14, resistor R1 is replaced by a capacitor rather than R2. This means the equation for determining gain becomes:

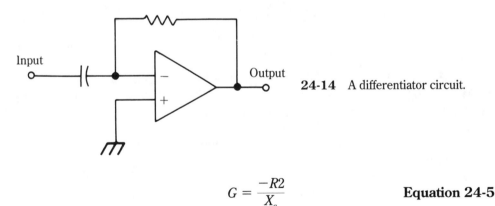

24-14 A differentiator circuit.

$$G = \frac{-R2}{X_c}$$ **Equation 24-5**

This can be algebraically rewritten as:

$$G = -2\pi FCR_2$$ **Equation 24-6**

With the capacitor in the input line, low frequencies are blocked before they get a chance to reach the operational amplifier itself. The higher frequencies are passed and amplified like an ordinary inverting amplifier.

Assume R2 is 10,000 Ω, and C is 0.01 μF ($1 \times 10^{-8}F$). Plugging these values into the equation you find that $G = 6.28 \times F \times 1 \times 10^8 \times 10,000 = -0.00628 \times F$.

This means the gain for a 60 Hz input would be equal to −0.00628 × 60, or about −0.4. There would be less than unity gain at 60 Hz.

If the input frequency is increased to 500 Hz, the gain becomes equal to −0.00628 × 500, or just over −3. Raising the input frequency to 2000 Hz, will increase the gain to − 0.00628 × 2,000, or about − 13. Obviously a differentiator is an active high pass filter.

Difference amplifier

If different signals are applied to each of the inputs of an operational amplifier with unity gain, the output will be equal to the difference between the input signals.

The circuit shown in Fig. 24-15 is a typical *difference amplifier.* Suppose 5 mVdc (0.005 V) is applied to the inverting input and 8 mVdc (0.008 V) is applied to the noninverting input of this circuit. What will the output be?

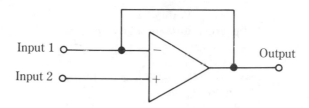

24-15 A difference-amplifier circuit.

First, ignoring the noninverting input, the output from the inverting input alone would be −5 mV (–0.005 V). Similarly, if you ignore the inverting input, the output from just the noninverting input would be + 8 mV (0.008 V).

Combining these two output signals, you have + 8 mV, and −5 mV, or an effective total of +3 mV—the difference between the two inputs.

It's also possible to build a difference amplifier with gain, although it isn't commonly done. If the difference amplifier had a gain of 5, the same 5 mV (inverting) and 8 mV (noninverting) inputs the output would be +15 mV (0.015 V).

There are almost countless other applications for operational amplifiers. It is unquestionably the most versatile kind of integrated circuit available.

Other op amps

The 741 isn't the only type of integrated circuit operational amplifier, though it is the most popular. See Table 24-1.

Some additional commonly used op amp devices are the 709 (see Fig. 24-16 and Table 24-2) and the 748 (see Fig. 24-17 and Table 24-3). These devices work in the same way as the 741, but with slightly different characteristics.

Another popular integrated circuit is the 747 (see Fig. 24-18) that includes two complete 741 operational amplifiers in a single 14-pin DIP package. Notice that the negative voltage supply terminal (pin 4) is shared by both op amps, but they each have a separate positive voltage supply terminal (pins 9 and 13). There must be the proper voltage on the appropriate pins for either op amp to function.

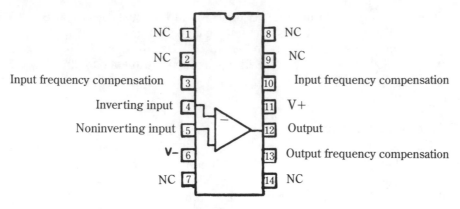

24-16   Pin-out diagram for a 709 op amp.

Table 24-2.
Specifications for a 709 op amp.

Maximum gain	45,000
Input offset voltage	1.0 mV
Input offset current	50 nA
Output resistance	150 Ω
Common-mode rejection	90 dB
Slew rate	0.25 V/μS

All specifications are for a ± 15-V power supply.

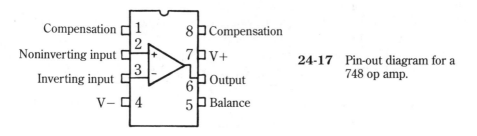

24-17   Pin-out diagram for a 748 op amp.

Table 24-3.
Specifications for a 748 op amp.

Maximum gain	200,000
Input offset voltage	1.0 mV
Input offset current	20 nA
Input bias current	80 nA
Output resistance	75 Ω
Common-mode rejection	90 dB
Slew rate	0.5 V/μS

All specifications are for a ∓ 15 V power supply.

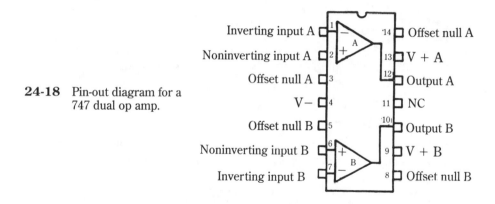

24-18 Pin-out diagram for a 747 dual op amp.

The 324 (Fig. 24-19) goes a step further. It consists of four complete 741 type operational amplifiers in a single package. In addition to saving space and costs, this particular integrated circuit can operate off a single polarity supply. Most operational amplifiers require both a positive and a negative voltage source with respect to ground.

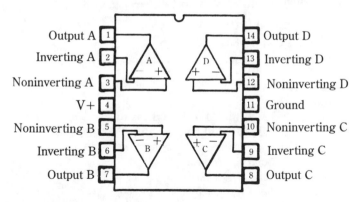

24-19 Pin-out diagram for a 324 quad op amp.

The Norton amplifier

Quite similar to the operational amplifiers you have been discussing, is the *Norton amplifier*. Its schematic symbol is shown in Fig. 24-20. Where ordinary operational amplifiers amplify the difference in the voltages applied to their inverting and noninverting inputs, the Norton amplifier amplifies the difference in applied currents. This device is also sometimes called a *current mirror*.

The Norton amplifier can be used in most op amp applications with little or no change in the basic circuitry, but this device requires only a single polarity power supply. It also has a slightly lower input impedance and a much higher output impedance than an ordinary operational amplifier. This might or might not be desirable, depending on the specific application.

The most common Norton amplifier IC is the LM3900, which contains four

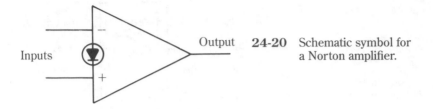

Output **24-20** Schematic symbol for
a Norton amplifier.

separate Norton amplifiers in a single package. The pin-out diagram for the LM3900 is shown in Fig. 24-21, and its specifications are listed in Table 24-4.

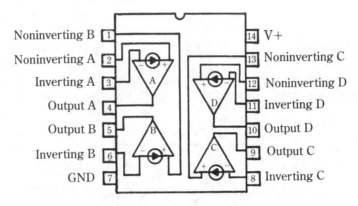

24-21 Pin-out diagram for a 3900 quad Norton amplifier.

**Table 24-4. Specifications
for a 3900 Norton amplifier.**

Input bias current	30 nA
Output resistance	8,000 Ω
Input resistance	1,000,000 Ω
Slew rate	0.5 V/μS

Other analog ICs

Operational amplifiers are certainly the most common analog integrated circuit devices, but there are countless others. Virtually any circuit can be miniaturized on an IC chip.

Some other common IC devices are oscillators, audio and RF amplifiers, voltage regulators (which hold a constant voltage output regardless of current drain or source voltage fluctuations—these devices will be discussed in the chapter on power supplies) and voltage comparators (which have an output that indicates whether the input voltage is greater than, less than, or equal to a reference voltage). And, of course, there are many, many others.

It would be impossible to even list all of the various circuits that are available in IC form, much less discuss them in any detail. You will just glance at a couple of typical examples.

The XR-2206 function-generator IC

Many types of equipment require some type of oscillator or waveform generator stage. A function generator (a circuit that can produce two or more standard waveforms) can also be used by technicians to test many types of amplifiers and other electronic equipment.

The XR-2206 is an IC function generator-stage. Only a few external resistors and capacitors are required to generate sine waves, triangle waves, pulse waves, square waves, and sawtooth (or ramp) waves. A single circuit can generate two or more simultaneous or switch-selectable waveforms. Figures 24-22 through 24-25 show XR-2206 function generator circuits. The circuit shown in Fig. 24-22 is a sine wave oscillator. Capacitor C sets the frequency range, with potentiometer R determining the actual output frequency. Frequency ranges for typical values of C are listed in Table 24-5.

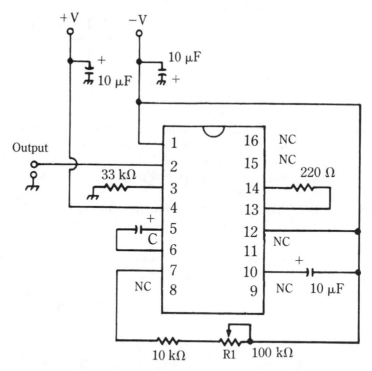

24-22 This circuit uses an XR-2206 function generator to produce sine waves.

The circuit shown in Fig. 24-23 will generate triangle waves instead of sine waves. Notice how similar this circuit is to the sine-wave oscillator circuit of Fig. 24-

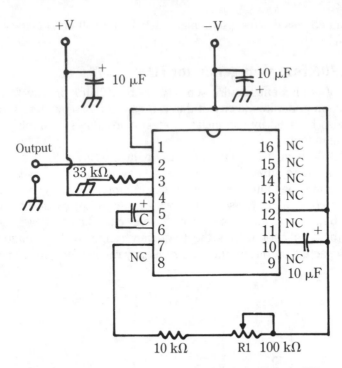

24-23   This circuit generates triangle waves from an XR-2206 function generator.

22. The only difference is the omission of the 220 Ω resistor between pins 13 and 14. These pins are left unconnected for the triangular-wave generator.

At further modification of the same circuit is shown in Fig. 24-24. In this circuit the output consists of square waves. Notice that this time the output is taken off of pin 11 and V—instead of between pin 2 and ground.

All three of these circuits have their frequency determined in the same way. The frequency ranges as defined by the value of C in Table 24-5 are valid for all three circuits.

A somewhat different generator circuit is shown in Fig. 24-25. This circuit generates sawtooth, or ramp waves. The output frequency is determined by the values of five passive components—capacitor C, and resistors R1 through R4. Two potentiometers allow fine tuning of the output frequency and waveshape. The output frequency can be determined according to the following formula:

$$F = \frac{2}{C}\left(\frac{1}{R1 + R2 + R3 + R4 +}\right) = \frac{2}{C}\left(\frac{1}{R2 + R4 + 2000}\right)$$

Assume a 1 μF capacitor is used for C. If both potentiometers (R2 and R4) are adjusted to their minimum setting, the output frequency will be equal to:

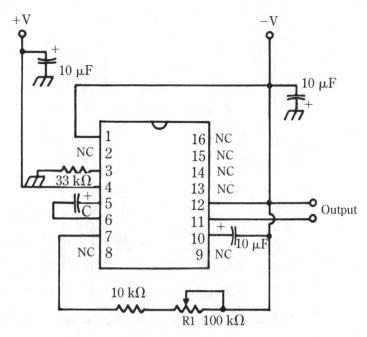

24-24 This square wave oscillator is built around an XR-2206 function generator.

Table 24-5. The component values to use for the XR-2206 oscillator.

C	F (minimum)	F (maximum)
1 μF	10 Hz	100 Hz
0.1 μF	10,000 Hz	1000 Hz
0.01 μF	1000 Hz	10,000 Hz
0.001 μF	10,000 Hz	100,000 Hz

$$F = \frac{2}{0.000001} \left(\frac{1}{0 + 0 + 2000} \right)$$

$$= 2000000 \left(\frac{1}{2000} \right) = 2000000 \times 0.0005$$

$$= 1000 \text{ Hz}$$

Turning both potentiometers up to their maximum setting decreases the output frequency to:

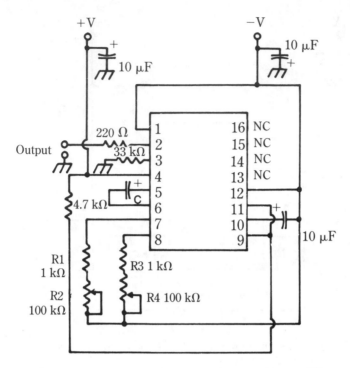

24-25 Circuit for generating sawtooth waves from an XR-2206 function generator.

$$F = \frac{2}{0.000001} \left(\frac{1}{100000 + 1000000 + 2000} \right)$$

$$= 2000000 \left(\frac{1}{202000} \right) = 2000000 \times 0.000005$$

$$= 10 \text{ Hz}$$

This circuit has wider output ranges for each value of C than the circuits in Figs. 24-22 through 24-24. The output ranges for this sawtooth wave generator are summarized in Table 24-6.

These basic waveform generator circuits can easily be combined into a multi-waveform function generator. A simple switching arrangement can make the necessary changes in the circuitry to allow the operator to select from the various output waveforms.

The XR-2206 can also be connected as a *voltage-controlled oscillator,* or *VCO,* in which the output frequency is determined by an externally applied control voltage. A number of other applications are also possible, although there is not the space to discuss them here.

**Table 24-6. The component values to use for
the XR-2206 sawtooth wave oscillator.**

C	F (minimum)	F (maximum)
1 µF	10 Hz	1000 Hz
0.1 µF	100 Hz	10,000 Hz
0.01 µF	1000 Hz	100,000 Hz
0.001 µF	10,000 Hz	1,000,000 Hz

XR-2208 operational multiplier

Another type of analog IC is the operational multiplier, which is closely related to the operational amplifiers discussed earlier. The XR-2208 Operational Multiplier is shown in block diagram form in Fig. 24-26. It consists of a four-quadrant multiplier/modulator, a high-frequency buffer amplifier, and an op amp in a single package. This device is used to perform mathematical operations such as multiplication, division, and square roots with a minimum of external components.

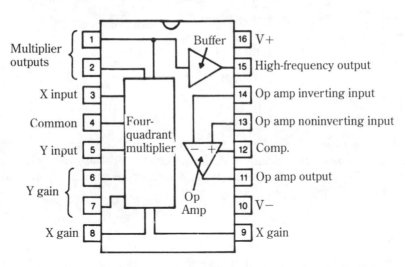

24-26 The pin-out diagram for the XR-2208 operational multiplier IC.

SAD 1024 analog delay

The SAD 1024 is used to delay analog signals. The SAD portion of its name stands for *serial-analog delay*. It uses what is known as the bucket brigade. Imagine ten people passing buckets of water from a well to a fire. Person number 1 passes the bucket to person number 2 and fills a new bucket. Person number 2 passes the first bucket to person number 3, gets the second bucket from person number 1, and so on. The buckets are passed down the line to the end person (person number 10). A bucket-brigade delay system uses a similar approach. Instantaneous signal levels are passed through a string of capacitors.

The SAD 1024 features two independent 512-stage analog delay sections. The clock controlled delay can cover a range from less than 200 μs (microseconds) to about 0.5 second. The analog signal fed through the delay can be anywhere within a wide range from 0 Hz (dc) to more than 200,000 Hz (200 kHz). The SAD-1024 analog delay introduces a very small amount of distortion (less than 1%) to the delayed signal.

This IC can be used for reverberation, chorus, and echo circuits in electronic music and sound-effects systems. It can also be used for variable-signal control of equalization filters, or amplifiers, time compression of telephone conversations or other analog signals, or voice-scrambling systems. The pin-out diagram for the SAD-1024 analog delay IC is shown in Fig. 24-27.

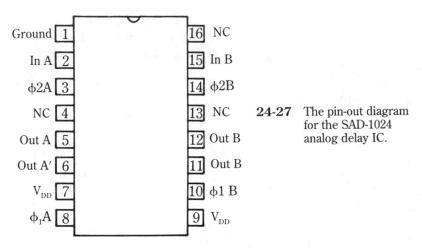

24-27   The pin-out diagram for the SAD-1024 analog delay IC.

And many more   This chapter has barely been able to scratch the surface. Other common analog ICs include FM (frequency modulation) demodulators, TV modulators, video games, temperature transducers, audio amplifiers, precision current sources, and many others.

Self-test

1. Which type of ICs are the least complex?

A SSI
B MSI
C LSI
D CPU
E None of the above

2. What is another name for *linear*?

A Integration
B Digital
C Analog
D Dip
E None of the above

3. Which of the following is the input of an op amp that will invert a signal 180 degrees?

A Inverting
B Noninverting
C Analog
D Digital
E None of the above

4. What is the gain of an inverting amplifier circuit when R1 equals 2.2 kΩ and R2 equals 18 kΩ?

A 8.2
B −0.12
C 1.2
D −8.2
E None of the above

5. Assuming a gain of 1, what is the output of an operational amplifier when 0.27 V is applied to the inverting input and −0.33 V is applied to the noninverting input?

A 0.06 V
B −0.60 V
C −0.06 V
D 0.60 V
E None of the above

6. Which of the following is not an op amp IC?

A 741
B 555
C 709
D 748
E None of the above

7. How many op amps are contained in a 324 IC?

A One
B Two
C Three
D Four
E None of the above

8. What is the name for op amp circuit using the inverting input with a resistor as the input component and a capacitor as the feedback component?

A Inverting amplifier
B Integrator
C Differentiator
D Difference amplifier
E None of the above

9. What type of circuit is the XR-2208?

A Op amp
B Function generator
C Analog delay
D Operational multiplier
E None of the above

10. What is the gain of a differentiator when the input frequency is 375 Hz? The resistor is 22 kΩ and the capacitor is 0.01 μF.

A −0.0005
B −0.5
C −2.2
D −0.035
E None of the above

25
Timers

Another popular and versatile type of integrated circuit is the *timer*. This device is given a chapter of its own because it is somewhere between analog and digital circuits. It can be used with either.

555 basics

The most common timer is the 555, which is currently being manufactured by several different companies. There are a number of other timer ICs, but they all work in basically the same way, so this discussion will be confined to the 555. The 555 integrated circuit is available in a number of package styles, but the 8-pin DIP version is the most frequently used. Figure 25-1 shows the pin-outs for this IC.

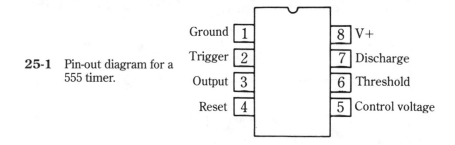

25-1 Pin-out diagram for a 555 timer.

Examine the 555 integrated circuit pin by pin. Pin 1 is simply the ground terminal for the circuit. The 555 must be used with a negative ground. No below ground voltages should be applied to this device. That is, the voltage on pin 1 should be the lowest voltage applied to any of the integrated circuit's pins.

Pin 2 is the *trigger* input. Normally this terminal is held at a constant value that is at least one-third of the source voltage. If the trigger voltage drops below the one-third source point, it will trigger the circuit (that is, the timing cycle will be initiated).

The output of the timer circuit is usually (but not always) taken from pin 3.

Pin 4 is labeled *reset*. As the name suggests, the signal on this terminal is used to return the device to its original rest state at the end of the timing cycle.

Pin 5 is left unused in most circuit applications, but in some special cases, it can be extremely useful. It is a *control voltage* input. The trigger voltage point can be determined by an external voltage applied to this terminal, rather than the normal one-third source voltage trigger point. When the voltage control mode is not used, pin 5 should be connected to ground via a 0.01 μF capacitor.

Pin 6 is called the *threshold* input. The voltage on this terminal tells the timer when to end its timing cycle. A *timing resistor* is connected from pin 6 to the positive terminal of the voltage source.

Pin 7 is called the *discharge* pin. This terminal is also used to determine the length of the timing cycle. A *timing capacitor* is connected from pin 7 to ground.

Pin 8 is used to power the circuit. The positive terminal of the voltage source is connected to this pin. In schematic diagrams this pin is usually labeled either $V+$, or *VCC*. The voltage source for a 555 timer IC should be between 5 and 15 V, with 15 V generally preferred for best results.

Monostable multivibrators

Figure 25-2 shows the most basic circuit that is built around a 555 timer IC. This circuit is called a *triggered monostable multivibrator*. Monostable means the circuit has one stable output state (at about ground level in this case). When this circuit is triggered, the output jumps to an unstable state (near the source voltage level in this circuit). The output holds this unstable level for a specific time period determined by the values of R_t and C_t. Then the output snaps back to its original, stable state. This process is shown graphically in Fig. 25-3.

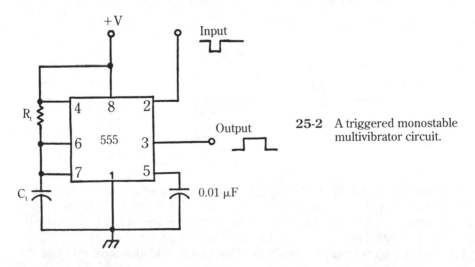

25-2   A triggered monostable multivibrator circuit.

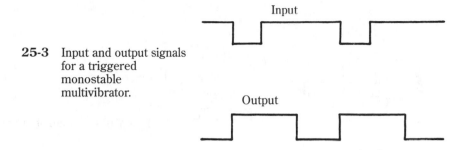

25-3   Input and output signals for a triggered monostable multivibrator.

The time of the output pulse is determined by the following formula:

$$T = 1.1(R_t \, C_t)$$ **Equation 25-1**

T is the time period in seconds, R_t is the resistance from pin 6 to Vcc in meg-ohms (one $M\Omega$ equals 1,000,000 Ω), and C_t is the capacitance between pin 7 and ground in microfarads (μF).

For reliable operation, keep R_t within the range of about 0.01 $M\Omega$ (10,000 Ω) to 10 $M\Omega$ (10,000,000 Ω). Keep C_t 100 pF (0.0001 μF) and 1000 μF.

Using Equation 25-1, you can see that this circuit can produce timing periods from 0.0000011 second (1.1 microseconds) to 11,000 seconds (about 183 minutes or just over 3 hours). Clearly, this is quite a wide range device. This timing pulse will always be the same length regardless of the time of the triggering signal—provid-ing, of course, that the trigger signal is less than the desired output pulse.

A triggered monostable multivibrator is an extremely useful circuit in countless timing and control applications. For example, a photographer might want a strobe light to flash on for 0.2 second when the shutter is opened. The photographer would use his shutter pulse to trigger a monostable multivibrator. In designing this cir-cuit, first arbitrarily select a convenient capacitance value. You'll use a 0.5 μF ca-pacitor. Because $T = 1.1(R_t C)$, then $R = T/(1.1 \times C_t)$. Filling in the known vari-able for the example, you find that $R_t = 0.2/(1.1 \times 0.5) = 0.2/0.55$, or about 0.36 $M\Omega$ (360,000 Ω).

Using a 0.5 μF capacitor and a 360,000 Ω resistor, the light can be turned on for precisely 0.2 second each time the unit is triggered. Of course, the precision of the timing period will depend on the tolerance of the timing components. If both the resistor and the capacitor have 20% tolerances, the final value might be off from the calculated value by as much as 40%. Obviously, low tolerance components should be used.

What component values would you use to generate a timing period of 1 second?

Astable multivibrators

Figure 25-4 shows a similar 555 timer circuit. The two most important differences between this and the previous circuit are that R_t has been split into two separate resistors (R_a and R_b), and there is no input for a trigger signal. This circuit is self-triggering. This circuit output fluctuates between the same two states as the trig-

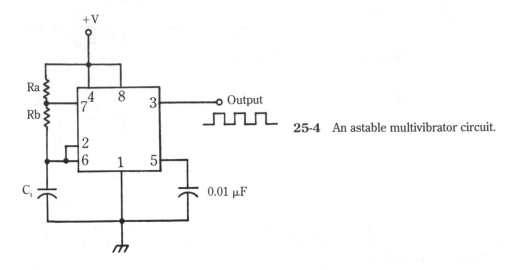

25-4   An astable multivibrator circuit.

gered monostable version, but in this circuit, neither output state is stable. This circuit is called an *astable multivibrator.*

The time the output is at its high voltage state is determined by C_t, R_a, and R_b:

$$T_1 = 0.693 \times (R_a + R_b) \times C_t \qquad \text{Equation 25-2}$$

Although the time the output is in its low, or grounded, state depends on only C_t and R_b

$$T/_2 = 0.693\ R_b\ C_t \qquad \text{Equation 25-3}$$

Obviously the total time of the complete cycle is simply the sum of the high time and the low time, or:

$$T = T_1 + T_2 \qquad \text{Equation 25-4}$$

These three equations can be combined and rewritten as:

$$T = 0.693 \times (R_a + 2R_b) \times C_t \qquad \text{Equation 25-5}$$

The output *oscillates* between the two states at a fixed rate. The circuit is also called a *rectangular wave oscillator.*

As Fig. 25-5 indicates, the output waveform is rectangular in shape. Notice that for this circuit, T_1 is always at least slightly longer than T_2. The relationship between the high-state time and the low-state time is called the *duty cycle*. For example, if the signal is in the high state for three quarters of each cycle, the duty cycle is 75%. If both times are equal, the duty cycle is 50%, and the signal is called a *square wave*. A perfect square wave can't be achieved with this particular circuit, but you can come quite close by using a very small value for R_a and a large value for R_b.

Because you have a repeating cycle, you can speak of it in terms of frequency. The frequency of a repeating cycle is the number of complete cycles that take place within a second. The frequency can be found by taking the reciprocal of the time required to complete one cycle. In algebraic terms this is:

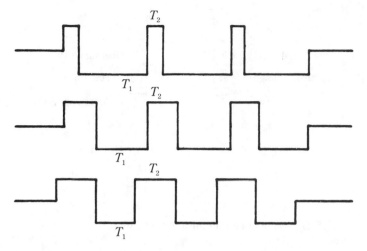

25-5 Rectangular waves.

$$F = \frac{1}{T}$$ **Equation 25-6**

Combining Equation 25-6 with Equation 25-5, you can define frequency as:

$$F = \frac{1.44}{(R_a + 2R_b)C_t}$$ **Equation 25-7**

F, of course, is the frequency in hertz for both equations.

Essentially the same range limitations are placed on the resistor and capacitor values as with the triggered monostable multivibrator circuit.

Most circuits built around the 555 integrated circuit timer are variations on one or the other of these two basic circuits. But these basic circuits can find applications in a vast number of electronics systems. A dual 555 in a single 14-pin DIP IC is also available. This component is called the 556. It can be used in place of two separate 555 ICs.

The pin-out diagram for the 556 dual timer IC is shown in Fig. 25-6. Except for the different pin numbers, the 556 timer is used in exactly the same circuits as the 555 timer. It is the exact equivalent of two 555 chips enclosed in a single housing. The two timer sections in a 556 IC are entirely independent. One can be used in the monostable multivibrator mode, and the other in the astable multivibrator mode, or both can be operated in the same mode. The time period of one timer of a 556 chip has nothing to do with the time period of the other timer section. The only pins the two timer sections have in common are the power supply pins—pin 14 (V +) and pin 7 (ground).

Another multiple timer IC is the 558 (sometimes the number is given as 5558). This chip is a quad timer chip. It contains four 555 timer sections in a single semiconductor package. The timer sections in the 558 are somewhat simplified in comparison with the standard 555 (or 556) timer. To keep the chip size and pin count down,

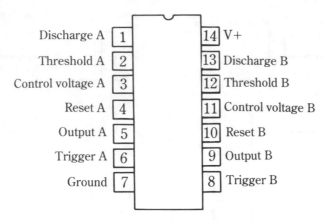

25-6 Pin-out diagram for a 556 dual timer.

not all of the ordinary timer functions are brought out to the pins of the 558. The pin-out diagram for the 558 quad timer IC is shown in Fig. 25-7.

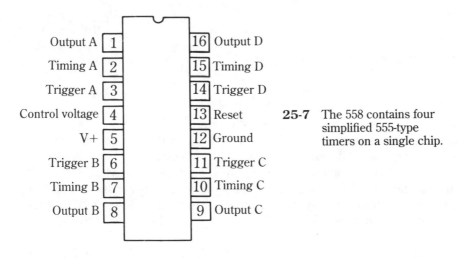

25-7 The 558 contains four simplified 555-type timers on a single chip.

Because the circuit designer does not have direct access to all of the ordinary 555 timer functions, the 558 can not be used in all standard 555 applications or circuits. This chip is intended primarily for use in the monostable multivibrator mode. It cannot normally be used in standard astable multivibrator circuits.

The 7555 CMOS timer

Another variation of the basic 555 timer IC is 7555, shown in Fig. 25-8. Notice that this chip is pin for pin compatible with the standard 555 timer IC. The 7555 can be directly substituted for the 555 in virtually any timer circuit. No circuit changes will be required to support such a substitution.

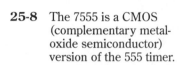

25-8 The 7555 is a CMOS (complementary metal-oxide semiconductor) version of the 555 timer.

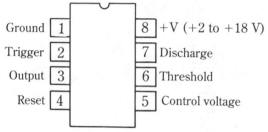

The difference between the 555 and the 7555 lies is in the way the timer circuitry is implemented within the chip itself. The standard 555 timer IC is a purely linear device. It's HIGH/LOW output can be interfaced with digital gates (see chapters 26–28), but its internal circuitry is still analog. The 7555 timer, on the other hand is a digital device. It uses CMOS digital circuitry to achieve exactly the same operational functioning as the linear 555 chip. The CMOS logic family is discussed in chapter 28.

The chief advantage of the 7555 timer over the standard 555 timer is the CMOS device features very low power consumption, and it can operate from a wider range of supply voltages. The power supply voltage for the 7555 can be anything from +2 V up to +18 V.

The digital CMOS circuitry of the 7555 also permit the timer to remain stable with longer timing periods than the standard 555 can cope with reliably. In other words, you can use larger timing capacitor and resistor values than are permitted for the standard 555 chip.

The 7555 can be directly substituted for a 555 in almost any timer circuit. The reverse is not always true. Usually a standard 555 chip can be substituted for a CMOS 7555 IC, unless the power supply voltage or the timing period is outside the normal acceptable limits for the 555. In some cases, such a substitution might cause problems, because the power supply can not reliably provide the higher current levels consumed by the 555's linear circuitry.

Precision timers

For the vast majority of practical timing applications, the 555 and its variations (the 556, the 558, or the 7555) will do the job very well, very easily, and conveniently. But some precision applications might call for a more precise timer than the 555. A number of precision timer ICs have been released. They are roughly similar to the 555 in concept, but in most cases, different circuitry and timing equations are needed.

Two typical precision timer chips are the 322 (shown in Fig. 25-9) and the 3905 (shown in Fig. 25-10). Both of these ICs are wide-range precision monostable timers. These two chips are essentially similar in their internal circuitry. The 3905 is a somewhat simplified version of the 322. A simplified block diagram of the internal circuitry used in both the 322 and the 3905 is shown in Fig. 25-11.

In the real world, compromises are inevitable. To achieve the improvements in the 322 and the 3905, something had to be given up. These devices can normally be

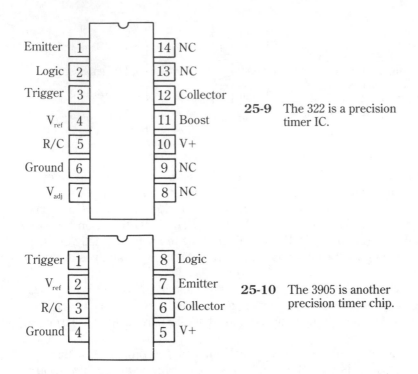

25-9 The 322 is a precision timer IC.

25-10 The 3905 is another precision timer chip.

used in monostable multivibrator applications only. They are not designed for use in astable multivibrator circuits. This is really less of a shortcoming than it might seem at first. The special advantages of the 322 and the 3905 (greater timing precision, and longer possible timing periods) generally are needed only in monostable multivibrator circuits. They are rarely relevant for astable multivibrator circuits, so the less expensive, and easier to find 555 (or one of its variants) will be suitable for the job.

There are other advantages of the 322 and the 3905 too. They feature greater immunity to any noise in the power supply lines than the 555. These devices can also be operated from a much wider range of supply voltages. The 322 and the 3905 are rated for supply voltages ranging from +4.5 V up to +40 V.

An on-chip voltage regulator circuit is used as a reference for the timing comparators. This gives greater accuracy in determining the timing period, because the actual supply voltage has virtually no effect, even if it changes or fluctuates during the timing period.

The output stages of the 322 and the 3905 precision timers can produce higher voltages than that of the 555. The output stage in these improved timer chips is also more generally flexible and versatile. It can be wired in either common-collector or common-emitter configurations, depending on the specific requirements of the intended application.

Another versatile feature of these precision timer ICs is that the stable (normal) output state might be externally selected with the *logic* pin. If this pin is made HIGH, the output will normally be LOW, and will go HIGH only during the timing period, just like the standard 555 timer. Holding the logic pin LOW produces the opposite

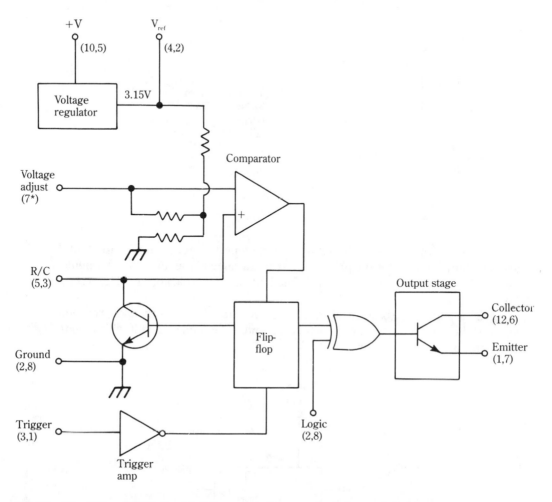

25-11   The 322 and 3905 precision timers use the same basic internal circuitry.

result. In this case, the normal (nontriggered) output state will be HIGH, and the output will go LOW only during the timing period. This action is shown in Fig. 25-12. This feature avoids the need for an external inverter stage in some applications.

Still another advantage of the 322 and the 3905 precision timer ICs is that their timing period equation is even simpler than the equation used for the standard 555 timer. In fact, this new timing equation is literally as simple as it can possibly be. The timing period is simply equal to the product of the timing capacitance and the timing resistance. That is:

$$T = RC$$

No constant is required as with the 555. The resistance must be in ohms and the capacitance must be in farads. If you prefer to work with capacitances in microfarads (μF), you must convert the resistance into megohms (1 MΩ = 1,000,000 Ω).

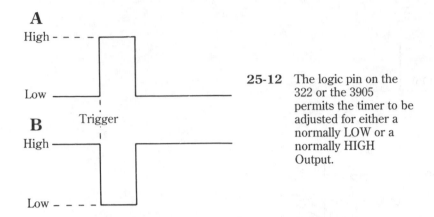

A

High - - - -

Low

B Trigger

High

Low - - - -

25-12 The logic pin on the 322 or the 3905 permits the timer to be adjusted for either a normally LOW or a normally HIGH Output.

These precision timers can be set up for timing periods extending up to several hours. The only limitation is the practical maximum value of the timing components. For example, you'd probably have a hard time finding a 2 F capacitor or a 330 MΩ resistor.

Fig. 25-13 shows the basic monostable multivibrator circuit for the 322 precision timer with the output in the common-collector mode. A similar circuit with a common-emitter output is shown in Fig. 25-14.

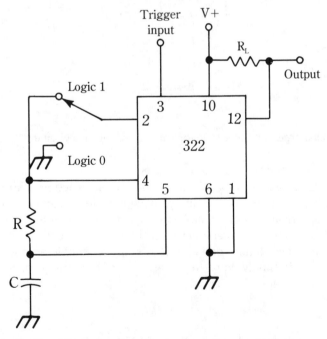

25-13 Basic common-collector output monostable multivibrator circuit for the 322.

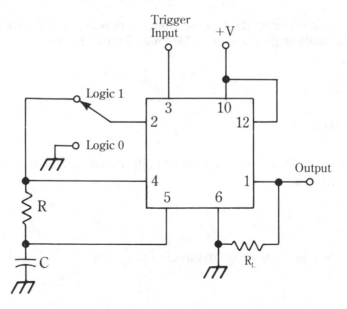

25-14 Basic common-emitter output monostable multivibrator circuit for the 322.

Self-test

1. How many stable output states does a monostable multivibrator have?

A None
B One
C Two
D Three
E None of the above

2. What is the name of the multivibrator circuit with no stable output states?

A Astable
B Monostable
C Bistable
D Unstable
E None of the above

3. How long is the output pulse of a 555 monostable multivibrator circuit when the timing resistor is 560 kΩ and the timing capacitor is 0.22 μ F?

A 135,500 seconds
B 0.13 second
C 0.03 second
D 52 seconds
E None of the above

4. What is approximately the longest timing period that can be achieved with a standard single-stage 555 monostable multivibrator circuit?

A 1 second
B 1 hour
C 183 seconds
D 11,000 seconds
E None of the above

5. Assume an astable multivibrator circuit built around a 555 IC uses the following component values

R_a = 33 kΩ
R_b = 27 kΩ
C_t = 0.1 μF

What is the output frequency of this circuit?

A 534 Hz
B 437 Hz
C 166 Hz
D 240 Hz
E None of the above

6. Which of the following is the correct formula for finding the output frequency of a standard 555 astable multivibrator circuit?

A $F = 1.44(R + R_b)C_t$
B $F = 1.44/((R_a + 2R_b)C_t)$
C $F = 1.1(R_a + RR_b)C_t$
D $F = 0.693(R_a + 2R_b)C_t$
E None of the above

7. What is the wave shape at the output of a standard 555 astable multivibrator circuit?

A Rectangular wave
B Sine wave
C Ramp wave
D Square wave
E None of the above

8. How many timer stages are there in the 558 chip?

A One
B Two
C Four
D Eight
E None of the above

9. Which of the following is not a variation on the basic 555 timer?

A 558
B 7555
C 3905
D 556
E None of the above

10. In a monostable multivibrator circuit built around the 322 precision timer IC, using a timing resistor of 270 kΩ and a timing capacitor of 33 μf, what is the timing period?

A 8910 seconds
B 8.9 seconds
C 0.089 seconds
D 303 seconds
E None of the above

26
Digital gates

You have read about analog integrated circuits. The other major category of integrated circuits is called *digital*. All input and output signals are converted into numerical (digital) equivalents and these numbers are treated by the integrated circuits in various ways.

The binary system

Ordinarily, you do our counting in the *decimal system*, which has ten digits (0, 1, 2, 3, 4, 5, 6, 7, 8, and 9). If you need to represent a number larger than 9 (the highest digit), you add additional columns of digits. The unit value of each column is equal to the *base* (number of digits in the system) times the unit value of the previous column. In other words, the number 2873 represents in decimal $2 \times 1000 + 8 \times 100 + 7 \times 10 + 3 \times 1$.

In working with electronic circuits, it is more convenient to use the binary system, which has only two digits (0 and 1). If a specific voltage is present, you can say you have a 1. If the voltage is absent, you have a 0. There are no intermediate values.

This process can be reversed. A present voltage could be called a 0, and no voltage could represent a 1. This would be called *negative logic*. There is really no difference, except sometimes it is more convenient conceptually to think in terms of negative logic.

Because there are no digits greater than 1 in the binary system, obviously you will need more than one column of digits to represent any number greater than one. The values of these columns are increased in the same way as in the decimal system, except each new column has a unit value of two times its predecessor. The first column is times 1, the second column is times 2, the third column is times 4, the fourth column is times 8, and so forth.

For example, the binary number 1101 consists of $1 \times 8 + 1 \times 4 + 0 \times 2 + 1 \times 1$, or 13 in decimal. Table 26-1 compares counting in the decimal and binary systems.

**Table 26-1. Counting in binary,
decimal, octal, and hexadecimal.**

Binary	Decimal	Octal	Hexadecimal
00001	1	1	1
00010	2	2	2
00011	3	3	3
00100	4	4	4
00101	5	5	5
00110	6	6	6
00111	7	7	7
01000	8	10	8
01001	9	11	9
01010	10	12	A
01011	11	13	B
01100	12	14	C
01101	13	15	D
01110	14	16	E
01111	15	17	F
10000	16	20	10
10001	17	21	11
10010	18	22	12
10011	19	23	13
10100	20	24	14
10101	21	25	15
10110	22	26	16
10111	23	27	17
11000	24	30	18
11001	25	31	19
11010	26	32	1A
11011	27	33	1B

Other number systems

Notice that Table 26-1 has two additional columns labeled octal and hexadecimal. These additional systems are used as an intermediate level between decimal and binary. Although binary numbers are quite simple for electronic circuits to handle—even when the values are quite large, they are rather unwieldy for humans. For instance, a binary number like 11010001101 is quite difficult to remember.

Unfortunately, human operators must put numbers into the circuits in the first place, and understand the numbers the circuit comes up with as output. Electronically converting a decimal input directly into a binary number is rather a complex job.

A compromise is reached if the operator learns to use the octal or the hexadecimal system, which can be quickly converted to binary, because the bases of these systems are squared multiples of the binary base (2).

The octal system is built on the base of eight. Only digits 0–7 are used. Notice in Table 26-1 that each column in an octal number corresponds to three columns in a binary number.

Similarly, the hexadecimal system uses a base of sixteen: the digits 0, 1, 2, 3, 4, 5, 6, 7, 8, 9, A, B, C, D, E, and F. A single column in a hexadecimal number corresponds to four columns in a binary number.

Simple switching circuits can easily convert octal or hexadecimal numbers to binary numbers. Of course, it is always important to identify which system you are working in. The number 111 represents seven in binary, seventy-three in octal, one hundred and eleven in decimal, or two hundred and seventy three in hexadecimal. Obviously, confusion can be a major problem if everything isn't carefully marked.

Sometimes numbers are identified by a *subscript* of their base. In the above example you could write 111_2 (binary), 111_8 (octal), 111_{10} (decimal), or 111_{16} (hexadecimal).

In most practical electronics work, you probably won't have to convert from one system to another very often, but you should have an idea of how it is done. And in digital circuits it is essential for you to understand the basics of the binary system.

Binary addition

The rules for combining or adding binary numbers are really quite simple. For any given column of digits, there are only two possibilities. If you are adding two binary digits, there are only four possible combinations. If both digits to be added are 0s, the total is 0. If one digit is a 0, and the other is a 1, the total is a 1. (Notice that this covers two possible combinations, simply by exchanging the position of the digits. $0 + 1 = 1 + 0$.) If both digits are 1s, then the total is 0, with a 1 carried into the next column. Here are some examples:

$$0 + 0 = 0$$
$$1 + 0 = 1$$
$$0 + 1 = 1$$
$$1 + 1 = 10$$
$$1010 + 1100 = 10110$$

Now you add $11 + 10$; $101 + 1110$; $1111 + 10$; $1001 + 111$.

To combine binary digits in various ways in electronic circuits, digital *gates* are used. Following are descriptions of the various types of gates used.

AND gates

One basic digital gate is called an *AND gate.* There are generally four AND gates in a single integrated circuit package (called a *quad AND gate*). The schematic symbol for an AND gate is shown in Fig. 26-1. Notice that there are two inputs and a single output. The output will be a 1 if, and only if, both inputs are 1s.

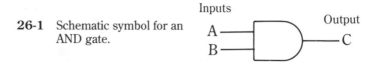

26-1 Schematic symbol for an AND gate.

Because there are two inputs, each with two possible signals, there are four possible input conditions. If both inputs are 0s, the output will be 0. If input A is 0 and input B is 1, the output will be 0. If input A is 1 and input B is 0, the output will still be 0. But if both input A and input B are 1s, the output will be a 1. These are the only four possible input combinations.

The relationship between the inputs and outputs of a gate are usually shown in a chart called a *truth table*. The truth table for a standard, two-input AND gate is given in Table 26-2A.

Although most AND gates have just two inputs, you will sometimes find AND gates (and other types of gates) with more inputs. They work in essentially the same way. For instance, Table 26-2B gives the truth table for a four-input AND gate. Notice that the output is a 1 if, and only if, all four inputs are 1s. If any input is a 0, the output is a 0.

Now, look at the truth table in Table 26-2C. It is the exact opposite of that for an AND gate. The output is 0, if and only if, both inputs are 1s. If either or both of the inputs goes to 0, the output goes to 1. Not surprisingly, this is a *Not AND* gate, or, more properly, a *NAND gate*. The schematic symbol for a NAND gate resembles that of an AND gate, except a small circle is added at the output to indicate *state inversion*. See Fig. 26-2.

In digital circuits, this small circle always represents state inversion. If the signal was a 1, it becomes a 0. If it was a 0, it becomes a 1.

OR gates

Another basic digital gate is the *OR gate*. Its schematic symbol is shown in Fig. 26-3, and its truth table is given in Table 26-3A.

Again, there are two inputs, and a single output that is dependent on the logic states of both inputs. As long as at least one of the inputs is a 1, the output of an OR gate will be a 1. The output will be a 0 only if both inputs are 0s.

A variation of the basic OR gate is the *exclusive OR gate*. As the name implies, the output is a 1 if one or the other input is a 1, but not if the inputs are both 1s, or both 0s. In other words, the output is a 1 if the two inputs are different. The output is a 0 if the two inputs are at the same level. The schematic symbol for an exclusive OR gate is shown in Fig. 26-4, and its truth table is shown in Table 26-3B.

Table 26-3C is the truth table for the inversion of an OR gate. This device (shown schematically in Fig. 26-5) is called a *NOR gate*. The output will be a 1 only if neither input is a 1. If either or both inputs are at a logic 1 level, the output will be a 0.

Table 26-2. Truth Tables: AND Gate (A); Four-Input AND Gate (B); Two Input NAND Gate (C).

A   **Two-input AND gate**

In		Out
A	B	C
0	0	0
0	1	0
1	0	0
1	1	1

e.g. C is true only if A + B are both true

B   **Four-input AND gate**

A	B	C	D	Out
0	0	0	0	0
0	0	0	1	0
0	0	1	0	0
0	0	1	1	0
0	1	0	0	0
0	1	0	1	0
0	1	1	0	0
0	1	1	1	0
1	0	0	0	0
1	0	0	1	0
1	0	1	0	0
1	0	1	1	0
1	1	0	0	0
1	1	0	1	0
1	1	1	0	0
1	1	1	1	1

C   **NAND gate**

In		Out
A	B	C
0	0	1
0	1	1
1	0	1
1	1	0

Inputs

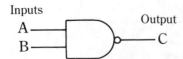

A —
B —
Output
— C

26-2 Schematic symbol for a NAND gate.

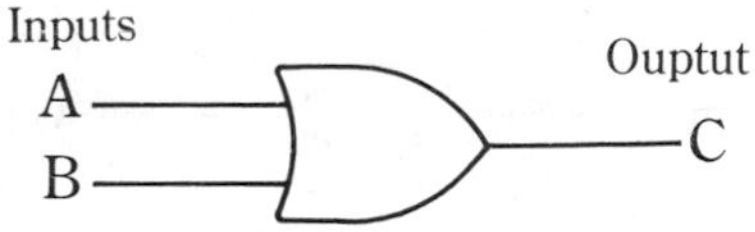

26-3 Schematic symbol for an OR gate.

Inputs
A
B
Ouptut
C

**Table 26-3. Truth Tables OR Gate (A);
exclusive-OR Gate (B); NOR Gate (C).**

A OR gate			**B Exclusive-OR gate**		
In		**Out**	**In**		**Out**
A	B	C	A	B	C
0	0	0	0	0	0
0	1	1	0	1	1
1	0	1	1	0	1
1	1	1	1	1	0

e.g. C is true only if either A or B or both are true

C is true only if either A or B is true — but not if both true

C NOR gate		
In		**Out**
A	B	C
0	0	1
0	1	0
1	0	0
1	1	0

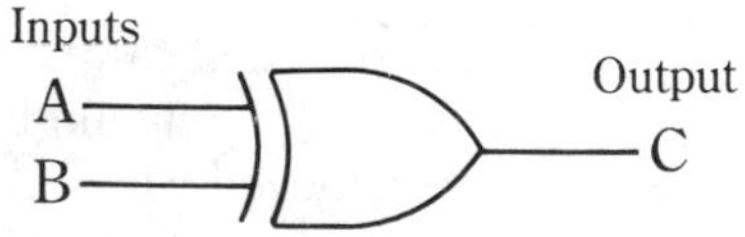

26-4 Schematic symbol for an exclusive-OR (X-OR) gate.

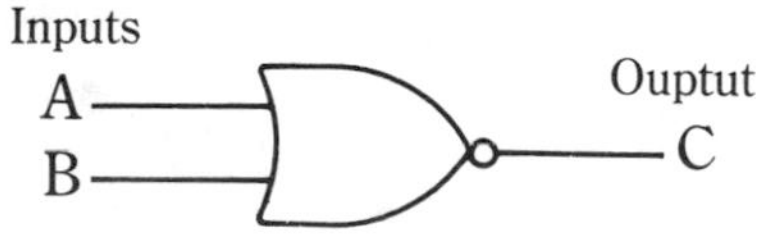

Inputs
A
B
Output
C

26-5 Schematic symbol for a NOR gate.

Inputs
A
B
Ouptut
C

Buffers and inverters

Figure 26-6 shows two single input/single output digital devices. Figure 26-6A is an inverter. Obviously, this device inverts or reverses its input. If the input is a 0, the output is a 1, or if the input is a 1, the output is a 0. These are the only possible states. The device shown in Fig. 26-6B might not seem particularly useful at first, since its output is the same as its input. This device does not affect the logic state in any way.

The output of any digital gate can feed only a limited number of inputs to other digital gates (or other devices). This is called the *fan-out*. The typical fan-out for *TTL*

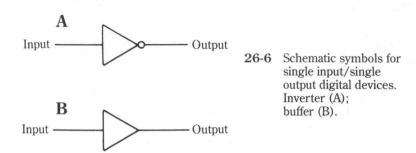

A

Input ————————▷o———— Output

B

Input ————————▷———— Output

26-6 Schematic symbols for single input/single output digital devices. Inverter (A); buffer (B).

gates (see chapter 28) is usually about 10. That is, each gate output can be used as the input for up to 10 other gates.

But, what do you do if you need to drive more than 10 gates from a single output? (This problem often crops up when digital gates need to be *interfaced* with other devices that might have a *fan-in* greater than one.)

Here is where the device in Fig. 26-6B comes in. It is called a *buffer*. A buffer typically has a fan-out of about 30. By passing the desired signal through a buffer, the buffer can drive many more gates (or other devices). Inverters also act as buffers in addition to their state reversal function.

Combining digital gates

These seven basic digital gates (AND, NAND, OR, exclusive OR, NOR, inverters, and buffers) can be combined to perform almost any logic function. There are a number of specialized digital IC gates for other logic functions, but these are not always readily available. Fortunately, any function can be built up from combinations of the basic gates.

Assume you need to generate the logic pattern shown in the truth table in Table 26-4. The output should be a 1 if, and only if, input A is a 1, and input B is 0. This could be achieved with either of the circuits shown in Fig. 26-7.

Table 26-4.
A Nonstandard truth table.

a	b	Output
0	0	0
0	1	0
1	0	1
1	1	0

Of course, more than two inputs can be used for certain applications. For example, Fig. 26-8A shows a three-input AND gate. The output will be a 1 only if all three inputs are 1s. If any of the inputs is a 0, the output will be a 0. This particular function is sometimes available in a single, ready-made package. If a single package unit is used, it is usually shown schematically as shown in Fig. 26-8B.

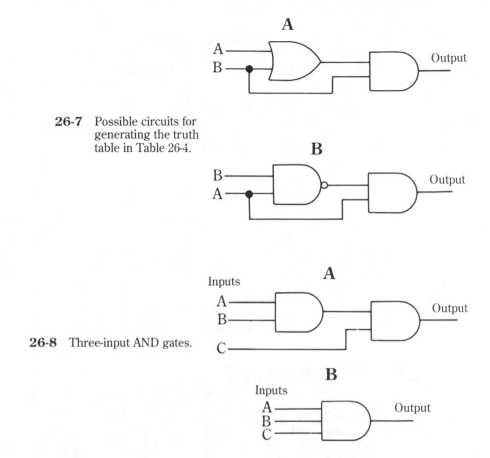

26-7 Possible circuits for generating the truth table in Table 26-4.

26-8 Three-input AND gates.

Table 26-5 shows a complex four-input logic function. There are a number of ways this can be achieved using the basic gates described in the previous sections. One possible solution is shown in Fig. 26-9. There are nine AND gates, four NOR gates, and five exclusive OR gates used in this circuit. Because there are usually four identical two input gates in each integrated circuit package, this circuit would require a minimum of five IC packages, with some of the available gates either left unused, or used in a separate circuit.

Can you come up with a simpler circuit to generate the same truth table? Can you determine the truth table for the logic circuit shown in Fig. 26-10?

Majority logic

In some applications, the straight yes or no approach of digital gates isn't entirely appropriate. For example, you might want the circuit to check for seven different conditions and respond if and only if most of them are HIGH. For example, in a seven-input majority logic AND gate the output goes HIGH when any four or more inputs go HIGH. If four or more inputs are LOW, the output stays LOW too. This logic is called *majority logic*. It is sort of like a digital democracy. Each signal gets one vote.

Table 26-5.
A complex four-input truth table.

a	b	c	d	Output
0	0	0	0	1
0	0	0	1	0
0	0	1	0	0
0	0	1	1	1
0	1	0	0	0
0	1	0	1	0
0	1	1	0	1
0	1	1	1	0
1	0	0	0	0
1	0	0	1	0
1	0	1	0	1
1	0	1	1	0
1	1	0	0	0
1	1	0	1	1
1	1	1	0	0
1	1	1	1	0

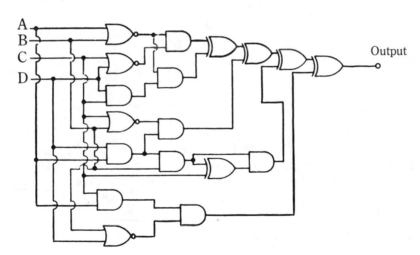

26-9   Possible circuit for generating the truth table in Table 26-5.

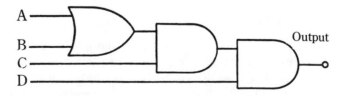

26-10   Another complex digital gating circuit.

Some dedicated majority logic ICs have appeared on the market from time to time. Usually, however, a majority logic circuit is made up individual standard gates or a multiplexer-demultiplexer circuit is used (see chapter 27). Often, the majority logic process will be implemented in software (computer programming) rather than hardware (an actual, physical circuit).

Table 26-6 shows the truth table for a three-input majority logic AND gate. This simple majority logic function is fairly easy to implement using standard digital gates. A typical circuit is shown in Fig. 26-11.

Table 26-6. Truth table for a three-input majority-logic AND gate.

Inputs			Output
A	**B**	**C**	
0	0	0	0
0	0	1	0
0	1	0	0
0	1	1	1
1	0	0	0
1	0	1	1
1	1	0	1
1	1	1	1

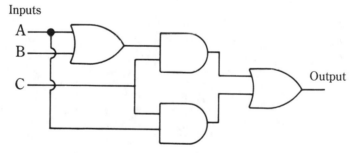

26-11   A simple three-input majority logic circuit can be constructed from standard digital gates.

Things get more complicated, of course, when the number of inputs is increased. For example, Table 26-7 is the truth table for a five input majority logic AND gate. It is up to you to devise the most efficient gating circuit to generate this truth table.

Most majority logic gates are either AND or NAND gates. A majority-logic NAND gate can easily be created simply by inverting the output of a majority logic AND gate. There wouldn't appear to be much point in a majority logic OR gate, because the output of an OR gate goes HIGH when at least one of its inputs is HIGH and is LOW only when all inputs are LOW.

Table 26-7. Truth table for a
five-input majority-logic AND gate.

Inputs					Output
A	B	C	D	E	
0	0	0	0	0	0
0	0	0	0	1	0
0	0	0	1	0	0
0	0	0	1	1	0
0	0	1	0	0	0
0	0	1	0	1	0
0	0	1	1	0	0
0	0	1	1	1	1
0	1	0	0	0	0
0	1	0	0	1	0
0	1	0	1	0	0
0	1	0	1	1	1
0	1	1	0	0	0
0	1	1	0	1	1
0	1	1	1	0	1
0	1	1	1	1	1
1	0	0	0	0	0
1	0	0	0	1	0
1	0	0	1	0	0
1	0	0	1	1	1
1	0	1	0	0	0
1	0	1	0	1	1
1	0	1	1	0	1
1	0	1	1	1	1
1	1	0	0	0	0
1	1	0	0	1	1
1	1	0	1	0	1
1	1	0	1	1	1
1	1	1	0	0	1
1	1	1	0	1	1
1	1	1	1	0	1
1	1	1	1	1	1

Self-test

1. What is 1001 0111 binary in decimal?

A 16

B 151

C 157

D 215

E None of the above

2. What is 169 expressed in binary?

A 1001 1010
B 1100 1001
C 1010 1001
D 1010 1101
E None of the above

3. Which of the following describes the action of an OR gate?

A The output goes HIGH when one or more of the inputs is HIGH.
B The output goes HIGH only when all of the inputs are HIGH.
C The output goes HIGH only when all of the inputs are LOW.
D The output goes LOW when one or more of the inputs is LOW.
E None of the above

4. How can an NAND gate be made from more basic gates?

A An AND gate with inverters at the inputs
B An AND gate with an inverter at the output
C An OR gate with an inverter at the output
D An AND gate with a buffer at the output
E None of the above

5. If the input to an inverter is a logic 1, what is the output?

A 0
B 1
C undetermined

6. What is the base of the octal numbering system?

A 2
B 8
C 10
D 16
E None of the above

7. Which of the following does not produce a 1 at the output when all inputs are 1s?

A Buffer
B AND gate
C Exclusive OR gate
D OR gate
E None of the above

8. Which of the following input combinations will produce a logic 1 output from a three input NOR gate?

A 010
B 111
C 110
D 000
E None of the above

9. If both inputs of an OR gate are inverted, the result will be the same as which basic gate type?

A NAND
B AND
C NOR
D X-OR
E None of the above

10. How many inputs can a majority logic gate have?

A Five
B Any number
C Any odd number
D Any even number
E None of the above

27

Other digital integrated circuits

Gates aren't the only type of digital devices. Some additional digital ICs are discussed in this chapter.

Flip-flops

In chapter 25, you read about monostable multivibrators (which have one stable state) and astable multivibrators (which have no stable state). As might be expected, there is a third type of multivibrator, which has two stable states. A trigger signal is required to switch the output from one state to the other. This device is called a *bistable multivibrator* or *flip-flop*. Sometimes it is also called a *latch* or a *one-bit memory*. Figure 27-1 shows the circuit for a very simple flip-flop. It primarily consists of two inverters connected from input to output.

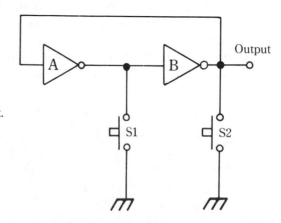

27-1   A simple flip-flop circuit.

If S1 is closed, the input to inverter B is pulled down to ground potential; for a positive logic system, this is a logic 0. Inverter B reverses the logic state to a 1. Besides being the output of the circuit, this signal is also fed back to the input of inverter A, which inverts the logic state back to a 0 and feeds this signal back into the input of inverter B. The circuit output will be a logic 1, even if S1 is reopened. The circuit latches itself into a logic 1 output state.

Similarly, if S2 is closed, the input of inverter A is forced to a logic 0. The output of inverter A is a 1, which is input to inverter B. This second-state inversion would produce a circuit output of 0. And of course, this signal is fed back to the input of inverter A, to latch the circuit into the 0 state, even after S2 is reopened. Either output state (1 or 0) will be held stable until the reverse state is initiated.

Closing both switches simultaneously forces the input and output of both inverters to logic 0. It is impossible to predict what the output of the flip-flop would be when the switches are opened. This is a *disallowed state*, and should be avoided.

Figure 27-2 is an adaptation of the basic circuit that prevents the disallowed state from occurring. For the best results, the switch should be a momentary-action, center-off SPDT switch.

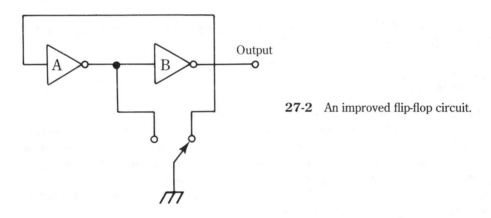

27-2   An improved flip-flop circuit.

The disadvantage of both of these circuits is that they can be *triggered* (forced to change states) only by mechanical switches. Relays could provide some form of automatic operation, but they are expensive and bulky. Fortunately, there is a better way of electronically triggering a flip-flop. Figure 27-3 is an improved version of the flip-flop circuit that can be triggered by the outputs of other logic gates.

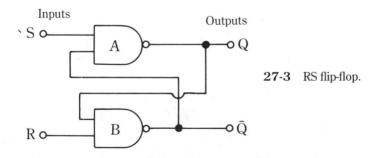

27-3   RS flip-flop.

Notice that there are two outputs from this circuit, labeled Q and $\overline{Q}$ (read these labels as "*Q*" and "*not $\overline{Q}$*"). The line over $\overline{Q}$ indicates that it is the *complement* (the opposite logic state) of Q. If Q is 1, $\overline{Q}$ must be 0, and if Q is 0, then $\overline{Q}$ must be 1.

This circuit also has two logic inputs. One is labeled *S* (*set*), and the other is labeled *R* (*reset*). Assume that when you first turn on this circuit, both S and R are at logic 1, and output Q is 0. Of course output $\overline{Q}$ must be a logic 1.

Gate A inputs are S (logic 1), and $\overline{Q}$ (logic 1). Gate A is a NAND gate, so its output (Q) remains at logic 0.

Similarly, gate B inputs are R (logic 1) and Q (logic 0), so its output ($\overline{Q}$) remains a logic 1. The circuit will be latched in this state as long as both inputs (S and R) are kept at logic 1.

If S is changed to a logic 0, the situation changes. The output of Gate A (Q) is changed to a logic 1, and this in turn converts the output of gate B ($\overline{Q}$) to a logic zero.

Even if S is now changed back to a logic 1 state, the output states will not change, because of the way they are fed back to the inputs of the gates. The circuit is latched into this new state, and any changes in the signal on the S input will have no effect on the output signal.

Now, if R is changed to a 0, it will change gate B output ($\overline{Q}$) back to a logic 1, and gate A output (Q) back to a 0. Further changes in the R input will have no effect on the output.

In other words, a 0 input at S sets the flip-flop (Q = 1, and $\overline{Q}$ = 0), while a 0 input at R resets the flip-flop (Q = 0, and $\overline{Q}$ = 1). A 1 at both inputs will latch the circuit in its present state.

Notice that a 0 at both inputs is once again a disallowed state. The output will be unpredictable under this condition. The truth table for this circuit is shown in Table 27-1. This type of circuit is called a *set-reset flip-flop*, or an *RS flip-flop*.

Table 27-1.
Truth table for the RS flip-flop.

Inputs		Outputs	
R	**S**	**Q**	**$\bar{Q}$**
0	0	Disallowed state	
0	1	0	1
1	0	1	0
1	1	No change—previous state	

JK Flip-flops

Another common type of bistable multivibrator is the *JK flip-flop*. The *JK* is used simply to distinguish this type of flip-flop from the RS type. It doesn't stand for anything in particular. The inputs and outputs of a JK flip-flop are illustrated in Fig. 27-4. Again, there are two outputs—Q and $\overline{Q}$. $\overline{Q}$ is always the complement of Q. This device also has a total of five inputs—*preset, preclear, J, K,* and *clock.*

The preset and preclear inputs work much like the set and clear inputs on an

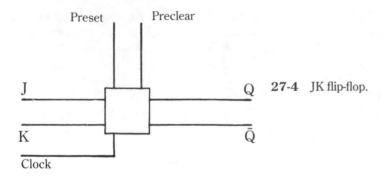

27-4   JK flip-flop.

RS flip-flop. A 0 input on the preset terminal immediately forces the Q output to a 1 state (and $\bar{Q}$ to a 0 state). Similarly, a 0 input on the preclear terminal immediately forces the Q output to a 0 state (and $\bar{Q}$ to a 1 state). If both of these inputs are logic 1, the output will be determined by the other three inputs.

Putting both preset and preclear in a logic 0 condition simultaneously is a disallowed state. These inputs and outputs are summarized in the truth table given in Table 27-2.

Table 27-2. Truth table for
the preset and preclear inputs of a JK flip-flop.

Inputs		Outputs	
Preset	**Preclear**	**Q**	**$\bar{Q}$**
0	0	Disallowed state	
0	1	1	0
1	0	0	1
1	1	Determined by clocked inputs	

The J and K inputs are *clocked inputs*. This means they can have no effect on the output until the clock input receives the appropriate signal.

There are two basic types of clocking—*level clocking* and *edge clocking*. In a level clocking system, the clock is triggered by the logic state of the input signal. It might be designed to trigger on either a 1 or a 0 (but not both). The clock will remain activated for as long as the input is held at the appropriate logic level.

Edge clocking, on the other hand, is triggered by the transition from one state to the other. Either the 0 to 1 (positive edge) transition, or the 1 to 0 (negative edge) transition can be used (but not both), depending on the design of the specific circuit. Obviously an edge triggered clock is activated for a much shorter time period than a level triggered device. For most digital work, edge triggering is usually preferred.

Clocked circuits have a number of advantages, especially within large systems. First, by triggering all of the subcircuits in a large system with the same clock signal, all operations can be forced to stay in step with each other throughout the system, preventing erroneous signals.

Another frequent source of errors in an unclocked circuit is noise on the input lines. The input is most likely to be noisy during the transition from one state to the other, especially when mechanical switches are used. No mechanical switch is perfect, and some bounce will be exhibited every time a switch is used. The slider will rapidly make and break contact many times before finally settling into position. See Fig. 27-5. For most applications this simply doesn't matter, but in digital circuits, it can be quite troublesome. If a flip-flop operated on each and every input pulse, the output would be very noisy and erratic and many undesired operations would take place. In this chapter, can read about a *bounceless switch*.

27-5 Mechanical switch bounce as it appears to a digital circuit.

In spite of this problem, if the inputs only change state during the time the circuit is held inactive by the clock, and the circuit is activated only when the inputs have had time to stabilize (only a tiny fraction of a second is needed), very clean, reliable outputs can be achieved. In the truth table shown in Table 27-3, *T* means the clock is triggered, and *N* means the clock is not triggered.

**Table 27-3. Truth table for
the clocked inputs of a JK flip-flop.**

Inputs			Outputs	
J	**K**	**Clock**	**Q**	**Q̄**
0	0	N	No change	
0	0	T	No change	
0	1	N	0	1
0	1	T	No change	
1	0	N	1	0
1	0	T	No change	
1	1	N	Output states reverse (0 becomes 1 and 1 becomes 0)	
1	1	T	No change	

Notice that if both the J input and the K input are at logic 1, the output will reverse states each time the clock is triggered. For example, assume the Q output starts out as a 0 ($\bar{Q} = 1$). On the first clocking pulse, Q will be a 1 ($\bar{Q} = 0$). The second clocking pulse will change Q back to a 0 ($\bar{Q} = 1$). The third clocking pulse will change Q to a logic 1 again ($\bar{Q} = 0$), and so forth.

D flip-flops

The JK flip-flop is quite useful and versatile, and there is no disallowed state for its clocked inputs (preset and preclear aren't used in most applications). But the JK flip-flop requirement for two inputs in addition to the clock is sometimes inconvenient in certain applications. The problem can be solved by using a *D flip-flop* (D stands for *data*). Figure 27-6 shows how a D flip-flop can be made from a JK flip-flop and an inverter. The J and the K inputs will always be at the opposite logic states. If J is a 1, K is a 0, and vice versa. Table 27-4 shows the truth table for a D type flip-flop.

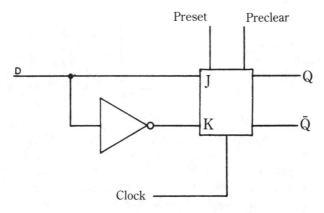

27-6   Converting a JK flip-flop to a D-type flip-flop.

Table 27-4. Truth table for a D-type flip-flop.

D (Input)	Ouputs	
	Q	Q̄
0	0	1
1	1	0

The D input is functioned only when the clock is triggered.

D type flip-flops also have preset and preclear inputs that function in the same way as in a JK flip-flop. Figure 27-7 shows a flip-flop circuit that is operated solely off of the clocking input. The D input is fed by the Q output. The clock is edge triggered so there is only time for a single state change during each clock pulse.

Assume that the Q output starts off at a logic 1 state. This means Q̄, and therefore D, must be at logic 0. Nothing happens until the clock is triggered; then the circuit looks at the data on the D input. Since this is a 0, the truth table tells you that Q should become a 0, and Q̄ should become a 1. This 1 is fed back to the D input, but by this time, the clocking pulse is gone, so the flip-flop waits until the next trigger

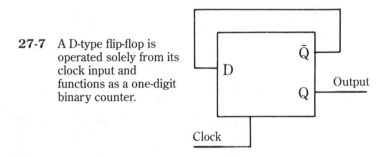

27-7 A D-type flip-flop is operated solely from its clock input and functions as a one-digit binary counter.

signal arrives. When the clock is triggered, the 1 on the D input changes Q back to 1, and $\overline{Q}$ back to 0, and you're back where you started.

The input and output signals are shown in Fig. 27-8. Notice that both the input and the output are square waves (the input can be any rectangular wave, but the output will always be a square wave). Notice also that it takes two complete input (clock) cycles to produce one complete output cycle. Two input cycles producing one output cycle means the output frequency is exactly one-half of the input frequency. For this reason, this type of circuit is often called a *frequency divider*. It divides the clock frequency by two.

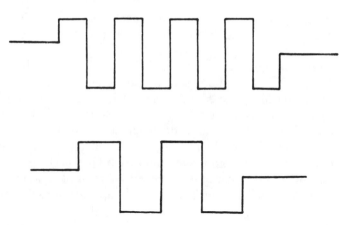

27-8   Input and output signals for the circuit shown in Fig. 27-7.

Counters

The circuit in Fig. 27-7 is also a binary counter. It counts the incoming clock pulses in binary form. But, because the binary system has only two digits, and this counter can only handle a single column of digits, the count starts over after every second input pulse. 0–1–0–1–0–1–0–1–. . . . The counting range can be extended to handle larger numbers by adding extra flip-flop stages, as shown in Fig. 27-9. This is a three-stage counter. Look at how it works.

The preset and preclear inputs of each of the three flip-flops are tied together. The preset inputs are connected to the voltage source, so they are always held at

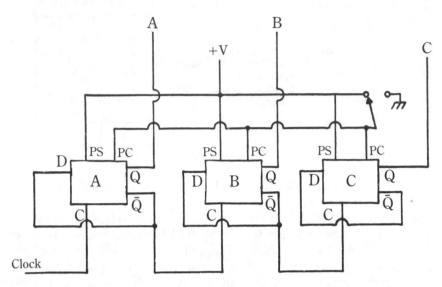

27-9   A three-stage counter.

logic 1. Remember, each integrated circuit package must also have its power supply pins connected to the power supply, even if this is not shown on the schematic diagram. It is always assumed.

The three preclear inputs go to a single SPDT switch, preferably with momentary contacts for one NC connection (shown in the diagram), and one NO connection. In its normal position, the switch sends a logic 1 to the preclear inputs. Remember that according to the truth table (Table 27-4) if both the preset and the preclear inputs are logic 1, the output is determined by the clock and D. But, if the switch is moved to its other position, a logic 0 is applied to the preclear inputs, forcing each of the Q outputs to a 0 state (and each $\overline{Q}$ to a 1), regardless of their previous states. This switch clears the counter by forcing it to a 000 output state. Start at this initial 000 condition and look at what happens on each clocking pulse. Assume the clock inputs trigger on the positive edge (that is, the 0 to 1 transition).

Clock pulse 1   Flip-flop A has a 1 on its D input (because D is tied to $\overline{Q}$). When the first clock pulse is received QA goes to 1, and $\overline{Q}$A goes to 0.

Flip-flop B is clocked by the 0 to 1 transition of the flip-flop A $\overline{Q}$ output. Since $\overline{Q}$A went from 1 to 0, no clocking pulse will be applied to flip-flop B. Its state will remain the same.

Similarly, flip-flop C is clocked by $\overline{Q}$B. This signal is not changed during this clock pulse, so flip-flop C's output state is also unchanged. Reading the outputs from right to left (that is, C B A), you have a binary count of 001, or 1.

Clock pulse 2 $\overline{Q}$A is a 0. This pulse is fed back to DA. When the clock pulse is received, QA goes to 0 and $\overline{Q}$A goes to 1.

Because $\overline{Q}$A has a 0 to 1 transition, flip-flop B is also triggered. Originally its outputs were QB = 0 and $\overline{Q}$B = 1, so the 1 on the DB input changes QB to a 1, and $\overline{Q}$B to a 0.

Flip-flop C is not triggered. Reading the outputs from right to left, you now have a binary count of 010, or 2.

Clock pulse 3 This pulse behaves like clock pulse 1, changing QA to a 1, and $\overline{QA}$ to a 0, and →leaving flip-flop B and C unchanged. The outputs now read 011, or 3.

Clock pulse 4 DA is being fed a 0, so QA changes to a 0 and $\overline{QA}$ changes to a 1. The 0 to 1 transition of $\overline{QA}$ triggers flip-flop B. DB is being fed a 0 from $\overline{QB}$, so QB becomes 0 and $\overline{QB}$ becomes 1.

Because $\overline{QB}$ has a 0 to 1 transition on this pulse, flip-flop C is also triggered. QC changes to a 1 and $\overline{QC}$ changes to a 0. Reading the outputs from right to left, you now have a binary count of 100, or 4.

Clock pulse 5 Once again, only flip-flop A is triggered. Notice that on any odd-numbered pulse, only the first flip-flop is triggered. QA switches to a logic 1 and $\overline{QA}$ goes to a logic 0. The output is now 101, or 5.

Clock pulse 6 QA changes from 1 to 0, and $\overline{QA}$ changes from 0 to 1. This triggers flip-flop B, causing QB to go from 0 to 1 and $\overline{QB}$ to change from 1 to 0. Flip-flop C is not triggered, so its output states remain the same. The binary count is now 110, or 6.

Clock pulse 7 As with the other odd numbered clock pulses, only flip-flop A is triggered. You now have a binary count of 111, or 7.

Clock pulse 8 QA changes from 1 to 0, and $\overline{QA}$ changes from 0 to 1, triggering flip-flop B. QB changes from 1 to 0, and $\overline{QB}$ changes from 0 to 1, triggering flip-flop C. QC also changes from 1 to 0, and $\overline{QC}$ goes from 0 to 1.

The binary count after the eighth clock pulse is 000 once more. The counter is reset for another series of count pulses. This pattern will continue repeating as long as there are incoming clock pulses.

Figure 27-10 compares the input (A clock) and the three output signals (A, B, and C). Notice that A is one half the original clock frequency. B is one half of the A frequency, or one quarter of the original clock frequency. C is one half of frequency B, or the original clock frequency divided by eight.

If, for example, the clock frequency was 6,000 Hz, output A would be 3000 Hz, output B would be 1500 Hz, and output C would be 750 Hz. In other words, a counter can be used as a higher level frequency divider.

The count (or frequency division) can be extended to any column of binary digits. That is, you can build counter circuits that count to 2, 4, 8, 16, 32, 64, 128, or so forth.

This maximum number of counts is called the *modulo* of the counter. Remember, a binary counter actually counts from 0 to one less than the modulo. For instance, a modulo-four counter would count 00 – 01 – 10 – 11 – 00 – 01 – 10 – 11 – 00 – 01 – 10 – 11 ... Or, in decimal, the count would translate to 0 – 1 – 2 – 3 – 0 – 1 – 2 – 3 – 0 – 1 – 2 – 3 ...

Suppose you needed a counter with a modulo that was not part of the regular binary series. For example, assume you need a counter with a modulo of six. Because six in binary is 110, this count cannot be achieved with a whole number of flip-flops. We need some way to set the count back to 000 after 101 (five).

Figure 27-11 shows the solution. The output to the NAND gate is fed back to

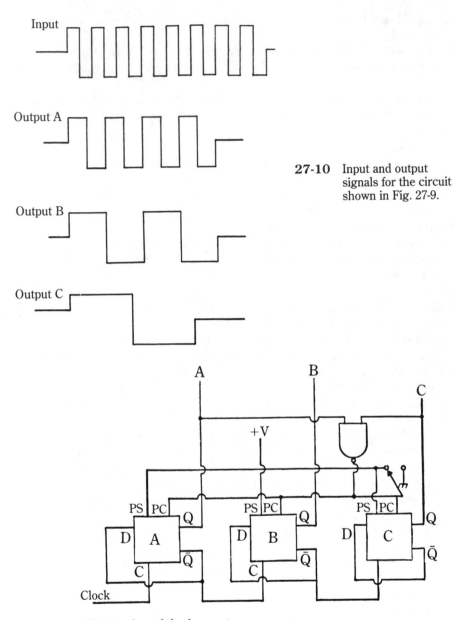

Input

Output A

27-10 Input and output signals for the circuit shown in Fig. 27-9.

Output B

Output C

A   B   C

+V

PS PC   Q   PS PC   Q   PS PC   Q

D   A   D   B   D   C

C   Q̄   C   Q̄   Q̄

Clock

27-11 A modulo-six counter.

the preclear input. Remember, a logic 0 level on this input line will force all of the Q inputs back to 0.

The output of the NAND gate will be a logic 1, unless both QA and QC are 1's. Whenever this happens, the gate's output goes to logic 0, allowing the preclear input to set the counter back to 000. As Table 27-5 shows, the binary count 101 (five) is the earliest point in the cycle where A and C are both at logic 1. Notice that the

**Table 27-5. Truth table
for the circuit of Figure 27-11.**

Binary output	Gate output	Decimal number
000	1	0
001	1	1
010	1	2
011	1	3
100	1	4
101	0	5
110*	1	6
111*	0	7

*These states never occur.

state of B doesn't matter in this case. Theoretically, the preclear input could also be triggered by binary 111 (seven) but this count will never be reached.

By using the appropriate number flip-flops and the correct gates, you can produce a counter for any whole number you choose. Because the ordinary counting system is built on a base of ten, a modulo 10 counter is particularly useful. Such a device would consist of a string of four flip-flops that are reset after a count of 1001 (nine).

This particular counter configuration is available in a number of single integrated circuit packages (with only slight variations from type to type). A modulo 10 binary counter is often called a *BCD* (short for *binary-coded decimal*). Table 27-6 shows the sequence of output states for a BCD.

Table 27-6. Output of a BCD circuit.

Binary input	Decimal equivalent
0000	0
0001	1
0010	2
0011	3
0100	4
0101	5
0110	6
0111	7
1000	8
1001	9
1010	Disallowed state
1011	Disallowed state
1100	Disallowed state
1101	Disallowed state
1110	Disallowed state
1111	Disallowed state

BCDs are especially useful to convert binary data to decimal numbers in display circuits. The four-line counter output is fed into another integrated circuit device called a *decoder/driver*, which rearranges the signals into seven outputs. These seven outputs are connected to the segments of an LED seven-segment display which will light up the decimal equivalent to the binary number at the decoder's input.

The driver section boosts the fan out so it can comfortably handle the relatively high current drawn by the LEDs. Some decoder/drivers are designed for use with common cathode displays. Others are intended for use with common anode displays. See Fig. 27-12 for the standard arrangement of these devices. The decoder and the driver section are usually (but not always) contained within a single integrated circuit package.

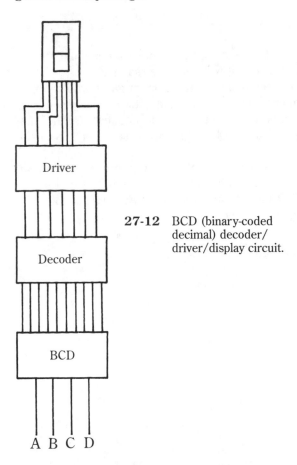

27-12 BCD (binary-coded decimal) decoder/driver/display circuit.

Shift registers

Figure 27-13 shows a variation on the basic counter circuit. This circuit does not necessarily count in a sequential manner, because on any specific clock pulse, the initial D input can be made either a 0 or a 1 via an external signal. Assume the

27-13   An SIPO (serial-in, parallel-out) shift register.

outputs all start at 000. On clock pulse 1, you will feed a logic 1 to the input. For all other clock pulses 0's will be inputted. Look at what happens on each clock pulse.

Clock pulse 1   Notice that the $\overline{Q}$ outputs are not used (although they could be, if complementary outputs were needed for some specific function). Remember, there is only time for a single state change to take place during a given clock pulse. This timing is essential for proper operation, and this is why only edge clocking is used for this type of circuit. In this example you will assume positive edge clocking (0 to 1 transition) is used, although negative edge (1 to 0 transition) can be employed just as well with only minor circuit changes.

QB is a 0, so DC is a 0, leaving QC at a 0 level when the clock pulse is received. Similarly, the initial 0 state of QA is fed to DB, holding QB at 0. But the input DA is being fed a logic 1 from an external source. This changes QA to 1. The binary output from right to left (C B A) is 001.

Clock pulse 2   QB is still 0, so there is still no change in flip-flop C output state on this clock pulse. QA is a 1, and this is fed into DB, so QB is now changed to a 1. DA (system input) is now receiving only 0s, so QA reverts back to logic 0. The binary output (from right to left) is now 010.

Clock pulse 3   QB, and thus, DC, is a logic 1, so QC changes to a 1. QA/DB is a logic 0, so flip-flop B output goes back to 0. DA is receiving only 0s, so QA remains at logic 0. The binary output (from right to left) is now 011.

Clock pulse 4   QB/DC is at logic 0, so QC also becomes a 0. The first two stages (A and B) also have 0's on their inputs, so their outputs stay at the 0 level.

The output has returned to the 000, initially cleared condition. The outputs will all stay 0 until another logic 1 is inputted to the system.

Notice how the 1 is shifted through the outputs. This circuit is called a *shift register*. This type of circuit is useful when selected digital information needs to be applied to different subcircuits at slightly different times in a specific sequence. It could be called a *digital delay circuit*.

One digit of digital information (a single 0 or 1) is called a *bit*. Bits are usually grouped into *words*, or *bytes*. Bytes are typically four, eight, or sixteen bits long. All bytes within a given system, of course, will be of equal length. Most shift registers are designed to hold a full byte of digital information.

The shift register shown in Fig. 27-13 accepts data in a serial fashion (one bit after another), and the output is parallel (all bits are available simultaneously). It is called a *SIPO* (serial-in, parallel-out) shift register.

The SIPO is probably the most commonly used type, but other combinations are also possible. Figure 27-14 shows a *SISO* (serial-in, serial-out) circuit. Figure 27-15 is a *PISO* (parallel-in, serial-out) device, and Fig. 27-16 is a *PIPO* (parallel-in, parallel-out) circuit. It is also possible to build a shift register with both serial and parallel inputs and outputs.

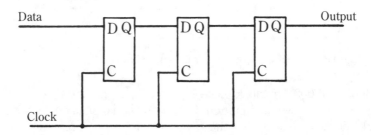

27-14   An SISO (serial-in, serial out) shift register.

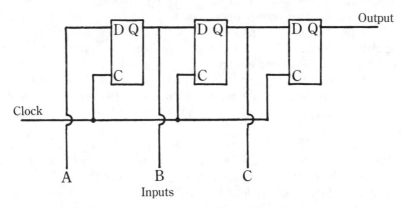

27-15   A PISO (parallel-in, serial out) shift register.

There are many other digital devices, but most are made up of combinations of the basic devices described in the last two chapters. Many complex digital functions (incorporating several of these basic devices) are available in single chip MSI and LSI integrated circuits.

Multiplexers and demultiplexers

In some moderate to complex digital circuits, you will need one signal at a given point in the circuit part of the time, but other signals will be needed instead at other times. To build such a system, you obviously need some way to select between two or more possible inputs. A subcircuit that has been designed for this purpose is called a *multiplexer*. Sometimes the name is abbreviated as *MUX*.

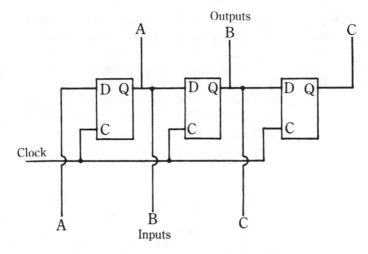

27-16   A PIPO (parallel-in, parallel out) shift register.

Figure 27-17 shows how a simple multiplexer circuit can be constructed from eight NAND gates. Use two standard quad NAND gate ICs. This circuit has four

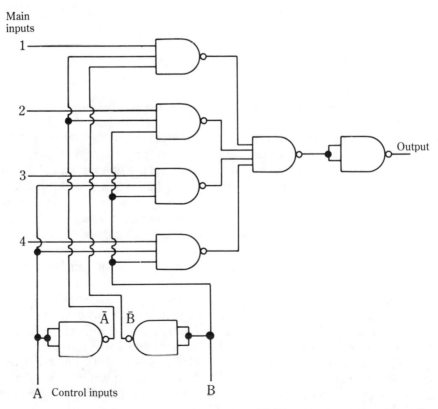

27-17   Standard NAND gates can be combined to create a 1-of-4 multiplexer.

main inputs (labelled 1 through 4 in the diagram), and two control inputs (labelled A and B in the diagram). The logic signals fed to the control inputs determine which of the main input signals will reach the output. Only one of the main input lines is active at any given instant. These main inputs are often called *data inputs* to more clearly distinguish their function from that of the control inputs.

The truth table for this multiplexer circuit is given in Table 27-7. Notice how only one of the four main input lines is of significance for any specific combination of signals for control inputs A and B. In a truth table, a small *x* means *don't care*. That is, the logic state of this input signal is irrelevant in determining the output logic state (or states).

Table 27-7. Truth table for a 1-of-4 multiplexer.

Control inputs		Main inputs				Output
A	B	1	2	3	4	
0	0	0	x	x	x	0
0	0	1	x	x	x	1
0	1	x	0	x	x	0
0	1	x	1	x	x	1
1	0	x	x	0	x	0
1	0	x	x	1	x	1
1	1	x	x	x	0	0
1	1	x	x	x	1	1

x = "don't care."

Because the control inputs can be used to select any of the main inputs for the output, this type of circuit is occasionally called a *data selector*, although *multiplexer* is generally the preferred name. As you can see, the multiplexer acts like a sort of digital traffic manager, determining just what signal goes where in the circuit.

The circuit shown in Fig. 27-17 is a 1-of-4 multiplexer, because any one of four main inputs can be selected. The same principle is commonly expanded to make 1-of-8 and 1-of-16 multiplexers. Dedicated multiplexers in all three of these sizes are readily available in IC form.

Some practical multiplexers invert the selected main input signal before feeding it to the output. In some cases, this might be desirable, in other applications it could be a problem (easily solved by adding an external inverter stage to the output line). In many applications such signal inversion won't really make any practical difference one way or the other. The truth table for an inverting 1-of-4 multiplexer is given in Table 27-8. Compare this truth table to the one shown in Table 27-7.

Multiplexers can take the place of complex gating circuits. For example, consider the truth table that is shown in Table 27-9. At best, it would be a definite nuisance to generate this truth table using separate gates. A typical circuit for accomplishing this task is shown in Fig. 27-18.

Table 27-8. Some multiplexers
invert the logic signal before the output.

Control inputs		Main inputs				Output
A	B	1	2	3	4	
0	0	0	x	x	x	1
0	0	1	x	x	x	0
0	1	x	0	x	x	1
0	1	x	1	x	x	0
1	0	x	x	0	x	1
1	0	x	x	1	x	0
1	1	x	x	x	0	1
1	1	x	x	x	1	0

x = "don't care"

Table 27-9. This truth table
would be difficult to implement
using standard gates.

Inputs				Output
A	B	C	D	
0	0	0	0	0
0	0	0	1	0
0	0	1	0	1
0	0	1	1	0
0	1	0	0	0
0	1	0	1	1
0	1	1	0	1
0	1	1	1	0
1	0	0	0	0
1	0	0	1	1
1	0	1	0	1
1	0	1	1	1
1	1	0	0	0
1	1	0	1	0
1	1	1	0	1
1	1	1	1	0

A 1-of-16 multiplexer can accomplish the same thing with a much simpler circuit. The 74150 is a typical 1-of-16 multiplexer IC. This chip is part of the TTL logic family (see Chapter 28). The 74150's pin-out diagram appears as Fig. 27-19.

Because there are 16 main or data inputs, four control inputs (labelled A through D) are needed to uniquely identify each input address. These control inputs will correspond to our logic inputs in the truth table of Table 27-9.

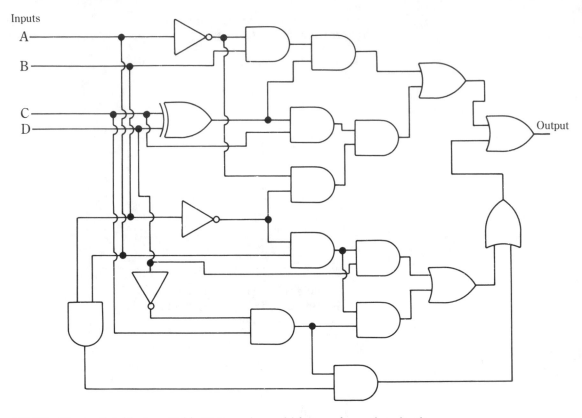

27-18   The truth table from Table 27-9 requires a fairly complex gating circuit.

The 74150 happens to be an inverting type multiplexer, so you feed each data input with the opposite logic state desired for the appropriate combination of control inputs. For example, when the control inputs are set to 1010, you want an output of logic 1, so you apply the opposite value (logic 0) to data input 10 (pin 21).

The complete circuit for generating the truth table in Table 27-9 is shown in Fig. 27-20. Notice how much simpler and easier to follow this diagram is compared to the discrete gating circuit of Fig. 27-18.

Literally any truth table can be readily generated using a multiplexer in this manner. In some cases, it might actually be more convenient to use separate standard gates. There would be little point in using a multiplexer to simulate a four input NAND gate, for example. But a multiplexer can really come in very handy when complex and/or unusual truth tables are called for.

A 1-of-16 multiplexer like the 74150 can, by definition, generate 2^{16} different truth tables. In other words, there are 65,536 possible combinations of inputs and outputs. Even allowing for the fact that some of these combinations are trivial or useless, this is still an incredibly versatile device.

A multiplexer can also be helpful when unusual counting sequences are called for. Multiplexers have many additional applications. For example, they can be used

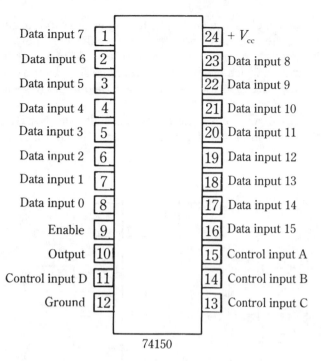

27-19 The 74150 is a 1-of-16 multiplexer in IC form.

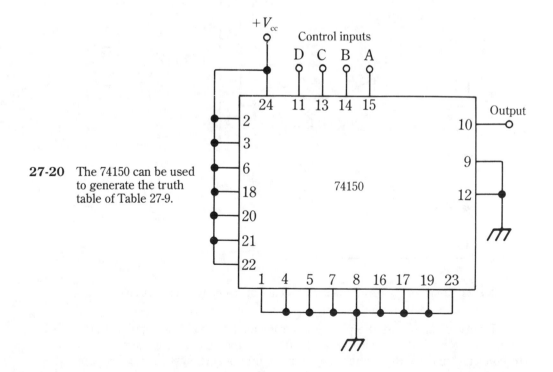

27-20 The 74150 can be used to generate the truth table of Table 27-9.

to scan a series of switches or a keyboard, testing each switch to see if it is open or closed, and telling the rest of the circuitry to respond accordingly.

The opposite of a multiplexer is a *demultiplexer.* Where the control inputs of a multiplexer determine which of several inputs will be seen by the output, the control inputs of a demultiplexer determine which of several output lines will respond to the single input.

A simple 1-of-4 demultiplexer circuit can be built from NAND gates, as shown in Fig. 27-21. The action of this circuit is outlined in the truth table given in Table 27-10.

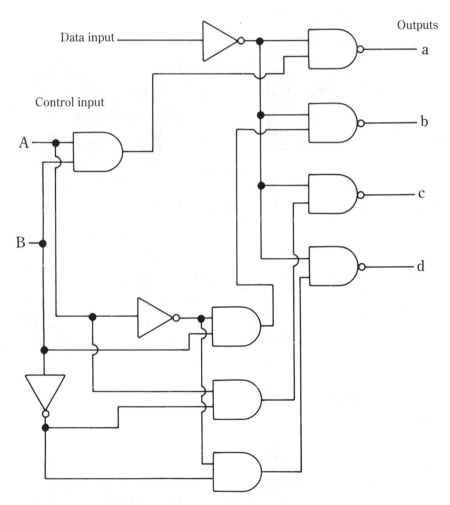

27-21   Standard gates can be used to construct a demultiplexer circuit.

Figure 27-22 shows the pin-out diagram for the 74154, a popular TTL 1-of-16 demultiplexer IC. It is much like a 74150 in reverse. A demultiplexer is often used to convert binary numbers into some other numbering system. For example, the

Table 27-10. Truth table for a 1-of-4 demultiplexer.

Control inputs		Main input	Outputs			
A	B		a	b	c	d
0	0	0	0	1	1	1
0	0	1	1	1	1	1
0	1	0	1	0	1	1
0	1	1	1	1	1	1
1	0	0	1	1	0	1
1	0	1	1	1	1	1
1	1	0	1	1	1	0
1	1	1	1	1	1	1

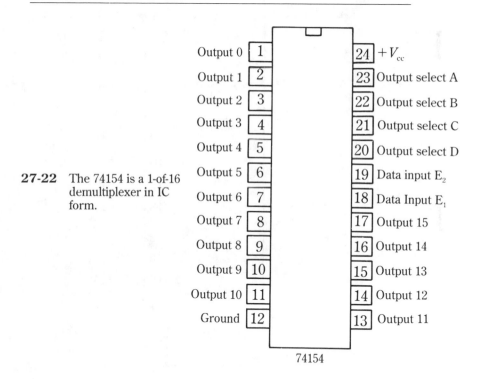

27-22   The 74154 is a 1-of-16 demultiplexer in IC form.

Output 0	1	24 $+V_{cc}$
Output 1	2	23 Output select A
Output 2	3	22 Output select B
Output 3	4	21 Output select C
Output 4	5	20 Output select D
Output 5	6	19 Data input E_2
Output 6	7	18 Data Input E_1
Output 7	8	17 Output 15
Output 8	9	16 Output 14
Output 9	10	15 Output 13
Output 10	11	14 Output 12
Ground	12	13 Output 11

74154

74154 can convert four digit binary numbers into single digit hexadecimal (base 16) numbers with the simple circuit shown in Fig. 27-23. For any combination of control inputs (any binary number from 0000 to 1111), one and only one of the sixteen outputs will be at logic 0. The other fifteen outputs will be held HIGH (logic 1), as outlined in Table 27-11.

The name *demultiplexer* is often abbreviated as *DEMUX*. Because these devices are frequently used for decoding digital data, they are often called *decoders*. By the same token, multiplexers might sometimes be referred to as *encoders*.

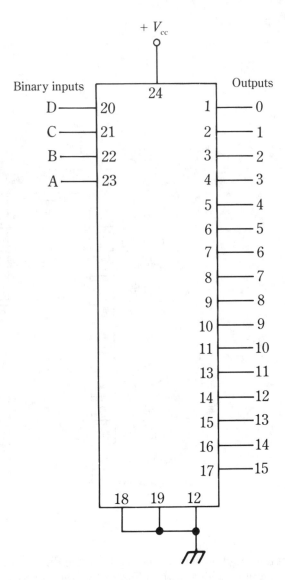

27-23 A demultiplexer can be used as a binary-to-hexadecimal converter.

**Table 27-11. A 1-of-16 demultiplexer can be used
as a binary to hexadecimal converter. (See Fig. 27-23.)**

Control inputs				Outputs			
a	b	c	d	0 1 2 3	4 5 6 7	8 9 A B	C D E F
0	0	0	0	0 1 1 1	1 1 1 1	1 1 1 1	1 1 1 1
0	0	0	1	1 0 1 1	1 1 1 1	1 1 1 1	1 1 1 1
0	0	1	0	1 1 0 1	1 1 1 1	1 1 1 1	1 1 1 1
0	0	1	1	1 1 1 0	1 1 1 1	1 1 1 1	1 1 1 1
0	1	0	0	1 1 1 1	0 1 1 1	1 1 1 1	1 1 1 1
0	1	0	1	1 1 1 1	1 0 1 1	1 1 1 1	1 1 1 1
0	1	1	0	1 1 1 1	1 1 0 1	1 1 1 1	1 1 1 1
0	1	1	1	1 1 1 1	1 1 1 0	1 1 1 1	1 1 1 1
1	0	0	0	1 1 1 1	1 1 1 1	0 1 1 1	1 1 1 1
1	0	0	1	1 1 1 1	1 1 1 1	1 0 1 1	1 1 1 1
1	0	1	0	1 1 1 1	1 1 1 1	1 1 0 1	1 1 1 1
1	0	1	1	1 1 1 1	1 1 1 1	1 1 1 0	1 1 1 1
1	1	0	0	1 1 1 1	1 1 1 1	1 1 1 1	0 1 1 1
1	1	0	1	1 1 1 1	1 1 1 1	1 1 1 1	1 0 1 1
1	1	1	0	1 1 1 1	1 1 1 1	1 1 1 1	1 1 0 1
1	1	1	1	1 1 1 1	1 1 1 1	1 1 1 1	1 1 1 0

Main Input grounded for constant logic 0 input.

Self-test

1. What is another name for a flip-flop circuit?

A Digital gate
B Astable multivibrator
C Bistable multivibrator
D Monostable multivibrator
E None of the above

2. What happens when a D flip-flop is triggered?

A The output reverses states
B The output goes LOW
C The output goes HIGH
D The output remains unchanged

3. Which of the following is not a standard type of flip-flop?

A JK
B RS
C QS
D D
E None of the above

4. What basic type of circuit is used to build up a multistage counter?

A AND gates
B Amplifiers
C Flip-flops
D Schmitt triggers
E None of the above

5. What is the maximum count of a five-stage counter in binary?

A 10000
B 11111
C 100000
D 111111
E None of the above

6. What is the maximum count of a five-stage counter in decimal?

A 32
B 16
C 50
D 31
E None of the above

7. If the Q output of a counter stage is at logic 1, what will be the value at $\overline{Q}$?

A 0
B 1
C Either 0 or 1

8. A counter stage can divide an input frequency by what factor?

A 0.2
B 3
C 4
D 2
E None of the above

9. Which of the following is a disallowed state for a BCD circuit?

A 0000
B 1001
C 0111
D 1101
E None of the above

10. How many outputs does a six-stage PISO shift register have?

A One
B Two
C Five
D Six
E None of the above

28
Logic families

There are a number of basic methods for designing digital circuits. Digital integrated circuits are classified according to their logic families or basic design approach. The earliest digital integrated circuits were usually either *RTL* or *DTL* devices. *RTL* stands for *resistor-transistor Logic*. Each digital gate is basically comprised of a resistor and a transistor. *DTL*, on the other hand, is *diode-transistor Logic*. Each digital gate is made up primarily of a diode and a transistor. Remember, these are not discrete components, but their electrical equivalents etched into the semiconductor material of an integrated circuit chip.

Although RTL and DTL ICs are relatively easy to manufacture, they are rather slow and inefficient in their power consumption. An improved digital family is *TTL*, or, as it is sometimes called, T^2L. This abbreviation stands for *transistor-transistor logic*.

TTL

Thanks to mass production and relative simplicity of design, TTL integrated circuits are quite inexpensive (as low as 25¢ a chip from some suppliers). These devices are also capable of high-speed operation—20 MHz (20,000,000 Hz) is typical, but some are capable of frequencies as high as 125 MHz (125,000,000 Hz). They have a fair fan-out capability (10 is the average), and a reasonable immunity to noise.

TTL devices are probably the most popular and widely available logic family on today's market. This fact is due to both the number of manufacturers making TTL ICs, and the large number of different devices available. Table 28-1 lists some of the more popular TTL devices and their basic functions.

Usually TTL integrated circuits will be numbered in the 74XX format, but there are a few exceptions. Often you will see devices numbered 54XX. These two numbering systems are generally interchangeable. For instance, a 7404 is essentially identi-

Table 28-1. TTL devices and their functions.

7400	Quad two-input NAND gate
7402	Quad two-input NOR gate
7404	Hex inverter
7408	Quad two-input and gate
7410	Triple three input NAND gate
7414	Hex Schmitt triggers
7417	Hex buffer
7432	Quad two-input or gate
7442	BCD to 1 of 10 decoder
7447	BCD to seven segment decoder/driver
7473	Dual JK level-triggered flip-flop
7474	Dual D edge-triggered flip-flop
7486	Quad exclusive-OR gate
7490	Decade counter
7493	Binary counter(+16)
7495	Four bit PIPO shift register
7412	Monostable multivibrator
74150	1-of-16 data selector

cal to a 5404. The only significant difference is that the 54XX devices can operate over a wider temperature range. This range is usually only needed for military and satellite applications. For most general applications the 74XX/54XX line of devices is ideal. The power supply for standard TTL integrated circuits must be a tightly regulated 5 V. The supply voltage should not be allowed to vary more than ±0.5 V.

When a TTL gate switches output states (goes from a 0 to a 1, or vice versa) it draws a large surge or current. This can cause sharp high-voltage spikes to appear on the power supply lines. These spikes could damage the delicate semiconductor chips.

To protect the integrated circuits, connect *despiking capacitors* between the +5 V and ground lines as close to the IC packages as possible. One 0.01 μF to 0.1 μF capacitor is usually sufficient to protect up to four gate packages, if they are closely placed.

For TTL circuits a logic 1 is usually defined as a signal of about + 2.4 V, while logic 0 is about half a volt above absolute ground. Any signal between these levels is undefined, and may be interpreted by the gate as a 0, a 1, or simply noise.

One important factor to bear in mind is that an unconnected TTL input will usually pull itself up to a logic 1 level. This can cause quite a bit of confusion to a circuit designer who is not prepared for it. Also, under certain conditions, an unused gate can influence the output of another gate within the same IC package. All unused inputs should be tied to one or the other fixed logic state, preferably unused inputs should be connected to ground.

There are a number of subfamilies within the basic TTL group. These trade off various advantages and disadvantages.

Low-power TTL

Regular TTL integrated circuits use a relatively small amount of power individually, but in moderate to large systems, using many IC packages, the power drain can very quickly add up. To alleviate this problem, special *low-power TTL* ICs are sometimes used.

These low-power units are numbered in the same way as regular TTL devices, but with a *L* added in the middle to indicate the low-power status. For example, a low-power version of the 7400 quad NAND gate is the 74L00. A low-power TTL integrated circuit generally consumes only about a tenth as much power as the standard TTL version.

This lower power consumption does not come without disadvantages, however. Low-power TTL chips can operate only about a tenth as fast as regular TTL devices.

High-speed TTL

Another subfamily of TTL devices is *high-speed TTL*. A high-speed 7400 would be in the number series beginning with 74H00. This class of devices operates about twice as fast as regular TTL, but consumes about twice as much power.

High-speed TTL integrated circuits are becoming less and less common. They are being replaced by Schottky devices, which are generally superior, although somewhat more prone to noise problems.

Schottky TTL

By using *Schottky diodes* in the circuit design, you can achieve a better speed-power trade off. Schottky diodes are very fast switching diodes, so *Schottky TTL integrated circuits* can be used for very high frequency operation. Generally, a Schottky TTL gate can operate about three and one-half times as fast as a regular TTL gate, but the power consumption is only doubled. A Schottky version of the 7400 would be numbered 74S00.

Low-power Schottky

By combining the design techniques of low-power TTL and Schottky TTL devices, integrated circuits can be manufactured that operate at the same speed as regular TTL, but at only a fifth of the power consumption. Obviously devices in this group are called *low-power Schottky TTL*. A low-power Schottky equivalent to the 7400 would be numbered 74LS00. The various TTL subfamilies are compared in Table 28-2. Regular TTL is used as the comparison standard.

CMOS

Another major category of digital devices are called *CMOS integrated circuits*. CMOS standards for *complementary metal-oxide silicon*. You have already read about metal-oxide semiconductors in the section on MOSFETs.

CMOS integrated circuits can operate at very high speeds with relatively low power consumption. CMOS devices are also much more tolerant of power sup-

Table 28-2. Comparing the TTL subfamilies.

Subfamily	Gating speed	Power consumption
Standard TTL	1	1
Low power	0.1	0.1
High power	2	2
Schottky	3.5	2
Low power Schottky	1	0.2

ply fluctuations than are TTL units. In fact, a CMOS IC can be powered by any voltage between +3 and +12 V. Surges to +15 V can be withstood but should be avoided, if possible.

However, CMOS integrated circuits tend to be somewhat more expensive than their TTL counterparts. Also, they are susceptible to damage from bursts of static electricity. Anything that touches a CMOS chip should be properly grounded, and that includes the human hand. Newer units are less sensitive than older devices, but problems can still arise unless precautions are taken.

There are two common numbering systems for CMOS integrated circuits. One system reflects TTL numbering. A 7400 TTL quad NAND gate could be replaced by a 74C00 CMOS unit. The equivalent in the other numbering systems would be a CD4011. The CD40XX system is more common and the 74CXX devices are gradually disappearing from the market. Table 28-3 compares a 7400 TTL quad NAND gate with a CD 4011 CMOS quad NAND gate.

Table 28-3. Comparing a 7400 TTL quad NAND gate with a CD4011 CMOS quad NAND gate.

Device	CD4011	7400
dc supply voltage range	−0.5 to +15 V	+4.5 to 5.5V
power dissipation	500 mW	60 mW

Device	CD4011 +5 V supply	CD4011 +10 V supply	7400
Propagation delay time, ns (nanoseconds)	50	25	
Output voltage	——	——	22
Logic 0	0	−0.5	0.4
Logic 1	4.95	9.95	2.4

ECL

A somewhat less common but still important logic family is *ECL*, or *emitter-coupled logic*. ECL devices are capable of extremely high switching speeds. The output transistors in an ECL gate are biased so that they are in their active region at all times. The transistors are not driven into saturation, so there is no problem with stored base charge. This operation minimizes propagation delay times.

The ECL family of digital devices differs from the more common TTL and CMOS devices in that the logic levels are negative. A logic 1 is represented by a voltage of about -0.9 V. A logic 0 is approximately -1.75 V. This voltage level can introduce some problems when interfacing ECL devices with circuitry from the other logic families.

The power supply for ECL devices should also be negative (with respect to ground). Acceptable supply voltages range from -8 V to -3 V. The preferred nominal supply voltage is -5.2 V.

The high-speed capability of ECL devices comes at a price, of course. Power dissipation is quite high because the transistors within the gate are continuously on. However, at high operating speeds, the differences in current consumption between ECL and other logic families becomes increasingly less significant. Of course, when the high speed capabilities of ECL devices is taken advantage of, great care must be taken in circuit layout. Interconnecting leads must be kept as short as possible. Shielding might be necessary in some circuits.

The inputs to ECL gates exhibit very high impedances. This means that to ensure reliable and predictable operation, no inputs should be left floating. All inputs must be fed an unambiguous logic 1 or logic 0.

ECL outputs are open emitters. A small external load resistor is required. The value will typically be between 250 to 500 Ω. The load resistor provides a current path for the output transistor.

Combining logic families

In some practical digital applications, it might be necessary to mix devices from different logic families in a single circuit, or circuits designed around different logic families might need to be interfaced. There are many possible reasons for such needs. For example, some unusual devices might be available in one logic family. Often different portions of a large system might have different requirements, calling for different logic families. For instance, in a circuit using many ICs, keeping the power consumption is likely to be a major design consideration. However, a few stages within the system might need to operate at very high switching speeds. It would probably be wasteful and needlessly expensive to use high-speed ICs throughout the entire circuit, if the high frequencies will only be encountered by just a few of the devices. The solution is to mix logic families.

Unfortunately, devices from differing logic families cannot just be directly wired to one another, willy-nilly. Different power supply voltages are generally used, and the logic levels might be defined at different voltage levels. Input and output currents might also be badly mismatched, creating potential fan-in/fan-out problems. Such problems might be severe enough to actually render the entire circuit inoperable.

By using a few simple tricks, the various logic families and sub-families can usually be successfully used together within a single circuit.

The addition of a simple pull-up resistor, as shown in Fig. 28-1 allows a TTL gate to drive a CMOS device without much trouble. The supply voltage must be suitable for the TTL ICs in the circuit—that is, a well-regulated +5 V. CMOS gates will work with this lower than normal voltage, although optimum performance generally requires a higher supply voltage. The value of this pull-up resistor is not terribly critical. A fairly low value is generally used. In most cases a 1 kΩ (1000 Ω) resistor will serve very well for this purpose.

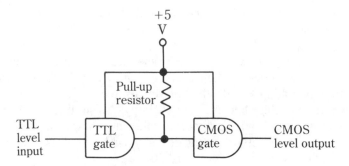

28-1   A pull-up resistor can allow a TTL (transistor-to-transistor logic) gate to drive a CMOS device.

Some CMOS devices can drive TTL gates directly, as shown in Fig. 28-2, provided that the same +5 V power supply voltage is used for both devices. Remember, TTL gates are very fussy about their supply voltage. It must be a well regulated +5 V, or the TTL chip could be permanently damaged.

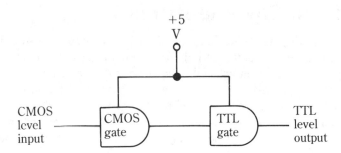

28-2   Some CMOS gates can drive TTL ICs directly.

Although this direct CMOS to TTL connection will work in some cases, a CMOS driving TTL situation will usually involve problems stemming from the ability of the CMOS device to supply and sink the required currents for each of the logic states. Typical TTL specifications call for current no more than −1.6 mA for a logic 0, and no more than 40 μA for a logic 1. In cases where a single device doesn't have enough current handling capability, two or more identical devices can be paralleled, as shown in Fig. 28-3.

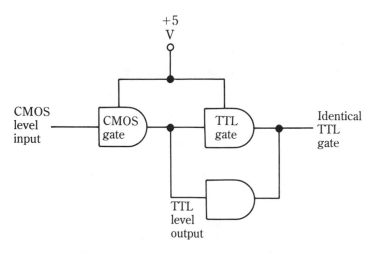

28-3   Paralleling identical gates allows CMOS gates to drive TTL devices.

A few CMOS devices are specially designed especially to make interfacing with TTL components easier. The CD4009A hex inverter and the CD4010A hex buffer are ideal for use as logic-level shifters. Such devices are particularly handy when the CMOS portion of the circuit uses a higher supply voltage than the TTL devices can handle. CMOS devices generally offer their peak performance with a supply voltage in the +9 to +15 V range. These logic-level shifting chips have multiple power supply terminals for precisely this purpose. One terminal can take a relatively high voltage (9 to 15 V) from the CMOS portion of the circuit, while the +5 V required for TTL can be fed into the same intermediate chip at the second supply voltage terminal. Figure 28-4 demonstrates the use of a CD4010A buffer in interfacing a CMOS gate with a TTL device. Of course, the CD4009A could be substituted directly into this circuit in place of the CD4010A if you wish to invert the CMOS output signal before feeding it into the TTL gate.

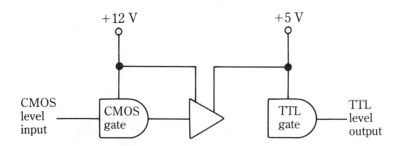

28-4   Special buffers, such as the CD4010A, can be used to interface CMOS and TTL gates.

Self-test

1. Which of the following is not a standard logic family?

A VMOS
B ECL
C TTL
D RTL

2. Which of the following logic families is considered obsolete?

A TTL
B CMOS
C DTL
D ECL
E None of the above

3. Which of the following logic families is most commonly used by hobbyists?

A DTL
B TTL
C ECL
D RTL
E None of the above

4. What is a good supply voltage for TTL devices?
A −5.2 V
B +12 V
C +25 V
D +5 V
E None of the above

5. What is a good supply voltage for CMOS devices?
A −5.2 V
B +12 V
C +25 V
D +5 V
E None of the above

6. Which of the following TTL devices use Schottky diodes in its internal circuitry?

A 74LS04
B 7400
C 74L17
D 74H86
E None of the above

7. Which of the following logic families has the highest switching speed capability?

A RTL
B TTL
C CMOS
D ECL

8. Which of the following can damage CMOS devices?

A Static electricity
B A 0 V input
C A supply voltage greater than 5 V, but less than 10 V
D A 1 V input
E None of the above

9. Which of the following is not a CMOS device?

A 74C10
B CD4001
C 7404
D 4049
E None of the above

10. Which of the following logic families can stand the greatest degree of variation in the supply voltage?

A TTL
B CMOS
C Schottky TTL
D DTL
E None of the above

29

Microprocessors

Digital circuits were initially designed primarily for use in computers. Microprocessors made the modern personal computer practical. The first computers used tubes and relays, and they filled buildings. When the transistor was developed, a computer only filled a large room. Using low-level ICs, a computer the size of a desk could be built. Today, there are small, desk-top computers, thanks to specialized LSI integrated circuits. The most sophisticated and complicated ICs developed have been microprocessors.

CPUs and microprocessors

The "brain" of a computer is the *CPU*, or *central processing unit*. This circuit accepts digital data and commands. It interprets what the commands mean, and performs mathematical operations on the data. In modern equipment, the CPU circuitry is almost always contained within a single IC chip, known as a *microprocessor*. A microprocessor is sometimes called a "computer on a chip," but this isn't entirely accurate. In most cases, a number of additional external ICs are required to form a practical, operating computer.

A CPU contains a number of *registers*, or temporary storage spaces. In its simplest form, one number for a given equation is put into one register. When the next number is entered, it is mathematically combined with the number already in the register. The register now contains the result, which is fed out to the display. This type of register is called the *accumulator register*. The circuitry for performing these mathematical functions is called the *arithmetic logic unit*, or *ALU*. Most calculators use CPUs that are not much more than simple ALUs.

The CPUs used in computers, however, are much more complex and have a number of additional registers. These registers are used for various purposes, such as storing an instruction, or keeping track of its place in the program. The 8080 (a popular CPU IC), for example, contains no fewer than 13 registers.

In addition to the registers, there must also be some way to get data in and out of the CPU to be manipulated. This transfer is done via the *data bus*. The data bus might consist of four, eight, or sixteen lines for digital data. Eight is probably the most popular number.

Each line carries a single *bit* (a 0 or a 1) of information. All of the lines together simultaneously carry a *word* of digital information. For a given system, the words are always of the same length. An eight-bit, or a 16-bit word is called a *byte*. A four-bit word is called a *nibble*. The size of the words used determines the amount of information they can communicate. A nibble, for example, has only 16 possible combinations. An eight-bit byte can have over 65,000 different values. Obviously, the greater the word length, the more powerful and versatile the computer.

Usually the same data bus is used for both inputting and outputting data. This means there must also be an additional single line input for telling the CPU whether data should be coming in or going out at any given instant.

There is also a *memory address bus* that is used to find a specific location in the memory. This typically consists of sixteen lines, so over 65,000 separate memory locations that can be individually addressed. Newer microprocessors use 16-bit commands and 32-bit address lines for very large memory access.

Commands and data

A CPU or microprocessor "understands" a number of commands. These commands must be in binary form. Examples are 01001101 or 11101001. Notice that the commands are in the same form as the data. The CPU uses position (what came just before the current number) and timing to distinguish between commands and data. A command tells the CPU what to do with the data. For commands, each binary number has a specific operational meaning for the CPU. A series of commands to perform a specific task is called a *program*.

If you program the microprocessor directly using the binary number commands, this is called *machine-language programming*. Translation programs are available to allow you to program the microprocessor with a more convenient (more English-like) set of commands.

The next step up from machine-language programming is *assembly-language programming*. Each binary number command is replaced with an easy to remember *mnemonic*. For instance, you use the instruction "ADD A,B" for adding values A and B instead of using the binary 10011100. Each assembly-language command corresponds to exactly one machine-language command.

For higher-level languages, such as BASIC, or Pascal, each user-entered command can be translated into a sequence of several machine-language commands in sequence. A special program (a *compiler* or an *interpreter*) is used to convert the English- like commands into the binary form understandable by the CPU.

Components of a computer

A CPU or microprocessor is not a full computer in itself. Figure 29-1 shows the basic structure of a typical computer. Two or more of these stages can be included on a single IC, or separate ICs are used for each stage. Some stages are made up of multiple ICs. This fact tends to be especially true for the memory stage. For your purpose here, you will consider the various stages of the computer to be separate and distinct. There are four primary sections of a computer:

Processor (CPU)
Memory
Input port
Output port

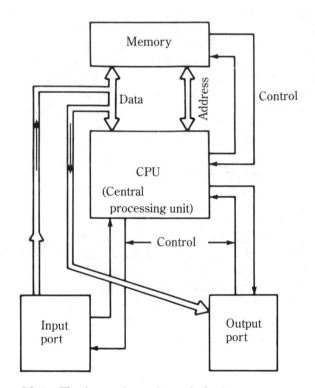

29-1   The four main sections of a basic computer.

The processor does the actual computing. A microprocessor is just a miniaturized processor. The memory stores the program commands and data used in executing the program.

The input and output ports allow the microprocessor to communicate with the outside world (anything that is not an integral part of the computer itself). The input port permits data from some external device (such as a keyboard or a *modem,* which is a device that can transfer data over telephone lines) to be fed into the computer.

Similarly, the output port lets the computer feed its results out to some external device (such as a printer or monitor screen).

Communication between the various internal sections of the computer are accomplished via buses. These buses are digital signal lines that can carry coded binary data either serially (one bit at a time) or in parallel (several bits at once). The binary data on a bus can represent numerical values, alphanumeric characters, or machine-language commands. The only difference is in how the CPU is instructed to interpret the binary information.

The *data bus* connects the microprocessor to everything else. The data bus goes from the CPU to the memory, the input port, and the output port. Data flows from the input port to the CPU, from the CPU to the output port, or in either direction between the memory and the CPU. The CPU determines which piece of data goes where.

The *address bus* goes from the CPU to the memory. The data on this bus determines which portion of the memory the microprocessor wants to use. The concept is simple enough to understand if you think that the CPU needs to know the address to find a friend's home.

The third bus is used for system control and synchronization. The signals on this bus keep the various sections of the computer functioning simultaneously. For example, the input port uses this bus to let the microprocessor know when there is some incoming data available from an external input device. Without proper synchronization, the data exchanged between the various sections of the computer would become garbled and meaningless.

Memories

Even with several registers, the CPU cannot store very much data, and it can only handle a single program instruction one step at a time. For this reason, some sort of external memory storage device is required for the computer to be useful.

There are two basic types of memory. They are called *ROMs* and *RAMs*. A ROM is a *read-only memory*. Data is permanently stored in this type of device. The data is either placed in the memory by the manufacturer, or it can be entered by the user if a *PROM* (*programmable read-only memory*) is used. In either case, once the data is entered, it cannot be altered.

There is also a device called an *EPROM*, or *erasable programmable read-only memory*, which can be cleared by exposing it to a strong ultra-violet light. It can then be reprogrammed like an ordinary PROM.

No data in any kind of ROM can be altered in any way while it is in use in the computer. If power is removed, the data in the ROMs will not be lost.

The other type of memory is called *RAM*, or *random access memory*. A better name would be *read-write memory*, since data can be written into a read out of this type of memory by the CPU. If power is interrupted, the data in a RAM will be lost.

With either type of memory, the data can be randomly accessed. This means the CPU can call out any number on the memory address bus, and look at (or, in the case of a RAM, write) the data at the specified memory location in the same amount of time for any memory address.

ROMs are typically used to store *language-translation programs*, and RAMs are used to store individual programs and data. Computer languages are discussed in the next section.

The total number of possible memory locations (both ROM and RAM) is limited by the number of possible address codes the CPU generate. In complex programs, or where amounts of data must be stored, there quite often will not be enough memory space available. Another problem is that a frequently used program must be re-entered each time it is used. This re-entering is time consuming and inconvenient at best.

For all of these reasons, some sort of external memory storage system is highly desirable, if not absolutely essential for certain programs. The earliest systems used punch cards, or punched paper tape. See Fig. 29-2. The presence or absence of a given punched hole identifies the logic state of a specific bit. This approach is often adequate, or, at least, better than nothing, but it is nonerasable, bulky, and inconvenient to store; the holes can snag and tear.

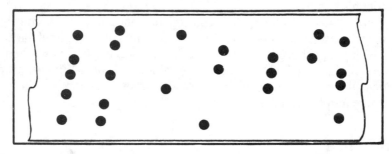

29-2   Computer punch tape.

External memory can also be stored on magnetic tape by digitally encoding a tone. Large commercial systems generally use reel-to-reel tape on large reels, and smaller systems often use ordinary cassettes, which are less expensive, easier to use, and more convenient to store.

The biggest disadvantage of using magnetic tape for storing data (especially cassettes) is that it is relatively slow, and not randomly accessible. The computer must search through the entire tape until it locates the data it needs. Much computer time is wasted searching for data.

Floppy discs are the answer to this problem. These are flexible discs that are coated with magnetic particles. Data can be recorded by these particles, in much the same way as ordinary magnetic tape. The surface of the disc is divided into *sectors* that are randomly accessible. Floppy discs offer large amounts of storage space and allow for extremely fast data transfer.

Thanks to improved technology, improved reliability, and steadily decreasing costs, hard disc drives are becoming more common in many small computer systems. A hard disk stores much more data than a floppy disk. The storage capacity of a hard disk is measured in megabytes (millions of bytes).

In the first personal computers, memory was rather expensive, so the maximum possible number of memory spaces was not included in most small computer systems. Today memory costs have come down considerably, although inexpensive models will tend to have much less memory than their more expensive relations.

Memory space in a computer is usually given as so many kB (kilobytes). Often, the abbreviation *k* is used instead of *kB* when computer storage or memory is discussed. Ordinarily *k* stands for 1000, as with resistor values. But in digital circuits, the value of 1 kB is actually 1024 (2^{10}) bits.

The CPU must address each individual memory location uniquely. The early personal computers all used simple, direct 16-bit address buses. This means the maximum number of memory locations that could be addressed was limited to an absolute maximum memory size of 64 kB. Inexpensive models often had just 4 kB or 16 kB of memory.

Later-model computers use memory bank switching techniques to permit much larger memory sizes. In fact, today a 64 kB would probably be sneered at as if it was just a cheap toy. Today, 256 kB is considered to be about the minimum memory size. Many commercial programs (software) call for a minimum of 640 kB. Many modern personal computers have memory sizes ranging up to several megabytes. (*Mega* means 1,000,000, or 1,048,576 (2^{20}) in a digital system.)

Memory size usually increases in powers of two, although this is not always the case. Several standard memory sizes past and present are listed in Table 29-1.

Table 29-1.
Memory k (kilobyte) sizes.

Stated size	Number of bytes
1 kB	1,024
2 kB	2,048
4 kB	4,096
8 kB	8,192
16 kB	16,384
32 kB	32,768
64 kB	65,536
128 kB	131,072
256 kB	262,144
384 kB	393,216
512 kB	524,288
640 kB	655,360
1 MB	1,024,000
2 MB	2,048,000

Today's CPUs and personal computers

Electronics technology has been advancing almost at an exponential rate, especially in personal computers. Overall prices have consistently dropped over time, or at least held fairly constant (even in times of general inflation), and the computing power offered has steadily increased. The computers that seemed incredibly power-ful and state of the art a few years ago are now treated like pathetic, obsolete toys. There is a great temptation to trade up for the latest, most powerful model. Some people hesitate buying a computer system because they are concerned that a newer, more-powerful model will soon appear on the market. But that will always be the case. If you want (or need) a personal computer, you're simply going to have to buy one sometime, and not worry about tomorrow's products.

Often obsolescence is merely in the mind of consumers and sales people. Be-fore junking your old computer for a newer, more-powerful model, stop and think about whether or not you really need the extra power. If the old model suits your needs adequately, it is not really obsolete at all for your purposes. Many people can still get along just fine with a dinosaur computer with a mere 16 kB of memory. The machine did the job in the past; why can't it keep doing the same job now?

Finding commercial software for most older personal computers is difficult if not impossible. There was no standardization, so when a model was discontinued, there was no more economic incentive for anyone to spend the time writing new programs for the orphaned machine.

Today, virtually all computers fit one of two competing standards—the IBM system, and the Macintosh (Apple) system. The IBM system is generally considered the norm, and software for this type machine is the most widely available. There are many IBM "clones" available on the market. These are computers from different manufacturers that more or less duplicate the IBM system. Although these clones are often advertised as fully IBM compatible, because of licensing and copyright restrictions, there are often a few slight differences. Usually these deviations from the true IBM standard don't make much practical difference, and they will probably be invisible to the user. However, in a few commercial programs, such internal changes in the computer structure could cause problems ranging from the mildly annoying to the severe, even to the absolutely catastrophic.

There are a few Mac (Macintosh) clones too, but they are less common than the IBM clones. Commercial software for the Mac system is widely available, though there usually isn't quite as many choices offered as there are for IBM compatibles.

The most frequent areas of improvement in the newer personal computers are in the area of memory size and computational speed. There is one subtle disadvan-tage in today's large memory sizes. In the old days, when memory was always at a premium, computer programmers learned to keep their software as neat and tight as possible, to cram as much into the small available memory size as possible. Today, there is little or no inherent penalty for beginning programmers using sloppy, memory-wasteful techniques. Modern commercial programs often use far more memory than they really need to. This forces consumers to buy (and pay for) more memory than they should need to use the programs. There are inexpensive, "bare-bones" word processing programs that call for a minimum of 384 k. This mem-ory requirement is ridiculous, because much more powerful word processing pro-grams once fit comfortably into 64 kB or even 16 kB. Of course, many newer word processing programs do use the extra memory space for a reason—extra, high-level commands, spelling correction dictionaries, style analyzers, indexing, and so forth. But for a simple, no-frills program to demand such memory indicates incredibly sloppy programming techniques.

Modern programmers seem to have a harder time keeping the memory requirements of their software down to a reasonable level, even though their predecessors managed the trick considerably smaller maximum memory sizes. Also, complex, modern programs with many special features can often get totally out of hand in memory consumption.

There are two basic ways to increase the operating speed of a computer. One way is to increase its data bus size, and the other method is to increase the frequency of the system clock oscillator.

The data bus size determines how many bits the CPU can look at simultaneously. Notice that this considers solely the data bus here—not the memory address bus. Early personal computers used 8-bit CPUs. That is, each word looked at by the CPU was eight bits (one byte) long. The CPU could only look at one byte at a time. With eight bits, up to 256 unique values can be represented. What if you needed to work with values larger than this? Two or more bytes are used in succession to identify different portions of the total number. For example, if two byte values were used, the CPU would first perform the programmed operation on the MSB (most-significant byte); then on the LSB (least-significant byte). This works, of course, but it obviously takes about twice as long to perform an operation on a two byte number as it does to perform the same operation on a single byte value. This will obviously slow down the overall computation speed of the system.

A 16-bit CPU, on the other hand, looks at words comprising sixteen bits at a time. This permits up to 65,536 unique values before a second word is required. A 32-bit CPU goes even further.

To add to the confusion, some modern CPUs are hybrids. For example, an 8/16 bit CPU is actually an 8-bit CPU with extra registers (internal memory space). It can perform operations on sixteen bit words, but it still uses eight-bit commands. This is a slight improvement, but it is not the same thing as a true 16-bit CPU, though that phrase might be used in advertisements for the computer.

Although the CPU performs one task for each clock cycle, it is obvious that increasing the clock frequency will increase the overall computing speed of the computer. The internal circuitry of the CPU chip determines the maximum usable frequency. If a CPU is forced to operate at too high a frequency, it could over-heat and become damaged. Even if it isn't permanently damaged, it will be more prone to unreliable operation.

In recent years, many new, improved CPUs have been placed on the market. A few offer new, more powerful commands, but usually the improvements are in bus size, and maximum operating frequency. A lot of recent advertising makes a big deal of which CPU is used in the particular model computer being marketed. At the time of this writing, 386 and 486 microprocessors are extremely popular. Each comes in a number of variants, identified by a suffix letter or two, such as 486DX or 386SL.

The basic 486 chip is a somewhat more recent development than the basic 386 chip, but some specific 386 CPUs might actually be faster and more powerful than some specific 486 CPUs.

The market changes so rapidly that there is little point in describing specific CPU devices in detail here. Such information is highly likely to be obsolete by the time you read it. If you are particularly interested in the specifics of the latest CPU chips, you should subscribe to some of the technical computer magazines, which have timely articles and news on such topics in almost every issue. Fortunately, most real computer users don't have to bother at all.

That's right, for the average personal computer user, it doesn't matter what CPU their computer uses. Few practical computer applications push even older CPUs even close to their operational limits. For example, if a word processing program keeps up with your typing speed, you won't notice any improvement if it can operate at even higher speeds. The practical limit is the user's input (typing) speed. For many programs, the latest, fastest CPU might be measurably faster, but the difference might be practically meaningless. For instance a data sorting procedure might take 48 seconds with one CPU, and just 35 seconds with the latest, top-of-the-line, high-speed CPU. That is an improvement of about 27%. But in human terms, will the 13 seconds saved really matter? Possibly, but probably not.

In other words, what counts is not that the latest and most advanced technology is used in your computer. As soon as you buy the present top-of-the-line model, someone is going to come out with something even "better." What really matters is whether or not the computer does the jobs you need it to do. It is foolish to pay extra for features and speed you will never get any practical advantage from.

Self-test

1. Which section of the computer is the "brain," which executes the actual program instructions?

A CPU
B Memory
C Input port
D Bus
E None of the above

2. What is the difference between a data byte and a command byte?

A Data bytes always begin with 1.
B Command bytes have fewer bits than data bytes.
C Command bytes have more bits than data bytes.
D There is no inherent difference.

3. How many bits are in a standard byte?

A 1
B 4
C 8
D 16
E None of the above

4. What is a program?

A A series of commands to perform a specific task
B A language for communicating with a computer
C A single instruction
D A data file
E None of the above

5. Which language can a CPU understand directly, without an interpreter or compiler?

A BASIC
B Machine language
C Assembly language
D Pascal
E English

6. What is a bus?
A A digital signal line
B An input device
C An output device
D A type of memory storage
E None of the above

7. Which type of memory is not user programmable?

A RAM
B ROM
C EEPROM
D Floppy disk
E None of the above

8. What happens to data stored in ordinary RAM when power is interrupted?

A Nothing
B It is inverted
C It is stored in the CPU registers
D It is moved to ROM
E It is lost

9. Which of the following is not a type of long-term data storage?

A Floppy disk
B Magnetic tape
C Punch cards
D Modem
E None of the above

10. Assuming a simple, direct 16-bit address bus, what is the maximum normal memory size. (No bank switching is used.)

A 16 kB
B 64 kB
C 256 kB
D 1 MB
E None of the above

30
Sensors and transducers

Sensors and *transducers* are devices that allow electronic circuits to communicate with the outside world. Sensors sense some external condition and produce an electrical signal or alter an existing electrical signal in response. Transducers transduce, or change, one form of energy into another. If a sensor produces an electrical signal in response to some external condition it is actually a transducer as well as a sensor.

In chapter 22 you read about light-sensitive devices, or photosensors. These devices convert light energy into electrical energy (or a related electrical parameter). The LEDs discussed in chapter 19 are also transducers, but they work in the opposite manner. They normally convert electrical energy into light energy. Under certain special circumstances, LEDs can be used as light sensors. In this chapter, you examine a number of sensors and transducers that detect or produce energy other than light.

Hand capacitance and resistance

An interesting way to control an electronic circuit is by using the operator's body itself as a component. Like all matter, your hand exhibits a certain amount of resistance. The value is relatively high, but it is quite variable. For example, lie detectors work on the principle that skin resistance changes with emotional states. A typical application is the *touch switch*. This switch is simply two conductive plates separated by a small distance (see Fig. 30-1).

If a touch plate is included in a circuit like the one illustrated in Fig. 30-2, ordinarily no current will flow. But if the operator touches both plates simultaneously (see Fig. 30-3), current can flow through the operator's finger to complete the circuit. For safety, use this device only with low power dc circuits!

Because the resistance of the human hand is relatively high, it can be used as the dielectric of a capacitor. The circuit board shown in Fig. 30-4 is used for this

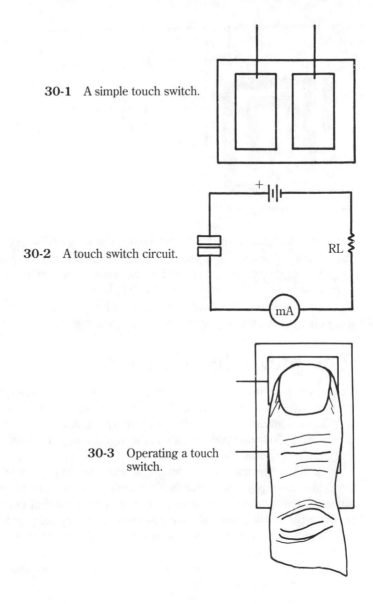

30-1   A simple touch switch.

30-2   A touch switch circuit.

30-3   Operating a touch
switch.

purpose. The dark areas are strips of copper foil mounted on an insulating board
(this is called a *printed circuit board*). This board can be connected into a circuit like
an ordinary capacitor. Ordinarily the dielectric would be air and the insulating board
material. But, if the operator laid his or her hand across the various copper traces,
the hand becomes the dielectric (because the hand's resistance is much lower than
that of the board, the board can now be ignored). By changing the pressure of the
hand on the board, the operator can vary the amount of contact area between the
dielectric (hand) and the plates (copper traces). Of course, changing the pressure
affects the effective capacitance.

Another interesting body capacitance effect takes place without any physical

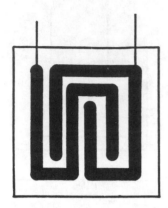

30-4   A hand capacitance
board.

contact at all. If a hand is brought near an *antenna* (discussed in another chapter), the antenna will act like a *proximity* detector that senses the distance to the hand.

An intriguing application of this effect is an instrument called the *Therimin*. The Therimin is a musical instrument that can be played without touching it. It generally has two antennas—one controlling pitch, and the other controlling volume. By moving your hands around the two antennas, you can produce different tones.

The piezoelectric effect

In the chapter on crystals, you learned how the piezoelectric effect can cause a crystal slab to oscillate. Remember that electrical stress along the X axis causes the mechanical stress to be produced along the Y axis. Similarly, a mechanical stress along the Y axis will produce an electric signal along the X axis. This means a crystal can also be used as a *pressure sensor.*

Perhaps the most common application for this effect is the *ceramic cartridge* used in record players. Records are plastic discs with a spiral groove cut into their surface. This groove undulates in a specific pattern that corresponds to the music recorded. A *needle*, or *stylus*, is connected to the crystal element in the crystal. This needle rides in the grooves cut on the record. As the needle is forced back and forth by the fluctuations of the groove different mechanical stresses are put upon the crystal. This stress is converted to an electrical signal that can be amplified and treated by the rest of the circuitry.

Thermistors

Another electronic sensor device is the *thermistor.* The thermistor is a resistor whose value changes in response to temperature. Some obvious applications for thermistors include electronic thermometers and thermostats for precise temperature control. They can also be used in fire alarms. The schematic symbol for a thermistor is shown in Fig. 30-5.

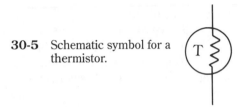

30-5 Schematic symbol for a thermistor.

Microphones

Microphones are transducers that convert audio energy (sound waves) into electrical energy. There are a number of basic methods for accomplishing this. Most microphone types are represented by the schematic symbol shown in Fig. 30-6. The word *microphone* is often shortened to *mike*.

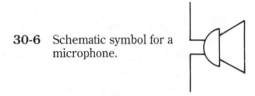

30-6 Schematic symbol for a microphone.

Carbon microphones

Perhaps the simplest type of microphone is the carbon microphone. The basic construction of this device is shown in Fig. 30-7. Basically a carbon microphone consists of a small container filled with carbon granules. This container has a carbon disc at either end. One of these discs is rigidly held in a fixed position, while the other is movable. The movable disc is connected to the *diaphragm*. Sound is caused by fluctuations in air pressure. This pressure moves the diaphragm (and thus the movable carbon disc) back and forth. This puts greater or lesser pressure on the carbon

30-7 Basic construction of a carbon microphone.

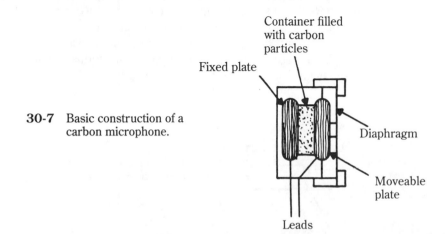

particles within the container. The changes in the density of these particles changes their effective resistance. If this assembly is in series with a dc voltage source (such as a battery, as shown in Fig. 30-8) the voltage drop will vary along with the changes in the resistance of the carbon particles, which is, in turn, caused by the changing sound pressure. Thus, the voltage output varies in step with the sound waves striking the diaphragm. We have an electrical equivalent of the acoustic energy.

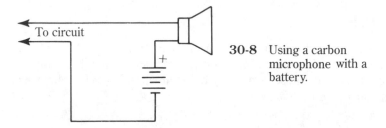

30-8   Using a carbon microphone with a battery.

The primary advantages of carbon microphones are that they are relatively low cost and they provide the highest level output of all commonly available microphone types. However, there are also a number of significant disadvantages. Carbon microphones require an external voltage source. They have a rather narrow frequency response, and their noise and distortion levels are higher than any other microphone type. Carbon microphones are typically used in telephone handsets.

Crystal microphones

Another type of microphone uses the piezoelectric effect. Figure 30-9 shows the basic construction of a *crystal microphone*. The sound pressure put on the diaphragm produces a mechanical stress on the crystal element. This, of course, generates a voltage in step with the mechanical stress. While resonant crystals are generally made of quartz, crystal microphones usually use elements made of Rochelle salt. The crystal microphone offers a number of advantages. It requires no external voltage, has a fairly high output level and a fair frequency response. However, it is rather fragile. Also, the Rochelle salt element can absorb moisture, ruining it. These two problems can be dealt with by replacing the Rochelle salt with a somewhat more rugged ceramic element. In this case we have a *ceramic microphone*. Crystal and ceramic microphones are good for general communications applications, but the frequency response is not adequate for high-fidelity use.

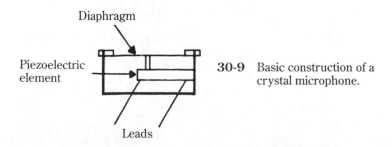

30-9   Basic construction of a crystal microphone.

Dynamic microphones

Dynamic microphones are probably the most popular type of general-purpose microphones. The basic structure of a dynamic microphone is shown in Fig. 30-10. The diaphragm is connected to a small coil which is suspended, so both can move freely in response to sound pressure. The coil moves within the magnetic field of a permanent magnet. Of course, this induces a voltage in the coil that varies in step with its movement. The output voltage of a dynamic microphone is fairly low, but its frequency response is quite good.

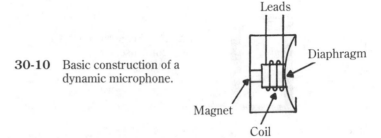

30-10 Basic construction of a dynamic microphone.

Ribbon microphones

Another type of microphone that is similar to the dynamic type is the *ribbon microphone*. Figure 30-11 shows the basic construction of this device. In this type of microphone a corrugated-aluminum ribbon is moved through the magnetic field of the permanent magnet. A small voltage will be induced in the ribbon via this process. The output voltage is extremely low, and usually has to be fed through a step-up transformer to reach a usable level. This transformer is often contained within the case of the microphone itself. Even with the transformer the output from a ribbon microphone is very low, but the frequency response is excellent. Ribbon microphones are also quite rugged.

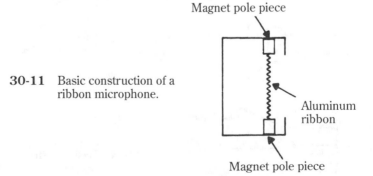

30-11 Basic construction of a ribbon microphone.

Condenser microphones

Figure 30-12 shows the basic construction of a condenser microphone. Two small plates are separated by a small amount. One is rigid, and the other is flexible (acting as the diaphragm). The movement of the diaphragm varies the distance between the two plates, which changes the capacitance between them. A small circuit within the microphone's case converts this varying capacitance to a varying voltage.

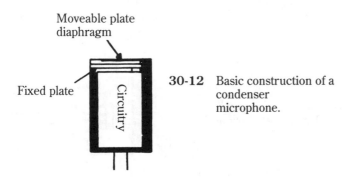

Moveable plate
diaphragm

Fixed plate

Circuitry

30-12 Basic construction of a
condenser
microphone.

Condenser microphones offer very low distortion and an excellent frequency response. However, they are rather expensive, and require their own dc voltage source.

Speakers

A *speaker* is the opposite of a microphone. It converts electrical energy back to sound pressure in the air. In fact, a small speaker can be used as a low quality dynamic microphone. The schematic symbol for a speaker is shown in Fig. 30-13. Often, the term *speaker* is shortened to *SPKR* in print. The basic construction of a speaker is shown in Fig. 30-14.

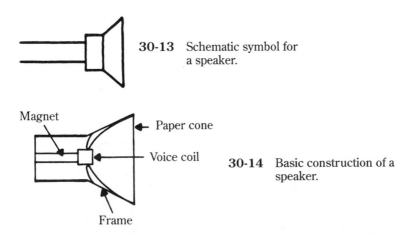

30-13 Schematic symbol for
a speaker.

Magnet

Paper cone

Voice coil

30-14 Basic construction of a
speaker.

Frame

The electrical signal is applied to a coil of wire called the *voice coil*. Because the voice coil is suspended within the magnetic field of a permanent magnet, it will move back and forth in step with the applied signal. The voice coil is connected to the center of a paper cone. The outer rim of this cone is firmly attached to a sturdy frame. The cone is forced to move in and out along with the voice coil. See Fig. 30-15. This movement of the cone changes the air pressure in step with the original electrical signal. These pressure fluctuations are perceived by the ear as sound.

30-15 Movement of a speaker cone. No signal (A); Positive signal (B); Negative signal (C).

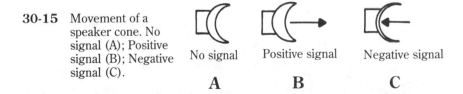

No signal   Positive signal   Negative signal

A   **B**   **C**

No speaker is perfect. All exhibit some signal loss and distortion. Also, the frequency response will always be less than ideal. A small speaker cone will reproduce high frequencies fairly well but won't be able to handle low frequencies. Conversely, a large cone area is required to reproduce low frequencies, but it won't be able to vibrate fast enough for high frequencies.

This limitation is why most high-fidelity speaker systems will contain more than one speaker within a single cabinet. A small *tweeter* reproduces the high frequencies, and a large *woofer* handles the low frequencies. A *crossover network* is used to separate the signals. The crossover network is basically a highpass filter that blocks low frequencies from the tweeter. A large amplitude low frequency signal could damage the delicate voice coil of a tweeter. High frequencies applied to a woofer will essentially be ignored by the speaker itself. Many speaker systems include a third speaker between the woofer and the tweeter. Not surprisingly, this third speaker is called a *midrange speaker.*

Some speakers are designed as *full-range* units and are capable of reproducing frequencies from about 100 to 15,000 Hz, but these full range units are not as accurate as separate woofer-tweeter combinations.

Although a speaker frequency range is definable, the frequency response is not. The frequency response of a speaker can be greatly affected by the shape and size of the cabinet, the material the cabinet is made of, the positioning of the speaker within the cabinet, the size and furnishings of the room the speaker is used in, and even the position of the speaker system within the room. Obviously all these variables provide infinite possible combinations.

Speakers are often tested for frequency response in echoless chambers so room variables do not enter the picture. Although the frequency responses obtained in this manner can usually be used for comparison purposes, it must be remembered that the test conditions are far removed from real world conditions. It is entirely possible for one speaker to produce better results than another in an echoless chamber, yet not sound as good in an actual listening environment.

Figure 30-16 shows the ideal frequency-response curve for a theoretically per-

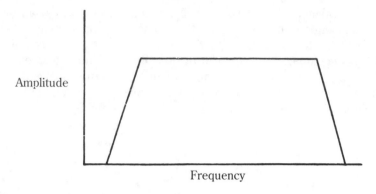

30-16   Ideal speaker frequency-response curve.

fect speaker. Notice that all frequencies are reproduced equally. This is called *flat response*. A typical frequency-response curve for a practical speaker is shown in Fig. 30-17. Notice that certain frequencies are reproduced at a higher level than others. Also remember, this curve would look quite different if the same speaker was placed in a different environment, but the basic shape will probably be more or less the same.

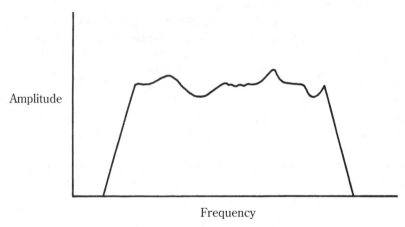

30-17   Typical frequency-response curve for a practical speaker.

Self-test

1. Which of the following terms describes a sensor that reacts to light?

A Photosensitive
B Photoemissive
C Transducer
D Piezoelectectric
E None of the above

2   What type of power should be used for a touch switch-circuit?

A Any dc
B ac
C Low-power dc
D Doesn't matter

3. Which of the following correctly describes the piezoelectric effect?

A A mechanical stress along the X axis produces an electrical signal along the Y axis
B A mechanical stress along the Y axis produces an electrical signal across the X axis
C A mechanical stress along either axis produces an electrical signal across the opposite axis
D Pressure affects the dielectric value, and thus, the capacitance

4. Which of the following microphone types requires its own dc voltage supply?

A Condenser
B Dynamic
C Ribbon
D Crystal
E None of the above

5. What bodily characteristic is used in lie detectors?

A Skin resistance
B Hand capacitance
C Heart rate
D Body inductance
E None of the above

6. What is a typical application of hand capacitance?

A Lie detector
B Optoisolation
C Pressure measurement
D Touch switches
E None of the above

7. Which of the following would be used for a pressure sensor?

A Photodiode
B Optoisolator
C Crystal
D Varactor
E None of the above

8. What electrical characteristic varies with temperature in a thermistor?

A Capacitance
B Resistance
C Gain
D Inductance
E None of the above

9. Which of the following is not a common type of microphone?

A Carbon
B Ribbon
C Condenser
D Dynamic
E None of the above

10. What type of speaker is best at reproducing low frequencies?

A Dynamic
B Tweeter
C Woofer
D Crossover
E None of the above

31
Experiments 3

Table 31-1 lists the equipment and parts needed for this collection of experiments.

Experiment 1 Inverting amplifier

Breadboard the circuit shown in Fig. 31-1 around a 741 operational amplifier in an 8-pin DIP package. If you use an op amp in another package configuration (perhaps a 14-pin DIP), the pin numbers in the diagram will not be correct. If you use the specification sheet for your op amp to determine the correct pin-out, the circuit will work.

Notice that three voltage sources are required. Two of these, +9 V (B2) and −9 V (B3), provide power for the operational amplifier, and the three V battery (B1) generates the input signal. R1 is a 10,000 Ω (brown-black-orange) resistor, and R3 is a 10,000 Ω potentiometer. For the first part of this experiment, use a 22,000 Ω (red-red-orange) resistor.

Connect your voltmeter between point A (+) and ground (−) and carefully adjust R3 for a reading of exactly 0.5 V. Now, reverse the polarity of the voltmeter and connect it between point B (−) and ground (+). The output voltage should be about 2.2 times the input voltage, and the polarity is reversed. Your exact measured value might vary from this somewhat due to resistor tolerances. Enter the output voltage in the appropriate space in Table 31-2. Repeat this procedure for input voltages of 1, 1.5, and 2 V. Enter each output voltage into Table 31-2.

Now replace R2 with a 33,000 Ω (orange-orange-orange) resistor and repeat the experiment.

Now substitute a 100,000 Ω (brown-black-yellow) resistor for R2 and go through the experiment again.

Finally, repeat the entire procedure using a 10,000 Ω (brown-black-orange) resistor for R2.

Table 31-1. Equipment and parts needed for the experiments.

VOM	Voltmeter, ac voltmeter
Breadboarding system	Oscillator, power supplies—+9 V*, −9 V*, +3 V*, and well regulated +5 V. The voltages marked "*" can be provided by batteries
1	270 Ω resistor (yellow-violet-brown)
2	1000 Ω resistor (brown-black-red)
2	4700 Ω resistor (yellow-violet-red)
4	10,000 Ω resistor (brown-black-orange)
1	22,000 Ω resistor (red-red-orange)
1	33,000 Ω resistor (orange-orange-orange)
1	100,000 Ω resistor (brown-black-yellow)
1	1,000,000 Ω resistor (brown-black-green)
1	0.01 µF disc capacitor
1	0.1 µF disc capacitor
1	1 µF 10-V electrolytic capacitor
1	10 µF 10-V electrolytic capacitor
1	1N914 diode
1	741 operational amplifier IC (8 pin DIP)
1	555 timer IC (8 pin DIP)
1	7400 TTL quad NAND gate IC
1	7402 TTL quad NOR gate IC
1	7474 TTL dual D type flip-flop IC
2	SPDT switches
1	SPDT momentary contact switch
1	DPDT switch
1	8 Ω miniature speaker
2	LEDs
1	10,000 Ω potentiometer

Now go back and calculate the gain for each input-output combination ($G = $ output/input). For the three smaller resistors the gain should be fairly constant for each input value. But when a 100,000 Ω resistor is used for R2, the gain is more than the operational amplifier can handle with these input levels. The theoretical gain in this case is 10 ($G = $ R2/R1), so an input voltage of 1 V should produce an output of 10 V. 1.5 V in should result in 15 V out, and 2 V in should give an output of 20 V. But the supply voltage is only ±9 V, so the amplifier saturates on these input levels. Remember, the output voltage can never be greater than the source voltage.

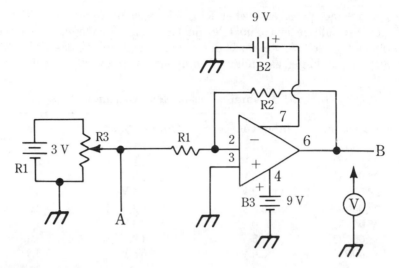

31-1   Inverting amplifier circuit.

Table 31-2. Worksheet for Experiment 1.

R2 Value	Input voltage	Output voltage	Gain
22,000 Ω	0.5 V	_____	_____
22,000 Ω	1.0 V	_____	_____
22,000 Ω	1.5 V	_____	_____
22,000 Ω	2.0 V	_____	_____
33,000 Ω	0.5 V	_____	_____
33,000 Ω	1.0 V	_____	_____
33,000 Ω	1.5 V	_____	_____
33,000 Ω	2.0 V	_____	_____
100,000 Ω	0.5 V	_____	_____
100,000 Ω	1.0 V	_____	_____
100,000 Ω	1.5 V	_____	_____
100,000 Ω	2.0 V	_____	_____
10,000 Ω	0.5 V	_____	_____
10,000 Ω	1.0 V	_____	_____
10,000 Ω	1.5 V	_____	_____
10,000 Ω	2.0 V	_____	_____

Also, notice the special case when R1 and R2 have the same value (G = 1). The output voltage should be approximately equal to the input voltage, except of course, the polarity is reversed. Any error is due to component tolerances. Table 31-3 lists the results you should get, assuming all components are exactly on value.

Table 31-3. Nominal results for Experiment 1.

R2 Value	Input voltage	Output voltage	Gain
22,000 Ω	0.5 V	−1.1 V	−22.2
22,000 Ω	1.0 V	−2.2 V	−2.2
22,000 Ω	1.5 V	−3.3 V	−2.2
22,000 Ω	2.0 V	−4.4 V	−2.2
33,000 Ω	0.5 V	−1.65 V	−3.3
33,000 Ω	1.0 V	−3.3 V	−3.3
33,000 Ω	1.5 V	−4.95 V	−3.3
33,000 Ω	2.0 V	−6.6 V	−3.3
100,000 Ω	0.5 V	−5 V	−10
100,000 Ω	1.0 V	−9 V	(−9)
100,000 Ω	1.5 V	−9 V	(−6)
100,000 Ω	2.0 V	−9 V	(−4.5)
10,000 Ω	0.5 V	−0.5 V	−1
10,000 Ω	1.0 V	−1.0 V	−1
10,000 Ω	1.5 V	−1.5 V	−1
10,000 Ω	2.0 V	−2.0 V	−1

Experiment 2 Noninverting amplifier

Breadboard the circuit shown in Fig. 31-2. Notice that it is quite similar to the one used in Experiment 1, except in this case you are working with a noninverting amplifier. The voltage supply values should be the same as in the previous experiment. Similarly, R1 and R3 retain the same values. You will also use the same sequence of resistors for R2.

For each step connect your voltmeter between point A (+) and ground (–) and adjust R3 to the desired input voltage. Then move the positive lead of the voltmeter over to point B and measure the output voltage. Enter this value into Table 31-4.

Repeat this procedure for each R2 value and input voltage listed in the table. The nominal (perfect tolerance) results are shown in Table 31-5.

Basically your results should be about the same as in the previous experiment, except for this circuit the output is the same polarity as the input.

Experiment 3 Difference amplifier

Build the circuit shown in Fig. 31-3. Notice that both the − and the + inputs are used in this circuit. This is a *difference amplifier*. All three voltage sources should

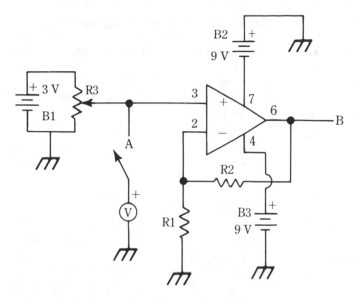

31-2 Noninverting amplifier circuit.

Table 31-4. Worksheet for Experiment 2.

R2 Value	Input voltage	Output voltage	Gain
10,000 Ω	0.5 V	_____	_____
10,000 Ω	1.0 V	_____	_____
10,000 Ω	1.5 V	_____	_____
10,000 Ω	2.0 V	_____	_____
22,000 Ω	0.5 V	_____	_____
22,000 Ω	1.0 V	_____	_____
22,000 Ω	1.5 V	_____	_____
22,000 Ω	2.0 V	_____	_____
33,000 Ω	0.5 V	_____	_____
33,000 Ω	1.0 V	_____	_____
33,000 Ω	1.5 V	_____	_____
33,000 Ω	2.0 V	_____	_____
100,000 Ω	0.5 V	_____	_____
100,000 Ω	1.0 V	_____	_____
100,000 Ω	1.5 V	_____	_____
100,000 Ω	2.0 V	_____	_____

have the same values as in the previous experiments. All resistors are 10,000 Ω (brown-black-orange).

Connect your voltmeter between point A (+) and ground (–) and measure the inverting input voltage. Nominally this value is 3 V (the value of B1), but your battery might not be exactly on voltage. Enter this voltage in the appropriate spaces in Table 31-6.

Table 31-5. Nominal Results for Experiment 2.

R2 Value	Input voltage	Output voltage	Gain
10,000 Ω	0.5 V	0.5 V	1
10,000 Ω	1.0 V	1.0 V	1
10,000 Ω	1.5 V	1.5 V	1
10,000 Ω	2.0 V	2.0 V	1
22,000 Ω	0.5 V	1.1 V	2.2
22,000 Ω	1.0 V	2.2 V	2.2
22,000 Ω	1.5 V	3.3 V	2.2
22,000 Ω	2.0 V	4.4 V	2.2
33,000 Ω	0.5 V	1.65 V	3.3
33,000 Ω	1.0 V	3.3 V	3.3
33,000 Ω	1.5 V	4.95 V	3.3
33,000 Ω	2.0 V	6.6 V	3.3
100,000 Ω	0.5 V	5.0 V	10
100,000 Ω	1.0 V	9.0 V	(9)
100,000 Ω	1.5 V	9.0 V	(6)
100,000 Ω	2.0 V	9.0 V	(4.5)

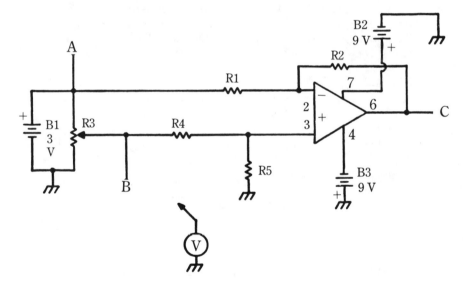

31-3   Difference amplifier circuit.

Now move the positive voltmeter lead to point B and adjust R3 for a noninverting voltage of 1 V.

Reverse the voltmeter leads and connect the meter between point C (–) and ground (+). Because the inverting input voltage is larger than the noninverting input voltage, the inverting input should predominate. Measure the output voltage and enter your result in Table 31-6.

Table 31-6. Worksheet for Experiment 3.

Inverting input	Noninverting input	Output
_____	1.0 V	_____
_____	1.5 V	_____
_____	2.0 V	_____
_____	2.5 V	_____

Repeat this procedure for each of the input values listed in the table. You do not have to measure point A each time—it should be a constant value.

When you are finished, each output voltage should equal the inverting input voltage minus the noninverting input voltage. Minor variations are due to unequal component tolerances.

Experiment 4 Integrator

Breadboard the circuit shown in Fig. 31-4. This is an *integrator*. B1 and B2 are the 9 V power supplies for the operational amplifier. R1 and R3 should be 10,000 Ω (brown-black-orange) resistors, and C1 is a 0.1 μF capacitor. R2 is 100,000 Ω (brown-black-yellow).

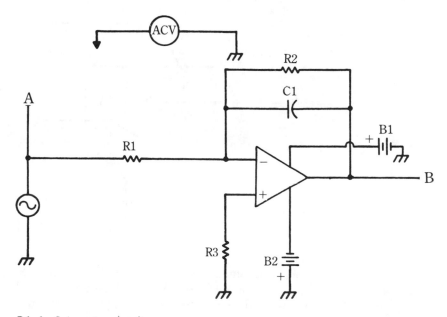

31-4 Integrator circuit.

Measure the input voltage at point A, then measure the output voltage at point B. Enter both values into Table 31-7. Do this for a number of input frequencies start-

Table 31-7. Worksheet for Experiment 4.

Frequency	Input voltage	Output voltage
A *(lowest)*	————	————
B	————	————
C	————	————
D	————	————
E *(highest)*	————	————

ing with the lowest frequency at the top of the table and increasing the frequency as you move down the column. The gain should decrease as the frequency increases.

If you have an oscilloscope it would also be interesting to compare the input wave shape with the output signal. There should be a dramatic difference.

Experiment 5 Differentiator

Repeat the previous experiment with the differentiator circuit shown in Fig. 31-5. R1 should be a 270 Ω (red-violet-brown) resistor. R2 and R3 should be 10,000 Ω (brown-black-orange) units. C1 is a 0.1 μF capacitor. The power supplies are ±9 V.

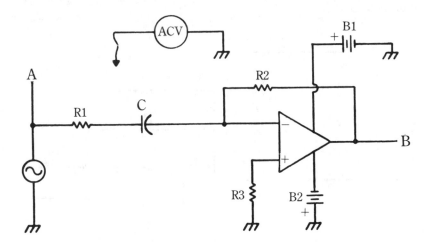

31-5   Differentiator circuit.

Your results in this experiment should be the opposite of what happened in Experiment 4. The gain should increase as the frequency increases. Record your results on Table 31-8. Again, it is interesting to compare the input and output signals with an oscilloscope.

Table 31-8. Worksheet for Experiment 5.

Frequency	Input voltage	Output voltage
A (lowest)	_____	_____
B	_____	_____
C	_____	_____
D	_____	_____
E (highest)	_____	_____

Experiment 6 A monostable multivibrator

Figure 31-6 shows a basic monostable multivibrator circuit built around a 555 timer IC. The pin numbers are given for an 8-pin DIP.

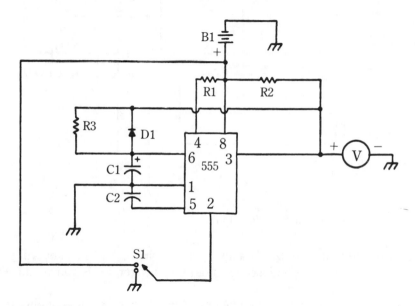

31-6   Monostable multivibrator circuit.

The power supply voltage can be anything between 5 and 15 V. A 9 V battery is fine. R1 and R2 are 4700 Ω (yellow-violet-red) resistors, and R3 is 1 MΩ, or 1,000,000 Ω (brown-black-green). D1 is 1N914 diode.

For the first part of this experiment, C1 will be a 1 μF electrolytic capacitor. Be sure to observe the correct polarity. C2 is a 0.01 μF disc capacitor.

S1 is an SPDT switch. Preferably, it should be a momentary-contact type. The diagram shows its normal position. Push the switch briefly and release it while watching the meter. The pointer should jump up from approximately zero to some value somewhat below the power supply voltage. After a definite period of time the pointer should move back down to near zero. This time should be approximately one second. The nominal value is actually 1.1 second.

Now, replace the 1 μF capacitor (C1) with a 10 μF electrolytic capacitor (observe polarity) and repeat the experiment. How long does the pointer stay at its high voltage position now?

Experiment 7 An astable multivibrator

Breadboard the astable multivibrator circuit shown in Fig. 31-7. Again the circuit is built around an 8-pin DIP 555 timer IC. B1 is 9 V, R1 is a 10,000 Ω potentiometer and R2 is a 1000 Ω (brown-black-red) resistor. C1 is a 0.01 μF capacitor. The speaker is a miniature 8 Ω unit.

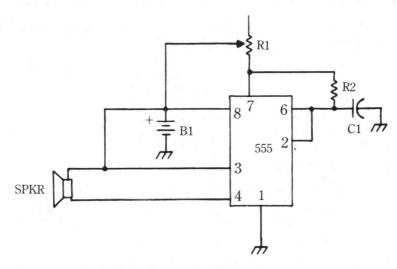

31-7   Astable multivibrator circuit.

You'll notice that as power is applied to this circuit, the speaker will sound a tone If you connected an oscilloscope across the speaker, leads you would see that the signal is a rectangular wave.

Move the knob on the potentiometer and listen to the sound coming from the speaker. The pitch (or frequency) should change as you vary the resistance of R1. R1 is the timing resistor, so changing its resistance changes the time constant of the circuit.

Experiment 8 The NAND gate

Connect pin 14 of a 7400 TTL quad two-input NAND gate integrated circuit to a well regulated 5 V power supply and connect pin 7 to ground.

Now, set up the rest of the circuit shown in Fig. 31-8. The pin labeled A in the diagram is pin 1. Pin B is pin 2, and pin C is pin 3. R1 is a 1000 Ω (brown-black-red) resistor, and S1 and S2 are SPDT switches.

When a switch is making contact with ground, it is providing a logic 0 input

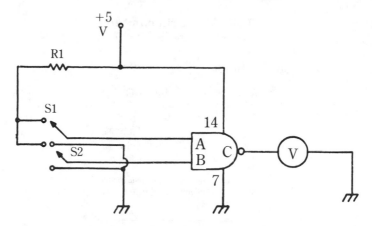

31-8 A NAND gate demonstration circuit.

to the gate. When a switch is making contact to +5 V through R1, it is a logic one. The meter pointer should be near 0 or between 2 and 3 V, corresponding to logic 0 and logic 1, respectively. Set the switches for each of the combinations shown in Table 31-9.

Table 31-9.
Worksheet for Experiment 8.

Switch 1	Switch 2	Output
0	*0*	_____
1	*0*	_____
1	*1*	_____
0	*1*	_____

You should get a logic 0 output only when both switches are in their logic 1 positions. For all other input combinations, the output should be a logic 1. This is a NAND gate.

Repeat the experiment for each of the gates in the package, using the pins listed in Table 31-10. You should get the same results from all of the gates, unless one of them happens to be defective.

Experiment 9 The NOR gate

Repeat the procedure of previous experiment with a 7402 TTL quad two-input NOR gate IC. See Fig. 31-9. The pin connections for each gate are listed in Table 31-11. Record your results in Table 31-12.

In this case, you should have a logic 1 output only when both switches are in

**Table 31-10. Pin numbers
for each gate in a 7400.**

Inputs		Output	
A	**B**	**C**	
1	2	3	Gate #1
4	5	6	Gate #2
9	10	8	Gate #3
12	13	11	Gate #4

Pin 14 is connected to the positive voltage source, and pin 7 goes to ground for
all gates.

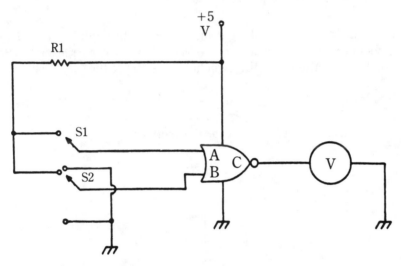

31-9 A NOR gate demonstration circuit.

**Table 31-11. Pin numbers
for each gate in a 7402.**

Inputs		Output	
A	**B**	**C**	
2	3	1	Gate #1
5	6	4	Gate #2
8	9	10	Gate #3
11	12	13	Gate #4

Pin 14 is connected to the positive voltage source, and pin 7 goes to ground for
all gates.

Table 31-12.
Worksheet for Experiment 9.

Switch 1	Switch 2	Output
0	0	_____
1	0	_____
1	1	_____
0	1	_____

their logic 0 positions. All other input combinations should result in a logic 0 output. This is a NOR gate.

Experiment 10 Combining gates

Breadboard the circuit shown in Fig. 31-10. IC1 is a 7400 quad NAND gate, and IC2 is a 7402 quad NOR gate. Before actually performing the experiment try to determine what the output state will be for each of the input combinations in Table 31-13.

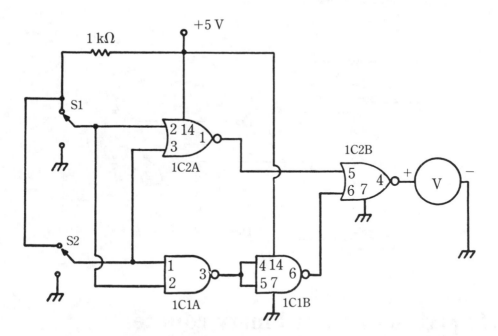

31-10   A complex gating circuit.

Try each of the switch combinations and see if you correctly predicted the output state. Does this circuit resemble one of the basic gate types?

Table 31-13. Worksheet for Experiment 10.

Switch 1	Switch 2	Predicted Output	Actual Output
0	0	_____	_____
0	1	_____	_____
1	0	_____	_____
1	1	_____	_____

Experiment 11 A bistable multivibrator

Breadboard the circuit shown in Fig. 31-11 using two of the NOR gates in a 7402 IC. S1 is a DPDT switch. Connect the positive voltmeter lead to point A. It should be either close to zero (logic 0) or between 2 and 3 V (logic 1). Changing the position of S1 should reverse the output state. Move S1 back and forth several times so you can see how it affects the output. Position the switch so there is a logic 1 output at point A. Now move the positive voltmeter lead to point B. You should read a logic 0 here. Now reverse the position of S1. Remember, this caused output A to change from 1 to 0. Output B should change from 0 to 1. A and B are always opposites. Either output state can be held indefinitely.

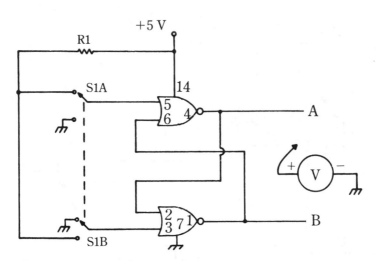

31-11   Binary bistable multivibrator.

Experiment 12 A binary counter

Construct the circuit in Fig. 31-12 using a 7474 TTL dual D-type flip-flop IC. S1 is an SPST momentary action switch. The figure shows its normal position. R1 and R2 are 1000 Ω (brown-black-red) resistors.

When you first turn on the circuit one or both of the LEDs might be lit. Press

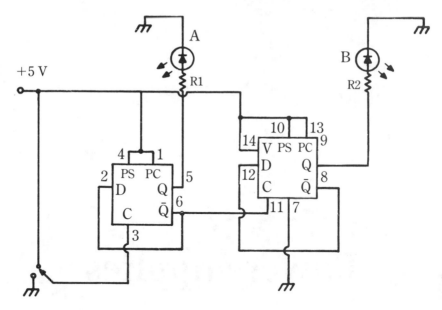

31-12 Binary counter.

S1 several times until both LEDs are dark. Assume that a dark LED is called a 0, and a lit LED a 1. LED A is the ones column. LED B is the twos column. You are starting out with a displayed count of 00 (both LEDs are dark).

If you press the switch once LED A will light up, and LED B will stay dark. You now have a binary count of 01, or one.

The second time you press the switch, LED A should go out and LED B should come on. Our binary count has changed to 10, or two.

The third time you press the switch both LEDs should be lit for a binary count of 11, or three.

On the fourth switch pressing both LEDs will go out again, and the count will start over.

What is the modulo for this counter? If you are unsure, read the section on counters in chapter 27.

32
Power supplies

Virtually all electronic circuits require some source of voltage. This means *power supply* circuits are extremely important. Fortunately, they are fairly easy to understand.

If a circuit requires an ac voltage, the power supply can simply be a transformer connected to the ac house current. If a circuit requires dc and has low power requirements, batteries can be used. Most practical circuits, however, are dc operated, and require power levels that would make battery operation uneconomical. These devices need a power supply circuit.

Actually, power supply circuits are somewhat misnamed. They do not truly supply power. They are more properly *power converters*. Generally they convert ac voltages into dc voltages.

There are a number of methods of accomplishing this function, using diodes. When diodes are used for this type of application, they are generally called *rectifiers*. Rectifiers can be either tube diodes or semiconductor diodes. The process itself is called *rectification*.

Half-wave rectifiers

A single diode power supply circuit is called a *half-wave rectifier*. This circuit is shown, in its most basic form, along with its input and output signals, in Fig. 32-1.

The terminals of the ac voltage source reverse polarity twice each cycle. For one-half of each cycle, terminal A is positive with respect to terminal B (ground). For the other half of each cycle, terminal A is negative with respect to terminal B. When terminal A is positive, the diode is forward biased, the current can flow through it to the output.

However, when terminal A is negative, the diode is reverse biased, so no current can flow. Only the positive half cycles appear at the output.

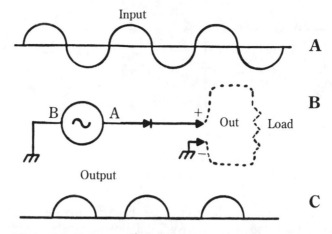

32-1   A half-wave rectifier circuit.

Of course, this is not true dc power at all. Half of the time there is no output voltage. The rest of the time the voltage is either in the process of rising from zero to the maximum level, or from the maximum level back to zero. It never holds any constant value.

A closer approximation of true dc can be achieved by placing a large capacitor across the diode output, as shown in Fig. 32-2.

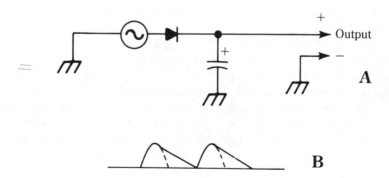

32-2   A half-wave rectifier circuit with a filter capacitor.

When the output voltage rises from zero to its peak value, the capacitor is charged. When the voltage drops off, the capacitor starts to discharge through the load. If the capacitor is large enough, it will not be fully discharged before the next positive half cycle starts.

In other words, the capacitor will be charged, partially discharged, charged, partially discharged, and so on. The final output signal will resemble Fig. 32-2B.

The larger the capacitance, the slower the discharge rate, and therefore, the shallower the discharging angle in the output waveform. Electrolytic capacitors with values of several hundred to a few thousand microfarads are typically used. Figure 32-3 shows an improved, practical half-wave rectifier circuit.

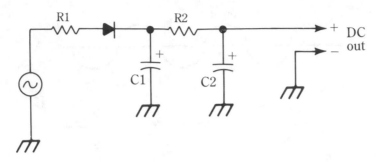

32-3   A practical half-wave rectifier circuit.

R2, C1, and C2 make up a lowpass filter that smooths the output signal more efficiently than a single capacitor. There will still be some voltage fluctuations (called *ripple*), but they won't be nearly as pronounced.

R1 is a surge resistor that is used to protect the diode from any sudden increase in the current drawn through the circuit. The surge resistor typically has a fairly small value, so normally the voltage drop across it is also rather small. But an increase in the current drawn through the resistor will cause its voltage drop to rise too, because according to Ohm's law, voltage equals current times resistance ($E = IR$). This brings the voltage applied to the diode down to a level it can comfortably dissipate.

Sometimes R1 is also fused for additional protection. A surge resistor is not only for protection against circuit defects. Often it is also needed for normal operating conditions.

Assume no source voltage at all is being applied to the circuit. Any residual charge on the capacitors will be soon discharged through R2 and the load circuit. The capacitors are completely discharged.

Now when power is first applied, the capacitors will draw a large amount of current until they are almost completely charged. There isn't sufficient time for them to be completely charged during a single cycle, so it takes a few cycles for ordinary operation to begin. Of course, this extra current drain will increase the voltage drop across the other components. Again, the surge resistor is used to protect the diode.

Often, for more efficient protection, a *thermistor* (a temperature- sensitive resistor) is used for R1. When the circuit is first turned on, the components, including the thermistor, are cool. The thermistor has a high resistance when it is cold. This means that when power is first applied, there is a relatively large voltage drop across the thermistor, leaving only a relatively small voltage to pass through the diode.

As current passes through the circuit, the components start to dissipate heat. The increased temperature causes the resistance of the thermistor (and thus, its voltage drop) to fall to a low value, and from then on it acts like any ordinary surge resistor.

Full-wave rectifiers

In the half-wave, rectifier half of each input cycle is completely unused. Of course this means power is wasted. Figure 32-4 shows a more efficient power supply circuit called a *full-wave rectifier*. Notice that a full-wave rectifier must be used with a center-tapped transformer.

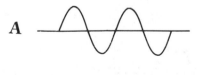

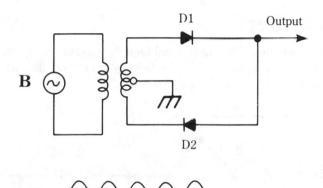

32-4 A full-wave rectifier circuit.

Remember that if the center tap of a transformer secondary is grounded, the lower half of the secondary winding will carry a signal that is equal to, but 180 degrees out of phase with the upper half signal. This means that when D1 is passing a positive half cycle, D2 is blocking a negative half cycle. And, when D1 is blocking a negative half cycle, D2 is passing a positive half cycle. One of the diodes is conducting and one is nonconducting at all times. This means the output win resemble Fig. 32-4C.

Notice that in addition to wasting less input power, the output of a full-wave rectifier is easier to filter, because there is less time for the filter capacitor to discharge before it is charged again. See Fig. 32-5. Notice that both the positive and the negative output lines need their own filter and they are isolated from the ac ground.

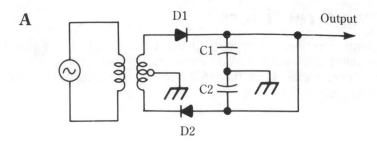

B

32-5   A full-wave rectifier circuit with filter capacitors.

Bridge rectifiers

A *bridge rectifier* circuit (see Fig. 32-6) combines the advantages of both full-wave rectifiers and half-wave rectifiers. Like the full-wave rectifier, the bridge rectifier uses the entire input cycle, and is easy to filter.

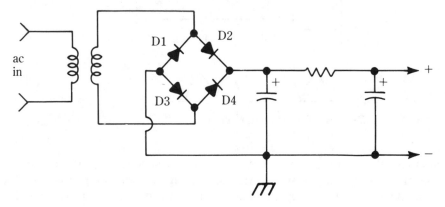

32-6   A bridge rectifier circuit.

Like the half-wave rectifier, the bridge rectifier does not require a center-tapped transformer. Although a bridge rectifier requires four diodes, it is still usually more economical for semiconductor circuits than most center-tapped transformers. The circuit normally requires less space and produces less heat. Bridge rectifiers using tube diodes are not practical.

Also, like the half-wave rectifier, one of the bridge rectifier output lines might be at ground potential. At any point of the input cycle, two of the diodes in the bridge are conducting and two are reverse biased. For the positive half wave the circuit effectively looks like Fig. 32-7. Figure 32-8 shows the equivalent circuit for the negative half cycle.

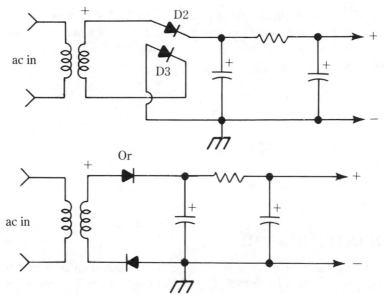

32-7 Equivalent circuits for a bridge rectifier circuit during a positive half cycle.

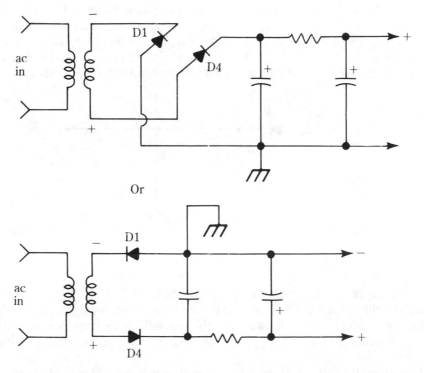

32-8 Equivalent circuits for a bridge rectifier circuit during a negative half cycle.

Bridge rectifiers might consist of four separate diodes, or they can be encapsulated in a single package, as shown in Fig. 32-9. The encapsulation is usually done simply to conserve space. Electrically, such a single unit bridge is the exact equivalent of four discrete diodes.

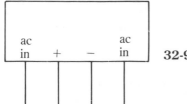

32-9   A typical bridge rectifier.

Voltage regulation

One problem with all of the power-supply circuits discussed so far is that the output voltage is dependent, to a large extent, on the amount of current drawn by the load circuit. If the current drawn by the load circuit increases for any reason, the voltage drop across the components within the power supply circuit will also rise, resulting in a lower output voltage. Of course, just the opposite happens if the current drawn by the load decreases—the output voltage will increase.

A partial solution is shown in the half-wave rectifier circuit in Fig. 32-10. R2 is replaced with a coil called a *choke*. This coil will oppose any change in the current passing through it. It also acts as a better filter than an ordinary resistor, this means lower ripple in the output signal. Also, because the dc resistance of a coil is extremely low, there will be very little wasted voltage drop across the choke.

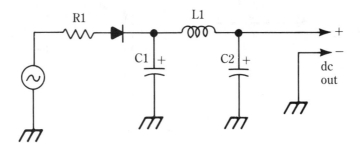

32-10   A half-wave rectifier circuit with a filter choke.

Using a zener diode across the output, as shown in Fig. 32-11 is another basic method of voltage regulation. Zener diodes are discussed in detail in chapter 18.

The best voltage regulation can be achieved with the voltage regulator circuit shown in Fig. 32-12. This circuit is actually a simplified basic circuit. There are a number of variations possible.

Transistor Q1 is what is known as a pass transistor. Q2 and Q3 make up a difference amplifier that is set up to detect any error in the output voltage. The two

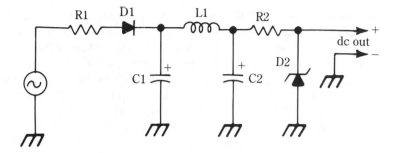

32-11   A half-wave rectifier circuit with a zener diode regulation.

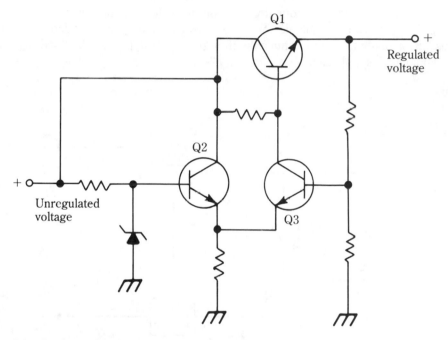

32-12   A simplified voltage-regulator circuit.

resistors between the output and the ground form a voltage divider (explained in the next section of this chapter).

If the output voltage increases even slightly because of an decrease in the current drawn by the load circuit, the voltage on the base of Q3 is increased. Notice that the voltage on the base of Q2 is held constant by the zener diode. Ordinarily these two base voltages should be equal. But you now have a condition where Q3 base is at a higher voltage than Q2 base. This difference is the difference detected by the difference amplifier.

The increase in base voltage on Q3, of course, increased the emitter and collector currents. This means the voltage drop across the shared emitter resistor must also increase ($E = IR$). because the emitter of Q2 is now at a somewhat higher voltage, the difference between the base and emitter voltages has decreased. This has the same effect as reducing Q2 base voltage. This means Q2 output must also decrease. The current through Q1 will be forced to compensate for the difference. In practice, the voltage drop across the emitter resistor will remain virtually constant, because the changes take place so quickly. In effect Q1 increases its resistance, which has the effect of dropping the output voltage back to the desired level. If the output voltage drops for any reason, the opposite reactions will take place.

The difference amplifier is extremely sensitive, and even changes of less than one-tenth of one percent can be quickly corrected by some circuits. Obviously, the voltage regulator also serves as a superior ripple filter, because ripple is a fluctuation in the output voltage, and is thus corrected by the circuit like any other output error.

Because voltage regulator circuits are so frequently needed, a number of IC versions are available for frequently used output voltages, such as 5, 12, or 15 V. Voltage regulators are available for either positive or negative ground operation. These are usually not interchangeable, though the two types can be used together for two polarity supplies, as shown in Fig. 32-13.

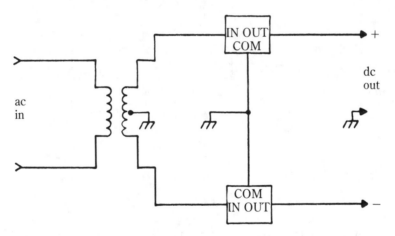

32-13   A regulated two-polarity power supply circuit.

Voltage regulator ICs have three output leads, input (the unregulated voltage), output (the regulated voltage), and common (the reference point for both the input and the output). The device usually resembles a somewhat oversized transistor. Figure 32-14 shows two typical package styles. The input voltage can vary over a large range without affecting the output voltage. One five V regulator will accept input levels up to 35 V.

Voltage regulators are limited in the amount of current that can be safely drawn from them. If greater current is required, several voltage regulators can be used in parallel.

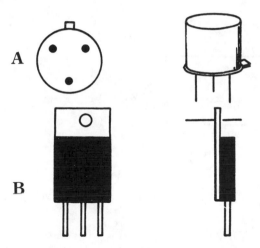

32-14   Typical voltage-regulator IC packages.

Voltage dividers

If you connect a string of series resistors across the voltage supply, varying voltages can be tapped off between each resistor pair. This device is called a *voltage divider*, because it divides the output voltage into a series of smaller voltages. Figure 32-15 shows a typical voltage divider.

Suppose all four resistors are 1000 Ω units, and 12 Vdc is being applied between +V and ground. because you have four 1000 Ω resistors in series, the total resistance is 4000 Ω. The current through the string is determined by Ohm's law. $I = E/R = 12/4000 = 0.003$ A or 3 mA.

Similarly, you can calculate the voltage drop across each of the resistors. $E = IR = 0.003$ A $\times 1000\ \Omega = 3$ V.

M1, of course, will read the full 12 V. R1 drops 3 V, so M2 will read 9 V. R2 drops 3 more V, so M3 reads 6 V. R3 drops an additional 3 V, bringing M4's reading down to 3 V. The remaining 3 V is dropped across R4, so the ground point will be at zero V.

The resistors in a voltage divider do not necessarily have to be of equal values. For instance, R1 might be 1000 Ω, R2 might be 270 Ω, R3 4700 Ω and R4 680 Ω. The total series resistance would be 6,650 Ω. This means the current flowing through the voltage divider with a 12 V input would be equal to 12/6,650 or 0.0018 A (1.8 mA).

Under these conditions M1 would read 12 V. R1 drops 1.8 V, so M2 will read 10.2 V. The voltage drop across R2 is about 0.5 V, leaving 9.7 V to be read on M3. R3 drops 8.5 V, so M4 must read 1.2 V. This remaining voltage is dropped by R4.

The voltage divider is an inexpensive circuit, but it is difficult to regulate because the load circuits are in parallel with the resistors, affecting the voltage drops. Voltage divider resistors are usually given the smallest values possible to minimize this effect.

A voltage divider circuit serves another important function. It provides a dis-

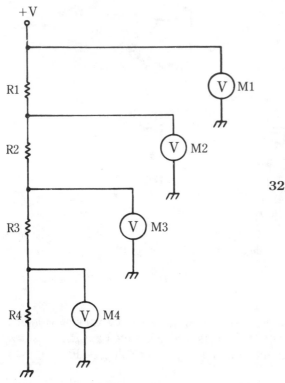

32-15   A voltage-divider circuit.

charge path for the heavy filter capacitors when the equipment is turned off. If these capacitors are not allowed to discharge to ground, a technician working on the circuit could receive a dangerous shock. Resistors used for this purpose are called *bleeder resistors*.

Voltage doublers

Sometimes you need a dc voltage that is more than the input ac voltage. Figure 32-16 shows a basic *voltage-doubler* circuit. During the negative half cycle, D1 is forward biased and D2 is reverse biased, charging C1. When the polarity is reversed during the positive half cycle the input voltage passes through the now forward biased D2, just as in an ordinary half-wave rectifier. D1 is reverse biased, and effectively out of the circuit. In addition to the regular input voltage, D2 also passes the voltage from C1 that is discharging through the diode. The charge on C1 is close to the input voltage, so the signal passing through D2 is essentially twice the input signal.

Besides being fed to the output, this voltage also charges C2. On the next negative half cycle, when D2 is again cut off, C2 partially discharges through the output, so the output voltage remains more or less constant. C3 is simply for filtering the output ripple.

The exact output is dependent on the values of C1 and C2. The larger they are, the closer the output will equal twice the input's peak voltage. These two capacitors

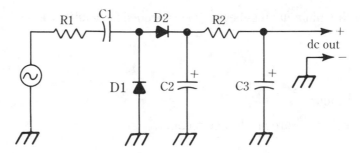

32-16 A voltage-doubler circuit.

should also be of equal capacitance value for the best results. If they are unequal, some ac might leak through, increasing ripple.

Self-test

1. What kind of power is supplied by batteries?

A ac
B dc
C Pulsating dc
D Static electricity
E None of the above

2. What component can be used as a half-wave rectifier?

A Capacitor
B FET
C Inductor
D Diode
E None of the above

3. What is the biggest disadvantage of a half-wave rectifier circuit?

A Expensive
B Wastes half of each input ac cycle
C Complicated circuitry
D Unreliable
E None of the above

4. What is the purpose of the output capacitor in a power supply circuit?

A Filter out ac ripple
B Act as a voltage divider
C Increase the output current capability
D Protect the diode from sudden current surges
E None of the above

5. What is the minimum number of diodes needed for a full-wave rectifier?
A One
B Two
C Three
D Four
E More than four

6. How many diodes are in a bridge rectifier?
A One
B Two
C Three
D Four
E More than four

7. What is the advantage of using a bridge rectifier for full-wave rectification?
A Cheaper
B Simpler circuitry
C No center tap is required on the transformer
D A lower input voltage can be used
E None of the above

8. How many leads does a typical voltage regulator IC have?
A One
B Two
C Three
D Four
E More than four

9. Refer to Fig. 32-15. Assume each of the resistors has a value of 33,000 Ω (33 kΩ), and V+ equals 18 V. What will the voltage reading on meter M2 be?
A 18 V
B 14 V
C 13.5 V
D 4.5 V
E None of the above

10. What is the purpose of a zener diode in a power supply circuit?
A Rectification
B Voltage regulation
C Filtering
D Current amplification
E None of the above

33
Amplifiers

The amplifier is probably the most widely used circuit type in all of electronics. Practically every complex electronics system or large circuit includes at least one amplifier stage. In a sense, every transistor used in every circuit is an amplifier. Even when a transistor is used for switching, its operation uses its amplification properties.

Defining amplifiers

In simple terms, an amplifier is a circuit that performs amplification, or increases the amplitude of a signal. In an ideal amplifier, the signal at the output should be absolutely identical to the signal at the input, except for the change in amplitude. Any other changes in the signal are due to distortion effects within the amplifier circuitry. All practical circuits distort to some extent, but in many cases the amount of distortion is negligible. In other cases, it can be quite severe.

Normally, an amplifier increases the signal amplitude, but there are occasional exceptions. Some amplifiers have negative gain. That is, the output signal is at a lower amplitude than the input signal. Strictly speaking, this is an attenuator rather than an amplifier, but because the same sort of circuitry is involved an active attenuator circuit is commonly called an amplifier.

Some amplifiers have unity gain—there is no change in the amplitude. The output signal (ideally) is identical to the input signal. This type of amplifier is called a *buffer* or a *buffer amplifier*. It is used primarily for impedance matching and stage isolation purposes.

Most amplifier circuits are divided into classes that define their basic operating characteristics. Some amplifier classes have much stronger distortion effects than others. Each amplifier class is suitable for a different set of applications. In this chapter we will take a look at the major amplifier classes. The most popular amplifier classes are:

Class A
Class B
Class AB
Class C

Additional classes run from D through H. These higher classes are used for special purposes, and some special considerations are involved when designing a circuit in one of these classes. In some cases, patents are involved. For this reason, this book concentrates on Classes A through C. These are the amplifier classes you are most likely to encounter in general electronics work, whether professionally or on the hobbyist level.

For convenience, this book concentrates solely on transistor amplifiers. The internal circuits in ICs work in pretty much the same way. Tube circuits are also basically similar, even though they are rarely encountered in modern electronics.

Biasing

The basic differences between the various amplifier classes are in how the transistor is *biased*. *Biasing* is the balancing of circuit polarities for correct operation of the transistor. For an NPN transistor to conduct, the collector must be more positive than the base, but the emitter must be negative with respect to the base. For a PNP transistor, these polarities are simply reversed.

An input signal (usually applied to the base—see chapter 20) will cause the biasing to fluctuate from its fixed levels. An input signal could throw off the fixed bias enough to cut off the transistor. The fixed bias point in the circuit must be selected with this in mind.

Class A amplifiers

In Class A amplifiers, the transistor is biased so that it conducts during the entire input cycle. This gives you an amplifier with very high linearity (low distortion), but low efficiency. The power supply to an amplifier circuit only supplies a finite amount of power. Any power consumed by the amplifier circuit itself, cannot be used in the actual amplification of the signal. In a Class A amplifier, most of the supplied power is consumed by the circuit. As much as 75 to 80 % of the supplied power is wasted in this manner. This limits how much amplification can be performed by the circuit. A typical Class A amplifier circuit is shown in Fig. 33-1.

Generally, low-power Class A amplifiers are biased at the center of the transistor's load line. That is, if you graph the operation of the transistor, the circuit bias point will be selected right in the middle of the linear portion of the graph.

Ideally, larger Class A power amplifiers are also biased at this mid-point, but as the power increases, there are some additional factors to be considered. It is an inescapable fact that for any transistor to deliver power, it must dissipate power.

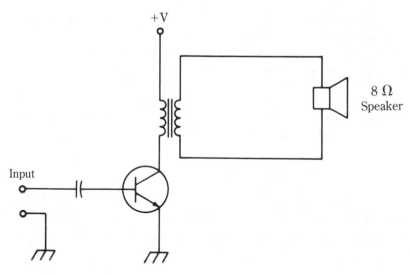

33-1 A Class A amplifier circuit.

Regardless of the size of the load (output circuit), the product of the collector current and the collector-to-emitter voltage at any point of the load line must be less than the maximum power dissipation rating of the transistor. If this maximum power dissipation rating is exceeded, the transistor could self-destruct from excess heat. In low-power circuits, this isn't much of a problem. In a larger power amplifier (Class A), the load line should be selected so that it always falls below the maximum permissible power dissipation curve of the transistor used in the circuit.

Class A power amplifiers are widely used to feed power to a loudspeaker in small radios. Most loudspeakers have a relatively low impedance (typically between 4 and 16 Ω), some sort of impedance matching network is usually employed to present a reasonable load to the collector circuit. In most cases, a transformer is used for this task (as shown in Fig. 33-1).

Theoretically, when a transformer is used between the loudspeaker and the transistor in a Class A amplifier, the maximum efficiency of the overall circuit is 50 percent. In other words, the maximum power that can be delivered to the load is equal to half the power supplied to the circuit. The other half of the supplied power is dissipated by the transistor, even if there is no input signal. For example, if you bias the transistor so that it draws a maximum of 10 W from the power supply, only 5 W can be applied to the output signal fed to the load. The other 5 W is dissipated as waste heat.

An efficiency of 50 % certainly sounds very low. But the situation is usually worse than theory suggests. In a practical Class A amplifier circuit, the actual efficiency is 20 to 40 % lower because of losses and limits of the transformer, and the amplifier circuitry itself.

In some cases, matters can be worse yet. If the transformer is eliminated, and the load is placed directly in the collector circuit, the load has to dissipate dc power too. This limits the maximum theoretical efficiency to a mere 25%, and that 25% can be reduced by the same 20 to 40% mentioned earlier. The net result is an amplifier circuit with an efficiency of about 20%—with luck.

In short, the Class A amplifier circuit has a very high price for its excellent linearity. It wastes an incredible amount of power. Fortunately, there are more efficient ways to amplify ac signals.

Because the Class A amplifier offers the lowest possible distortion, it is used in some audiophile stereo amplifiers, where price (and wasted power) is no object. Generally, Class A amplifiers are normally used only in very low-power applications.

Class B amplifiers

A Class B amplifier conducts for just half of each input cycle, and rests the remainder of the time. This is shown in Fig. 33-2. Because the amplifier is active for only half of each cycle, the circuit has much greater efficiency than with a Class A amplifier.

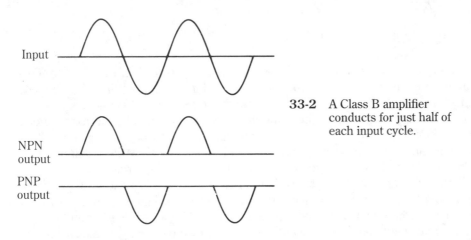

33-2   A Class B amplifier conducts for just half of each input cycle.

To allow the Class B amplifier to conduct during just one-half of each cycle of the input signal, the transistor is biased at the point where the *quiescent* (or idling) *current* is equal to zero. If there is no input signal, the transistor does not conduct—it just sits there, theoretically consuming no power.

If you assume that an NPN transistor is being used in a Class B amplifier circuit, the only time current will flow through the transistor is when the positive portion of the cycle is applied to the base. During the negative half cycle, the transistor is cut off, and no current flows. A PNP transistor works exactly the same way, of course, except that all of the polarities are reversed. A PNP transistor in a Class B amplifier circuit conducts only during negative half-cycles and is cut off during positive half cycles. A simple Class B amplifier circuit is shown in Fig. 33-3.

This approach to amplification is very efficient compared to a Class A amplifier. Unfortunately, chopping off half of each cycle of the input signal is an extreme form of distortion. This chopping is clearly undesirable in virtually any audio application, especially if any degree of high fidelity is desired. In order for a Class B amplifier to reproduce signals with reasonable accuracy, two parallel transistors must be used

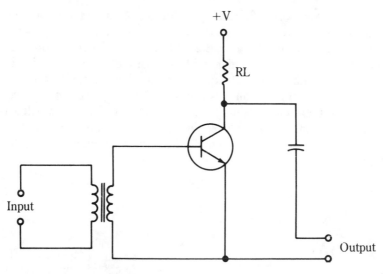

33-3   Class B amplifier circuit.

in the circuit. One of these parallel transistors is an NPN type to amplify the positive half cycles, and the other is a PNP device to amplify the negative half cycles. The two amplified half cycles are then recombined across the load to produce an amplified version of the original total input waveform. This kind of dual Class B amplifier circuit is known as a *push-pull amplifier.* A typical push-pull circuit is shown in Fig. 33-4.

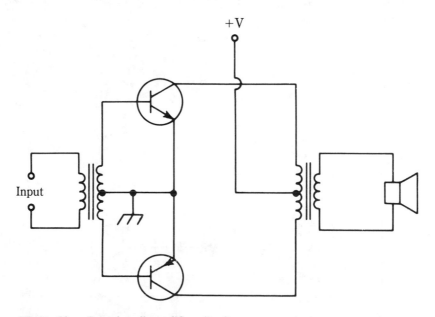

33-4   Class B push-pull amplifier circuit.

This circuit is far more efficient than the Class A amplifier. It can deliver up to 78.5% of the supplied power to the load. Because of the push-pull arrangement, the power that can be delivered to the load over the complete cycle is double the *maximum instantaneous power* each individual transistor can pass during the cycle.

Because the two halves of each output cycle are being amplified by two separate transistors, any (even minute) differences between the electrical characteristics of the transistors (or their surrounding components) will result in some degree of non-linearity (distortion) at the point where the two halves of the cycle are recombined. This is called *crossover distortion*. A signal with greatly exaggerated crossover distortion is shown in Fig. 33-5.

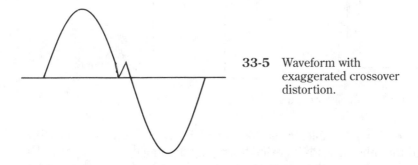

33-5 Waveform with exaggerated crossover distortion.

Crossover distortion in Class B audio amplifiers can be reduced by adding a source of bias to the circuit, as shown in Fig. 33-6. This modification actually turns the circuit into a Class AB amplifier, because the collector current now flows for more than half of each cycle. Class AB amplifiers are discussed in the next section of this chapter.

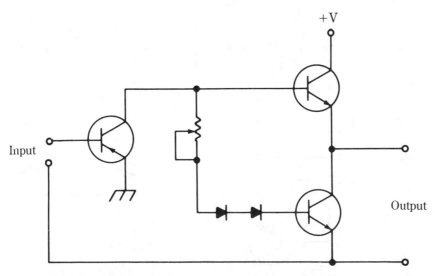

33-6 Crossover distortion in class B push-pull amplifiers can be reduced by adding bias.

Because of its inherent high efficiency, the Class B, push-pull direct coupled amplifier (as shown in Fig. 33-7) is probably the most popular audio amplifier configuration in use today. Early designs (before the mid-1970s) tended to rely more on

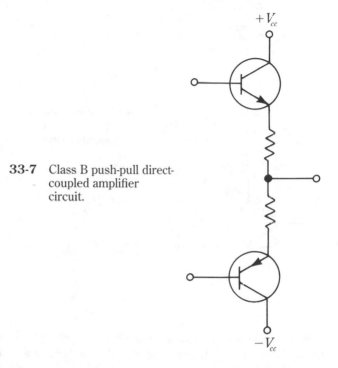

33-7 Class B push-pull direct-coupled amplifier circuit.

capacitor coupling (as shown in Fig. 33-8). The chief advantage of capacitor coupling is that such circuits require only a single ended power supply. In addition, because capacitors block dc voltages, speaker damage in the event of problems is unlikely using this approach.

Designers and audiophiles soon learned that capacitor coupling is scarcely ideal. The capacitors tend to cause poor low-frequency response, and instability. Direct coupling is not susceptible to these particular problems, so it is now more popular. However, a direct coupled amplifier does require a dual-polarity power supply, and fairly complex circuitry to protect the speakers against the possibility that one of the output transistors might short out and dump the full supply voltage (dc) across the speaker's voice coil, burning it out.

In dealing with Class B amplifiers, you might come across the term *output symmetry*. This term refers to how equally the positive and negative half cycles are reproduced. An amplifier with poor output symmetry will have *asymmetrical* (unequal) *clipping*, and reduced output power for a given distortion rating.

Although a Class B amplifier is highly efficient (up to nearly 80% efficiency) when it delivers its maximum power, it isn't quite as efficient at lower amplification levels. Under normal music listening conditions, an audio amplifier is generally driven to deliver full (or nearly full) output power for only a very small fraction of

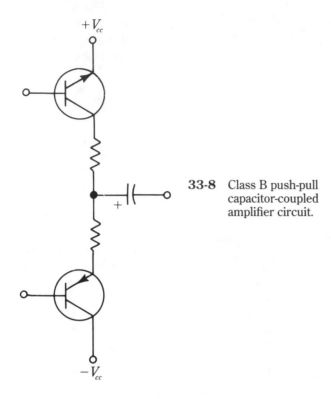

33-8   Class B push-pull
capacitor-coupled
amplifier circuit.

the time it is operating. Signal levels vary quite a bit in most music. Under practical music listening conditions, the actual efficiency of a typical Class B amplifier might be as low as 20%.

Class AB amplifiers

The Class AB amplifier is a neat compromise between the high linearity (low distortion) of a Class A amplifier and the efficiency of a Class B amplifier. Practical push-pull circuits are almost always Class AB amplifiers.

As you should recall from the earlier sections in this chapter, in Class A operation, the transistor is biased at the center of the load line, but in Class B operation, the transistor is biased at the point where no collector current flows when there is no input signal. In Class AB operation the transistor is biased to operate somewhere between Class A and Class B.

In the AB mode, some small amount of collector current flows when the transistor is idling (no input signal). This decreases the efficiency of the circuit somewhat. A Class AB amplifier can be useful where good and reliable (low distortion) reproduction of the input signal is required.

The chief advantage of Class AB amplifiers is that this mode of operation significantly reduces crossover distortion. This type of distortion occurs in a Class B push-pull amplifier because no (or very little) *conduction* takes place through a transistor unless a specific minimum voltage is exceeded. Virtually no current will flow through the base-emitter junction of a germanium transistor if less than 0.2 or 0.3 V is applied. For silicon transistors, the minimum signal voltage is between 0.5 and 0.7 V.

As the ac input signal passing through a Class B push-pull amplifier goes from positive to negative (or vice versa), it obviously passes through zero. There is a brief delay between the time one transistor is cut off and the other is turned on. This slight delay (along with any electrical mismatch between the two transistors) is what causes crossover distortion. If a transistor is cut off at a rapid rate, large transient voltages can develop in the circuit. This could cause the transistor to break down. Such problems are pretty much avoided in Class AB operation. One transistor is not cut off until after the other transistor has already started to conduct.

Class C amplifiers

In a Class C amplifier, the transistor is biased so that it conducts for less than one-half of each cycle. Only a very small portion of each input cycle is passed through the amplifier. This severely distorts the signal at the output, but in some applications (especially RF—radio frequency amplifiers) linearity isn't nearly as important as low power consumption (high efficiency). The output of a Class C amplifier is in the form of pulses, and the efficiency can be as high as 85%.

Using an NPN transistor, a Class C amplifier is created by biasing the base so that with no input signal, it is negative with respect to the emitter. This keeps the transistor cut off except when the input signal exceeds the bias voltage, which happens only during the peaks of the input cycle.

The Class C mode of operation is absolutely worthless for audio applications, because of its extreme distortion. But Class C amplifiers are widely used in IF (intermediate frequency) and RF (radio frequency) circuits. A resonant LC circuit (see chapter 10) is normally placed at the output of a Class C amplifier in a receiver. Each time a signal is applied to the amplifier input, a full cycle is generated across the LC tank, provided that the circuit is tuned for the frequency (or a multiple of the frequency) of the signal applied to the amplifier input.

Comparing the classes

Each of the amplifier classes can be compared with regard to efficiency and distortion (linearity).

Class	Efficiency	Linearity
A	poor	very good
B	very good	fair
AB	good	good
C	excellent	poor

The best trade off depends on the specific desired application.

Self-test

1. What does an amplifier do to an input signal?

A alters the wave shape
B changes the frequency
C alters the level
D changes the linearity

2. Which of the following is not a standard amplifier class?

A Class A
B Class B
C Class BC
D Class AB
E Class C

3. What kind of amplifier circuit is subject to crossover distortion?

A Class A
B Push-pull
C Class C
D RF

4. What type of gain does a buffer amplifier have?

A Unity
B Zero
C Positive
D Negative
E None of the above

5. Which of the following best describes a Class A amplifier?

A High efficiency and low distortion
B High efficiency and high distortion
C Low efficiency and high distortion
D Low efficiency and low distortion
E None of the above

6. Which of the following best describes a Class C amplifier?

A High efficiency and low distortion
B High efficiency and high distortion
C Low efficiency and high distortion
D Low efficiency and low distortion
E None of the above

7. How is a Class B amplifier biased?

A In the middle of the load line
B So it is cut off for less than half of each cycle
C So it is cut off for half of each cycle
D So it is cut off for more than half of each cycle

8. Which class of amplifier offers the highest efficiency?

A Class A
B Class B
C Class AB
D Class C

9. Which class of amplifier offers the lowest distortion?

A Class A
B Class B
C Class AB
D Class C

10. What type of transistors should be used in a push-pull circuit?

A Both NPN
B Both PNP
C One NPN and one PNP
D It doesn't matter as long as both transistors are the same type.

34
Oscillators

An *oscillator* is a circuit that produces an ac signal at some specific frequency (which might or might not be variable). Often a distinction is made between oscillators and *waveform generators*. Under these definitions, the ac signal produced by an oscillator is a sine wave (see Fig. 34-1) that is a very pure signal that (theoretically) contains only the nominal frequency. A generator, on the other hand, can produce other, more complex waveforms that contain the nominal (or *fundamental*) frequency and one or more of harmonics of this frequency (or whole number multiples) in some specific proportion. Common complex waveforms will be discussed later in this chapter.

 34-1 A sine wave.

For your purposes here, you can consider the terms *oscillator* and *generator* to be. Define an oscillator as any electronic circuit that produces a repeating ac signal.

Sine-wave oscillators

Producing a truly pure, harmonic-free sine wave is actually quite difficult, but practical circuits can generate reasonable approximations that are adequate for most applications.

Most sine-wave oscillators are built around parallel resonant LC circuits. The fundamental operations of an LC oscillator are illustrated in Fig. 34-2.

When the switch is closed, as in Fig. 34-2A, the voltage through coil L1 will

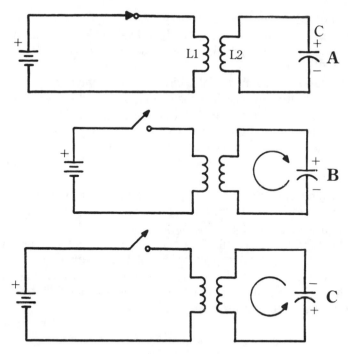

34-2   Oscillation in a parallel resonant LC circuit.

rapidly increase from zero to the source voltage. Of course this means the current through the coil must also be increasing. A change of current through a coil will generate a magnetic field. This magnetic field will induce a similar voltage in L2, which will be stored in capacitor C.

When the current through L1 reaches a stable point and stops increasing, the magnetic field collapses. This means no further voltage is induced into L2 and C can discharge through the coil in the opposite direction from the original voltage.

This discharge voltage will cause L2 to induce a voltage into L1, which in turn induces the voltage back into L2, charging C in the opposite direction. This action is shown in Fig. 34-2B.

Once the induced voltage in the coils collapses, this whole discharge-charge process repeats with the polarities again reversed, as in Fig. 34-2C.

Theoretically, this cycling back and forth between the capacitor and the coil will continue indefinitely. In real-world components, however, the coil and capacitor will have some dc resistance that will decrease the amplitude on each oscillation. The signal will look like Fig. 34-3. This is called a *damped* sine wave.

34-3   A damped sine wave.

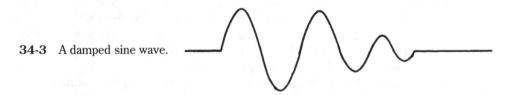

In addition to dc resistance, any energy that is tapped out of the circuit to be used by another circuit will subtract from the available energy within the LC circuit. Eventually, a point will be reached where the signal level is too weak to feed back and sustain oscillation.

For this reason, practical oscillator circuits always include some kind of amplifier stage. The output of the oscillator is continuously fed back to the input of the amplifier, and so it will be maintained at a usable level. Positive feedback is used.

It might at first seem that the amplifier would keep increasing the output amplitude infinitely, but all amplifiers have natural limitations that will prevent any further increases beyond some specific output level. This level is usually linked with the source voltage.

In an oscillator, this *saturation point* will be reached within a few cycles of power on, and the amplitude of the output signal will remain essentially constant. This is just a natural characteristic of amplifiers.

Another natural characteristic of amplifier circuits is often taken advantage of to start the oscillation in the first place. All amplifiers have some internal noise that will produce a tiny output signal even if the input is perfectly grounded. This noise is fed back through the amplifier until it is increased to a level that can start oscillations within the LC circuit, or *tank*.

The frequency of the oscillations will be determined by the resonant frequency of the specific coil-capacitor combination used. The formula is:

$$F = \frac{1}{6.28 \sqrt{LC}} \qquad \qquad \textbf{Equation 34-1}$$

where *F* is the frequency in hertz, *L* is the inductance in henries, and *C* is the capacitance in farads. The value 6.28 is two times pi (π).

As you should remember from chapter 8, this resonant frequency is the frequency at which the reactance of the capacitor equals the reactance of the coil. As frequency is increased, inductive reactance increases, while capacitive reactance decreases. Only at resonance (equal reactances) can oscillations be sustained.

The most common types of sine wave oscillators are the *Hartley oscillator*, the *Colpitts oscillator*, and the *crystal oscillator*.

The Hartley oscillator

A typical Hartley oscillator circuit is shown in Fig. 34-4. You should notice that the coil L has a center tap. In effect, it acts like two separate coils in close proximity. A current through coil AB will induce a signal into coil BC. For obvious reasons, this type of circuit is also known as a *split-inductance oscillator*.

When power is first applied to this circuit, R2 places a small negative voltage on the base of the transistor, allowing it to conduct. Internal noise will build up within the transistor amplification stages as discussed above. When this noise signal reaches a usable level, current from the collector will pass through R4, R2, and R1, finally reaching coil section AB. This rising current will induce a voltage into coil section BC. This voltage is stored by capacitor C1.

C2 is selected so that it has an extremely low impedance at the oscillating fre-

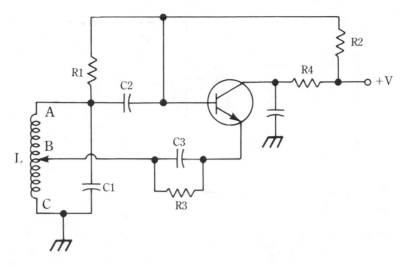

34-4 A Hartley oscillator.

quency, so the base of the transistor is effectively connected directly to C1. (The base voltage provided by R2 is quite low, so it can be ignored once oscillations begin. It is only needed to initiate oscillation.)

As C1 charges, it will increase the bias on the transistor. This in turn will increase the current through coil section AB and the induced voltage through coil section BC. At the same time, the charge on both C1 and C2 is increased.

Eventually the voltage from C1 will equal the R1/C2 voltage, but with the opposite polarity. In other words, these voltages cancel out. At this point the transistor is saturated. Its output will stop rising, so coil section AB's magnetic field collapses. C1 starts to discharge through coil section BC, allowing C2 to discharge through R1, cutting off the transistor until the next cycle begins.

It takes some finite time for C1 to discharge through coil section BC, so as the current through the coil increases, it builds up a magnetic field. Once the capacitor is discharged, the coil will tend to oppose the change in current flow, so it continues to conduct, charging the capacitor in the opposite direction, and the entire process is repeated.

In some applications the low impedance of the transistor might load the tank circuit excessively, increasing power loss, and possible decreasing stability. This problem can be readily dealt with simply by using a high impedance active device in place of the transistor. Figure 34-5 shows a Hartley oscillator built around an FET instead of an ordinary bipolar transistor. Except for the increased impedance, operation is the same.

The Colpitts oscillator

The *Colpitts oscillator* is very similar to the Hartley oscillator, except it is a *split-capacitance* instead of a split-inductance device. A typical Colpitts oscillator is shown in Fig. 34-6.

With the two capacitors in series, they will work like a single capacitor as far as

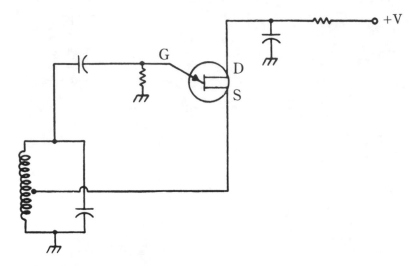

34-5 A FET Hartley oscillator.

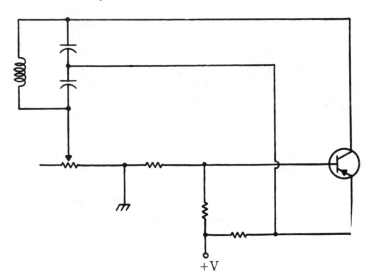

34-6 A Colpitts oscillator.

the LC resonant circuit is concerned, but the center tap (the connection between the two capacitors) provides a feed-back loop to the transistor's emitter.

If the two capacitors are of equal value, the total effective capacitance within the LC network (determining the frequency of oscillation) will be equal to one half the value of either capacitor separately. If they are unequal, the total can be calculated using the standard formula for capacitors in parallel; that is:

$$\frac{1}{C_T} = \frac{1}{C_1} + \frac{1}{C_2}$$

Equation 34-2

In actual practice the two capacitors are usually of unequal values, because the strength of the feed-back signal is dependent on the ratio between these two capacitances. By changing both of these capacitors in an inverse fashion the feed-back level can be varied while the resonant frequency remains constant. That is, when C1 increased, C2 would be decreased by a like amount, or vice versa.

This points up one of the chief problems of the Colpitts oscillator. When the frequency is changed, you don't want the feed-back signal to vary, or the level of the output signal will not be constant. In fact, oscillations might not be sustained. This limitation means you have to change both capacitances simultaneously.

True, you could make the coil adjustable instead of the capacitors, but in most applications this would be just as impractical. Generally speaking, it is preferable to use an adjustable capacitor rather than an adjustable coil.

A common solution to this problem with the Colpitts oscillator is illustrated in Fig. 34-7. This method keeps the C1, C2 ratio constant, but the resonant frequency is readily adjustable. The total effective capacitance in the tank circuit is determined by the following formula:

$$C_T = C_3 + \cfrac{1}{\left(\cfrac{1}{C_1} + \cfrac{1}{C_2}\right)} \qquad \text{\textbf{Equation 34-3}}$$

34-7 A variable frequency tank circuit for a Colpitts oscillator.

The Colpitts oscillator is a very popular circuit because it offers very good frequency stability at a reasonable cost. It also appears in a number of variations. One common variation of the Colpitts oscillator is the *Ultra-Audion oscillator,* which is well suited for VHF (very high frequency) applications. A typical Ultra-Audion oscillator circuit is shown in Fig. 34-8.

Internal capacitances within the transistor itself provide the feed-back paths. These internal capacitances give the effect of small capacitors from the base to the emitter and from the emitter to the collector. In the VHF range, these internal capacitances have quite a low reactance. For operation at lower frequencies, larger external capacitors could be added, as shown in Fig. 34-9. The Ultra-Audion oscillator is frequently used as the local oscillator in television tuners.

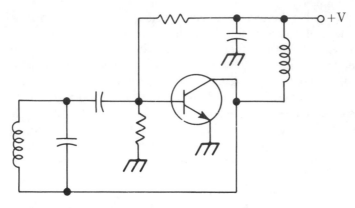

34-8   An Ultra-Audion oscillator.

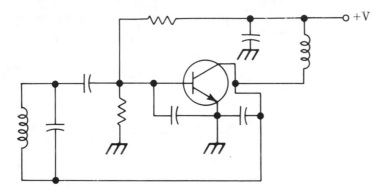

34-9   A low-frequency Ultra-Audion oscillator.

Another variation of the Colpitts oscillator is shown in Fig. 34-10. This is the *Clapp oscillator*, which is tuned by a series resonant LC circuit, rather than the more common parallel resonant tank. Feedback for the Clapp oscillator is provided by the voltage divider formed by the two smaller capacitors in the emitter circuit. Figure 34-11 shows a practical Colpitts oscillator that can be used as the variable-frequency ac source in the experiments. The parts list is given in Table 34-1.

Crystal oscillators

Although the frequency stability of LC oscillators like the Hartley and the Colpitts can be quite good, it often isn't precise enough for certain critical applications, such as in broadcast transmitters (see chapter 37), or the local color oscillator in a television receiver.

Oscillators built around quartz crystals can be extremely accurate and stable, especially if a constant temperature is held. Most broadcast stations (both television and radio) keep their carrier frequency oscillators in special crystal ovens to maintain a precisely constant temperature.

The fundamental operation of a crystal is described in chapter 11. Remember

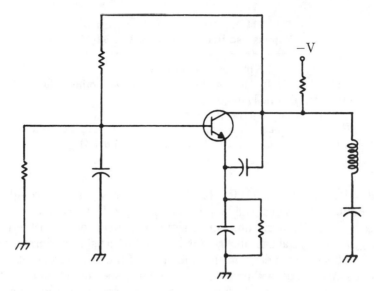

34-10 A Clapp oscillator.

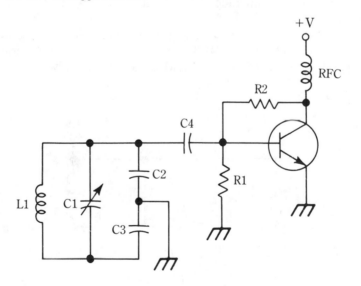

34-11 A practical Colpitts oscillator.

that the thickness of the crystal slab determines the resonant frequency. Obviously, there is a practical limit to just how thin a crystal can be cut. Because the thinner the crystal, the higher the resonant frequency, this puts a theoretical upper limit on the frequency of a crystal oscillator.

Higher than normal frequencies can be obtained from crystal oscillators by passing the signal through special circuits called *frequency doublers*. For example, if a 5 MHz (5,000,000 Hz) oscillator signal is fed through a frequency doubler, the

**Table 34-1. Parts list for the practical
Colpitts oscillator circuit of Fig. 34-11.**

C1	365 pF variable capacitor
C2, C3, C4	0.1 μF capacitor (experiment with other values)
RFC	radio frequency choke
L1	0.1 μH coil
Q1	almost any npn transistor (2N3904, 2N2222, etc.)
R1, R2	experiment should be at least 1,000 Ω

output will be 10 MHz (10,000,000 Hz). If this signal is passed through a second frequency doubler, the signal will be raised to 20 MHz (20,000,000 Hz).

For crystal oscillator circuits, an ac signal is applied between the plates, and thus, through the crystal slab itself. This causes the crystal to vibrate, due to the piezoelectric effect. Depending on the thickness of the crystal slab, it will be mechanically resonant (most willing to vibrate) at some specific frequency.

If the applied ac voltage is equal to the crystal resonant frequency (or some exact harmonic of it) the amplitude of the vibrations will be quite large, and oscillations will be sustained.

Figure 34-12 shows the schematic diagram for an oscillator using a crystal as a parallel resonant circuit. Figure 34-13 shows an example of the series resonant mode. The parallel resonant form is somewhat more commonly used.

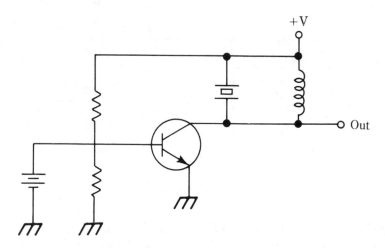

34-12   A parallel resonant-crystal oscillator.

There are a number of other types of sine wave oscillators, but these are the basic types you are most likely to encounter.

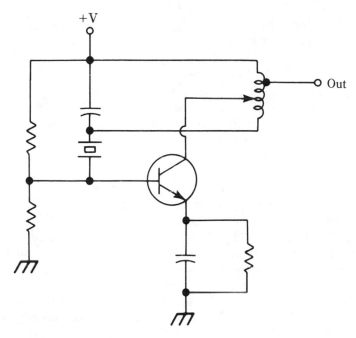

34-13   A series resonant-crystal oscillator.

Triangular-wave oscillators

Figure 34-14 shows another common waveshape, the triangular wave. This is also sometimes called a *delta wave*. The triangular wave is somewhat similar to a sine wave in that the instantaneous level rises from a minimum value to a maximum value, then immediately reverses direction, without ever holding any constant value. But where the signal level in a sine wave varies according to a sinusoidal curve, the level in a triangular wave varies linearly.

34-14   A triangular wave.

The primary result of this is the presence of the odd harmonics. Besides the nominal fundamental frequency, the signal contains weaker frequencies at three times the fundamental (third harmonic), five times the fundamental (fifth harmonic), seven times the fundamental (seventh harmonic), and so forth. No even harmonics (those divisible by two) are included in a triangular wave signal. The amplitude of the various harmonics relate to the amplitude of the fundamental according to the following formula:

$$A_{\mathrm{h}} = \frac{A_{\mathrm{f}}}{h^2}$$ **Equation 34-4**

where A_{h} is the amplitude of the harmonic in question, h is the number of the harmonic, and A_{f} is the amplitude of the fundamental frequency. If you arbitrarily assign A_{f} a value of 1, the amplitude of the third harmonic would be $\frac{1}{3^2}$ or $\frac{1}{9}$, the amplitude of the fifth harmonic would be $\frac{1}{5^2}$ or $\frac{1}{25}$, the amplitude of the seventh harmonic would be $\frac{1}{7^2}$ or $\frac{1}{49}$, and so forth.

A triangular waveform (or any other complex waveform) can be created by combining sine waves of appropriate frequencies and amplitudes—one for the fundamental and one for each of the harmonics. This is called *additive synthesis*.

Similarly, if you filter a triangular wave (or any complex waveform) so there is no output except for the fundamental frequency, or one specific harmonic frequency, you would be left with a sine wave. This is called *subtractive synthesis*.

If you wanted to build up a 200 Hz triangular wave using additive synthesis starting with an 18 V peak-to-peak fundamental at 200 Hz, you'd need to add a 2 V peak-to-peak 600 Hz sine wave (third harmonic), a 0.72 V peak-to-peak 1000 Hz sine wave (fifth harmonic), a 0.37 V peak-to-peak 1400 Hz sine wave (seventh harmonic), a 0.22 V peak-to-peak 2200 Hz sine wave (eleventh harmonic). For most purposes, any additional harmonics beyond the eleventh would be at too low a level to be of significance.

As a second example, a triangular wave with a 1450 Hz, 30 V peak-to-peak fundamental would have a 4350 Hz, 3.33 V peak-to-peak third harmonic, a 7250 Hz, 1.20 V peak-to-peak fifth harmonic, a 10,150 Hz, 0.61 V peak-to-peak seventh harmonic, a 13,050 Hz, 4.07 V peak-to-peak ninth harmonic, and a 15,950 Hz, 0.25 V peak-to-peak eleventh harmonic.

Notice that the fundamental frequency is always at a much higher voltage than all of the harmonic frequencies taken together. For many applications, the harmonics can simply be ignored, and the triangular wave used in place of a sine wave. This substitution is often made because it is generally easier and more economical to produce a stable triangular wave than a comparable sine wave.

Figure 34-15 shows a typical triangular-wave oscillator circuit built around two 741 operational amplifier integrated circuits. IC1 is a square-wave oscillator (discussed in chapter 24). An integrator converts a square-wave signal into a triangular wave. The fundamental frequency of the output signal is determined by C1 and R3. Of course, a single 747 dual op amp could be used in place of the two separate 741s.

Rectangular-wave oscillators

The ac voltage at the output of a sine wave oscillator or a triangular wave oscillator varies continuously between its minimum and maximum values. A rectangular wave, however, consists of only two voltage levels—a maximum level and a minimum level. Ideally, the voltage switches instantly from one level to the other. A real world circuit

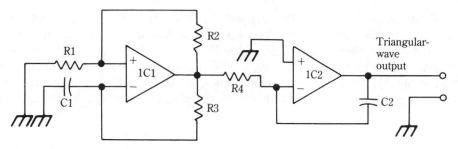

34-15   A triangular wave oscillator.

takes some finite time to change states. This time is measured as a specification called the *slew rate*.

The relationship between the high-voltage time and the low-voltage time is called the *duty cycle*. The duty cycle is important because it determines the harmonic content of the signal. A rectangular wave with a duty cycle of 1:X will contain all harmonics, except those that are multiples of X. The amplitude of each harmonic relates to the amplitude of the fundamental by the formula:

$$A_h = \frac{A_f}{h}$$ **Equation 34-5**

For example, consider a rectangular wave with a 1:3 duty cycle (see Fig. 34-16), and a 10 V peak-to-peak 1000 Hz fundamental. The second harmonic would be equal to $2 \times F = 2 \times 1000$ or 2000 Hz. The amplitude equals $A/h = 10/2$ or 5 V peak-to-peak. The third harmonic is absent. The fourth harmonic is at 4×1000 or 4000 Hz with an amplitude of 10/4 or 2.5 V peak-to-peak. The fifth harmonic is at 5×1000 or 5000 Hz, and its amplitude equals 10/5 or 2 V peak-to-peak. The sixth harmonic is absent. The seventh harmonic is 7×1000 or 7000 Hz. Its amplitude is 10/7 or 1.43 V peak-to-peak. The eighth harmonic is at 8×1000 or 8000 Hz, and its amplitude equals 10/8 or 1.25 V peak-to-peak. The ninth harmonic is missing, and so forth. Every third harmonic is absent from the signal.

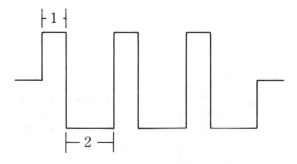

34-16   A 1:3 duty-cycle rectangular wave.

Similarly, take a rectangular wave with the same fundamental, but a duty cycle of 1:5. The harmonics would be 2000 Hz (second harmonic), 3000 Hz (third harmonic), 4000 Hz (fourth harmonic), 6000 Hz (sixth harmonic), 7000 Hz (seventh harmonic), 8000 Hz (eighth harmonic), 9000 Hz (ninth harmonic), 11,000 Hz (eleventh harmonic), 12,000 Hz (twelfth harmonic), and so forth. Every fifth harmonic is missing from the signal. Figure 34-17 shows a rectangular wave with a 1:5 duty cycle.

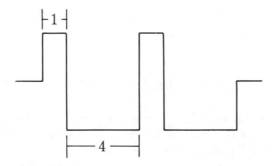

34-17   A 1:5 duty cycle rectangular wave.

A special case is a rectangular wave with a 1:2 duty cycle (see Fig. 34-18). This is called a *square wave*. Sometimes other rectangular waves are called square waves, but it is incorrect to do so. A square wave has only the odd harmonics, like a triangular wave, but the proportions are different. Table 34-2 compares the harmonic content of a triangular wave and a square wave with 500 Hz, 50 V peak-to-peak fundamentals. You can see that the square wave has considerably more harmonic content than the triangular wave.

Rectangular-wave oscillators are also called multivibrators. You read about multivibrators in other chapters. Remember that there are three basic types of multivibrators—the monostable, the bistable, and the astable. In this context you are only concerned with astable multivibrators. Figure 34-19 repeats the astable multivibrator built around a 555 timer integrated circuit discussed in chapter 25. Figure 34-20 shows another rectangular-wave oscillator, using discrete transistors. When power

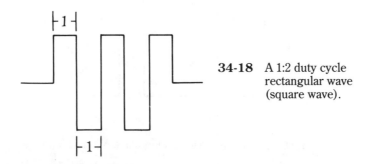

34-18   A 1:2 duty cycle rectangular wave (square wave).

Table 34-2. Comparing a square wave with a triangular wave.

Harmonic number	Frequency	Amplitude in a square wave	Amplitude in a triangular wave
First (fundamental)	500 Hz	50 V	50 V
Third	1.500 Hz	16.67 V	5.56 V
Fifth	2.500 Hz	10.00 V	2.00 V
Seventh	3.500 Hz	7.14 V	1.02 V
Ninth	4.500 Hz	5.56 V	0.62 V
Eleventh	5.500 Hz	4.54 V	0.42 V
Thirteenth	6.500 Hz	3.85 V	0.30 V
Fifteenth	7.500 Hz	3.33 V	0.22 V
Seventeenth	8.500 Hz	2.94 V	0.17 V
Nineteenth	9.500 Hz	2.63 V	0.14 V
Twenty-first	10.500 Hz	2.38 V	0.11 V
Twenty-third	11.500 Hz	2.17 V	0.09 V
Twenty-fifth	12.500 Hz	2.00 V	0.08 V

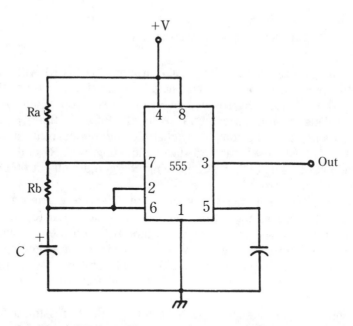

34-19 A 555 timer IC rectangular wave oscillator.

is first applied to this circuit, one of the two transistors will start conducting a little faster than its mate, because no two transistors are ever exactly identical. It could be either transistor, and it doesn't really matter which one it is. For this discussion, you will assume Q1 conducts first.

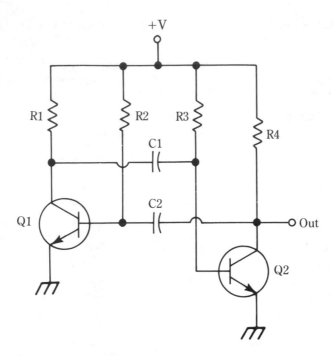

34-20   A transistor rectangular wave oscillator.

As Q1 draws more current, the voltage dropped across R1 will increase (because $E = IR$), pulling the collector of Q1 negative, and charging C1 so that the end connected to the collector of Q1 is negative and the end connected to the base of Q2 is positive. This keeps the second transistor cut off for the time being.

At some point C1 will be completely charged, and the current flow to the base of Q2 will stop. Q2 will now be able to start conducting, with R4 and C2 mirroring the earlier action of R1 and C1. As C2 is being charged, it will cut off Q1, allowing C1 to discharge.

This entire process continues back and forth indefinitely. The way the transistors are wired, they are either cut off completely, or saturated (operating at their maximum level) with a very small transition time between the two states (that is, the slew rate). For most applications, you can consider the switching time to be effectively zero, so the output is always either high, or low—switching back and forth at some specific frequency. Of course, when one transistor is cut off the other is always saturated.

The output could be taken from the collector of either transistor. In fact, in some cases, both outputs are used. Obviously these two outputs are mirror images, or complements of each other—when one is high, the other is low, and vice versa.

Four components determine the frequency of this oscillator. These are R1, C1, R4 and C2. R1 and C1 determine the one time for Q1 (time = $R \times C$), and R4 and C2 control Q2. If these two lines are equal (that is, R1 = R4, and C1 = C2) the output will be a square wave.

The rectangular wave oscillator shown in Fig. 34-21 works in essentially the same way, except most of the circuitry is contained within the digital NOR gates. Notice that there is only one capacitor. It is used for both halves of the cycles. This means the on and off times are automatically equal. The output of this circuit is always a square wave.

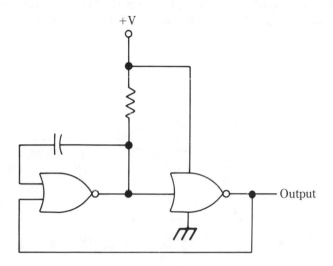

34-21 A digital square-wave oscillator.

Another simple square-wave oscillator circuit is shown in Fig. 34-22. This circuit uses four sections of a hex inverter integrated circuit, such as a TTL 7404, and a single external capacitor—just two parts!

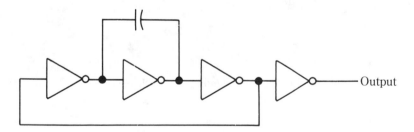

34-22 Another digital square-wave oscillator.

The capacitor can have a value anywhere between about 300 pF (giving an output frequency of about 1 MHz—1,000,000 Hz) to about 300 μF (giving an output frequency of about 1 Hz). There is one important precaution to be taken with this circuit. The capacitor must *not* be an electrolytic type. In this circuit, current must be able to pass through the capacitor in both directions, and, of course, this can't be done with an electrolytic capacitor. If the polarity across an electrolytic capacitor changes, damage will result both to the capacitor itself, and (possibly) to other components in the circuit (in this case, the integrated circuit).

Semiconductors (such as transistors and integrated circuits) are particularly well suited for use in multivibrator circuits because they can switch from cutoff to saturation and back very rapidly, giving excellent slew rates. Tubes can also be used, but not as efficiently.

An operational amplifier, or a timer IC is a perfect point for designing a rectangular wave oscillator. Figure 34-23 shows a square-wave oscillator built around a 741 operational amplifier.

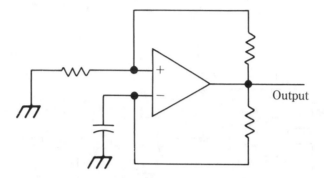

34-23   An op amp rectangular wave oscillator.

Sawtooth-wave oscillators

Figure 34-24 shows an *ascending sawtooth*, or *ramp wave*. The signal starts as a minimum level, builds linearly up to a maximum, then drops back down to its original minimum level and starts over. An ascending sawtooth wave is also sometimes called a *positive sawtooth wave*.

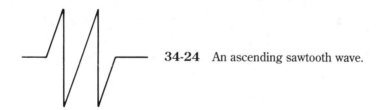

34-24   An ascending sawtooth wave.

This waveform contains all of the harmonics (second, third, fourth, fifth, sixth, seventh, and so forth). The amplitudes of the successive harmonics decrease in an exponential manner. The mirror image of this waveform is also sometimes used. A *descending sawtooth wave* is shown in Fig. 34-25. This waveform is also called a *negative sawtooth wave*.

Generally speaking, positive sawtooth waves are more commonly used. If the type is unspecified (that is, when a circuit is called just a "sawtooth wave oscillator")

34-25 A descending sawtooth wave.

it is usually safe to assume a positive sawtooth wave is intended. Figure 34-26 is a simple sawtooth-wave oscillator built around a unijunction transistor. This circuit is extremely versatile, because all of the components can be virtually any value. Typically, however, R2 and R3 are between 15 Ω and 1500 Ω. These resistors do not have to be of identical values.

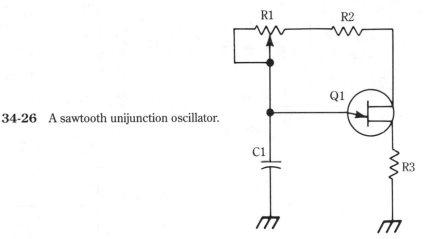

34-26 A sawtooth unijunction oscillator.

The frequency of the oscillator is determined by R1 and C1. C1 charges through R1 at a specific fixed rate (that is, the time constant). At some point, enough voltage is built up at the emitter for the transistor to fire. When this happens, the capacitor quickly discharges, and the whole cycle starts over again. This process produces a positive sawtooth wave signal on the emitter of the transistor.

It is also possible to take a signal off of either base. B1 produces a string of negative spikes, as shown in Fig. 34-27A. The B2 signal is shown in Fig. 34-27B. This is the mirror image of B1's output—a string of positive spikes. All three signals are produced at the same frequency, of course.

Function generators

A *function generator* is actually nothing more than an oscillator that can produce more than one type of waveform. The sawtooth oscillator in Fig. 34-26 is actually a

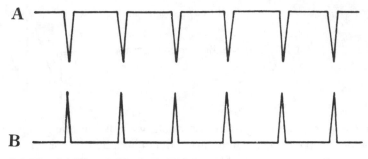

34-27 Additional signals for Fig. 34-26.

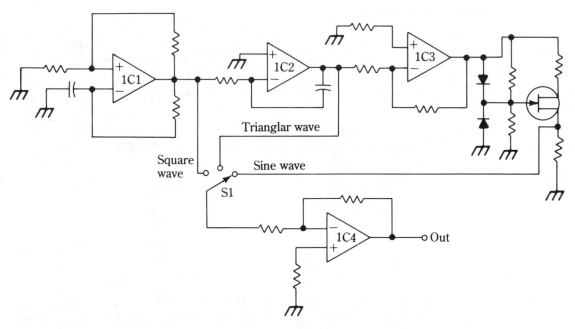

34-28 An op amp function generator.

function generator of sorts. Most function generators produce triangular waves and rectangular waves, and perhaps one or two other waveforms, such as a sine wave or a sawtooth wave.

Figure 34-28 shows a simple function-generator circuit built around operational amplifiers. IC1 is a square wave oscillator. IC2 is an integrator that converts the square wave to a triangular wave. IC3 is a filter that converts the triangular wave to a sine wave. S1 selects which of the three waveforms will appear at the output. IC4 is simply an amplifier. This circuit could be built using a single 324 quad operational amplifier integrated circuit. Another simple function-generator circuit is given in Fig. 34-29. This circuit uses discrete transistors.

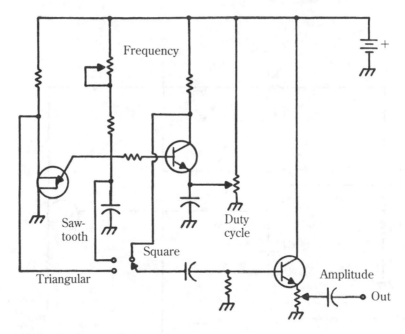

34-29 A transistor function generator.

Function generators are extremely useful devices for electronic testing because of the wide range of possible output signals that can be experimentally fed through other electronic circuits.

Voltage-controlled oscillators

All of the oscillator circuits described so far either have a fixed frequency output or an output whose frequency can be altered by manually adjusting some variable component (a potentiometer, or a variable capacitor). For many applications, it is preferable or even necessary to have some way to automatically adjust the frequency via purely electronic means. A *voltage controlled oscillator* (abbreviated *VCO*) is an oscillator whose output frequency will vary in step with an input voltage. This input voltage is called the *control voltage*, or *CV*.

Figure 34-30 shows a voltage controlled version of the sawtooth wave oscillator from Fig. 34-26. Q2 acts like a voltage variable resistor—the equivalent circuit is shown in Fig. 34-31.

R4 and R(Q2) take the place of R1 in the original circuit. Because the output frequency is determined by R1 (R4 and R(Q2)), varying R(Q2) will vary the output frequency. Because the effective resistance of Q2 is determined by the external voltage applied to its base, you have a voltage-controlled oscillator.

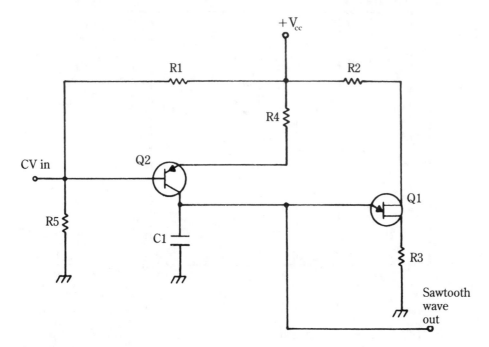

34-30   A VCO.

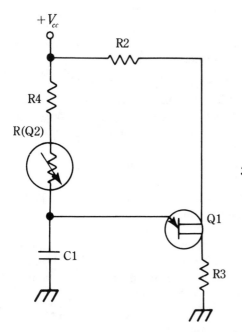

34-31   Equivalent circuit for Fig. 34-30.

Self-test

1. What is the simplest, purest type of waveform?

A Square wave
B Sine wave
C triangular wave
D Sawtooth wave
E None of the above

2. Which of the following is not a common type of sine wave oscillator?

A Hartley
B Crystal
C Multivibrator
D Colpitts
E None of the above

3. Which of the following is an oscillator with a spilt inductance?

A Hartley
B Crystal
C Multivibrator
D Colpitts
E None of the above

4. If the two capacitors in the LC tank of a Colpitts oscillator have values of 0.01 μF and 0.0047 μF, what is the total effective capacitance?

A 0.0147 μF
B 0.0053 μF
C 0.0032 μF
D 0.000047 μF
E None of the above

5. What is another name for a triangular wave?

A Delta wave
B Pulse wave
C Ramp wave
D Angular wave
E None of the above

6. In a rectangular wave with a duty cycle of 1:4, which of the following is a correct summary of the lower harmonic content?

A Fundamental, third, fifth, seventh, ninth
B Fundamental, second, fourth, fifth, seventh, eighth, tenth
C Fundamental, second, third, fifth, sixth, eighth, ninth, tenth
D Fundamental, second, third, fifth, sixth, seventh, ninth, tenth
E None of the above

7. What is the duty cycle of a square wave?

A 1:2
B 1:5
C 1:4
D 1:1.5
E None of the above

8. Is the fifth harmonic stronger in a triangular wave or a square wave?

A triangular
B Square
C They are equal
D It varies
E None of the above

9. What is the output frequency range of a square-wave oscillator made up of four inverter stages and a capacitor (Fig. 34-22)?

A 1 Hz to 100 kHz
B 10 Hz to 1 MHz
C 100 Hz to 100 kHz
D 1 Hz to 1 MHz
E None of the above

10. What is the name for a waveform that begins at a low level, smoothly builds up to a maximum level, then quickly snaps back down to the minimum and starts over?

A Descending sawtooth
B Delta
C Spike
D Ascending sawtooth
E None of the above

35
Filters

A *filter* is a circuit that removes selected frequencies from a signal. There are four basic types of filters: *lowpass, highpass, bandpass,* and *band-reject.*

Lowpass filters

Figure 35-1 shows the basic frequency-response graph of a typical lowpass filter. Notice that low frequencies are able to pass through to the output, but higher frequencies are increasingly blocked. The steeper the cut-off slope, the better the filter. The point marked X is the *cutoff frequency* of the filter. The simplest possible lowpass filter is shown in Fig. 35-2. This consists of just a resistor and a capacitor. The cutoff frequency is determined by the formula:

$$F = \frac{159,000}{RC}$$

Equation 35-1

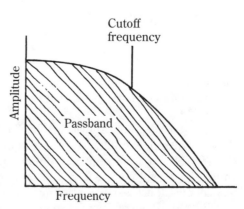

35-1 Frequency response graph for a typical lowpass filter.

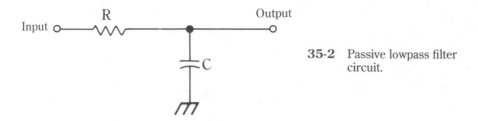

35-2   Passive lowpass filter circuit.

where *F* is the cutoff frequency in hertz, *R* is the resistance in ohms, and *C* is the capacitance in microfarads. Suppose you have a filter consisting of a 10,000 Ω (10 kΩ) resistor and a 0.1 μF capacitor. The cutoff frequency would be equal to 159,000/(10,000 × 0.1) = 159,000/1,000 = 159 Hz. Similarly, if *R* is 330 Ω and *C* equals 0.5 μF, then *F* equals 159,000/(330 × 0.5) = 159,000/165 or approximately 964 Hz.

A 0.0022 μF capacitor and a 470 Ω resistor would result in a cutoff frequency of 159,000/(470 × 0.0022) = 159,000/1.034 or about 153,772 Hz. Notice that reducing the value of the capacitance or the resistance increases the cutoff frequency. If you have to design a filter with a specific cutoff frequency, the formula can be rewritten as:

$$R = \frac{159,000}{FC}$$ **Equation 35-2**

As an example, suppose you need a lowpass filter with a cutoff frequency of 1,000 Hertz. First you arbitrarily select a convenient value for *C*. Use a 0.1 μF capacitor. Now you can calculate that the required resistance equals 159,000/(1000 × 0.1) = 159,000/100 = 1590 Ω. Unless the application is extremely critical, you can select the closest standard resistance value. In this case you would probably use a 1500 Ω resistor.

The same cutoff frequency could be just as easily achieved with a different combination of components. For instance, if you started with a 0.033 μF capacitor, the resistor would have to be equal to 159,000/(1000 × 0.033) = 159,000/33 = about 4818 Ω. A standard 4700 Ω resistor could be used. Notice that, for a given frequency, as one component increases in value, the other component must decrease in value.

Of course it is possible to start with an arbitrarily selected resistance value and solve for the capacitance using the formula in this form:

$$C = \frac{159,000}{RF}$$ **Equation 35-3**

But there are more readily available standard resistor values than standard capacitor values, so it is generally more convenient to select a convenient capacitor and solve for resistance.

The simple filter shown in Fig. 35-2 has two major disadvantages. Its cutoff slope is extremely gradual—quite a bit of the output signal could consist of frequencies above the cutoff frequency and because only passive components are used, the entire signal will be attenuated to some degree. Even the desired frequencies will be partially decreased. This attenuation is called *insertion loss*.

Both of these problems can be taken care of by using an active filter instead of a passive one. An active filter includes an amplifier stage to compensate for any insertion loss. The disadvantages of an active filter are increased cost and circuit complexity, and the need for a power supply.

Figure 35-3 shows a basic lowpass filter built around an operational amplifier. This circuit is also known as an integrator, and has been discussed earlier in this book. As you'll recall, the formula for determining the gain in an operational amplifier is:

$$G = \frac{R_2}{R_1}$$
Equation 35-4

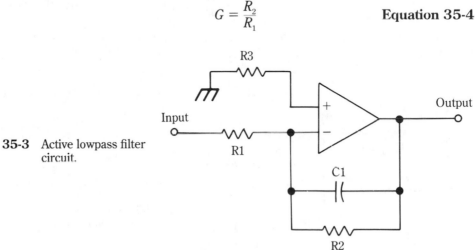

35-3 Active lowpass filter circuit.

But in this circuit the feedback component is a capacitor whose reactance decreases as the frequency increases, according to the formula:

$$X_c = \frac{1}{2\pi FC}$$
Equation 35-5

So you now have to combine the two equations:

$$G = \frac{X_c}{R_1} = \frac{1/(2\pi FC)}{R_1}$$
Equation 35-6

Assume R1 is 4700 Ω, and C is 0.022 μF (0.0000000022 F). The basic formula becomes $(1/(2 \times 3.14 \times F \times 0.0000000022))/4700$, or $(1/(0.00000014 \times F))/4700 = (7234316/F)/4700$.

At 10 Hz, the gain is equal to $(7234316/10)/4700 = 72341.6/4700$, or about 154. But if the frequency is increased to 1000 Hz, the gain becomes equal to $(7234316(1000)/4700 = 7234.316/4700$ or about 1.5. Increasing the frequency to 5000 Hz brings you into the region of negative gain—that is, the output level is less than the input level. The gain equals $(7234316/5000)/4700 = 1447/4700 =$ just over 0.3.

If the frequency is increased to 50,000 Hz, the gain goes down to $(7234316/50000)/4700 = 145/4700 =$ a gain of 0.03. The output level at this frequency is just 3% of its original input level. Although it is a definite step up over the simple passive filter previously described, the cutoff slope of this circuit still isn't particularly steep.

Figure 35-4 shows an improved active filter with a much steeper cutoff slope. The cutoff frequency for the circuit is determined by the following formula:

$$F = \frac{1}{2\pi\sqrt{R_2 R_3 C_1 C_2}}$$ **Equation 35-7**

and then gain is determined by:

$$\text{GAIN} = \frac{R_3}{R_1}$$ **Equation 35-8**

This, of course, is the gain for frequencies below the cutoff frequency—that is, within the *passband*. For unity gain, all three resistors should be equal.

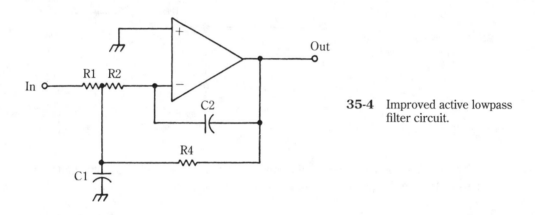

35-4 Improved active lowpass filter circuit.

Assume you have such a circuit with three 1,000 Ω resistors, C1 equals 0.5 μF, and C2 equals 0.02 μF. The cutoff frequency equals 1/(6.28 × (1000 × 1000 × 0.0000005 × 0.00000002)½) = 1/(6.28 × (0.00000001)$^{1/2}$) = 1/(6.28 × 0.0001) = 1/0.000628. The cutoff frequency is approximately 1,592 Hertz. The passband gain of course is 1 or unity.

Highpass filters

A highpass filter is the mirror image of a lowpass filter. Figure 35-5 shows the frequency response chart for a typical highpass filter. For a simple passive highpass filter, the components simply change places, as shown in Fig. 35-6. The equations are the same as the basic lowpass filter.

Similarly, a simple op amp highpass filter (differentiator) is the same as the lowpass version, except the resistor and the capacitor swap positions. See Fig. 35-7.

Figure 35-8 shows a steep-slope active highpass filter built around an operational amplifier.

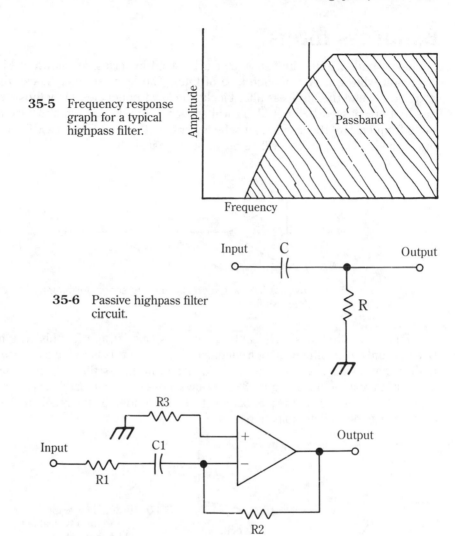

35-5 Frequency response graph for a typical highpass filter.

35-6 Passive highpass filter circuit.

35-7 Active highpass filter circuit.

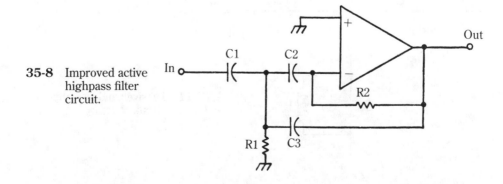

35-8 Improved active highpass filter circuit.

10. Assume a passive lowpass filter made up of a 390 Ω resistor and a 0.1 μF capacitor. What size inductor should be added for a center frequency of 12,000 Hz?

A 4 mH

B 0.000002 mH

C 1.8 mH

D 18 mH

E None of the above

36
Modulation

If one ac signal is used to control some aspect of a second ac signal, the process is called *modulation*. There are several types of modulation, each type having its own unique characteristics. Modulation is often used in such applications as electronic music and data storage. By far, its most common application is in radio transmission. This subject is discussed in the next chapter.

Amplitude modulation

Perhaps the most basic type of modulation is called *amplitude modulation*, which is often abbreviated as *AM*. As the name indicates, one ac signal modulates, or controls, the amplitude, or level, of a second ac signal. This process is shown in Fig. 36-1. The controlled signal is called the *carrier*, and the controlling signal is called the *program signal*. The program signal is superimposed onto the carrier signal.

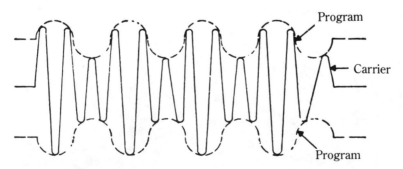

36-1   Amplitude modulation.

514

The process of amplitude modulation produces additional signals that are not present in either of the input signals. These newly created frequencies are called *sidebands*. For simplicity, first consider what happens when both the carrier and the program are pure sine waves. In this case there will be two sidebands. The frequency of the upper sideband is determined by the sum of the carrier and program frequencies. That is:

$$F_{us} = F_c + F_p \qquad \qquad \textbf{Equation 36-1}$$

where F_{us} is the frequency of the upper sideband, F_c is the carrier frequency, and F_p is the frequency of the program signal. All three quantities are in hertz, of course.

The lower sideband is the difference between the carrier and program frequencies. That is:

$$F_{is} = |F_c - F_p| \qquad \qquad \textbf{Equation 36-2}$$

F_{is}, of course, is the frequency of the lower sideband. The two vertical lines enclosing the right side of the equation indicate that the result is the *absolute value* of the equation. This means the answer will always be positive, even if F_p is greater than F_c. Negative frequencies, of course, are impossible.

If one, or both, of the input signals is a complex waveform, rather than a pure sine wave, the number of sidebands increases. Each individual harmonic will produce its own set of sidebands with each and every harmonic in the other input signal. Obviously, the resultant signal can be quite complex.

Assume you have a 150 Hz sine wave program signal amplitude modulating a 1000 Hz sine wave carrier signal. The output would consist of the 1000 Hz carrier with 150 amplitude fluctuations per second, plus weaker signals at 1150 Hz (upper sideband) and 850 Hz (lower sideband).

Now, suppose the program signal is changed to a 150 Hz square wave. A square wave, of course, contains all of the odd harmonics. The fundamental will behave in the same way as a simple sine wave. That is, the output will contain signals at 850 Hz, 1000 Hz, and 1150 Hz. The third harmonic (450 Hz) also will combine with the 1000 Hz carrier to produce additional sidebands at 550 Hz and 1450 Hz. The fifth harmonic (750 Hz) will produce sidebands at 250 Hz and 1750 Hz, and so on. If both the carrier and the program signals are complex waveforms, the number of sidebands becomes very large. Note that the sidebands don't necessarily (and probably will not) bear any harmonic relationship to either of the input signals.

AM is achieved by using a circuit called a *voltage-controlled amplifier* (or *VCA*). The program signal is used as an automatic volume control, determining the instantaneous level of the carrier signal. The basic setup is shown in Fig. 36-2.

The carrier signal is usually generated by a sine-wave oscillator, and the program signal can be virtually any electrical signal, such as the output from a microphone.

Balanced modulation

A variation of the amplitude modulation concept is *balanced modulation*. In a balanced modulator, the original input signals are suppressed—only the sidebands

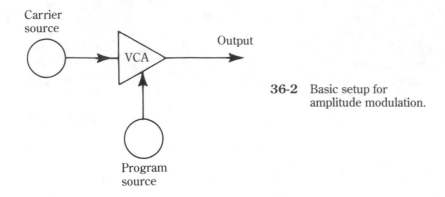

36-2 Basic setup for amplitude modulation.

are present at the output. For example, if the input signals are 700 Hz and 1300 Hz sine waves, the output would consist of just 600 Hz (lower sideband) and a 2000 Hz (upper sideband) sine wave. Balanced modulation is also sometimes called *ring modulation*.

Heterodyning

Closely related to amplitude modulation is the process known as *heterodyning*, or *nonlinear mixing*. If two signals whose frequencies are very close to each other are superimposed onto each other, they will mix into an apparently single tone with amplitude fluctuations equal to the difference between the two frequencies. For example, a 998 Hz signal and a 1004 Hz signal would combine with amplitude fluctuations at a 6 Hz rate.

Heterodyning is often used to fine tune two signals to the same frequency, by adjusting for the *zero beat* (no amplification fluctuations). Heterodyning is also widely used in radio receivers. This subject is discussed in the next chapter.

Frequency modulation

If one ac signal is used to control the frequency of a second oscillator (that is, a VCO), *frequency modulation* takes place. *Frequency modulation* is often shortened to *FM*. The basic setup for frequency modulation is shown in Fig. 36-3, and typical signals are shown in Fig. 36-4. As with amplitude modulation, sidebands are produced by the process of frequency modulation. But there are many more sidebands produced with frequency modulation. FM sidebands appear above and below the carrier frequency at multiples of the program signal.

For example, if the carrier signal is a 1000 Hz sine wave and the program signal is a 60 Hz sine wave, the first set of sidebands would appear at 940 Hz and 1060 Hz. The second set of sidebands would be at 880 Hz and 1120 Hz. The third set of sidebands would be at 820 Hz and 1180 Hz, and so forth.

Of course, if more complex input signals are used, the pattern of the sidebands will become much more dense and complicated. FM sidebands have different amplitudes, which are at their lowest levels at frequencies farthest from the carrier frequency.

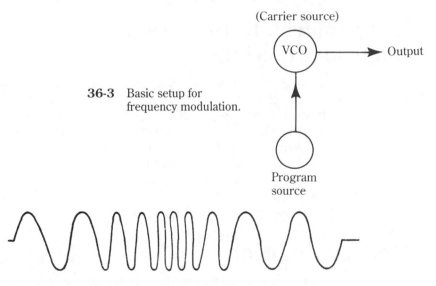

(Carrier source)

VCO → Output

36-3 Basic setup for frequency modulation.

Program source

36-4 Frequency modulation.

The number of sidebands produced in a given FM signal is determined by a factor called the *modulation index*. This quantity can be mathematically found with the following formula:

$$MI = \frac{\Delta F}{F_p}$$ **Equation 36-3**

MI is the modulation index, and ΔF is the peak frequency deviation. This is the difference between the modulated carrier frequency and the unmodulated carrier frequency. The value is constantly changing. For example, a 1000 Hz carrier signal that is modulated so that it fluctuates between 800 Hz and 1200 Hz has a peak frequency deviation of 200 Hz. F_p is the frequency of the program signal.

The frequency deviation at any instant is determined by the instantaneous amplitude of the program signal. A high program amplitude produces a large frequency deviation. For a given amount of input power, as the modulation index increases, the output amplitude of the carrier frequency decreases. If, for example, a 1000 Hz carrier signal has a peak frequency deviation of 200, and the program frequency is 40 Hz, the modulation index would be equal to 200/40, or 5. This means there would be five sidebands above the carrier frequency, and five below. In other words, a modulation index of 5 indicates a total of ten sidebands. In this example, the output would consist of frequencies at 800 Hz, 840 Hz, 920 Hz, 960 Hz, 1000 Hz (the carrier), 1040 Hz, 1080 Hz, 1120 Hz, 1160 Hz, and 1200 Hz.

FM equipment for radio applications tends to be somewhat more expensive and complex than AM equipment, but FM is considerably less susceptible to electrical noise. FM is more resistant to electrical noise because most electrical noise sources (such as lightning or motor brushcs) act like an AM signal, and contain very little frequency modulation.

Single-sideband, suppressed-carrier modulation

In an AM signal the upper and lower sidebands are exact mirror images of each other. Both sidebands together can carry no more information than either one separately. Similarly, the carrier frequency, because it is a constant, contains no information. Transmitting an extra set of sidebands and the carrier uses up power that gives no greater distance or program content. This power is, in effect, wasted.

It is possible to eliminate one of the sidebands and filter out the carrier and use all the power to transmit just one set of sidebands. This more efficient system is called *single-sideband*, *suppressed-carrier*, or *SSSC*. The carrier signal is regenerated in the receiver for *demodulation* (see the next chapter).

Phase modulation

Phase modulation is similar to frequency modulation except, with phase modulation, the phase of the carrier signal (with respect to some constant standard) is varied in step with the instantaneous amplitude of the program signal. See Fig. 36-5.

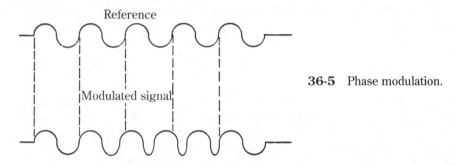

36-5   Phase modulation.

Phase modulation is generally used in conjunction with another form of modulation when too much information for a single carrier is required. An example of the use of phase modulation is the color information signal in color television. The sound information is transmitted via frequency modulation, the basic picture by amplitude modulation, and the color by phase modulation.

Pulse-code modulation

Pulse-code modulation, or *PCM* converts the program signal into a string of pulses representing a binary number. Pulse-code modulation has the highest resistance to noise of any form of transmission modulation because unless the noise coincidentally resembles one of the modulation pulses, it can easily be ignored by the receiving equipment.

Analog program signals are adapted for digital use by a circuit called an *A/D converter* (analog-to-digital converter). See Fig. 36-6.

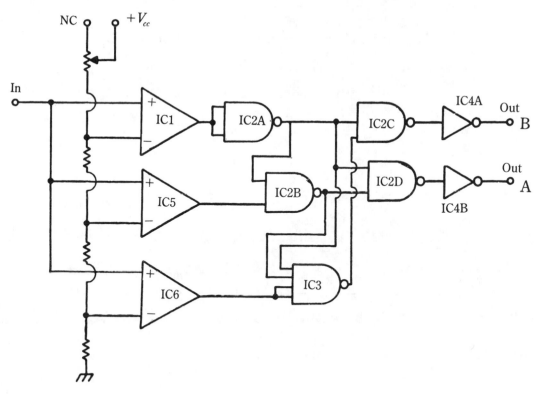

36-6 A simple A/D converter circuit.

The converter in Fig. 36-6 is a two-bit A/D converter. That is, it produces a two-bit digital number that corresponds to the level of the analog input voltage. IC1, IC5, and IC9 are op amp voltage comparators. If the input to one of these devices is equal or greater than a specific reference voltage, the output can be used as a digital logic 1. Otherwise, the output is a logic 0. IC2, IC3, IC6, and IC7 are 2 input NAND gates. IC10 is a 4 input NAND gate. IC4 and IC8 are simply buffers.

Table 36-1 lists some possible input/output combinations assuming IC1 has a reference voltage of 3 V, IC5 is referenced to 2 V, and IC9 has a reference voltage of 1 V.

Note that because there are only two digital bits in the output, there are only four possible output conditions. Practical A/D converters would, of course, require a much larger number of output bits to be really useful.

The digital numbers can be changed back into analog signals with a *D/A converter* (Digital to Analog converter). See Fig. 36-7.

A four-bit digital number is inputted to this circuit. The output will be one of sixteen voltage levels—each corresponding to a specific digital number. A is the most significant (largest value) digit, and D is the least significant (lowest value) digit.

This type of circuit is sometimes called an "R—2R ladder network," because the effective resistance increases in a step-like manner from the most significant to

**Table 36-1. Typical input
and output signals for the circuit shown in Fig. 36-6.**

Input voltage	Comparator outputs			Digital outputs	
	IC1	IC5	IC9	A	B
0.0	0	0	0	0	0
0.5	0	0	0	0	0
1.0	0	0	1	0	1
1.5	0	0	1	0	1
2.0	0	1	1	1	0
2.5	0	1	1	1	0
3.0	1	1	1	1	1
3.5	1	1	1	1	1

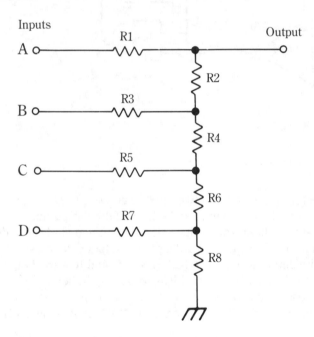

36-7   A simple D/A converter circuit.

the least significant digit. Resistors R2, R4, and R6 are all of equal value (R). Resistors R1, R3, R5, R7, and R8 are all twice the value of R2 ($2R$). Table 36-2 shows the outputs that could be expected from such a circuit for each of the possible digital inputs. The 1 state voltage for this example is 4 V. R is 1000 Ω, so 2R equals 2000 Ω. The output from this type of circuit is usually fed through an operational amplifier for buffering and amplification.

Pulse-code modulation is also useful for high-quality recording. All analog recording systems produce some degree of degradation of the signal because of mechanical and electrical imperfections. If a second recording is made from the first,

**Table 36-2. Typical input and output
signals for the circuit shown in Fig. 36-7.**

	Digital in			Analog out,
A	B	C	D	volts
0	0	0	0	0.00
0	0	0	1	0.25
0	0	1	0	0.50
0	0	1	1	0.75
0	1	0	0	1.00
0	1	0	1	1.25
0	1	1	0	1.50
0	1	1	1	1.75
1	0	0	0	2.00
1	0	0	1	2.25
1	0	1	0	2.50
1	0	1	1	2.75
1	1	0	0	3.00
1	1	0	1	3.25
1	1	1	0	3.50
1	1	1	1	3.75

the distortion and noise will increase. A third copy will have an even more degraded signal. Eventually a point will be reached when the signal is no longer usable.

Pulse-code modulation recordings, (or digital recordings), do not have this problem. The signal on the hundredth copy will be identical to that on the first. This is because only digital numbers are recorded. The noise and distortions are linear in nature, so they are simply ignored by the system.

Self-test

1. Which of the following is not a common type of signal modulation for transmission or storage of information?

A Amplitude modulation
B Frequency modulation
C Phase modulation
D Ring modulation
E None of the above

2. What is the frequency of the upper sideband produced when a 50,000 Hz carrier signal is amplitude modulated by a 4500 Hz program signal?

A 54,500 Hz
B 45,500 Hz
C 100,623 Hz
D 95,000 Hz
E None of the above

3. How many sidebands are produced in an AM signal, assuming sine wave inputs?

A One
B Two
C Four
D It varies with the strength of the original signals
E None of the above.

4. A 2500 Hz carrier is frequency modulated by a 300 Hz program signal with an amplitude that causes the carrier signal to vary between 1,900 Hz and 3100 Hz. How many sidebands will be produced?

A Eight
B Two
C Four
D Six
E None of the above

5. Suppose the amplitude of the program signal in the problem described in question 4 is raised so that the carrier signal is varied between 1000 Hz and 4000 Hz. How many sidebands are now produced?

A 2
B 5
C 12
D 10
E None of the above

6. What type of modulation is used to transmit analog data in digital equipment?

A Amplitude modulation
B Frequency modulation
C Pulse-code modulation
D Phase modulation
E None of the above

7 How can a digital signal be converted into an analog signal?

A With pulse-code modulation
B With an A/D converter
C With phase modulation
D With a D/A converter
E None of the above

8. If a 15,000 Hz carrier signal is amplitude modulated by a 2,000 Hz program signal, how many sidebands are produced? (Assuming sine waves.)

A 7
B 15
C 2
D 10
E None of the above

9. Two signals whose frequencies are close together are superimposed and mixed into a single tone with amplitude fluctuations. What is this effect called?

A Heterodyning
B D/A conversion
C Balanced modulation
D Linear mixing
E None of the above

10. What determines the number of sidebands in an FM signal?

A The difference between the carrier and program frequency
B The modulation index
C There are always two sidebands
D The square of the program frequency
E None of the above

37
Radio

Radio is a means of communicating information from one place (*transmitter*) to another (receiver) via *electromagnetic waves*, or *electrostatic waves*.

Radio waves

When ac current passes through a conductor, the conductor is surrounded by both an electromagnetic field (discussed in other chapters) and an *electrostatic field* (which is sometimes called just an *electric field*). These two types of fields are at right angles to each other, as shown in Fig. 37-1.

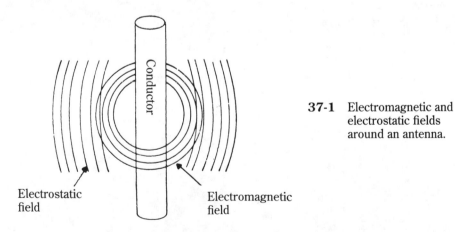

37-1 Electromagnetic and electrostatic fields around an antenna.

The magnetic and electrostatic lines of force are radiated outward from the conductor at the speed of light. Of course, the farther the lines of force get from the conductor, the weaker they are. Because the strength of these fields is determined by the instantaneous strength of the current through the conductor, the lines of force will vary in step with the applied ac signal. If the current through the conductor is high enough, and the ac frequency is within a specific range of frequencies, the lines of force can be transmitted over very long distances. These frequencies are called *radio frequencies*, or RF. The range of radio frequencies is called the *radio spectrum*.

The radio spectrum extends from approximately 20 kHz (20,000 Hz) to 30,000 MHz (30,000,000 Hz). It is divided for convenience into smaller bands, as listed in Table 37-1. These various bands have somewhat different characteristics that are discussed in the section on *antennas*.

Table 37-1. The radio spectrum.

Frequency	Name of band
15 to 30 kHz	Very low frequency—VLF
30 to 300 kHz	Low frequency—LF
300 kHz to 3 MHz	Medium frequency—MF
3 to 30 MHz	High frequency—HF
30 to 300 MHz	Very high frequency—VHF
300 to 3000 MHz	Ultra high frequency—UHF
3000 to 30,000 MHz	Super high frequency—SHF

Portions of the radio spectrum are legally designated for specific types of services. Some of the most important of these are listed in Table 37-2. There are occasional changes in the allocation of RF frequencies from time to time to accommodate changing needs in society. For example, the UHF television band once contained channels 14 through 83, but all channels over 69 have recently been deleted from television services and have been re-allocated to various special services. The high-frequency UHF channels were not used very much, because of their rather limited

Table 37-2.
Major services within the radio spectrum.

Frequencies	Services
535 to 1605 kHz	AM broadcast band
0.2 to 30 MHz	Short-wave band—international broadcasting and HAM (amateur) transmission
27 MHz	Citizen's band (CB)
54 to 88 MHz	Television channels 2 through 6
88 to 108 MHz	FM broadcast band
140 to 170 MHz	Emergency services (police, fire, ambulance, etc.)
174 to 216 MHz	Television channels 7 through 13
470 to 890 MHz	UHF television channels (14 through 83)

broadcasting range, and other portions of the RF spectrum were overcrowded, so the UHF frequencies from 806 MHz to 890 MHz were reassigned.

These frequency allocations refer to the carrier frequency sent out by the transmitter. The program information modulates this carrier frequency in some way, as described in this chapter. The frequencies contained in the modulated program having nothing to do with the RF carrier frequency. In most cases, the program will consist of audio frequencies (about 50 Hz to 15,000 Hz), and the carrier frequency will be well above the limits of audibility. The program signal is not necessarily audio, however. It might contain encoded video information (television pictures), computer data, or almost any other form of intelligence someone wishes to transmit over the airwaves.

The earliest and simplest type of radio communication consisted simply of turning a transmitted signal (an unmodulated waveform) on and off in an organized pattern and detecting the periodic presence or absence of the carrier frequency with a receiver in another location.

These on/off pulses were arranged in specific patterns consisting of short and long pulses in a system called the *International Morse Code*. The letters of the alphabet, numerals and a few punctuation symbols are each assigned a unique pattern of dots (short pulses) and dashes (long pulses). Each word in a message is spelled out, encoded and transmitted to a receiver where it is decoded and the original message can be read out. The Morse code is given in Table 37-3.

A short pause is inserted between consecutive letters, numerals, or symbols to avoid confusion, because the length of the code varies from letter to letter, ranging from one to eight marks (dots or dashes). The most commonly used letters are given the shortest and easiest codes. For example, an E is represented by just a single dot, and a T is encoded as a single dash.

The international distress code, SOS, derives directly from Morse Code. Contrary to a number of popular legends, SOS does not stand for "Save our ship," or anything else. It has no meaning in itself. It was selected because of its distinctive, easy-to-memorize and easy-to-recognize Morse Code pattern (. . . - - - . . .) (three dots, three dashes, three dots). It is very easy even for an untrained operator to send the SOS code in an emergency.

In a Morse Code message, a slightly longer pause is used between words. Sentences are separated by punctuation codes, or a special STOP code.

Some people find the Morse Code relatively easy to learn using flash cards and a little practice. Others find it almost impossible to memorize the correct sequence of dots and dashes.

To a large extent, the Morse code is technically obsolete. It is still required to obtain certain short-wave operator licenses from the FCC (Federal Communications Commission). A few special frequencies in the short-wave band are set aside for Morse Code transmissions only, but these frequencies are little used in comparison to the short-wave voice transmission frequencies. This is not surprising, because any such code is an awkward and inconvenient way for human beings to communicate. You are used to the spoken word.

Still, knowing the Morse Code can be very helpful in certain specialized conditions. In an emergency situation where voice transmission is not possible for any

Table 37-3. Morse code.

Letters and numbers

A .-	*B* -...	*C* -.-.	*D* -..
E .	*F* ..-.	*G* --.	*H* ....
I ..	*J* .---	*K* -.-	*L* .-..
M --	*N* -.	*O* ---	*P* .--.
Q --.-	*R* .-.	*S* ...	*T* -
U ..-	*V* ...-	*W* .--	*X* -..-
Y -.--	*Z* --..		
1 .----	*2* ..---	*3* ...--	*4* ....-
5	*6* -....	*7* --...	*8* ---..
9 ----.	*0* -----		

Punctuation and special marks

. (period)	.-.-.-	
, (comma)	--..--	
? (question mark)	..--..	(also known as IMI)
: (colon)	---...	
; (semicolon)	-.-.-.	
- (hyphen or dash)	-....-	
" (quotes)	.-..-.	or .-.-.
((open parenthesis)	-.--.	
) (close parenthesis)	-.--.-	
' (apostrophe)	.----.	
$ (dollar sign)	...-..-	
Error sign	....	(8 dots)
Separation indicator	-...-	(also known as BT)
Starting signal	-.-.-	
Invitation to transmit	-.-	
Wait	.-...	(also know as AS)
End of message	.-.-.	(also known as AR)
End of work	...-.-	(also known as SK or VA)

reason, the Morse Code could be a literal lifesaver. It is also used in emergency communications other than radio. For example, someone trapped in a cave-in might be able to tap out a message calling for help when their voice couldn't be heard.

Submarine personnel are more likely to use Morse Code than almost anyone other people. Radio transmissions to or from a submerged submarine have special technical problems. High-frequency radio waves are effectively shielded by the ocean water. Very low frequencies (50 kHz, or lower) must be used to achieve any transmission distance at all. It is more difficult to adequately modulate such low carrier frequencies. Sufficient bandwidth can be a real problem. With a simple on/off code, there is no true modulation needed at all. An extremely narrow bandwidth will do the job. The signal can get through more reliably because of its inherent simplicity.

As a general rule of thumb, however, the Morse code is more a historical relic than a practical tool in virtually all modern electronics and radio work. Code requirements for amateur radio licenses have been reduced in recent years. There has been a long-raging debate over whether all code requirements should be eliminated altogether, because they are of only minimal value now. There is really no more reason for an amateur radio operator to know Morse code than anyone else. A freak occasion when the code is needed might come up, but it is unlikely. The arguments for retaining Morse code requirements for licensing seem to be basically elitist. Only someone who is really serious about amateur radio is likely to spend the time needed to learn Morse Code well enough to pass the tests, so "undesirables" are weeded out. Such an attitude is probably not valid for government licensing in a democratic society.

The main point of the Morse code when it was originally devised was to simplify the technical requirements of message transmission as much as possible. The technology just wasn't ready for more sophisticated message transmission. Today, spoken audio messages, or even television signals are less real fuss and bother to transmit than the inherently cumbersome Morse code. Although the Morse code is technically simple, and it is certainly better than no long distance communication at all, it is, by definition, a clumsy and unnatural method of sending messages.

Although the Morse Code system is technically simple, and is certainly better than no long-distance communication at all, it is inherently a rather cumbersome method of sending messages. More information can be efficiently transmitted if the carrier is modulated by the signal to be transmitted. This can be speech, music or digital data. Amplitude modulation and frequency modulation are the most common types of transmission used today.

Although it is definitely taking a back seat to the various modulation systems, Morse Code is still used in some applications because it is simple and inexpensive, and it produces no sidebands. The transmitted bandwidth is simply the carrier frequency. Either it is transmitted or not transmitted. Modulation always produces some sidebands. Usually these have to be limited so a reasonable number of usable frequencies can be fitted into the radio spectrum.

Transmitters

A *transmitter* is used to generate the RF carrier signal, amplify it, modulate it with the program signal, and prepare it for the *antenna* (antennas are discussed in the next section). The RF carrier signal is generated by an oscillator. It is always a sine wave, because harmonics would behave like additional subcarriers at different frequencies, wasting energy, and spreading over a large bandwidth. The frequency of the oscillator must also be very tightly controlled. If the carrier frequency is allowed to drift even slightly off value, reception will be difficult, or impossible. Also, the frequency shifted signal might interfere with other signals on nearby frequencies. Radio transmitters are legally required to be on frequency within a small fraction of 1%.

For the best possible accuracy, these RF oscillators are generally of the crystal type, and they're usually enclosed in special ovens for precise temperature control. *Phase-locked loops* (or PLLs) also are frequently used. A phase-locked loop is a circuit that continuously tests the output of a voltage-controlled oscillator. If the oscillator is on frequency, nothing happens. If, however, the frequency has drifted (either increased or decreased) the phase-locked loop generates an *error voltage* that pulls the VCO back on frequency. See Fig. 37-2.

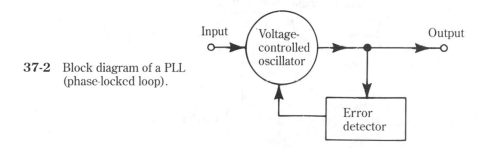

37-2 Block diagram of a PLL (phase-locked loop).

Harmonics and stray oscillations at undesired frequencies also must be stringently avoided in transmitter circuits. Lowpass and bandpass filters are often used to block out these undesired frequencies from the final output signal. The modulation index and bandwidth of broadcast signals are also tightly controlled and regulated by the FCC. For example, all significant sidebands and radiated energy for a station within the standard AM broadcast band must be within ±5 kHz (10,000 Hz) of the nominal carrier frequency. This, of course, limits the frequency of the program signal to 15 kHz. FM and television channels are allowed wider bandwidths, but there are restrictions.

RF amplifiers are generally similar to audio-frequency (AF) amplifiers, except greater care must be taken to prevent stray capacitances. Because capacitive reactance decreases as frequency increases, an unintentional capacitance between say, two nearby circuit leads could act as a short circuit, or a feedback path for an RF signal. This could cause parasitic oscillations, or other problems.

Figure 37-3 shows a basic block diagram for a typical AM transmitting station. The turntable or the microphone are the source for the program signal. They are combined in the desired proportions in the mixer. The equalization circuit compensates for any frequency inequalities in the system. No circuit has a completely flat frequency response. Certain frequencies might be overemphasized, and others might be cut back. Equalization is a way of forcing the frequency response to be flat, by amplifying the weak frequencies, and partially filtering (that is, attenuating) the overly strong frequencies. The program signal then passes through one or two stages of audio amplification.

Meanwhile, the crystal oscillator is generating the carrier frequency, and the phase-locked loop is helping it stay on the correct frequency. The unmodulated carrier is amplified by one or more stages of RF amplifiers. The program is superimposed on the carrier signal in the modulator and the modulated signal is amplified again. Finally the modulated signal is fed to the transmitting antenna where it is radiated into space.

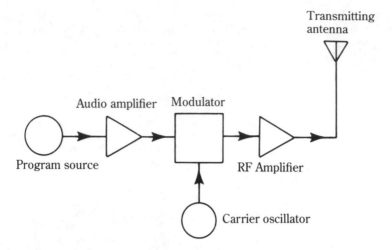

37-3   Block diagram of a typical AM broadcasting station.

Antennas

When the electromagnetic and electrostatic waves radiate out from the antenna, they can follow one or more different paths to the receiving station. Each type of path has different characteristics. The paths taken depend on a number of factors—especially the carrier frequency.

The three basic types of signal paths are called *ground waves*, *direct waves*, and *sky waves* (see Fig. 37-4.)

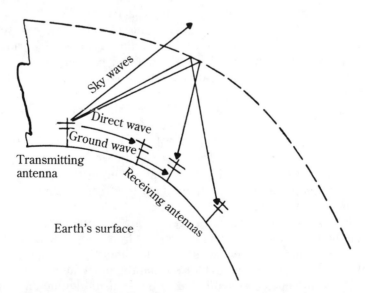

37-4   RF propagation—ground waves, direct wave, and sky waves.

Ground waves basically follow the curvature of the earth. Direct waves move in a direct, line-of-sight path from the transmitting antenna to the receiving antenna. Sky waves radiate upwards to the upper atmosphere. Some of the sky waves are absorbed by the atmosphere, some pass right through into space, and some bounce off the atmospheric layers and are reflected back to earth. The atmosphere has a number of different layers. The portion you are concerned with here is the *ionosphere*. Because an *ion* is an electrically charged particle, the ionosphere is an electrically charged portion of the atmosphere.

The ionosphere is also divided into layers. There is a D layer, an E layer, and two F layers. The D and E layers exist only during daylight hours. At night they disappear.

Each frequency band in the radio spectrum is *propagated* by different combinations of these basic wave types. *Propagation* is the way a given wave travels from the transmitting antenna to the receiving antenna.

VLF (very low frequencies) and LF (low frequencies) (20,000 Hz to 300,000 Hz) are propagated mostly by the ground wave. The direct wave is generally too weak to be very useful at these frequencies, and the sky waves are virtually all absorbed by the atmosphere. The signals in this range are quite reliable, and their reception is practically independent of time of day and weather conditions. However, the range is somewhat limited.

These low frequencies tend to travel farther over sea water, because it has an extremely high amount of conductivity, so they are often used for marine applications such as ship-to-shore communications, distress signals, and weather stations.

Medium frequencies (MF) travel along a combination of ground waves and sky waves. The sky waves are reflected back to earth by the lower portions of the ionosphere, so the reception point isn't very far from the transmitting antenna. At night, when the D and E layers disappear, the sky waves are reflected from a higher angle, so the signal is propagated farther.

A transmitting station in this frequency range has three basic service areas. The primary service area is served almost entirely by the ground wave. Any sky waves in this area are extremely weak if they exist at all. The primary service area is the closest to the transmitting antenna. The secondary service area is at some distance from the transmitting antenna. This area receives relatively little signal during the day, and the station might not be receivable at all. At night, however, this area receives a strong sky wave signal, and reception is usually quite reliable.

In between the primary and secondary service areas is a region called the *fading area*. Here the signal comes from both the ground wave and the sky waves. If at a given point these two signals are in phase, they will reinforce each other. If they are out of phase, they will partially or completely cancel each other. Because the phase of the sky wave changes from moment to moment, as the reflection angle shifts slightly, the signal will tend to fade in and out, making reception unreliable. The standard AM broadcast band is included in this frequency range.

HF (high frequencies) (3 MHz to 30 MHz) are propagated mostly by sky waves. This frequency range is often called the short-wave band. Citizens' Band radios (CB) operate at approximately 27 MHz. These frequencies are not bent at as sharp an angle by the ionosphere as MF signals, so the sky waves can reach over very long distances. This is why short-wave transmissions can be used for global communications. VHF (very high frequencies) and higher behave almost like a beam of light. They travel along a line-of-sight path. That is, the signal is propagated by the direct wave.

A few sky waves are reflected back by the upper atmosphere, but not many. Most of the sky waves pass right through the atmosphere and into space. The higher the frequency, the more easily the signal can pass through the atmosphere. Obviously, signals transmitted at these extremely high frequencies can rarely be received at large distance from the transmitting antenna.

The transmitting antenna, of course, is the conductor that radiates the signal. The receiver must also have an antenna. As the signal passes the receiving antenna, a small voltage is induced into the antenna. This voltage is extremely small. Even a 50,000 W transmitting station will rarely induce more than a microwatt (one millionth of a watt) into the receiving antenna. Often the induced signal will be only a thousandth of that.

Any antenna will be most sensitive to a specific frequency (or range of frequencies). Electrically, it behaves like a series resonant RLC (resistive-inductive-capacitive) circuit. The length of the antenna elements is the primary factor in determining the R, L, and C values, and thus, the resonant frequency.

Remember that in a series resonant circuit, the maximum amount of current flows at the resonant frequency. Thus a signal at or near the antenna resonance point will induce the strongest signal into the antenna.

Figure 37-5 shows one basic type of antenna. This antenna is called a *Marconi antenna*. It simply consists of a conductor of the appropriate length, and a

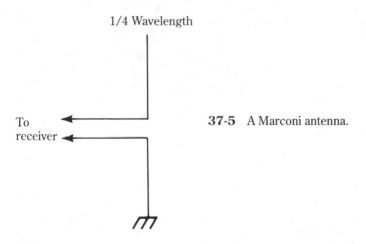

1/4 Wavelength

To
receiver

37-5   A Marconi antenna.

ground connection. Figure 37-6 shows the electrical equivalent circuit for a Marconi antenna.

A Marconi antenna will be resonant at a wavelength just under four times the length of the antenna itself. For this reason, it is often called a *quarter-wave antenna*.

Another common type of receiving antenna is the *hertz antenna*, shown in Fig.

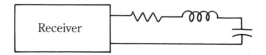

37-6 Equivalent circuit for Marconi antenna.

37-7. Because this type of antenna has two sections, it is also known as the *dipole antenna*. Its electrical equivalent circuit is shown in Fig. 37-8.

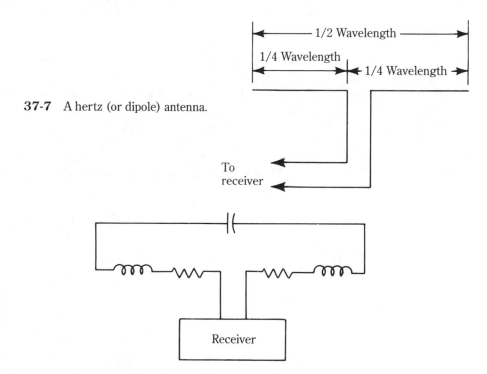

37-7 A hertz (or dipole) antenna.

37-8 Electrical equivalent circuit for a hertz antenna.

Notice that it consists essentially of two Marconi antennas. No ground connection is required. Each section is just under a quarter wavelength long, so the entire antenna is often called a *half-wave antenna*. Typically, the actual length is about 5% less than one-half wavelength.

Of course, most practical antennas are required to receive a number of different frequencies. For example, an FM receiver needs an antenna that is functional along the entire 88 to 108 MHz (megahertz) FM broadcast band. FM band is a range of 20 MHz. No antenna can be truly resonant over such a wide range. Generally you have to compromise by tuning the antenna to the center of the band, so either end of the band is reasonably close to the resonance of the antenna.

A practical formula for finding the correct average length of a half-wave antenna is as follows:

$$X = \frac{468}{F}$$ **Equation 37-1**

where X is the length of the antenna in feet and F is the frequency in megahertz (1,000,000 Hz). For example, assume you need an antenna for receiving the FM broadcast band. As mentioned above, it should be resonant to the center of the band, or 98 MHz. The antenna length should therefore be equal to 468/98, or about 4.8 feet. As a second example, find the length of an antenna for the standard AM broadcast band. The center of the band is 1070 kHz (1,070,000 Hz), or 1.07 MHz. A half-wave antenna for this frequency would have length of 468/1.07, or over 437 feet. Obviously this would be extremely impractical.

For low frequencies, a coil is often added in series with the antenna, increasing the effective inductance of the circuit, and decreasing the resonant frequency. Although this method is practical, a short antenna of this type will not pick up as great a signal as a true half-wave antenna would. Often portable AM broadcast band radios simply use a coil within the case of an antenna.

In a sense, an antenna acts like a bandpass filter, emphasizing only a specific set of frequencies from all of the signals that pass it. Like a bandpass filter, an antenna has a definite Q. A high Q antenna (such as a dipole made of thin wires) has a very narrow bandwidth. Increasing the diameter of the antenna elements decreases the Q, and increases the bandwidth. The Q can also be altered by changing the physical shape of the antenna.

Antennas generally do not receive signals equally well from all directions. A half-wave dipole receives signals along its length better than signals that strike its ends. Its reception area is a figure-8 pattern, as shown in Fig. 37-9.

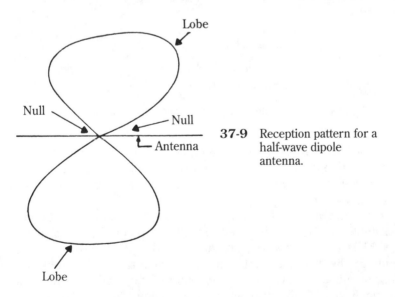

37-9   Reception pattern for a half-wave dipole antenna.

This pattern is true only for signals that have a wavelength that is twice the length of the antenna. At other frequencies, the reception pattern would be changed.

For example, most television antennas are one half wavelength in the middle of the band consisting of channels 2 through 6. For channels 7 through 13 the antenna is about one and a half (3/2) wavelength long. The reception pattern would resemble Fig. 37-10.

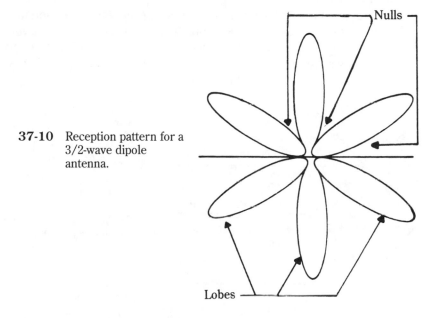

37-10 Reception pattern for a 3/2-wave dipole antenna.

You have effectively added two additional half-wave lengths to the original dipole and got six *lobes* (strong reception areas) and six *nulls* (weak reception areas). In general increasing the number of half-wave lengths increases the number of lobes and nulls. Bending the antenna would make it more sensitive to signals from one side, and reducing the sensitivity on the opposite side, as shown in Fig. 37-11.

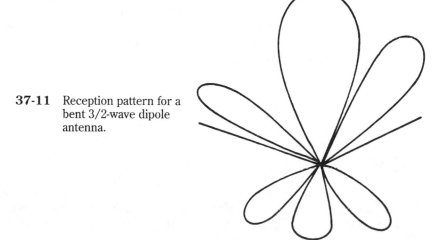

37-11 Reception pattern for a bent 3/2-wave dipole antenna.

Often the *gain* of an antenna will be spoken of in technical literature. Actually an antenna has no true gain—it does not amplify the signal. Gain, in this context, is used to compare the efficiency of a specific antenna with a basic dipole antenna. An antenna with a gain of two would receive twice as much signal as a standard, simple dipole antenna.

There are basically two ways of increasing the gain of an antenna. You could connect two or more dipole sections together or you could use *reflectors*. These are simply extra rods or wires that are placed around the antenna to effectively focus the passing wave onto the antenna itself. This arrangement is shown in Fig. 37-12. Television antennas often use this sort of arrangement.

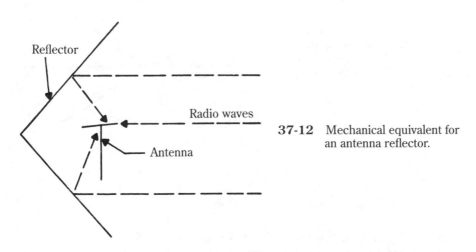

37-12   Mechanical equivalent for an antenna reflector.

The reflectors are not electrically connected to the receiver. They are called *parasitic elements*. The reflector is usually about 5% longer than the main antenna element, and is placed about 0.15 wavelength behind the antenna itself with reference to the desired signal source (see Fig. 37-13A). The signal that passes the antenna strikes the reflector and bounces back to the antenna.

Another type of parasitic element is called a *director*. This element is about 4% shorter than the main antenna element and is placed about a tenth of a wavelength in front of it. See Fig. 37-13B. The director focuses the passing signal onto the antenna, somewhat like an optical lens. Both types of parasitic elements can be used within a single system. This type of system is called a *Yagi antenna*.

Note that these parasitic elements increase directivity as well as gain. They make the antenna sensitive to essentially only those signals arriving from one specific side.

SWR

An antenna has its own specific impedance, that should match the impedance of the circuit it is connected to for maximum power transfer. The cable connecting the antenna and the circuit (either receiver, or transmitter—in this case, you are more

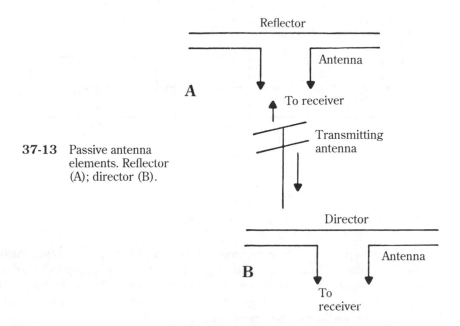

37-13 Passive antenna
elements. Reflector
(A); director (B).

concerned with the transmitter although the following discussion applies to receivers too, but to a lesser extent) is called the *transmission line*. It also has a specific inherent impedance. This value is constant regardless of the length of the cable. This impedance should closely match the impedances of the antenna and the transmitter. For a receiver the impedances only have to be approximately equal.

When these impedances are mismatched, *standing waves* will appear along the line. These standing waves consist of power that is reflected back from the antenna to the transmitter. Obviously, this power is serving no useful purpose, and is therefore wasted. This effect is measured by a factor called the *standing-wave ratio*, or *SWR*. An ideal SWR would be 1:1. This would mean all of the power from the transmitter is reaching the antenna. Unfortunately, this ideal is impossible to achieve in practical circuits. Generally a 1:1.1 to a 1:1.5 SWR is the best that can be reached. Usually anything better than 1:2.5 is good enough for most applications.

In receiving only application, SWR is generally ignored beyond just roughly matching up the nominal impedances. You would not get very good results from a 75 Ω antenna connected to a 300 Ω receiver, but if the actual impedance of the antenna were 250 Ωs, there would probably be no noticeable problems in a 300 Ω system.

Receivers

A *receiver* accepts the signal from an antenna. The receiver amplifies and demodulates the signal to retrieve the original message or other program signal. Most modern receivers are of the *superheterodyne* type. This name is often shortened to *superhet*. Figure 37-14 shows the block diagram for a superheterodyne receiver.

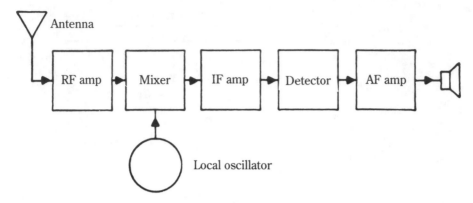

37-14 Block diagram for a superheterodyne receiver.

Not all superheterodyne receivers include an RF amplifier stage. Inexpensive AM radios almost always leave this stage out, but it is generally included in better receivers. A superheterodyne receiver is very sensitive. That is, it can operate on a fairly weak input signal. The RF amplifier boosts the incoming signal from the antenna, so even weaker signals can be received.

The RF amplifier stage also helps eliminate a problem called *image interference*. A latter stage in the receiver converts the RF signal to a lower, intermediate frequency (IF). Another station operating at a frequency that is twice this intermediate frequency could interfere with the desired signal.

The RF amplifier is tuned to the desired frequency and is set up to reject this potentially interfering frequency. Strong signals can be applied directly to the mixer without an RF amplifier stage, but a weak signal could be drowned out by internal noise in the mixer circuit. An RF amplifier stage increases sensitivity.

Figure 37-15 shows the schematic diagram for a typical RF amplifier stage in an AM receiver. Capacitor C1 is tunable, so the C1-L1 parallel tank circuit can be made resonant at the desired frequency. L1 is wound around a ferrite rod, and acts as the antenna. No external antenna is required for this circuit. The ferrite core in the coil gives it a very high Q, so the resonant circuit is quite selective, rejecting nearby frequencies almost totally.

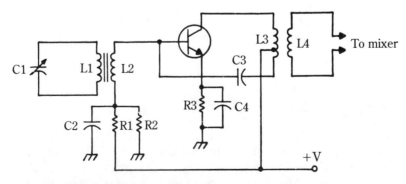

37-15 A typical AM RF amplifier stage.

Coil L2 is wound around the same core as L1, so the signal flowing through L1 is induced into L2. This signal is, in turn, applied directly to the base of the transistor. The signal is also sent through C2 to ground, and from ground through C4 to the emitter of the transistor. R3 helps prevent thermal runaway. Resistors R1 and R2 form a voltage divider to forward bias the transistor.

The amplified RF signal from the collector of the transistor passes through L3 where it induces the signal into L4, which is the input of the next stage in the receiver. This might be either another stage of RF amplification (uncommon) or the mixer.

C3 is vitally important in an RF amplifier. Internal capacitances within the transistor itself could provide a feedback path for the RF signal from the collector back to the base. This could cause the stage to break into oscillation. C3 feeds back a small out of phase signal to compensate for this. This component is called a neutralizing capacitor.

Next the RF signal is fed to a mixer stage where it is combined with the signal from a local (contained in the receiver) oscillator. The local oscillator actually modulates the received RF signal again, producing two sidebands—the received frequency plus the oscillator frequency, and the received frequency minus the oscillator frequency. Everything except this lower sideband is filtered out.

If the received signal is 1000 kHz, and the local oscillator frequency is 545 kHz, the lower sideband would be 455 kHz. This is the intermediate frequency (IF). It is constant for all stations received by the radio. The oscillator is tuned so that it is always equal to:

$$F_o = RF - IF \qquad\qquad \textbf{Equation 37-2}$$

where F_o is the frequency of the local oscillator, *RF* is the radio frequency of the desired signal, and *IF* is the intermediate frequency. The *IF* used in most AM radios is 455 kHz. FM radios usually have an IF of 10.7 MHz. There are two major advantages of converting the RF signal to a lower IF signal. One is that it is easier to amplify lower frequencies without the circuit breaking into oscillation. The components in an IF amplifier can usually be less expensive types than are typically found in comparable RF amplifiers.

The other advantage is that each stage in the radio does not have to be returned each time you want to receive a different frequency. All of the stages past the mixer can be tuned to a single frequency and left there. The process of turning the IF stage is called *alignment*. Only the RF amplifier and the local oscillator have to be retuned for each received frequency. The retuning is usually done via a single, ganged two-section capacitor.

Figure 37-16 shows a typical mixer/oscillator circuit. Component Q1 is the mixer and Q2 is the oscillator. Figure 37-17 is a similar circuit that uses a single transistor for both the mixer and the oscillator functions. L3, C3, and C4 are the oscillator frequency determining tank circuit. L2 provides feedback compensation. R3 is fairly small and R4 is rather large. The large resistance of R4 prevents energy fed through C5 from feeding back into the power supply.

Coils L4 and L5 couple the signal into the first IF amplifier.

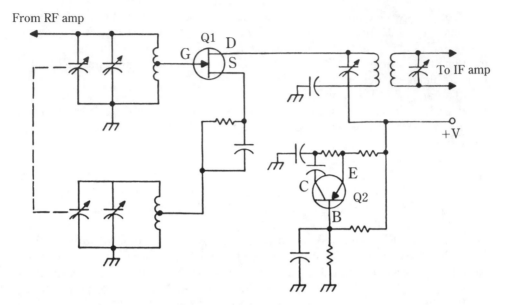

37-16   A typical mixer/oscillator circuit.

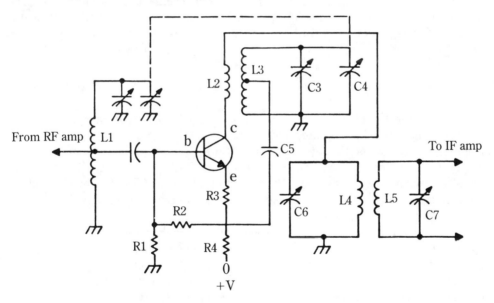

37-17   A single transistor mixer/oscillator circuit.

IF amplifiers are usually quite similar to RF amplifiers, but neutralizing capacitors generally aren't needed, thanks to the lower frequencies involved.

Usually two or more stages of IF amplification are included, at least in better quality receivers. A tuned coupling circuit like the one shown in Fig. 37-18 is included between each pair of stages (including the mixer/first IF amplifier connection, and the connection between the final IF amplifier and the demodulator stage).

Each of these coupling tanks is tuned to the IF. Each additional stage increases the Q of the filtering action, reducing the bandwidth that is allowed to pass through the circuit. By making the bandwidth as narrow as possible, potential interference problems from adjacent frequencies are minimized. In other words, the selectivity of the receiver is increased by increasing the number of IF stages.

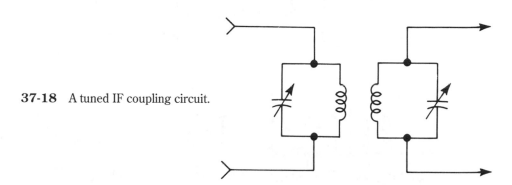

37-18 A tuned IF coupling circuit.

Because only a single frequency is allowed to pass through the IF stages, they only have to be tuned during alignment. In operation they are left alone.

Next comes the *detector,* or *demodulator.* This stage recovers the original program information from the carrier. The mixing process converted the RF carrier to the IF, but the signal modulation remained unchanged. Semiconductor diodes are usually used for signal demodulation.

Figure 37-19 shows a simple AM detector circuit. The incoming signal resembles Fig. 37-20A. The diode will conduct only when it is forward biased by a signal with the correct polarity. This means the output consists of series of pulses whose amplitude varies in step with the original program signal. See Fig. 37-20B.

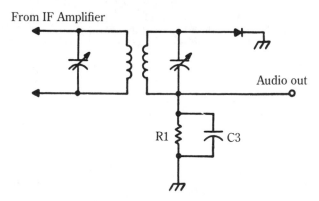

37-19 A simple AM detector circuit.

These variable amplitude current pulses produce similar voltage pulses across R1. These pulses charge C3. In between pulses this capacitor discharges slightly.

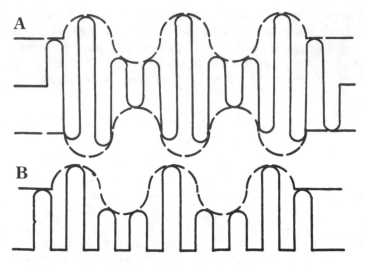

37-20   Input and output signals for the circuit of Fig. 37-19.
Incoming signal (A); output signal (B).

This produces a smoothing action on the output voltage, as shown in Fig. 37-21. This
capacitor is relatively large, so it will discharge only a small amount between pulses.
The output voltage follows the amplitude of the carrier. It is an ac signal superimposed on a dc signal (see Fig. 37-22). The dc voltage is proportional to the strength
of the received signal. The dc voltage is often tapped off to adjust the gain of previous
amplifier stages to boost weak signals, or attenuate overly strong signals. The ac
portion of the output, of course, is a replica of the original program signal.

37-21   AM detector output when smoothed by a large filter
capacitor.

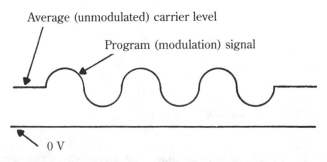

37-22   The output of an AM detector is an ac signal
superimposed on a dc voltage.

Figure 37-23 shows a circuit for separating the ac and dc signals. Direct current cannot pass through the capacitor C4, so only the ac portion of the signal can be taken off of R2, R3 and C5 from a frequency sensitive voltage divider. At dc the value of R3 is much lower than the reactance of C5, so most of the voltage drop is across the capacitor, and can be taken off at the junction between these two components. At audio frequencies, however, the situation reverses. The resistance of R3 resistance is considerably larger than the reactance of C5 at audio frequencies, so most of the ac signal is dropped across R3, and never reaches the R3-C5 junction. There will be some ac at the dc output, but it will be negligible value.

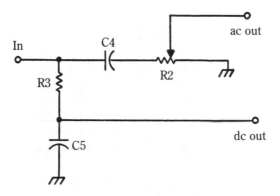

37-23 A circuit for separating ac from dc.

FM is somewhat more difficult to demodulate. This is due to both the type of modulation, and the allowable bandwidth (150 kHz for standard FM broadcasting). Look at three typical FM demodulator circuits.

The simplest type of FM detector is shown in Fig. 37-24. This detector is called a *slope detector*. Notice that it is quite similar to the basic AM detector.

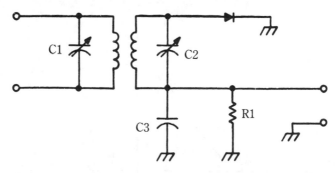

37-24 A slope detector.

The resonant circuits in the slope detector are tuned slightly away from the IF. To see why this is necessary, first assume it is tuned for resonance at the IF. Assume the IF is 500 kHz (usually FM receivers use an IF of 10.7 MHz). The signal passing through the resonant circuits will be at its maximum level when it is at the IF. Increasing the frequency to 510 kHz would cause the output level to drop off somewhat. Similarly, decreasing the frequency to 490 kHz, would also decrease the output level. The frequency response graph is shown in Fig. 37-25. This circuit would be of no use in recovering the program signal because it can only detect if the frequency has changed—it has no way of determining whether the frequency has increased or decreased.

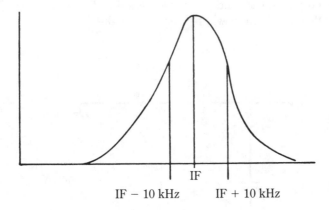

37-25   Frequency-response graph for a tuned slope detector.

But, if the circuit was detuned slightly, so that it was resonant at 510 kHz, the frequency response graph would look like Fig. 37-26. Now the unmodulated carrier would produce a specific output voltage. Increasing the frequency would increase the output. Decreasing the frequency would decrease the output level. The circuit can now determine the direction of frequency change, and the program signal can be recovered. Because the unmodulated carrier frequency falls midway down the

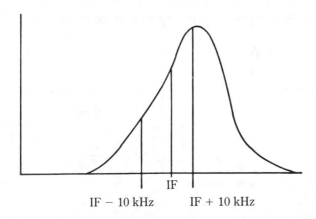

37-26   Frequency-response graph for a slightly detuned slope detector.

slope, this is called a *slope detector*. Although this circuit is fairly simple, it is rarely used because its fidelity is not very good. It is certainly unsuitable for music. Sometimes, however, it is used for narrowband communications systems, where fidelity isn't particularly important.

Another type of FM detector is the *FM discriminator*, or the *Foster-Seeley discriminator*. This circuit is shown in Fig. 37-27. Actually this circuit detects changes in phase rather than changes in frequency, but it really amounts to the same thing because when the frequency changes, its phase will also have to shift.

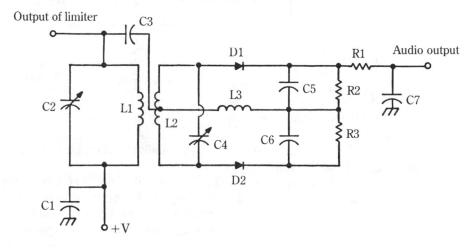

37-27   An FM discriminator circuit.

The FM discriminator is quite sensitive to amplitude variations in the signal. Every effort is made at the transmitting station to hold the carrier at a constant amplitude level, but as the signal is propagated, it might pick up some fluctuations. For this reason this type of detector is always preceded by one or more *limiter* stages. A limiter is a circuit that removes amplitude variations from a signal. It limits the amplitude to a single level.

The FM signal is demodulated by the discriminator by creating an output voltage that is proportional to the difference between the phase at the output of the limiter, and the phase across coil L2. The amount of conduction through the diodes is controlled by these phase relationships.

The voltage drop across R2 is proportional to the conduction level of D1. Similarly, the voltage of RS is determined by the conduction of D2. As the input signal deviates from one nominal IF, one diode will start to conduct more than the other, which causes the voltage drop across the resistors to change. The voltage drop across R2 and R3 will vary in step with the original program signal.

The most commonly used type of FM detector is the *ratio detector*, shown in Fig. 37-28. This circuit does not require a preceding limiter stage. Notice that this circuit is extremely similar to the FM discriminator, except the polarity of one of the diodes is reversed, and a large electrolytic capacitor (C3) is bridged across the output resistors. This is a *stabilizing capacitor*. It keeps the total voltage dropped across the resistors constant, even though the voltage dropped across them individually

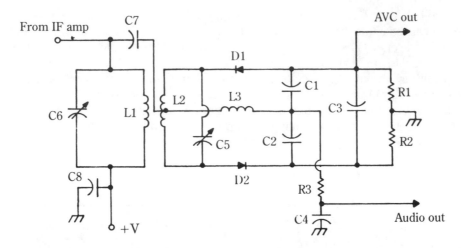

37-28   An FM autodetector circuit.

varies. This provides a self-limiting action. Any amplitude variations in the input signal are blocked from the output.

Direct-current voltages can flow through this circuit in only one direction (see Fig. 37-29). The ac signal paths are shown in Fig. 37-30. At the IF (unmodulated carrier), the voltage across C1 equals the voltage across C2. If the voltage across C1 is 10 V, the voltage across C1 equals the voltage across C2 which equals 5 V. R1 and R2 each also drop 5 V.

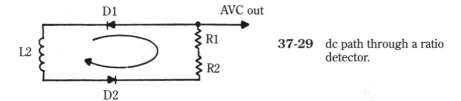

37-29   dc path through a ratio detector.

If the input signal drops in frequency, the impedance of the circuit becomes primarily capacitive, so the current through the secondary of the transformer leads the voltage. This means the voltage applied to D1 is greater than the voltage applied to D2. The voltage across C1 is therefore greater than the voltage across C2 (the voltages across C3 and its resistors remain unchanged). The audio output signal will be a negative voltage that is equal to one half the difference between C1 voltage and C2 voltage.

The opposite happens if the input signal increases in frequency. The voltage across C2 is greater than the voltage across C1, and the output signal is positive. The output voltages of the ratio detector tend to be somewhat lower than those of the FM discriminator. This means more audio amplification is required. R1 and R2 vary in response to the average level of the carrier signal, producing the AGC signal.

Regardless of the method of demodulation, you now have an audio signal that is the same as the original program signal. You merely need to boost the signal level

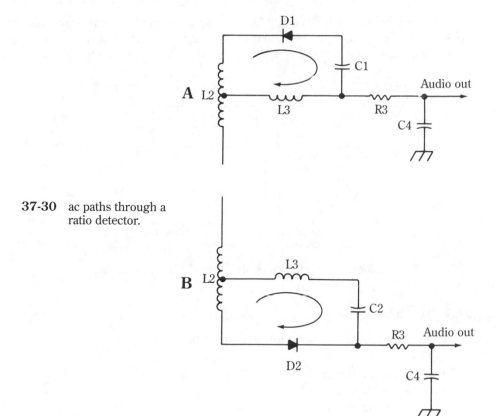

37-30 ac paths through a ratio detector.

with an audio amplifier, like the one shown in Fig. 37-31, and apply the output signal to a speaker, and you can hear the original program.

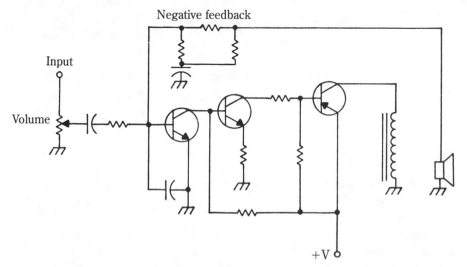

37-31 A typical audio amplifier circuit.

Often a single radio is designed for AM and FM reception. Although these two types of signals require entirely different circuitry for certain stages, some stages can be shared, as shown in the block diagram of an AM/FM receiver in Fig. 37-32.

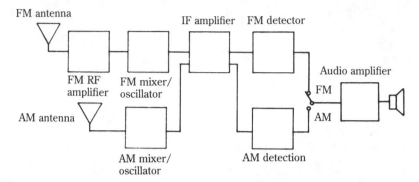

37-32   Block diagram for a typical combined AM/FM receiver.

Self-test

1. What type of waves are used for radio transmission?

A Sine waves
B Electromagnetic waves
C Asymmetrical waves
D Symmetrical waves
E None of the above

2. What is the name for the portion of the radio spectrum that includes 500 kHz?

A LF
B MF
C HF
D VHF
E None of the above

3. Which of the following is used in the simplest form of radio transmission?

A AM
B FM
C PCM
D Morse code
E None of the above

4. What is used to hold a VCO on frequency?

A Phase-locked loop
B Heterodyning
C Modulator
D Bridge rectifier
E None of the above

5. Which of the following is not a basic type of signal path for radio transmission?

A Ground waves
B Direct waves
C Reflected waves
D Sky waves
E None of the above

6. What should be the length of a half-wave antenna to receive signals at 65,000,000 Hz?

A 15.4 feet
B 7.2 feet
C 8.6 feet
D 0.0086 inches
E None of the above

7. How many lobes does a half-wave dipole antenna exhibit?

A One
B Two
C Four
D Six
E None of the above

8. What should be the length of a reflector for a half-wave antenna for 100 MHz signals?

A 4.45 feet
B 4.68 feet
C 4.91 fcct
D 9.36 feet
E None of the above

9. What causes SWR?

A Heterodyning with nearby frequencies
B Reflected duplicate signals
C Mismatched impedances
D Incorrect positioning of the antenna
E None of the above

10. What is used for demodulating AM signals?

A IF coupling circuit
B Slope detector
C Foster-Seeley discriminator
D Diode
E None of the above

38

Stereo

If an audio amplifier system puts out an audio signal to just a single speaker (or the identical signal to a number of speakers), the system is a *monaural* or *mono* system. All of the sounds will come from a single source.

Real-world sounds, however, arrive at your ears from various directions. Listening to a monaural recording of an orchestra or chorus has been compared to listening through a hole in the wall, as shown in Fig. 38-1. Although the sound might be extremely good, with little or no distortion, it will still lack something in realism.

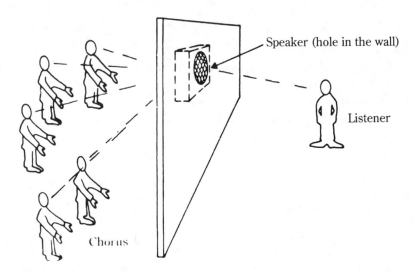

38-1 Effective acoustic image of a monaural sound system—the hole-in-the-wall effect.

A *stereophonic* (or just *stereo*) system, on the other hand, provides two sound sources (a right speaker and a left speaker) for a more realistic effect. A stereo system can apparently place the instruments or voices in a number of different positions (see Fig. 38-2).

At first glance, it might seem that stereo would be like two holes in the wall, with nothing in between. But that is only the case if there is no duplication between the signals fed to the two speakers. This rarely occurs in practice.

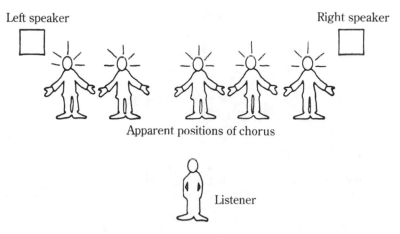

38-2   Effective acoustical image of a stereo sound system.

Suppose you are making a recording of a trio consisting of a violin, a cello, and a viola. You'll assume each of the instruments plays a brief solo passage, each at the same volume, so that if this was a monaural recording, the speaker would be fed a constant 3 W of power. For your stereo recording you will record the violin so that its full 3 W is applied to the right speaker. Similarly for the cello solo, the left speaker will receive 3 W from the amplifier, and the right speaker will receive no signal. So far, you have just the two holes in the wall effect.

The viola solo, however, is recorded so that each of the speakers receives 1.5 W of signal (for a total of 3 W). Because the sound coming from the two sources is identical, your ears will interpret this sound as coming from a nonexistent single source between the two speakers. The acoustic image would appear to resemble the drawing in Fig. 38-3. In this way, various sounds can be apparently placed anywhere between the two speakers simply by controlling the amount of signal fed to each speaker. Of course, this is purely an illusion, but it is quite effective in simulating realism. Virtually all modern high fidelity (realistic reproduction) sound systems are stereophonic.

Until recently, there were three basic signal sources for home stereo systems. These were magnetic tape recordings, phonograph records, and stereo FM radio broadcasts. In the last few years, a fourth stereo signal source has appeared and gained enormous popularity. This source is the *CD*, or *compact disc*. Each of these signal sources are discussed in the remainder of this chapter.

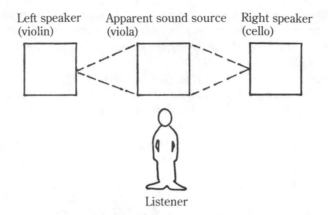

Left speaker   Apparent sound source   Right speaker
(violin)   (viola)   (cello)

Listener

38-3   How a stereo system can create more than two
apparent sound sources.

Tape recording

Magnetic tape consists of a thin strip of plastic that is coated with magnetic particles. Ordinarily, the particles are randomly placed, as shown in Fig. 38-4. If this strip of tape is passed through a strong magnetic field, the particles will be aligned into a coherent pattern as shown in Fig. 38-5. The degree of alignment depends on the strength of the magnetizing field. By varying the magnetic field at an audio frequency (40 to 16,000 Hz), an audio signal can be recorded on the tape. The particles will retain their alignment after the magnetic field is removed.

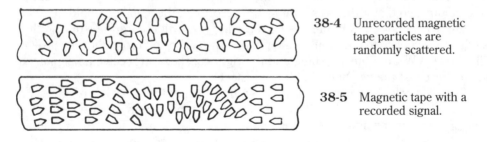

38-4   Unrecorded magnetic
tape particles are
randomly scattered.

38-5   Magnetic tape with a
recorded signal.

The varying magnetic field is produced by a device called a *recording head*. An electrical voltage resembling the audio signal to be recorded is fed to a coil in the recording head. Of course, the varying current through the coil will cause it to generate a varying magnetic field. Now, if the recorded tape is passed over a second type of head, called a *playback head*, the varying magnetic fields of the passing magnetic particles will induce a voltage into a coil within the playback head. This voltage will be a replica of the original signal. Thus, a signal can be recorded onto a strip of magnetic tape and played back. Often a single head is used for both the recording and playback functions, but most high quality tape decks use separate heads.

The highest frequency that can be recorded depends on the *gap* in the protective shield on the head and the speed at which the tape is moving past the head.

(The protective shield is essentially a small window that allows the head to look at only a small portion of tape at any given instant.) The highest possible frequency can be increased by either narrowing the head gap, or by increasing the speed of the tape.

The three most common tape speeds are 1-⅞ inches per second (*ips*), 3-¾ ips, and 7-½ ips. Some professional recording machines operate at 15 ips or 30 ips. Obviously, the faster the speed, the more tape that will be used up within a given time, so there is a tradeoff is involved between frequency response and tape economy.

A recorded tape can be reused by rescrambling the magnetic particles and recording the new signal. This rescrambling is done with a third kind of head called an *erase head*. The signal provided to the head is called *bias* and is produced by an internal *bias oscillator*. The bias is a very high-level, high-frequency signal. The high level of the magnetic field level forces the magnetic particles out of alignment, but the frequency is too high to give them time to realign themselves, so they end up randomly placed. The erase head is always separate from the record and playback heads.

Figure 38-6 shows the arrangement of these three heads in a tape recorder. They are always in this order. A two-headed machine (record and playback heads combined into one) is shown in Fig. 38-7.

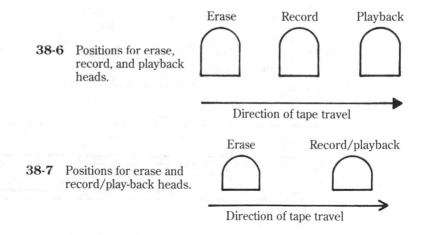

38-6   Positions for erase, record, and playback heads.

38-7   Positions for erase and record/play-back heads.

The simplest way of putting a signal on a strip of tape is to use the entire width of the tape, as shown in Fig. 38-8. This method is called *one-track mono recording*. Typically, most tapes are ¼ inch wide. The tape is wound onto a spool or reel. The free end is connected to a second reel (this is called a *reel-to-reel recorder*). As the tape is pulled off the first reel by the *capstan* and *pincher wheel*, it is moved past the heads, and wound onto the second reel. See Fig. 38-9. The capstan turns at a fixed speed and controls the speed of the tape past the heads.

Once the tape is wound entirely onto the second spool, it is rewound back onto the first wheel and fed through the system once again for playback. Of course, this takes extra time, and is a bit inconvenient. Also, the full width of the tape usually isn't needed to record the signal, so tape is wasted.

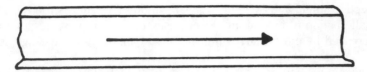

38-8 One-track monaural recording—the tape is usually 1/4-inch wide.

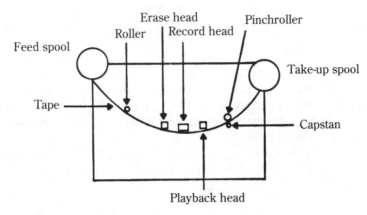

38-9 A reel-to-reel tape recorder.

Some recorders are designed so that when the tape is wound completely onto the second spool, the two spools are reversed, and the tape is sent past the heads again in the opposite direction, recording a second program. See Fig. 38-10.

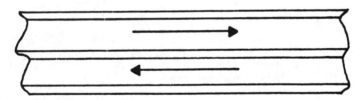

38-10 Two-track monaural recording.

Alternatively, you can record both tracks simultaneously with the two halves of a stereo program. See Fig. 38-11. On playback, the signal from the upper track is fed to the right speaker, and the lower track signal is sent to the left speaker. These two systems are called *two-track mono*, and *two-track stereo*. Of course the

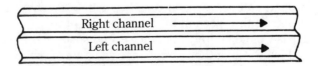

38-11 Two-track stereo recording.

stereo form must be rewound between recording and each playback as with one-track mono.

Two-direction stereo recording can be achieved by recording on four tracks. There are two ways of doing this. In Fig. 38-12 the tracks of the two programs are interweaved. This helps cut down on *crosstalk* (part of one channel bleeding over into the other channel, blurring stereo imaging). This system is the one used in most four-track stereo reel-to-reel recorders.

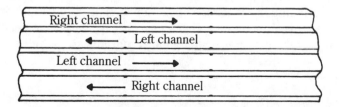

38-12   Four-track stereo recording, as used in reel-to-reel
systems.

The other method, shown in Fig. 38-13, is used in *cassettes*. A cassette, as shown in Fig. 38-14, is basically a miniature reel-to-reel system with both reels enclosed in a single plastic housing. The tape is only ⅛ inch wide, and it only moves at 1-⅞ ips,

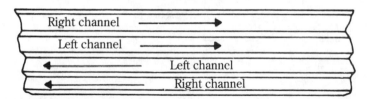

38-13   Four-track stereo recording, as used in cassette systems.

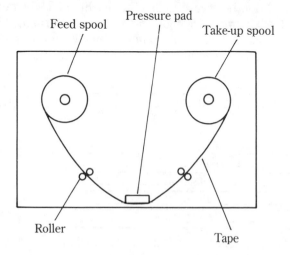

38-14   A tape cassette.

so fidelity isn't quite as good as with reel-to-reel systems, but modern circuitry can compensate for these disadvantages. Modern cassette systems can often produce sound that is virtually as good as a comparable reel-to-reel system.

Ordinarily, cassettes use the two-track mono method of recording shown in Fig. 38-10. Two-track recording is what cassettes were originally designed for. The four-track stereo arrangement of Fig. 38-13 allows cassettes recorded on stereo machines to be directly interchangeable with mono machines.

A third type of system is the *eight-track stereo cartridge* (see Fig. 38-15). The tape is wound into an endless loop, with the ends spliced together with a piece of aluminum. When this aluminum band passes a special sensor, the head is mechanically moved to play the next track, so the four programs are played sequentially.

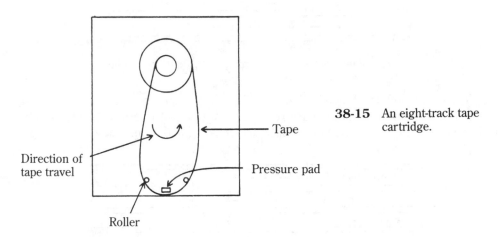

Direction of tape travel

Roller

Tape

Pressure pad

38-15 An eight-track tape cartridge.

At the end of the fourth program, the head is returned to the first program position and starts over. This action continues for as long as the cartridge is plugged into the machine. Another sensor shuts the machine off when the tape is removed. It is impossible to rewind an eight-track stereo cartridge. Usually a switch is provided to allow you to select programs manually, and sometimes a fast-forward control is included. The arrangement of tracks is shown in Fig. 38-16. Notice that

Program 1	Right channel
Program 2	Right channel
Program 3	Right channel
Program 4	Right channel
Program 1	Left channel
Program 2	Left channel
Program 3	Left channel
Program 4	Left channel

38-16 Eight-track stereo recording as used in cartridge systems.

all of the tracks run in the same direction. The motion of the tape can never reverse direction.

Eight-track cartridges were once popular in car stereo systems because the driver could just plug in a tape and let it play over and over while he or she concentrated on driving. Some automobile cassette players will automatically reverse direction at the end of a tape for the same kind of advantage; others have to be manually turned over at the end of a side.

Phonograph records

A *phonograph record* is a vinyl disc with tiny grooves cut into its surface. The disc is placed on a platter (called a *turntable*) that rotates at a fixed speed. A sapphire or diamond needle, or *stylus*, rides in the grooves. The grooves force the stylus to move back and forth, and these variations from a straight path are used to generate an electrical signal. Figure 38-17 shows the way these undulating grooves are cut into a disc, or record. The grooves in this drawing are greatly exaggerated for clarity. The farther a groove varies from its nominal, no-signal path, the higher the amplitude of the recorded signal.

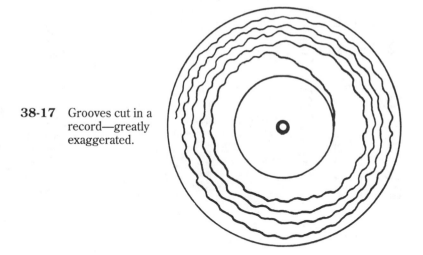

38-17 Grooves cut in a record—greatly exaggerated.

The device that converts the stylus mechanical motion into an electrical signal is called the *cartridge*. There are two basic types of cartridges used in record playing systems. These types are *ceramic cartridges* and *magnetic cartridges*.

Ceramic cartridges

A ceramic cartridge works by the piezoelectric effect (see chapter 11). The motion of the stylus places a mechanical stress along one axis of a piece of ceramic material. This causes a similar electrical stress to be generated across another axis in a different plane.

Ceramic cartridges have a number of advantages. They are inexpensive and sturdy, and they produce a strong output signal. These advantages are offset by certain disadvantages. The disadvantages include a relatively poor frequency response and a high *tracking force*. The tracking force of a cartridge is the amount of pressure (or the effective weight) exerted by the stylus on the record grooves. Obviously, a high tracking force will wear down the grooves (thus rendering the record useless) faster than a light tracking force would. Tracking force is measured in grams. Most ceramic cartridges exert a tracking force of about five grams.

Ceramic cartridges are found in most inexpensive record players. Better systems generally use magnetic cartridges.

Magnetic cartridges

In a magnetic cartridge, the stylus is attached to a piece of magnetic material, which is moved between two small coils. This moving magnetic field induces an ac voltage into the coils. See Fig. 38-18. This ac voltage is a reconstruction of the original signal encoded into the grooves of the record.

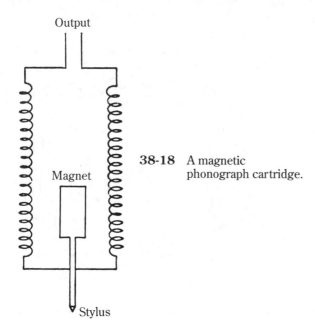

38-18   A magnetic phonograph cartridge.

Magnetic cartridges typically offer a much better high-frequency response and a lower tracking force (¾ to about 3 grams is the typical range) than ceramic cartridges. However, they are generally much more expensive, rather delicate, and they have a much lower output signal level. This type of cartridge requires an extra stage of *preamplification* before the signal is strong enough to be fed to the regular amplifier.

The grooves in a record can be cut at virtually any angle with respect to the surface of the disc. For example, the stylus could be forced by the undulations to move laterally (from side to side), or vertically (up and down). In either case, the output from the cartridge would be exactly the same. In practice, vertical groove records are fairly difficult to manufacture, so lateral motion is used for monaural recordings.

Now, suppose you had a groove with both lateral and vertical modulations. Because the two motions are in different planes (at 90-degree angles to each other) two separate signals could be detected by a single stylus, even if both motions take place simultaneously.

Because of the difficulty of cutting variable depth grooves, and to ensure compatibility with existing monaural equipment, both sets of groove undulations are rotated with respect to the disc, so that they are both at 45-degree angles to the surface of the disc (see Fig. 38-19). In this way, stereo signals can be recorded onto vinyl discs. A monaural cartridge would combine both signals into a single output.

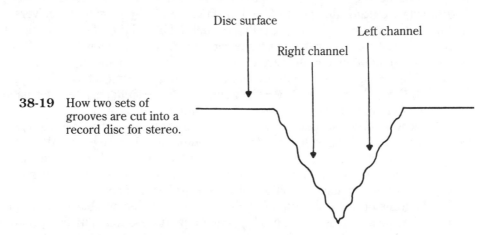

38-19 How two sets of grooves are cut into a record disc for stereo.

A stereo cartridge translates one set of stylus motions to information for the right channel, and the other set of stylus motions to information for the left channel. In practical equipment there is some leakage, or *crosstalk* between the two channels. That is, if a disc is recorded with only right-channel groove modulations, there will still be some (albeit, probably weak) signal in the left channel output too. Better cartridges usually have lower crosstalk.

Stereo broadcasting

Stereo signals can also be sent out over radio waves. Most stereo systems include a stereo FM receiver. Stations within the FM broadcast band are allowed a bandwidth that is more than twice what is needed for transmitting a monaural signal. Therefore they could transmit two separate signals—one for the right channel and one for the left channel—each with its own carrier, as shown in Fig. 38-20.

Unfortunately, a monaural receiver would only receive the right channel information. You need some method of broadcasting a high-fidelity stereo signal that is

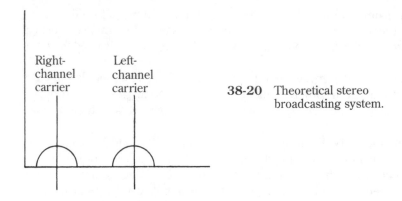

Right-
channel
carrier

Left-
channel
carrier

38-20   Theoretical stereo
broadcasting system.

fully compatible with existing monaural FM receivers. The solution is *matrixing*. Two signals are transmitted. One (the only one received by monaural receivers) is the full monaural signal (that is, the right-channel signal and the left-channel signal). The other signal consists of the difference between the two channels (L − R). These two signals are combined in the receiver to recreate a regular stereo output.

If the two transmitted signals are combined in phase with each other you have (L + R) + (L − R), or L + L + R − R, or simply 2L. The right channel information is canceled out, and you are left with the left channel.

Similarly, another section of the circuit combines the two transmitted signals 180 degrees out of phase with each other. The output in this case equals (L + R) − (L − R) = L + R − L + R = L − L + R + R = 2R. The left channel information is canceled out and the right channel remains. Thus, you have a way of transmitting a stereo signal that is fully compatible with monaural equipment.

At the transmitting station the (L − R) signal is produced by a balanced modulator (see chapter 36). This circuit produces no output unless a modulating program signal is applied. In other words, the unmodulated carrier for the (L − R) signal (which is 38 kHz higher than the (L + R) carrier) is never actually transmitted. This type of transmission is called *double-sideband, suppressed-carrier* (DSSC) transmission. Because the (L − R) carrier is not transmitted, it must be recreated at the receiver for demodulation to take place.

To let the receiver know when it is receiving a stereo transmission, a 19 kHz *pilot* signal is also transmitted. This signal is not modulated. It is simply either present (stereo transmission) or not present (monaural transmission). Note that this 19 kHz signal is exactly one half of the required suppressed-carrier frequency (38 kHz). The received pilot signal can be passed through a frequency doubler and reinserted into the (L − R) signal. Figure 38-21 shows the complete stereo FM signal as transmitted.

AM stereos

For the past few decades, stereo has been available in FM only. AM radios were designed for only monaural broadcasts. Stereo AM transmission was not allowed by the FCC. In recent years, portable FM radios, and high-quality FM receivers have

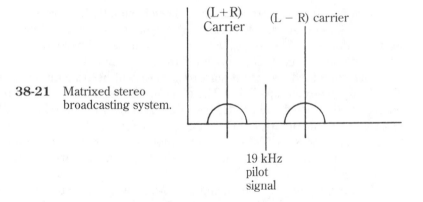

38-21 Matrixed stereo broadcasting system.

put a severe dent in the audience for AM radio. To boost their audiences (and their profits), AM broadcasters have been pushing for the FCC to legalize stereo broadcasting for AM as well as for FM. The idea is not new. AM stereo was proposed to the FCC in 1958. The FCC rejected the proposal in 1961, primarily because of technical problems.

Several manufacturers were not entirely willing to give up on the idea and they continued development of AM stereo transmission systems. The AM stereo proposal was again placed before the FCC in 1977. No fewer than five systems were proposed. They were developed by the following corporations: Belar, Harris, Kahn, Magnavox, and Motorola. The FCC was to consider the various proposed systems and select one as a standard.

Recently, the FCC voted to allow AM stereo transmission, but they decided not to declare a standard. The FCC ruling stated that the best of the various systems should be selected in the marketplace, rather than prematurely limiting the industry to a single standard.

There is certainly something to be said for this idea. Television transmission standards have recently been found to leave quite a bit to be desired, but they can't be easily changed without making millions of existing television sets obsolete.

On the other hand, the lack of a consistent standard might end up killing off AM stereo altogether. The public might be reluctant to invest in a receiver designed to receive only one of the incompatible systems, when the selected system might be the loser in the marketplace war. There would be a lot of obsolete and useless equipment when a competing system becomes the standard. If enough prospective buyers decide to wait until a clear winner emerges, there might be no winner at all. Sales might be poor enough to kill off all of the systems. This is essentially what happened with the competing quadraphonic sound systems (discussed in this chapter).

This situation might not be the case, however. Two incompatible videotape systems (VHS and Beta) managed to find a solid place in the market and shared the market until Beta eventually lost out to VHS. You will have to wait and see what happens with AM stereo over the next few years. Many industry experts consider the front runner to be the Magnavox system. This system was initially chosen as the standard by the FCC, before they decided to opt for no official standard. The Magnavox system combines phase modulation with the standard AM signal. Adapting broadcast transmitters would be relatively simple. Receivers will require major redesign. Of course, all of the proposed systems are completely compatible with existing monaural AM radios. Directional antennas will have to be precisely phased and positioned for correct reception of the stereo signal.

The combined monaural signal (L + R) will be transmitted as ordinary AM signals, to ensure compatibility. The phase of the carrier will be modulated with the difference (L − R) signal. A 5 Hz pilot signal is also transmitted to identify the signal as a stereophonic broadcast. The pilot signal could also be adapted in the near future to transmit low speed digital information such as station identification, time, weather, traffic information, and so forth. There are no firm plans in this area at the moment.

A broadcasting system using the Magnavox AM stereo system is shown in block diagram form in Fig. 38-22. The receiver's block diagram is shown in Fig. 38-23.

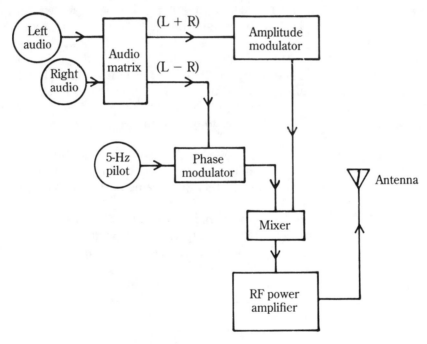

38-22   Block diagram showing the basics of an AM stereo transmitter using the Magnavox system.

Motorola's proposal is called *C-Quam,* or compatible quadrature amplitude modulation. Two separate carriers are transmitted at the same frequency in *phase-quadrature* and modulated with separate left and right signals. The transmitter outputs are tied together and fed to a common antenna. A standard monaural radio will demodulate the combined right and left signals together for a full monaural signal. In a stereo receiver, on the other hand, the two carriers are independently demodulated by two synchronous detectors.

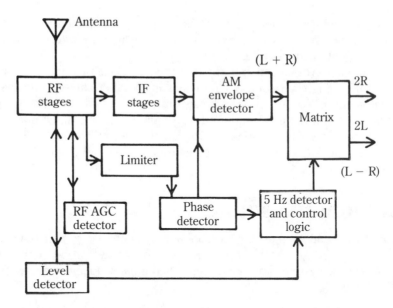

38-23   The block diagram for an AM stereo receiver using the
Magnavox system.

The Harris system uses carrier phase modulation, but in a somewhat different manner than used by Magnavox. Independent sidebands are used in the Kahn
system. Belar Electronics proposes a combination AM/FM signal for stereo
broadcasting.

Which system is the best? Well, they all have their own advantages and disadvantages. Which one will win out in the end (if any)? Only time can tell.

Simulated stereo

Sometimes it might be nice to be able to convert a monaural source into sound like
stereo. This conversion is often done by record companies when older, monophonic
recordings are re-released. Most systems for simulating stereo use some form of
filtering. The simplest approach would be a single simple crossover network that
routes the low frequencies (*bass*) to one channel and the high frequencies (*treble*)
to the other. This works, but doesn't sound very much like true stereo.

More advanced systems would use a series of bandpass filters whose outputs
are added or subtracted from each of the two channels. The filtered frequencies are
increased in volume through one speaker, and decreased in volume in the other.
This gives a vague, somewhat muddy stereo-like effect. Some sounds will appear to
come from the right, others from the left, and still others from somewhere between
the two speakers.

The best results are obtained with a method known as *comb filtering*. This
method is used professionally in recordings that are rechanneled to simulate stereo.
The monaural signal is delayed slightly and mixed with itself. Some frequencies

will partially cancel each other out, and others will be emphasized (increasing their apparent volume). This allows much better separation at more frequency points than using independent bandpass filters.

The frequency crossover points should not be too close together, or stereo separation will suffer. Slight variations in frequency from note to note would make each instrument seem to wander all over the apparent sound field. Generally speaking, best results are achieved if the audio range is split up into 1 kHz chunks. This can be accomplished by delaying the monaural signal by 0.5 milliseconds. Similar techniques are occasionally used to simulate quadraphonic sound (see below) from stereophonic sources.

Quadraphonic sound

Just as the two sound sources in a stereophonic system give greater realism than a single monaural speaker, an even greater sense of reality can be achieved with a four sound source system. This system is called a *quadraphonic system*, or simply *quad*. See Fig. 38-24.

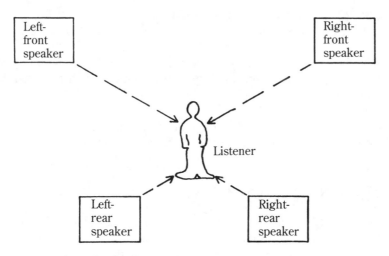

38-24 A quadraphonic sound system.

Unfortunately, quad has virtually died out in the marketplace because of incompatible competing systems. The two most common systems were CD-4 and SQ™ (SQ is a Trademark of Columbia Broadcasting System, Inc.).

CD-4

The CD-4 system is the simplest quadraphonic system. The four channels are simply kept separate on tape or on a record disc. Figure 38-25 shows the way a CD-4 signal would be recorded on tape. Notice that only a single direction of playback is possible. The full width of the tape is used for one program. CD-4 cannot be transmitted by radio.

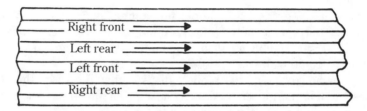

38-25 A four-track CD-4 recording tape.

SQ

SQ (stereophonic-quadraphonic) is a compatible matrixing system. An SQ program can be played on standard stereo equipment without losing any of the program signal. The left channel contains portions of both rear channels as well as the left front channel information. These signals are combined in the following formula:

$$L_T = L_F - j0.707L_r + 0.707R_r \qquad \textbf{Equation 38-1}$$

where L_T is the total left-channel signal, L_F is the left-front channel, L_r is the left-rear channel, and R_r is the right rear-channel. The letter j indicates a 90-degree phase shift.

Similarly, the contents of the right channel signal (R_T are given in the following formula:

$$R_T = R_F + j0.707L_r - 0.707L_r \qquad \textbf{Equation 38-2}$$

A phase shift circuit in the amplifier decodes the rear signals from the front signals. The rear channels are amplified by only about 0.707 times the amplification of the front channel. They are effectively at half the volume of the front-channel signals (the ear does not hear volume linearly).

For broadcasting purposes, the signal is further matrixed into standard stereo signals. That is, $L_T + R_T$ and $L_T - R_T$. Filling in from Equations 38-1 and 38-2, the formulas become:

$$\text{Lower Channel} = L_F + R_F - j0.707L_r \qquad \textbf{Equation 38-3}$$
$$- 0.707L_r + 0.707R_r + j0.707R_r$$

$$\text{Upper Channel} = L_F - R_F - j0.707L_r \qquad \textbf{Equation 38-4}$$
$$+ 0.707L_r + 0.707R_F - j0.707R_F$$

Although these formulas are rather lengthy and complex, it is easy for the circuitry to decode.

Compact discs

The compact disc or CD has gone from an audiophile novelty to one of the most popular recorded sound sources available. The recorded sounds are stored in digital form as pits and islands on a 5-inch metal disc. The pits and islands are detected by a small laser in the CD player. Additional circuitry converts the digital data back into analog form, or sound.

The CD offers an extremely wide dynamic range, as compared to other audio signal sources. There is virtually no background noise. The clarity of sound on a well-recorded CD is incredibly good and often difficult to distinguish from live musicians right in the room with you.

Another major advantage of compact discs is that the discs themselves are nearly indestructible. Large scratches can cause problems in playback, but for the most part, you really have to try pretty hard to damage a CD.

Self-test

1. What is the name for a system with two sound sources to allow more realistic effects?

A Monophonic
B Stereophonic
C Quadraphonic
D SQ
E None of the above

2. What determines the highest frequency that can be recorded on a magnetic tape recorder?

A The recording head gap
B The tape width
C The bias oscillator frequency
D The bias level
E None of the above

3. Which of the following is not a standard speed for tape recorders?

A 3.75 ips
B 33.33 ips
C 15 ips
D 1.875 ips
E None of the above

4. How many channels are used on a four-track stereo tape?

A One
B Two
C Four
D Eight
E None of the above

5. What two signals are transmitted by a stereo FM station?

A L and R
B 2L and 2R
C $(L - R)$ and $(L \times R)$
D $(L + R)$ and $(L - R)$
E None of the above

6. What is the frequency of the pilot signal in an FM stereo broadcast signal?

A 5 Hz
B 5 kHz
C 19 kHz
D 38 kHz
E None of the above

7. Which of the following types of cartridges requires a preamplifier?

A Magnetic
B Ceramic
C Crystal
D None of the above

8. What signal is received by a monaural FM receiver when tuned to a stereo FM station?

A L
B R
C (L + R)
D (L − R)
E None of the above

9. Which of the following is the matrixing system for quadraphonic sound?

A CD-4
B SQ
C Magnavox
D Kahn
E one of the above

10. Which of the following is the legal standard for AM stereo broadcasts?

A Magnavox
B Motorola
C Belar
D Kahn
E None of the above

39
Television

For most people, probably the most familiar and obvious type of electronic equipment is the television set. This ubiquitous entertainment device is considered virtually a necessity rather than a luxury in most modern cultures. Very few American homes do not have at least one set, and many homes have more than one. (These statistics can't help but cast some doubt on the popular pseudointellectual claim that "I never watch television.")

On the most basic level, television is essentially just an expanded form of radio. Along with the audio portion of the program, which is the same as an ordinary radio broadcast, a television signal also includes data for reconstructing visual images. The radio waves don't really care how the data they carry is used. Electrical waves are electrical waves. The same radio waves that carry modulated audio signals can carry modulated video data just as well. In fact, in the early days of television, it was often referred to as "radio with pictures."

The picture tube

The key component in a television receiver is the picture tube. The picture tube is a deluxe *CRT* (cathode-ray tube) as discussed in chapter 19. Some of the key concepts of CRTs are reviewed here. The CRT, or picture tube, displays the received video signals. The images are drawn dot by dot by the electron gun. Each dot making up the composite picture is called a *pixel* (picture element). To keep things fairly simple, consider only a black-and-white television for the time being. The main parts of interest in the picture tube are the *electron gun*, the *yoke* and *deflection plates*, and the phosphor-coated screen. There are several other important components in the CRT, of course, which are discussed in chapter 19, and aren't directly relevant to this discussion.

The electron gun fires a stream of electrons towards the front of the tube. Eventually each electron will strike the screen, which is coated with phosphors. When a phosphor is struck by an electron, it briefly glows, creating part of a visual image. This part of the image is a *pixel* (picture element).

Left to itself, the electron gun will never vary its aim. Every electron will strike the screen dead center, creating a nice clear image of a bright dot on a dark background, but nothing else. To create useful images, you need some way to aim electrons at other pixels on the screen.

Theoretically, the "obvious" solution would be to move the electron gun so it can take aim at each individual spot on the screen. This might be an obvious approach, but it is hardly a practical one. The mechanical system required would be very complex, very delicate, and very expensive. Because the distance from pixel to pixel is extremely short (in the millimeter range or less), the aim would have to be incredibly precise. The slightest mechanical inaccuracy would totally jumble the displayed image, making the entire television worthless. Because an electron can be deflected from its path by an electromagnetic field, which can be electrically varied without any mechanical components, it is much better to move and bend the electron beam after it has left the electron gun.

Deflecting electrons is precisely the purpose of the yoke and deflection plates. They create the electromagnetic fields that deflect the electron beam to the desired spot on the screen. There are two sets of deflection plates—the vertical plates that control up and down motion, and the horizontal plates that control right and left motion.

In an oscilloscope, the signal to be monitored is fed to the deflection plates, so the waveform is drawn on the screen. In a television, however, you want a complete picture each time. The electron beam must scan the entire screen over and over again, regardless of the content of the video signal. This screen image is called the *raster*. Two sawtooth wave signals are used to create the raster. The higher frequency control signal draws fine horizontal lines from left to right, then shoots back to the far left to start over again. At the same time, the second, lower-frequency sawtooth wave, is moving the lines from the top of the screen to the bottom. When the bottom of the screen is reached, the signal jumps back up to the top of the screen to start over.

In the U.S., 525 complete lines make up a full frame. But for various technical reasons, a full frame is not displayed at once. Instead, it is split up into two parts, called fields, each consisting of 262.5 lines. The first field draws all the odd lines, then the second field draws all the even lines. The two fields are therefore interlaced to display the complete frame. The frequency is fast enough that the persistence of vision of the human eye blends the fields into an apparently continuous picture. Sixty fields, or 30 complete frames, are displayed per second. (This number is used in the U.S. Some other countries use slightly different systems with other frequencies. The basic principles remain the same.)

There are 60 fields per second, so the vertical control signal must move from top to bottom at a frequency of 60 Hz. Because each of these 60 fields consists of 262.5 lines, a total of 15,750 horizontal lines must be drawn per second. The horizontal frequency therefore is 15,750 Hz.

So far, you just have the picture tube displaying a blank raster over and over.

The modulated video signal is fed to the electron gun. It must be perfectly synchronized with the vertical and horizontal control signals, so it "knows" which pixel the electron beam is currently aimed at. If the signal voltage at this instant is high, a bright white dot will be displayed at this point because the electron beam emitted from the gun is very strong. A very low voltage will create a black dot, because the electron beam will be very weak. In between voltages will generate dots of varying shades of gray. The final picture is made up of dots of varying intensity at each specific point on the screen.

A color television is a bit more complicated. There are three electron guns in a color CRT. They are aimed slightly off from one another, so they don't quite hit the exact same points on the screen. The screen is coated with three different types of phosphors that glow with different colors—red, green, and blue. One electron gun aims only at red dots, the second only at blue dots, and the third only at green dots. These colored dots are very closely spaced, and tend to blend into one another when they are simultaneously glowing. By lighting up these three color dots with various intensities, a whole rainbow of different colors can be displayed. The specifics of color television are discussed in a little more detail in this chapter. For the time being, concentrate on the somewhat simpler black-and-white (or monochrome) television system.

The television signal

The audio portion of a television program is transmitted via FM, but AM (amplitude modulation) is used to carry the picture data. The video signal is much more complex than the audio signal, because there is more information to be transmitted. You won't any special consideration to the audio portion of the television signal in this chapter, because it is essentially the same as a regular FM radio broadcast, as explained in the preceding chapter.

The video signal must contain the intensity (and color) information for each pixel in every frame. Remember there are 15,750 horizontal lines per second. Each line is composed of hundreds of pixel dots, so you can see that there is a lot of information for the video signal to carry. In addition, there must be synchronization signals so each line and field begins and ends at the right instant, or the displayed picture will be scrambled. Other housekeeping signals are also included in the video broadcast signal. As you can see, this is a very complex signal.

In the transmission signal, the black-to-white range might appear inverted from what common sense might suggest. The whitest possible signal that turns the electron gun most fully has the lowest amplitude of the transmission signal. The higher the signal amplitude, the weaker the drive voltage to the electron gun, and the "blacker" the resulting dot will appear. As you shall see, this is not an arbitrary arrangement at all.

Each horizontal line begins with a synchronization pulse. This alerts the television receiver to "begin a new line now." This synchronization pulse rides on top of a blanking pulse. The purpose of the blanking pulse is to give the horizontal signal time to retrace from right to left to begin a new line. The intensity signal during the blanking pulse is "blacker than black". This means, the blanking pulse amplitude is higher than the blackest picture element to ensure that the electron gun will be held completely off during the retrace, so it won't interfere with the displayed image. The synchronization pulse has an even higher signal amplitude. These signals are shown in Fig. 39-1. If the black-and-white image data were not "inverted" like this, the synchronization pulses would have to have the lowest signal amplitude. They could all too easily get lost in noise, or noise could confuse the television receiver into thinking it sees synchronization pulses where they should occur. The synchronization pulses are the most critical part of the video signal. Without them, no useful image can be displayed. Therefore, they are given the highest amplitude in the transmitted signal to ensure that they get through to the receiver's control circuitry properly. But to avoid the possibility of stray pixels being lighted up during the retrace, the synchronization and blanking pulses must appear to the electron gun as superintense black. This is why the brightest white will have a lower signal amplitude than a pure black.

After 262.5 horizontal lines have been displayed, there must be a vertical retrace to begin a new field. The vertical sync pulse is shown in Fig. 39-2. Because there is a longer retrace path for the lower frequency vertical signal, the vertical blanking pulse is longer than the horizontal blanking pulse. In fact, it lasts for a period equal to several horizontal lines. Most of the vertical blanking interval is made up of simple filler pulses. Other signals can be added to the television broadcast signal during the vertical blanking pulse. For example, this is where the data for closed-captioning decoders is hidden.

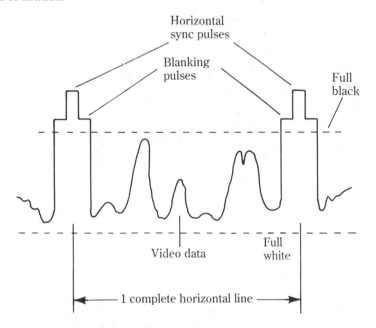

39-1   Television sync pulses are transmitted as "blacker-than-black" signals.

In the television receiver, the horizontal and vertical sync (synchronization) pulses must be separated from one another. A highpass filter of differentiating cir-

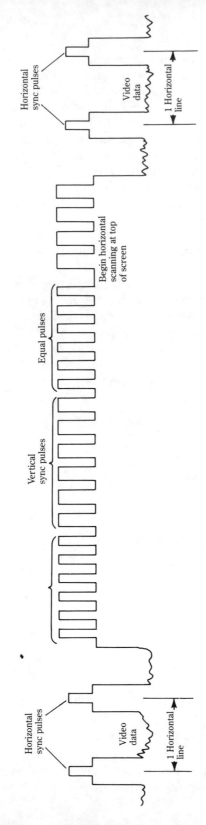

39-2 The vertical sync pulse begins a new field.

cuit directs the horizontal sync pulses to the horizontal oscillator circuit to lock its signal in phase with the received broadcast signal. A lowpass filter or integrating circuit strips off the horizontal components from the vertical sync pulses so the vertical oscillator can be properly synchronized.

The complete television broadcast signal takes up about 6 MHz, as shown in Fig. 39-3. To limit the bandwidth to reasonable limits, the lower sideband of the video signal is suppressed. The audio signal carrier is placed about 4.5 MHz above the video signal carrier. The audio signal has a bandwidth of ±25 kHz. In color television broadcasts, a *subcarrier* at 3.58 MHz carries the color information. This concept is explained more fully elsewhere in this chapter. This color subcarrier is simply ignored by a black and white television receiver.

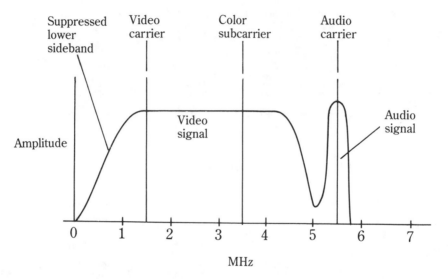

39-3   The complete television broadcast signal takes up about 6 mHz.

Table 39-1 lists the frequency assignments for the 68 standard television broadcast channels used in the United States. For historical reasons, channel 1 has long because been dropped from the available channel frequencies. On older television sets, the channel selector might go up to channel 83, but channels 70 through 83 are no longer used for television broadcasting, and those frequencies have been reallocated for other radio uses.

Inside a black-and-white television set

A television set is certainly a complex device. It is very easy for the electronics novice to become totally intimidated by the hundreds of components inside a typical television receiver. But, like any other piece of advanced electronic equipment, a television set comprises multiple subcircuits. If you divide the television receiver into functional sections and deal with each section individually, it is nowhere near as intimidating. Many of the subcircuits in a television set should already be some-

Table 39-1.
Standard broadcast television channels in the U.S.

Channel number	Video carrier	Audio carrier	Frequency range, MHz
VHF band			
2	55.25	59.75	54–60
3	61.25	65.75	60–66
4	67.25	71.75	66–72
5	77.25	81.75	76–82
6	83.25	87.75	82–88
Gap that includes FM radio band frequencies			
7	175.25	179.75	174–180
8	181.25	185.75	180–186
9	187.25	191.75	186–192
10	193.25	197.75	192–198
11	199.25	203.75	199–204
12	205.25	209.75	204–210
13	211.25	215.75	210–216
Gap			
UHF Band			
14	471.25	475.75	470–476
15	477.25	481.75	476–482
16	483.25	487.75	482–488
17	489.25	493.75	488–494
18	495.25	499.75	494–500
19	501.25	505.75	500–506
20	507.25	511.75	506–512
21	513.25	517.75	512–518
22	519.25	523.75	518–524
23	525.25	529.75	524–530
24	531.25	535.75	530–536
25	537.25	541.75	536–542
26	543.25	547.75	542–548
27	549.25	553.75	548–554
28	555.25	559.75	554–560
29	561.25	565.75	560–566
30	567.25	571.75	566–572
31	573.25	577.75	572–578
32	579.25	583.75	578–584
33	585.25	589.75	584–590
34	591.25	595.75	590–596
35	597.25	601.75	596–602
36	603.25	607.75	602–608
37	609.25	613.75	608–614
38	615.25	619.75	614–620

Table 39-1—Continued

Channel number	Video carrier	Audio carrier	Frequency range, MHz
39	621.25	625.75	620–626
40	627.25	631.75	626–632
41	633.25	637.75	632–638
42	639.25	643.75	638–644
43	645.25	649.75	644–650
44	651.25	655.75	650–656
45	657.25	661.75	656–662
46	663.25	667.75	662–668
47	669.25	673.75	668–674
48	675.25	679.75	674–680
49	681.25	685.75	680–686
50	687.25	691.75	686–692
51	693.25	697.75	692–698
52	699.25	703.75	698–704
53	705.25	709.75	704–710
54	711.25	715.75	710–716
55	717.25	721.75	716–722
56	723.25	727.75	722–728
57	729.25	733.75	728–734
58	735.25	739.75	734–740
59	741.25	745.75	740–746
60	747.25	751.75	746–752
61	753.25	757.75	752–758
62	759.25	763.75	758–764
63	765.25	769.75	764–770
64	771.25	775.75	770–776
65	777.25	781.75	776–782
66	783.25	787.75	782–788
67	789.25	793.75	788–794
68	795.25	799.75	794–800
69	801.25	805.75	800–806

what familiar to you from other chapters in this book. You aren't really dealing with any concepts here—the same old concepts are simply put together into new combinations.

A functional block diagram of a typical black-and-white television set is shown in Fig. 39-4. In some practical sets, some of these functions might be combined into a single sub-circuit, but the principle is still the same. You don't need to concern yourself with the way the actual circuitry is implemented here. You are simply interested in the general way it functions.

This television is a black-and-white television receiver here to keep things as simple as possible. A color television set has all the same functional stages shown

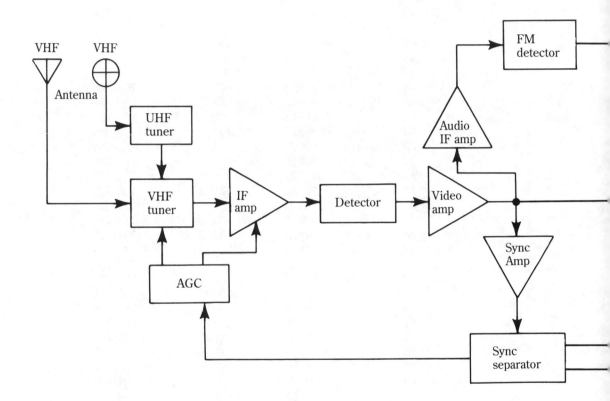

39-4   Functional block diagram for a typical black-and-white television.

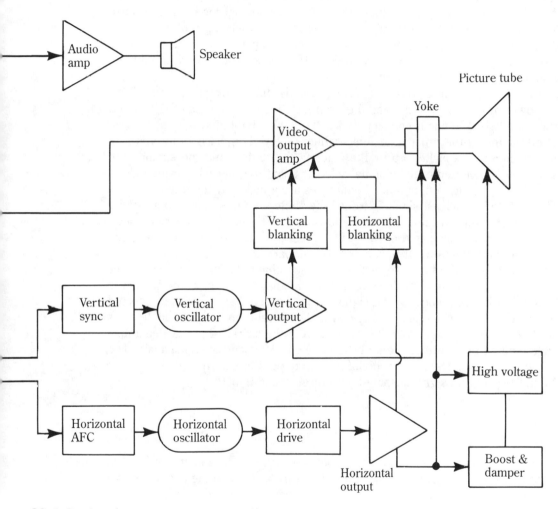

39-4 Continued.

here, plus some additional stages to add the displayed color at the picture tube. You will discuss color television later in this chapter.

The broadcast signal is picked up by an antenna. Television broadcasts occur in either of two bands—VHF (very high frequency) and UHF (ultra high frequency). There are some technical differences between these two bands because of their different frequency ranges. For example, UHF signals are generally line-of-sight phenomena (waves that travel in straight lines and are not easily deflected). VHF broadcasts will travel greater distances under most normal conditions. Because the dimensions of the antenna should correspond to the wavelength of the desired signal for maximum reception efficiency, it follows that separate antennae should be used whenever possible for VHF and UHF. UHF wavelengths are, by definition, much shorter than VHF wavelengths.

In the television set a pretuner circuit (or UHF tuner) drops received UHF signals down into the VHF range, so they can pass through the main VHF tuner. Naturally, the VHF antenna is connected directly to the VHF tuner.

The function of the tuner is to select which received signal is of interest. All available channels are picked up by the antenna the entire time the stations are broadcasting. The setting of the channel selector tells the tuner what frequency to look at. All signals centered on other frequencies are ignored by the tuner.

The selected broadcast signal is frequency shifted, so it is centered around a more convenient and constant frequency, known as the IF (intermediate frequency). Remember, the lower sideband of the video signal is suppressed, so the nominal center frequency is not at the true center of the signal.

The IF is always the same, regardless of which channel is selected. The tuner converts the original frequency down to the intermediate frequency. The IF serves as the new carrier frequency for the signal. This converted signal is then boosted by one or more IF amplifier stages. The converted signal then reaches the detector that starts to strip off the program information from the carrier (IF). A video amplifier stage (sometimes more than one) boosts the entire broadcast signal again. The video amplifier must have a very wide bandwidth to pass the entire signal.

Next, the program signal is split off into three separate paths:

Audio
Video
Sync

The audio signal, of course, is the sound portion of the program. It has its own carrier frequency 4.5 MHz above the video carrier frequency. So the audio IF is 4.5 MHz above the video IF. By passing the entire program signal through an appropriate high-pass filter, you can isolate the audio signal at its own if.

From here on, the audio path is much like an ordinary FM radio. There are one or more audio IF amplifier stages (with a much narrower bandwidth than the full signal IF amplifier stage(s) encountered earlier). Then the program signal is recovered via an FM detector stage. The resulting audio signal is then amplified by a circuit that is like any other AF (audio frequency) amplifier, and the sound is finally heard from the speaker.

The video path deals with the actual picture information contained in each horizontal line. The sync path is only for the synchronization pulses, which are always at a level that is blacker than black. It is easy enough to separate the video and sync signals on the basis of voltage. Any voltage that is lower than the maximum video black level is part of the video signal. All higher voltages are fed through the sync amp.

The video signal is amplified a bit more before being fed to the electron gun of the picture tube (CRT).

The blacker-than-black sync signals are separately amplified, then separated into their various types. The horizontal sync pulses are isolated from the vertical sync pulses. If these signals are too weak, a boost voltage is fed back through the *AGC* (automatic gain control) circuits to the tuner and IF amplifier to increase the amplification gain to the incoming signal. Similarly, if the sync pulses are too strong, the AGC circuits decrease the amplification in the tuner and IF amplifier. The AGC system ensures regularity of the signal strength by the time it reaches the picture tube.

The vertical sync signal adjust the frequency of the vertical oscillator, if necessary. The sync signal is amplified and fed to the vertical yoke of the picture tube. Also, a blanking signal is sent to the video output amplifier to make sure that the electron gun is fully turned off during the vertical retrace period.

The horizontal sync pulses are processed similarly, adjusting the frequency of the horizontal oscillator. These pulse are fed to the horizontal yoke of the picture tube. Also a horizontal blanking signal is fed to the video output amplifier to make sure that the electron gun is fully turned off during the horizontal retrace period.

The 15,750 Hz horizontal signal is also used by the high-voltage and boost and damper stages to create the very high voltages required by the CRT (picture tube) to operate. The ordinary power supply voltages used throughout the rest of the television circuitry are nowhere near strong enough.

In the following sections of this chapter, you will take a closer look at some of the key subcircuits throughout the television receiver as outlined here.

Tuner, IF, and video amplifier circuits

The video signal takes up quite a bit of frequency space, because it has to carry so much information. The full television broadcast signal takes up 6 MHz. More than 5 MHz of this channel allocation comprises the video signal itself. This means the RF (tuner), if, and video circuits in a television receiver must have very flat frequency response over a fairly wide bandwidth. Moreover, you are dealing with relatively high frequencies here—in the MHz region. This means all components in these sub-circuits must be suitable for use at such frequencies. Component leads must be kept as short as possible to minimize stray RF pick-up problems. Shielding of such circuits is also a must.

A block diagram of a typical television tuner stage is shown in Fig. 39-5. Everything shown in this block diagram was included in the single block labeled *Tuner* in Fig. 39-4. There might be some variations from this diagram in some practical television tuner circuits.

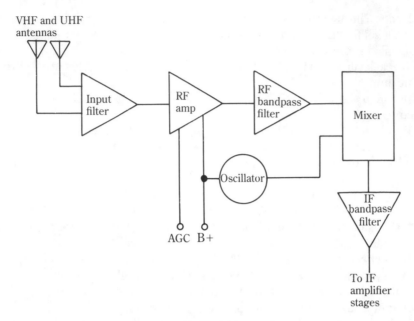

39-5   Block diagram of a typical television tuner.

The tuner stages accept the incoming signals from the antenna(s). Any signals outside the allocated television broadcast band are filtered out at the input to minimize potential distortion problems. An RF amplifier stage boosts everything coming in from the antenna. Notice that this amplifier stage must have a very wide bandwidth. This initial amplification stage is needed to boost relatively weak signals so they don't get lost in noise generated by later circuit stages. Any noise already in the signal by the time it reaches the antenna will get amplified right along with the desired signal. The amplification of this stage is usually fairly minimal.

The signal then passes through an RF bandpass filter, before being combined with the signal from a local tuning oscillator in the mixer stage. The frequency of this oscillator is varied for each television channel from 2 to 69. The output of the mixer stage is the 6 MHz television signal, now riding the IF carrier frequency, instead of the original RF carrier frequency. By changing the oscillator frequency to suit each available channel, the oscillator frequency to suit each available channel, the combination can be made to always result in the same IF frequency. No matter what channel is selected, the IF will always have the same value. This simplifies the design of all the later stages, which do not have to be retuned when the selected channel is changed.

Another bandpass filter removes anything but the desired 6 MHz IF signal. Any other channels broadcasting in the area will be ignored by the later stages in the television receiver.

Because of the wide differences in the frequencies involved, most television sets use separate tuners for UHF and VHF reception. In some modern sets, a single tuner is used, but with different resonant tanks to suit the desired frequency band.

Some recent models use digital circuitry in the tuner. In most cases, the digital circuitry is used primarily to display the selected channel numbers, and to use electronic switching inside of physically moving mechanical switches for channel selection. This makes special features such as remote control and channel programmability a lot easier and more practical to implement. Some deluxe modern television sets actually use digital circuitry in place of the analog mixer stage to select the desired channel. For your purposes here you only need to acknowledge that such circuits exist. You don't need to concern ourselves in the specialized details of how they function.

Early television sets used mechanical switches to select the 12 VHF channels. A rotary control (either resistive or capacitive) was used to select channels in the UHF band (which at that time extended up to channel 82). Because the tuning oscillator frequency was linearly adjustable over its full range, accurately tuning in the desired UHF channel was sometimes rather tricky and frustrating. Because of the wide frequency range covered by the UHF channel control (69 channels), a minor incorrect adjustment could make a major difference in the reception of the selected channel.

Mechanical TV tuners were once the only available choice. Today they have been replaced in most commercial television sets with all-electronic solid-state tuners. You will briefly look at each of these two basic types of television tuners.

In the mechanical tuner, different resonant networks are switched in and out of the circuit for each position of the channel selector dial. A simplified diagram of such a mechanical VHF tuner is shown in Fig. 39-6. Changing the position of the channel selector switch effectively alters the apparent length of three inductors. (Two inductances in series add.) This changes the resonant frequency in each of three resonant networks, permitting the tuner to select the desired broadcast channel.

Mechanical tuners are relatively simple, but, as with any equipment with moving parts, they are subject to wear and tear over time, especially if used roughly. They can also develop problems due to dirty contacts. Particles in the air can settle into all sorts of hard to reach nooks and crannies. Conductive particles can cause short circuits. More commonly, the particles are insulating and prevent good contact from being made in certain switch positions. Calibrating such a mechanical tuner is also a very tedious job. There are many adjustments to make, and most of them interact.

You can understand why all-electronic tuners were developed as soon as it was practical to do. As soon as all-electronic tuners were available, mechanical tuners began to be phased out. Although they are still being used occasionally, mechanical television tuners are rather rare today.

The key to the electronic tuner is the varactor diode, which is discussed in chapter 18. As you should recall, changing the reverse voltage applied across a varactor diode alters its effective capacitance in a predictable way.

By using a varactor diode in place of a regular capacitor in a resonant network, you can change the resonant frequency via an applied voltage. A dc voltage can easily be modified and controlled in many ways. Either analog or digital control of this voltage is possible. No mechanical switches need to be used at all.

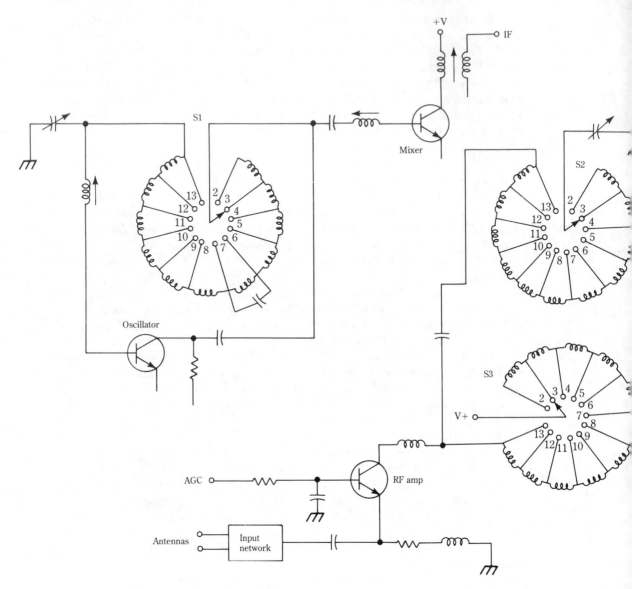

39-6   Simplified diagram of a mechanical VHF tuner.

The IF and video amplifier circuits in a television receiver are fairly straightforward, but considering the high (MHz range) frequencies involved, some special consideration must be made to the limitations of practical components in such circuits. These limitations especially apply to active components, such as transistors and ICs.

Of course you know that real-world inductors, capacitors, and resistors are not perfect. The inherent practical imperfections are more critical at high frequencies.

For example, any inductor has stray capacitances associated with it. Similarly, a capacitor exhibits stray inductances. Resistors also exhibit some internal capacitance and/or inductance. In ordinary low-frequency circuits, these stray inductances and capacitances are so small their effect is negligible, and they can reasonably ignored. Generally they make no noticeable difference in practical circuit operation.

But the resonant frequency of a small inductance and a small capacitance is a high frequency, so such stray inductances and capacitances can make a real difference if their resonant frequency is near the desired operating frequency range of the circuit. Some types of resistors have so much internal capacitance and inductance that they can not be used in high frequency circuits at all. For example, a wire-wound resistor essentially contains a coil.

Practical imperfections of this type certainly aren't limited to passive components. If anything, they can be even more of a problem in active components. Transistors and other active components also exhibit stray capacitances between their various terminals, as shown in Fig. 39-7. A transistor internal capacitances usually make up its chief imperfections when it is used at high frequencies. In some transistors, these internal capacitances might be large enough that such devices can only be used in low-frequency circuits. Other transistors are specially designed for operation up to the gigahertz range. (Refer to chapter 21.)

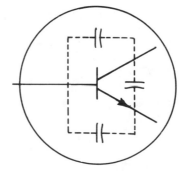

39-7   Transistors exhibit stray capacitances, which become important at high signal frequencies.

Consider how bipolar transistors function in RF circuits. For convenience, alpha and beta are normally treated as constant values in low-frequency circuits. It is normally assumed that the only way these values vary from their nominal values is in step with variations in the collector current. Actually, in any practical bipolar transistor, alpha and beta are not constants at all. They vary with frequency, becoming smaller as the frequency increases. This means that the input impedance of a particular circuit is much higher at low frequencies than at higher frequencies. In a high-frequency circuit, the lower impedance might cause excessive loading.

The manufacturer's specification sheet for any active component will list its rated operating frequency range. A transistor rated for operation up to 500 kHz obviously should not be used in an IF or amplifier circuit in a television receiver—not if you want the television set to function properly, that is.

Sync circuits

For most people, probably the most "mysterious" and confusing sections of a television set are the sync circuits. Actually, there is nothing all that unusual or complex about these subcircuits. They really aren't all that difficult to understand. The entire composite video signal is fed to the sync separator subcircuit. This circuit has two primary jobs. It first removes the sync pulses from their blacker-than-black blanking pedestals, discarding the rest of the composite video signal. Next, it separates the horizontal (line) sync pulses from the vertical (field) sync pulses. Often, the sync separator will also perform some amplification on the isolated sync pulses as well. In other television sets, a separate sync-amplifier stage can be used for this purpose. A fairly typical sync-separator circuit is shown in Fig. 39-8. You should be aware that many other circuit designs can be used to perform the same functions.

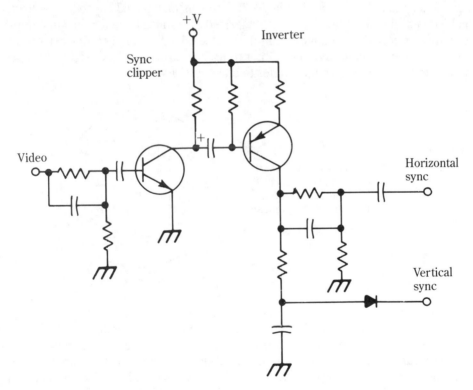

39-8   Typical sync separator circuit.

This circuit is divided into two stages, each stage built around one of the two transistors. The first transistor (Q1) pulls the desired sync pulses from the composite video signal. The video signal first passes through an RC network (R1, R2, C1, and C2), which performs some simple prefiltering. The output from the collector of this transistor is fed to the midpoint of a voltage divider string made up of resistors R3 and R4. Only the sync pulses manage to get through to the base of transistor Q2, the second stage of the circuit. This transistor mainly serves as an amplifier. The actual separation of the vertical and horizontal sync pulses is accomplished by a simple passive integrator network (R7, C4, and D1) and a simple passive differentiator network (R8, R9, C5, and C6).

An integrator is essentially a lowpass filter. Rapidly changing voltages are filtered out of the signal passing through such a network. The horizontal sync pulses are brief pulses separated by rather wide intervals, so they are pretty much filtered out by the integrator network. The vertical sync pulses, however, consist of several fairly wide pulses, separated by narrow gaps. This drives the output of the integrator network to a higher voltage during these pulses. This integration action is shown in Fig. 39-9.

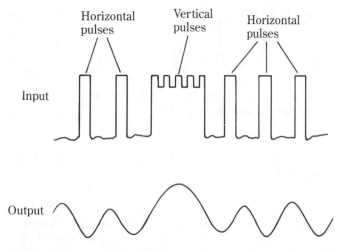

39-9 The integrator network produces a higher output voltage during the vertical sync pulses.

The differentiator network works in basically the opposite way, as shown in Fig. 39-10. This time slowly changing voltages are filtered out, while fast changes reach the output. Each horizontal pulse comes through the output of the differentiator network. The horizontal pulse frequency can be speeded up slightly during the vertical blanking interval, but this doesn't matter, because the horizontal sync pulses really aren't doing anything useful at this point, although there is no particular reason to try to turn them off or suppress them during the vertical blanking interval. Because the electron gun in the picture tube is turned off through the entire vertical blanking interval, the horizontal lines "drawn" during this period are not displayed, and are completely ignored.

High-voltage circuits

Never open the back of a television set without knowing exactly what you are doing. If you have any doubts, leave any repairs, modifications, or internal adjustments to

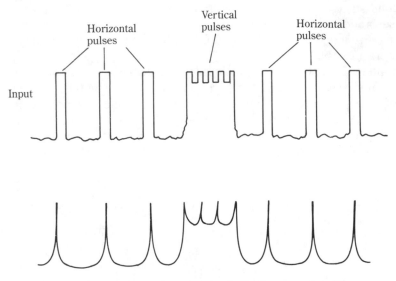

39-10 The differentiator network is basically the opposite of the integrator network.

qualified service personal. Please don't ever take any foolish chances. This book certainly will not qualify you to work on television circuitry. Making such an attempt without proper training can be extremely dangerous and possibly even fatal. It's simply not worth the risk. The chief danger involved is the extremely high voltages generated within the television set to drive the picture tube. These voltages are 50 kV (50,000 V) or more, and the voltages are at moderately hefty currents. This energy is more than sufficient to provide a fatal electrical shock to even a husky human being who gets a little careless.

Test equipment leads should never be applied directly to a high voltage circuit point. Most expensive test equipment would be grossly overloaded, and would probably be utterly ruined in a fraction of a second. At best, you won't get any meaningful readings. At worst, you could electrocute yourself.

Another danger that still gets talked about from time to time is X-ray radiation. This was a potential problem with early television sets (although there is still some debate as to just how significant the risk was in the first place), but it is certainly a negligible concern with modern equipment. In fact, the U.S. government has recently lightened the required antiradiation safeguards manufacturers must include in their sets. There are several people who have gotten quite paranoid about this turn of events, and they are making dire predictions of what could happen from letting manufacturers operate on trust. Actually, it's not a matter of trust at all, but a matter of science. In early TV sets, high-voltage tubes and supporting circuitry acted as X-ray radiation sources. But the old-type tube circuits have not been used for years. All modern television sets use solid-state (transistor or IC) circuitry, except for the picture tube. (More and more sets use LCD screens instead of CRTs, too). The picture tube was never a significant contributor to the X-ray problem. In addition, all CRTs are internally shielded, and the relevant regulations for these devices remain in effect.

Ultimately, the easing of the government regulations on the TV X-ray radiation issue just means that you will no longer have to pay for some useless shielding around solid-state high-voltage circuits that aren't generating any measurable amount of X-ray radiation to be shielded.

Radiation and radioactivity scares a lot of people. While some concern is certainly valid, much of what has appeared in the media more closely resembles blind paranoia from writers, reporters, and so-called consumer advocates (as opposed to genuine consumer advocates) who simply don't have any background education in the relevant scientific fields. Most people are not aware that everything on Earth is radioactive. This book is radioactive. So are the clothes you are wearing and the chair you are sitting on. Your own body is radioactive.

There is some (admittedly debatable) scientific evidence suggesting that some minimal level of background radiation might be necessary to sustain life. At the very least, you need to put the risk into perspective. If a given level of background radiation can cause a genetic mutation in 1 out of every 10,000,000 cases, it is not a reasonable risk to worry about. There is a greater risk that you will be killed or badly injured by having the television set fall on top of you.

Modern television sets simply don't emit significant amounts of X-ray radiation. It is scarcely measurable even a few inches away. Once you get a couple of feet from the set, any X rays from the television are more than swamped out by the natural background radiation level that is always all around us. Most of the energy modern television sets do radiate is in the form of light and heat—and even these forms of radiation are weaker than the similar emissions from an ordinary light bulb. Frankly, it would make more sense to worry about being hit be a meteorite than about X rays from a television set.

Incidentally, a lot of nonsense has been written about supposed radiation risks from computer monitors. All a computer monitor is is a slightly modified television set without a tuner. The modifications are simply to aid increases in frequency bandwidth for more detailed picture detail. By far, the biggest risk related to computer monitors is eyestrain from staring at imperfectly formed letters composed of dots of light for extended periods.

Even older television sets were adequately shielded to prevent harmful X-ray emissions. But if you ever work on an older television set, be very careful about removing any shielding within its works. It can be very dangerous to operate some old tube-type sets with certain shielding elements removed, even for testing purposes.

Because of the change in regulations, the high-voltage circuits in newer television sets will have considerably less shielding. This will make no noticeable difference (except perhaps a slight savings in cost) to the ordinary consumer. Using these sets is perfectly safe. But if you open up the set, it is more critical than ever to know what you are doing to avoid unnecessary shock or fire risks. In older sets, you could usually identify most of the high-voltage circuitry from the shielding. In newer sets with little or no shielding needed, there might be little to identify a potentially dangerous high-voltage circuit from any other circuit. Let me repeat the vitally important warning—do not attempt to service, modify, or internally adjust any television set unless you know what you are doing. When in doubt, never open the case of the television yourself. Turn all such work over to a qualified technician. Yes, it will be more expensive. But what price tag would you put on a serious injury, a fire, or even a death? It's not worth the risk.

Especially watch out for large electrolytic capacitors. They will be physically larger than most other components. They can store very large voltages, even when the set has been unplugged for some time. Yes, you can get a harmful or even fatal electrical shock from a television set that has been unplugged for a week. Any large transformers or coils can also be serious potential shock risks, although they are less likely to hold a charge when power is removed from the circuit.

Some electronics technicians test for the presence of high voltage by holding a well-insulated screwdriver near the high-voltage point to be tested. If the high-voltage is present, a spark will arc up to the nearby metal tip of the screwdriver. The handle of the screwdriver must be very well insulated, but even then this is a some-what risky procedure. Your hand could slip and momentarily touch something conductive at just the wrong instant. The spark could conceivably touch some-thing flammable and start a fire. Some of the spark energy could be deflected or re-flected to some other component in the circuit that could be damaged. Even under the best of circumstances, this procedure places a momentary overload on the high-voltage supply. That probably won't do any harm, but you never really know until it is too late.

To check out the high-voltage circuits properly in a television set, a high-voltage probe for a VOM or other voltmeter is the best choice. Such probes are commer-cially available from many manufacturers. Besides being a lot safer, this method provides much more information than the crude "screwdriver test" described above. You can actually measure the high-voltage value (although usually only with fair accuracy). It is often helpful in television servicing to check the changes in the high voltage as the brightness control is adjusted to a different setting.

Two common high-voltage related problems in television sets are *arcing* and *corona*. Arcing occurs when a spark jumps from one conductor to another, either getting into the wrong place, or applying a signal at the wrong time. A common cause of arcing is accumulated dirt that can form a low resistance path between a high-voltage point and ground. Sharp points can also shorten the electrical path, encouraging arcing. Old, partially dried insulation can lose some of its effectiveness, and allow arcing to occur. Generally speaking, arcing problems occur most fre-quently under relatively high humidity condition, or where air is thin.

The problem of corona is a little more complex. In essence, it is a more or less continuous high-voltage discharge into the surrounding air. The corona can usually be seen in a partially darkened area. It appears as a purplish glow. In most cases, it will not glow brightly enough to be visible under full lighting conditions. Often you will also be able to hear a fine hissing sound, and you might be able to smell ozone. Usually the corona source will be a sharp point, rather than a smooth large surface.

Corona problems can look rather frightening to a lay person. There is some shock and/or fire hazard whenever uncontrolled high voltage is present, of course. In most cases, corona will only be visible if the set is opened up, which the lay person should never do in any event. In some fairly rare cases, some corona might be visible outside the television set's cabinet. If a television set exhibits corona it should be turned off and unplugged, and qualified service personal should be called in as soon as possible—certainly before the set is put back into use. Why take foolish chances? If you notice an odd ozone smell, you should also disconnect the set and call in a qualified technician.

In most cases, however, the risk of corona is only to the television set circuitry. It's wasting much of the high-voltage energy. The current that is drawn off into the air, instead of driving the picture tube, as it is intended to do. Corona will usually lead to arcing sooner or later.

Usually corona problems are identified through operational symptoms, such as a weak or abnormally dark picture, due to the insufficient high-voltage current reaching the picture tube.

One last time, read again the so-important warning—television sets include some very high-voltage circuits, which can present a serious shock and fire hazard if you don't know exactly what you are doing. If you are not trained in television circuitry, never open up the cabinet of any television set. Always call in a qualified technician. Safety is definitely worth the expense. Dozens, if not hundreds, of careless, untrained experimenters and hobbyists unnecessarily injure or kill themselves every year through taking such foolish chances. Also be aware that color television sets use even higher voltages than black-and-white sets.

Color television

So far you have been concentrating entirely on black-and-white television sets, because they are a little less complex than color sets. But these days, color television is by far the norm. The vast majority of television sets in use today are color. In fact, very few black-and-white sets are manufactured any more. This situation is a bit of a shame, because a black-and-white set can be so much less expensive than a color set of similar quality. Often a black-and-white set would make more sense, especially for a second set for a child's room or for occasional use. Frankly, few television programs are really enhanced by color anyway. When you watch *Jeopardy*, for example, you probably don't much care what color Alex Trebek's tie is. But apparently this opinion is a minority opinion, because the today's electronics marketplace has decisively declared that color is virtually mandatory for modern television sets.

However, you have hardly wasted your time examining black-and-white television sets first here. In fact, if you have jumped ahead to get right into color television, it is very strongly recommended that you go back and read the first part of this chapter. Everything said about black-and-white television sets also applies to color television sets. The practical differences in electronics terms are that some specifications in a color television receiver might be a little tighter than in a comparable black-and-white set, and, of course, the color set also has some additional circuitry to provide the coloring of the displayed images. The special features of color are all you will read about in this section. To discuss the other subcircuits in a color television would essentially be a matter of simply repeating the information already given for black-and-white television sets. In a sense, a color television set is just a black-and-white television set, with a few special features added.

Light waves have frequencies, just like sounds, or ac waveforms. These light frequencies are very high. Within a limited range of the light frequency spectrum, you have visible light. Light waves within this frequency range can be seen. The specific frequency of the light wave (or waves) in question determines the perceived color. The lowest visible color frequency is red, and the highest visible color frequency is violet. Green is about in the middle of the visible spectrum. The eye is most sensitive to colors in this midrange.

When light waves of different visible frequencies coexist close to one another, they combine to form a new color. You can see this effect in a tricolor LED (see chapter 19). When a dc voltage of one polarity is applied across the tristate LED, it will give off a glow. Reversing the polarity of the dc voltage applied across the device will cause it glow green. If a moderate to high-frequency ac voltage is applied across the tristate LED, it will alternately glow red and green, but at too high a rate for the human eye to distinguish the individual flashes of colored light. The two colors will blend together due to the persistence of vision. The tristate LED will appear to have a yellow glow. Adding red light with green light produces yellow light.

Color television works in much the same way. All of the displayed colors are made up of tiny dots of just three colors—red, green, and blue.

If all dots are off, you get a black spot on the television screen. White is produced by a more or less equal combination of red, blue, and green. The combination is more or less equal because the human eye is not equally sensitive to all visible colors. It is much more sensitive to green, than to red or blue, so to get pure white, the blue and red light will have to be a bit stronger than the green light.

In essence, in black-and-white television, you are just using a white light control signal. The stronger (more negative) the signal, the brighter the white spot that will be displayed on the CRT screen. The weaker this white light control signal is, the darker, or grayer the displayed spot will be. If the signal is turned off (at its lowest, or most positive value), the phosphor on the screen won't light up at all, and you will see a black spot.

The white light control signal in television is often called the luminosity signal. It determines the overall intensity, or brightness of the picture.

Now, you know that pure white light can be made up of the proper combination of red light, blue light, and green light. (Don't concern yourself with the exact proportions here.) This approximation permits you to write simple algebraic equations for each of the three color signals—blue (B), red (R), and green (G):

$$B = W - (R + G)$$
$$R = W - (B + G)$$
$$G = W - (B + R)$$

These equations might appear to be too simple and obvious to be worth mentioning, but as you will soon see, they are very important equations for understanding the inner workings of color television.

The color picture tube has three electron guns, one for red, one for blue, and one for green. The inner surface of the screen is covered with three types of phosphors. The electron guns themselves are identical, except for their aim. One type glows red when it is struck by the electron beam. The second type glows green, and the third blue. Each electron gun is carefully aimed so it strikes only its own color phosphors.

Three types of information must be included in the color television broadcast signal. They are usually called brightness, hue, and saturation. The brightness is the ordinary black-and-white signal, or the luminance signal. This signal must be kept intact or black-and-white television sets won't be able to receive the color program. Black-and-white television receivers simply ignore the hue and saturation data.

The hue data defines which color or colors is to be displayed at any given point. That is, it controls which of the electron guns are permitted to fire at that pixel.

Saturation refers to how strongly colored the picture image will be. It controls the intensity to a single electron gun.

The brightness or luminance signal in a television broadcast is often identified as Y. When color television was developed, two standard phase-shifted signals were added. They are called the I and Q signals. The I signal was displaced 57 degrees from a reference signal, and the Q signal was displaced a further 90 degrees from the I signal, for a total 147 degrees from the reference signal. Most modern technical literature has redubbed the old I and Q signals X and Z for some reason. Probably someone felt X and Z go better with the already existing Y (luminance) signal.

This system is still in use in television broadcast signals, but it has been somewhat modified on the reception end. In most modern color television sets, the I and Q signals themselves are not used. Instead, the set works with three color difference signals:

$$R - Y$$
$$B - Y$$
$$G - Y$$

Remember Y is the overall brightness or luminance signal. It can be considered equivalent to white. R, B, and G, of course, refer to red, blue, and green, respectively. These color difference signals are derived from the broadcast signal by a special subcircuit called the color decoder.

In the broadcast signal, a special color subcarrier is placed 3.58 MHz above the standard video carrier. The X and Y signals are phase-shifted from the 3.58 MHz reference signal. The color difference signals can be derived from the X and Z (or I and Q) signals. The various phase relationships in the color signal are shown in Fig. 39-11.

The color subcarrier is both amplitude modulated and phase modulated to carry the required color information. A special technique known as frequency interlacing is used to prevent the color signal from interfering with the ordinary luminance (black- and-white) video information. A black-and-white television receiver behaves as if the color signal simply didn't exist at all.

The luminance (black-and-white) signal is not actually a continuous frequency band running from 0 MHz (the video carrier frequency) to 4 MHz. Instead, this signal is made up of many narrow frequency bands based on harmonics (whole

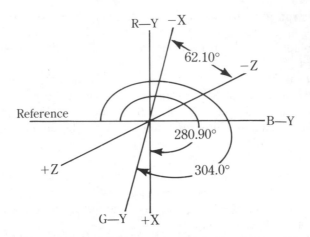

39-11 The color signals are phase shifted.

number multiples) of the horizontal line frequency (15,750 Hz). The color signal, which is fairly narrow in itself, is placed in one of the gaps between adjacent bands in the luminance signal, so it doesn't cause interference.

While the color subcarrier frequency is usually given as 3.58 MHz, it is actually 3.579545 MHz, but it is such an awkward value to write out, so the value is rounded off 3.58 MHz. Whenever you see a reference to 3.58 MHz in electronics literature, remember the actual frequency is a little lower than this. Fortunately, in most practical electronics work, you can safely ignore this difference, but you should be aware that it exists.

To receive the color information, a color television set must have a local oscillator with a frequency of precisely 3.579545 MHz. Moreover, this local oscillator signal must be perfectly in phase with the reference signal (color sub-carrier) transmitted from the broadcasting station. To permit such synchronization, an unmodulated color reference signal is transmitted on the horizontal blanking pulse, just after the horizontal sync pulse, as shown in Fig. 39-12. This reference signal consists of exactly eight cycles. It is called the color burst signal. In a black-and-white television broadcast, the color burst signal is not transmitted. The color television receiver senses its absence and turns off its color circuits with a special subcircuit known as the color killer. The color killer prevents random colors due to noise signals from being added to the black-and-white picture.

During the horizontal blanking pulse, the local color oscillator (in the television set) is locked onto the exact frequency and phase of the reference signal. This way the color circuits can properly translate the phase-modulated color information interlaced with the main luminance signal.

A functional block diagram of a color television receiver would essentially be the same as that for a black-and-white set, with the addition of the special color circuits shown in Fig. 39-12.

The full luminance (black-and-white) signal is applied to all three electron guns in the picture tube. In addition, each electron gun also receives the appropriate color difference signal. For example the red electron gun is fed the Y and R − Y signals.

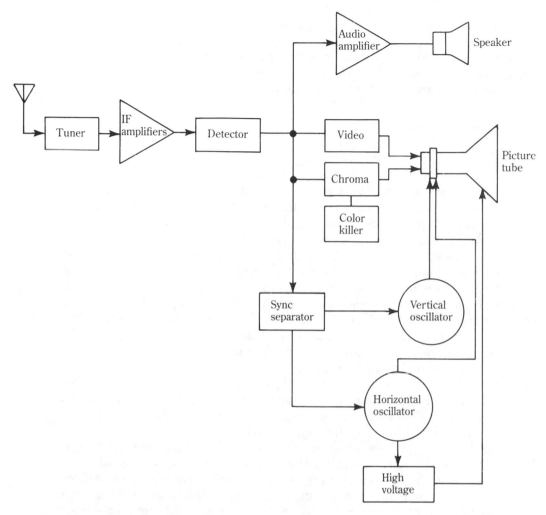

39-12   Functional block diagram of a typical color television set.

Notice that most of the full luminance (Y) signal is canceled out, except where it relates to red images in the picture. The R part of the R − Y difference signal also includes the appropriate hue and saturation information the electron gun needs to do its job properly.

Remember that any two or all three electron guns might fire at the same pixel with different intensities to form different colors. For example, to display a yellow spot, the red and green electron guns must fire at the spot, but not the blue electron gun.

The high-voltage circuits in a color television set also tend to be somewhat different than their counterparts in a black-and-white set, because a color picture tube typically requires higher operating voltages than a black-and-white picture tube of similar dimensions. This is just a minor difference in degree, not in kind. There is no particular difference on the theoretical level.

Some newer color picture tubes use a somewhat different design. For example, some modern color picture tubes manage to use a single electron gun to illuminate all three color dots (red, blue, and green) on the screen. You don't need to get into the details of how this is done here. The basic operating principles remain the same, even though they are carried out somewhat differently.

TV antennas

To receive a broadcast signal, a television receiver, like a radio, must have some sort of antenna, or *aerial*, tuned to the frequency band of interest. In an urban environment, with several broadcasting stations nearby, only a small antenna might be required. Often a simple indoor antenna that sits atop the television set will be adequate. In more remote locations, a large outdoor antenna mounted on the roof (or otherwise raised as high as possible) might be necessary.

If your main concern is to pull in distant stations from different directions, an omnidirectional antenna (sensitive to broadcast signals arriving from all directions) might be an appropriate choice. More often, especially in urban areas, a more directional and selective antenna might be required. If you have a powerful omnidirectional antenna, and there is a station 45 miles to the north broadcasting on, for example, channel 3, and a second station 190 miles to the southeast that is also broadcasting on that same channel, you could run into interference problems. You almost certainly won't be able to receive the more distant station—its signal will be swamped out by the more local station using the same frequency. But the distant station's signal might be strong enough to interfere with the signal from the nearby station. The television set won't be able to distinguish between the two signals, because they are both operating on the same frequency.

A more common problem with omnidirectional TV antennas, especially in urban environments, is multipath interference. This interference happens when the same signal takes different paths to the receiving antenna, arriving at different times. This effect is shown in Fig. 39-13. The main signal goes along a simple straight-line path from the transmitting antenna to the receiving antenna. But some of the signal goes on by and is reflected from some surface, such as a tall building or a mountain. This signal is then reflected back to the receiving antenna. This signal is the same

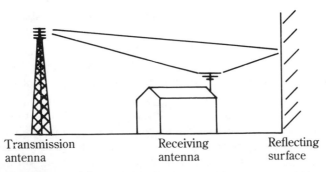

Transmission antenna   Receiving antenna   Reflecting surface

39-13   Multipath reception causes *ghosting*.

as the straight-path signal, of course, but obviously it takes a little longer for the reflected signal to reach the receiving antenna, so this signal is delayed slightly. The television will display multiple, out-of-phase images, or "ghosts." A more directional antenna can be selectively aimed to receive only the straight-path signal, ignoring any reflections from different angles.

The problem with directional antennas is one of convenience. Suppose your home is located as shown in Fig. 39-14. If you have a good, directional antenna, you will have to re-aim it each time you want to watch a program on a different channel. This nuisance is a minor one with an indoor antenna on top of the set, but it is obviously very impractical for an outdoor, roof-top antenna. One solution would be to mount separate directional antennas for each desired station. A switching network inside your home can be used to select the appropriate antenna for each desired channel. This would be expensive, and awkward. Your neighbors would probably consider it a real eye sore.

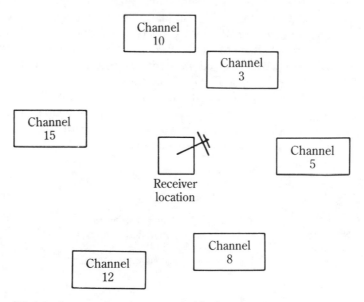

39-14   In some locations, a movable directional antenna is necessary.

Fortunately, a solution does exist: use an antenna rotor. The antenna is mounted on a movable mast, connected to a stepper motor. A control box inside (usually on top of the television set itself, or very nearby) determines when the motor moves and which way. This allows you to conveniently and easily re-aim the outdoor roof-top antenna from the television viewing area.

These days, a lot of people are watching television without an actual antenna. As far as the television set is concerned, the signal provided by the cable television company is from an antenna. The cable hookup takes the place of the ordinary antenna. Strictly speaking, the antenna isn't really removed from the system. It is just relocated. The cable company uses antennas (usually satellite dishes) to receive the various channels it carries into your home. You can read about cable television and satellite television a little more in this chapter.

A VCR, a video game, or a computer can also be hooked up to the antenna terminals of a television set. As far as the television receiver is concerned, it is receiving a broadcast signal from an antenna, no matter what the actual signal source is.

Most practical television aerials are some variation on the basic dipole antenna. The basic construction of a simple dipole antenna is shown in Fig. 39-15. The dipole consists of two equal-length conductive legs. At the center connection point between these two legs, or poles, the current of the received signal is at its maximum value, and the voltage at its minimum. At this point, the lead-in wire or cable is connected to the antenna to carry the signal into the actual television receiver. The length of each is equal to one-quarter of the wavelength of the signal to be received by the antenna. Because most practical antennas are used to receive multiple frequencies or channels (within a specific band), there is almost always some approximation and compromise involved in determining the one-quarter wavelength dimension. In television, you are dealing with rather wide frequency ranges. A simple dipole antenna can only operate efficiently at a single, specific frequency (equal to four times the length of each leg). It will operate if the frequency is a little off, but with decreasing efficiency, the further the received frequency is from the nominal resonant frequency of the antenna (determined by the leg length).

Because of the very wide range of frequencies used in television broadcasting, it is clear that a simple dipole antenna wouldn't do a very good job by itself. It would offer reasonable efficiency only for one or two channels. Fortunately, there are ways to modify the basic dipole antenna to get around such difficulties.

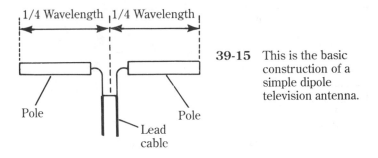

39-15   This is the basic construction of a simple dipole television antenna.

Normally a dipole has a bidirectional sensitivity pattern, that can be graphed as a figure 8, as shown in Fig. 39-16. This type of antenna ignores signals from either side, but it is equally sensitive to signals from the front or back. This isn't really a very desirable pattern for television reception because of the possibility of multipath interference and ghosting mentioned earlier. A dipole antenna can be made more sensitive in a single direction by adding a reflector, as shown in Fig. 39-17. In a practical television antenna, there are several legs in addition to the two halves of the dipole itself. These additional elements are of two types—directing elements and reflecting elements. A directing element is slightly longer than the one-quarter wavelength dipole, and a reflecting element is slightly shorter than one-quarter wavelength. These directing and reflecting elements are placed a little less than one-

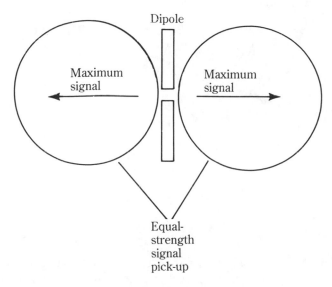

39-16 A dipole antenna has a bidirectional, figure-8 reception pattern.

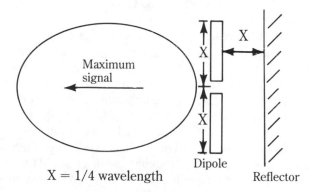

X = 1/4 wavelength

39-17 A dipole can be made more sensitive in a single direction by adding a reflector.

quarter wavelength from the dipole itself. These elements maximize the antenna sensitivity and make its directionality pattern as narrow as possible. This type of antenna is often called a Yagi antenna. There is considerable variation in the number and placement of the elements, but a basic Yagi antenna looks something like the one shown in Fig. 39-18.

Of course, two separate dipole antennas must be used for the two widely separated frequency bands used in television broadcasts—VHF and UHF. Sometimes entirely separate antennas are used for these two bands. Often, however, both dipoles will be mounted on a single unit for convenience and simplicity of installation. Naturally, because of the much higher frequencies involved, a UHF dipole is considerably smaller than a VHF dipole.

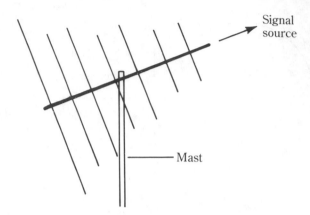

39-18   Basic Yagi antenna.

An antenna, by definition, deals with ac signals. Any practical electronic device that carries ac signals has some natural impedance. The exact impedance value will vary with the actual signal frequency, of course, but in practical usage, a rounded off, standardized impedance value is commonly given. Two standard impedance values are offered in television antennas—300 Ω and 75 Ω. Different types of wiring or cable is used for the two basic impedances.

A 300 Ω system usually uses twin-lead wire. The twin-lead wire is simply two separate wires bundled conveniently together with a common plastic insulation jacket. The two wires are spaced a specific distance apart, as shown in Fig. 39-19, separated by the insulation. Connections are usually made directly to a pair of screw terminals at the base of the antenna assembly and on the back of the television set. 300 Ω twin-lead is fairly inexpensive, flexible, and easy to work with. However, it is really only suitable for relatively short runs. It can be subject to RF pick-up and interference problems. In effect, the connecting wire between the antenna and television receiver acts like a secondary antenna itself. This adds to interference from unwanted signals, and lowers the gain and sensitivity of the antenna.

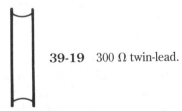

39-19   300 Ω twin-lead.

For long runs, or where interference is a problem, a 75 Ω *coaxial* system is used. For 75 Ω systems, coaxial cable is used. The signal conductor runs through the middle of the cable. It is surrounded by a thick layer of insulation. Then the ground wire is woven into a shielding jacket around this insulation. An additional layer of insulation is then placed around the shielding. A cross-sectional view of a coaxial cable is shown in Fig. 39-20.

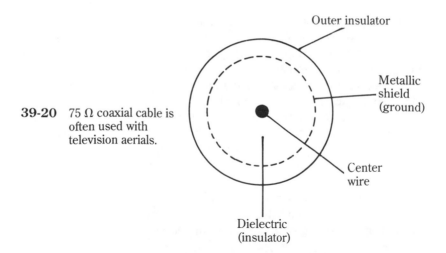

39-20 75 Ω coaxial cable is often used with television aerials.

Outer insulator

Metallic shield (ground)

Center wire

Dielectric (insulator)

Special screw-on connectors are used for 75 Ω television hookups. Do not use the 300 Ω screw terminals. Most modern television sets have a 75 Ω coaxial connector on their back panel in addition to (sometimes instead of) the standard 300 Ω screw terminals. If your television set doesn't happen to have a 75 Ω connector, you can use an external impedance transformer or *balun*. These devices cost just a few dollars, and are widely available from almost any electronics or television dealer or service shop.

Another add-on box that might be put in the antenna lead-in line is a splitter. Often, to simplify the physical wiring, both the VHF and UHF antennas will use the same lead-in cable. Before connecting the cable to the television set, you must separate these two signals again. Most television sets have separate inputs for VHF and UHF antennas. This separation can be accomplished with an inexpensive signal splitter. Often a signal splitter and an impedance matching balun will be built into a single unit.

In mounting a roof-top or other outdoor antenna, it is vitally important to be very safety conscious. Besides the obvious risk of falling, there is a very serious risk of electrocution, or at least a very bad shock, if there are any electrical power lines in the area. Unless you happen to have one of the relatively few homes that receives electrical power via underground cables, there will be overhead power lines to contend with. Yes, the power cables are insulated, but you are talking about huge amounts of power here. No practical insulator is perfect. To eliminate the electrocution or shock risk entirely would require several feet of insulation around each cable, which would be impossibly unwieldy. Remember, an antenna is made of metal, and it is therefore a conductor. If it touches an overhead power line, it will almost certainly be carrying deadly amounts of live current. The only solution to such problems is to use great care and common sense when working with an outdoor antenna. Take your time.

Every year, there are dozens of needless deaths and serious injuries related to the installation of outdoor television aerials. With the increasing usage of cable, the absolute numbers have dropped somewhat, but because you hear of fewer accidents of this type, it seems there is less safety awareness on the part of many of the remaining antenna installers. There appears to be an increase in the proportion of such needless accidents.

This problem is not the type of problem that can be solved by legislation, although there have been attempts. At least one bill has been introduced to Congress that would require adequate insulation on all receiving antennas designed for outdoor use. A nice idea, but obviously this bill was not written by anyone with any knowledge of electronics. Aside from the problem of "adequate" insulation—the antenna would have to be incredibly thick and bulky—such insulation would render the antenna totally useless as an antenna. It has to be conductive to the signals in the atmosphere in order to do its job. Incidentally, the inherent error was soon spotted, and this proposed bill didn't get very far.

If you install or otherwise work with any outdoor antenna, please be very careful. Always bear in mind that a moment of carelessness could cost your life. It's hardly worth taking any short cuts or unnecessary chances.

Cable television

It seems strange that just a decade or so ago, few homes were wired for cable television. Now, cable TV is rapidly becoming the norm, if it isn't already. With cable television, you don't need to bother with installing and aiming an antenna. The cable company takes care of that for you. Even stations some distance away that would be difficult or impossible to receive with an over-the-air antenna can be clearly received via cable. A cable television hookup typically offers the viewer of three to ten times as many stations as they can receive over the air in the same area. Many channels are available only on cable (or by satellite—see the next section of this chapter.)

Theoretically, a cable television hookup should give you perfect reception on all channels, without ghosts, interference, or other typical problems of broadcast reception. In practice, however, the cable company might be receiving a defective signal, or, more commonly, it's equipment might be out of alignment, so the system delivers than perfect signals to your home.

The cable company receives its signals from three basic sources. Local channels are received by a large master antenna. Larger cable companies will often have a separate antenna specifically tuned and aimed for each broadcast channel in the area. In principle the master antenna is just a deluxe version of an outdoor antenna used for television reception in a private home. A cable company can afford a larger and more efficient antenna than most individual television viewers. Often, there are strict regulations for antenna height in a private home, but a cable company will probably be permitted to mount a significantly taller antenna tower. The higher the antenna, the better the job it will do.

National cable channels, such as HBO (Home Box Office) and CNN (Cable News Network) are received via a satellite receiver. The originating station transmits a master signal to an orbiting satellite which then beams back the signal to receiving stations all over the country (and sometimes internationally). This system is similar to a home satellite television system, which will be described in more detail shortly.

Finally, most cable companies have their own studios to create their own local origin shows. Many states require cable systems to offer one or more public access channels. In most cable systems, the local studio space and facilities are small and rather limited. Still, they make it possible to produce unique, special interest programming (with little or no budget) that otherwise wouldn't get televised.

Programs from these three signal sources are assigned individual channels in the cable system. They are fed through the cable company's distribution system and on to the homes of their customers for viewing.

A simplified block diagram of a typical cable system is shown in Fig. 39-21.

Most cable systems use the standard broadcast frequencies for the VHF channels (2 through 13). Some local broadcast channels might be renumbered in the cable system. For example, local channel 5 might appear on cable channel 6. This channel shifting is done to prevent a strong direct reception signal from the channel 5 transmission antenna won't get mixed in with the slightly delayed channel 5 signal coming from the cable company's master antenna. If allowed to occur, this interference could result in severe ghosting in the displayed video images. The interfering over-the-air signal from channel 5 would be too weak to noticeably interfere with completely different programming coming through the cable system on channel 5. The interference is only critical only when both the cable channel and the broadcast channel are the same, with a minor phase shift between them.

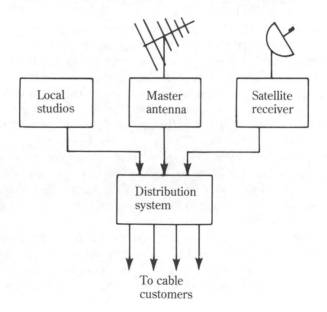

39-21 Block diagram of a typical cable television system.

Most cable systems do not carry UHF frequencies. Instead, additional channels over the 12 basic VHF channels (2 through 13) are carried on special cable-only midband or super-band frequencies. A cable ready television set can receive these nonbroadcast band channels directly via its built-in tuner. Otherwise, a separate converter box (supplied by the cable company) will be required to receive these channels. In some systems, the cable-only channels are identified by letters (channel A, channel B, channel C, and so forth), but it is more common to use standard numerical channel names, ranging from channel 14 up. It is important to realize that cable channel 14 is not on the same frequency as UHF broadcast channel 14. Theoretically hundreds of cable channels can be carried. Most practical cable systems today carry just two or three dozen channels.

Usually there is an attempt to avoid two broadcast stations in a given area from operating on adjacent channels. For example, if there is a channel 4 in your area, channels 3 and 5 are probably left unused. This minimizes the chances of adjacent channels from interfering with each other. This type of interference can be very effectively filtered out of a cable system, so adjacent channels are normally used. Most cable companies try to avoid leaving any blank channels.

Some are all of these cable channels might be scrambled. That is, they cannot be received properly without a special decoder box provided by the cable company. Even a cable-ready set will only display garbage from a scrambled signal. The signal is scrambled to discourage illegal theft of service. Scrambling is almost always done for special channels that require a special fee such as HBO or Showtime. If a cable customer only pays for basic service, he or she won't be able to receive these special pay channels. For an extra monthly fee, the cable company will supply the required decoder box so the pay channel can be received.

Many scrambling systems are in widespread use, and many are extremely sophisticated. There is no reason to go into the operational specifics of such scrambling systems here. In general, they all involve suppression or intentional distortion of the horizontal and vertical sync pulses in the video signal. When the receiving set can't find these sync pulses where it thinks they should be, it is unable to lock onto a stable picture. You'll just see a lot of distorted garbage and wavy lines on your television screen. The audio portion of the program might or might not be scrambled as well as the video signal.

Unfortunately, no security system is foolproof. Most cable subscribers will be stymied by the scrambling, but technical hackers seem to consider such a system a challenge to be defeated as a point of "honor." Every time the industry comes up with a new "unbeatable" scrambling system, someone else soon develops and often markets a decoder. These decoders are often sold in kit form to get through some loopholes in the laws. Strictly speaking, the sale of such devices is illegal. So is their use. The cable companies have found some very clever ways to locate and disable illegal decoders. If you are caught using one you could face a very stiff fine, and possibly a jail sentence. Still such methods of catching offenders are also far from perfect, so a great many people are willing to gamble.

For special pay channels, the argument for scrambling is very strong. Without scrambling there would be no way for these channels to make a profit, and therefore, they wouldn't exist at all, and nobody would get to watch them. Ordinary broadcast channels carries advertising that pays the bills for keeping them on the air. HBO, Showtime, and similar pay channels get their income from the viewers, who then get the advantage of not having to sit through countless sales pitches. What has fueled the debate over unofficial descrambling is that many cable companies have gone a bit overboard, and scramble basic cable channels too. Usually they only scramble some of their basic channels, which is very diffi-cult to defend logically.

Supposedly such scrambling is to discourage unauthorized people from tapping into the cable line without subscribing to the cable service at all. Some people try to defend such theft of service, the fact is, it is stealing, just as much as shoplifting in a store. The cable company makes its profits from selling its services. They have every right to expect payment for such services. Without such payment, they will eventually go out of business and everyone will have to go without their services. The question is, does scrambling the basic channels discourage the thieves? Almost definitely not. Anyone with the technical savvy to tap into the cable (at least without being crudely obvious about and getting caught almost immediately) will almost surely be quite capable of building or finding and hooking up an illegal descrambler.

What is really stupid is that many cable companies only scramble only some, but not all of their basic channels. If they are using scrambling to dissuade theft of service, why don't they scramble everything? Don't they care if the services of the unscrambled channels is stolen? Clearly, this is not why they are using scrambling.

In many areas cable companies can charge subscribers extra fees for renting them decoder boxes. Could some cable companies be scrambling some of their ba-sic channels to force basic subscribers to rent a decoder box whether they want a special pay channel or not? Doesn't this kind of business practice only encourage theft of service in the long run?

Besides the added expense of renting a decoder box, the basic subscribers are unnecessarily inconvenienced. If you have a cable ready television set or VCR (video cassette recorder), you won't be able to take advantage of it. Hopefully, the practice of scrambling basic channels soon becomes a thing of the past. It is unquestionably ineffective in preventing any significant degree of theft of service.

The scrambling issue becomes even more heated when you get to satellite tele-vision, which is discussed in this chapter.

Before signing on with a cable television service it is a good idea to find out just what is involved when any servicing is required on your hookup. Check with your local Better Business Bureau to find out if there are any serious complaints against the company. Almost any company, no matter how good, will receive a few complaints. If a lightning strike knocks the cable system off the air for three hours, someone will file a complaint, even though its not the cable company's fault, and they made the necessary repairs in record time. On the other hand, many cable companies have really lousy repetitions when it comes to service or billing prob-lems. Save yourself some grief and check it out before you sign anything. You might decide you'd do just as well without cable, especially in an urban area with many broadcast channels on the air.

Unfortunately, cable companies get away with a lot of sloppy business practice, because they usually have no competition within a given area. They compete with each other to get franchises, but once the Acme Cable Company gets the franchise in Typical Town U.S.A., the Brand X Cable Company won't be able to service homes in that area. So your only available choices are to deal with the Acme Cable Company or do without cable.

You should also consider just how much the cable hookup will really offer you.

many cable companies make a big deal of how many channels they offer, but absolute numbers of channels aren't too meaningful in themselves. Just what is on those channels? How many of them will you ever actually watch? In one area, the local cable company offered 38 channels, which sounded pretty good on the surface. But most families only watched about 12 of those channels. The 12 included the four broadcast channels in the area. The 38 cable channels included three home shopping channels (all advertising, uninterrupted by programming), and four classified ad/billboard channels that played music from local radio stations and displayed written notices, and/or the time. This cable company did not offer enough special programming to make their service worth the subscription fee for me. They also had a bad reputation for billing problems and errors, and for transmitting inferior signals. Apparently they did not properly calibrate their equipment often enough. All in all, many families are better off doing without cable. This discussion is included because many people today are technophiles. They want all the latest high-tech goodies whether they need them or not. Cable television seems almost like a status symbol in some circles. This is silly. If you get your money's worth out of the service, it's a good deal, if not, it's not. "Old-fashioned" broadcast television still works out just fine for many people. New is not always a practical improvement. Not even if everybody else seems to be buying it. The question is do you really want or need it?

Satellite television

A cable television system receives its national channel signals from a radio satellite. Couldn't individual viewers receive these satellite signals directly themselves? Yes. The necessary equipment is not cheap, but it is in the affordable range (at least, for individuals making good incomes), and it is commercially available.

A home satellite television system is called a *TVRO* (television receive only) system. This name is actually a lousy name, and not very informative. After all, an ordinary television set and a simple rabbit ears antenna could be said to make up a television receive only system. Nevertheless, this curiously inadequate name seems to have stuck. In usage, the term TVRO refers specifically to a system for receiving television signals from an orbiting RF satellite.

The television signals transmitted by the satellite do not originate aboard the satellite itself of course. They are beamed up from an Earth transmission station directed towards the satellite. The satellite are placed in a geosynchronous orbit. This means they hold a constant position with relation to the rotating Earth. The satellite is always at the same position in the sky. The originating Earth station beams the transmission signal up to the satellite, which shifts the signal to a different carrier frequency, and rebroadcasts it back towards Earth, as shown in Fig. 39-22. The high altitude of the satellite permits it to cover a much, much larger area than any transmission station on Earth. If the same signal is carried by two or three separate satellites, true world-wide coverage might be possible.

There are several special problems in receiving television transmissions from a satellite. These signals are relatively weak—typically just a few watts. The low power is necessary because an orbiting satellite must carry its own self-contained

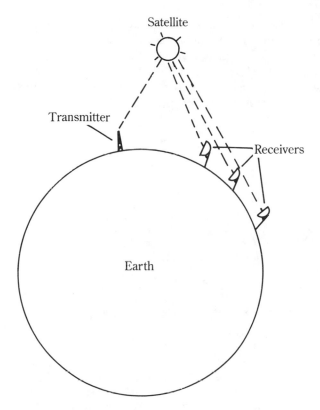

39-22 Satellite television can broadcast over very
wide geographical areas.

power supply. It can not afford to put out too much transmission power, or it would
be too heavy to be economically launched into orbit. A single satellite transmits a
dozen or more independent television channels, and its limited available power must
be divided among them.

The satellite transmissions are in the microwave region, with frequencies far
higher than UHF. Specialized circuitry is required to receive and utilize such high-
frequency signals. Even standard electronic components might not work. Specialized devices are required to work with microwave frequencies. Some of these
high-frequency devices are discussed in chapter 21.

In the TVRO system, these weak microwave signals must be captured by a
highly efficient antenna, amplified up to a usable level, then down-converted down
to a frequency range that can be recognized and used by the television set. For this
reason, TVRO systems are expensive, and they are likely to remain so. The required
antenna is quite large, and must be mounted outside. A small outdoor antenna is
simply out of the question.

The antenna for a TVRO system resembles a large dish set on edge, with a
center protrusion at the focal point, as shown in Fig. 39-23. The dish itself is actually
a reflector, and the central protrusion, called the "feedhorn assembly," is the antenna

proper. The reflecting dish focuses as much of the received energy to the antenna at the feedhorn assembly at the focal point as possible. Compared to an ordinary television antenna, a satellite dish is very highly directional, and must be very precisely aimed at the desired satellite. All TVRO satellite dishes are mounted on precision rotor bases for accurate positioning. It is necessary to physically re-aim the antenna to receive a signal from a different satellite.

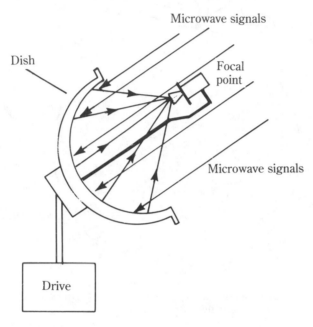

39-23   A special dish antenna is used to receive satellite television signals.

A drive assembly is used to physically aim the dish at the desired satellite. This assembly is a somewhat more sophisticated version of an ordinary antenna rotor. In the case of the satellite dish, precise, controlled motion is needed in all three dimensions. The satellite must be located along three planes—north-south, east-west, and up-down. Depending on the location of the receiving station on Earth, the satellite might appear to be higher or lower in the sky, even though its actual height, with respect to the surface of the Earth, is a constant. This effect comes from looking at the satellite from Earth at an angle.

In most TVRO systems the drive assembly is motor driven. In some systems, however, to keep the costs down a little, a manual drive assembly might be used. The user must turn several cranks or other controls to adjust the positioning of his antenna. Even with a cost-cutting manual-drive assembly, a TVRO system is still a fairly expensive item. It is difficult to cut corners in such a system and have it function at all, much less give good performance. The weak microwave video signals picked up from the dish antenna are then fed through a low-noise amplifier. The specifications for this amplifier must be very tight and precise. You are dealing with very weak, wide bandwidth signals with extremely high carrier frequencies (in the microwave region). A slight imperfection or inaccuracy in this amplifier, or even a tiny amount of noise will have a very drastic effect on the video signal when it is eventually displayed.

The amplified microwave signals are then fed through a down converter circuit, which readjusts them to a lower frequency the television receiver can handle. The received microwave signals are dropped down into the VHF band. As far as the television set is concerned, it is picking up the signal from an ordinary VHF broadcast station.

For the most part, these satellites were not originally intended for direct reception by individuals. Their primary purpose was to carry national channels (such as "super stations") to cable companies, and for network feeds to their affiliate stations across the country. But there are now more than three million TVRO systems in operation, so some special satellite-only broadcast channels have appeared.

Signal scrambling has been a bit of a political hot potato for satellite television, as well as for cable television. For example, the networks don't particularly like having the general public tap into their unedited news feeds. Of course the cable companies don't want viewers to be able to receive the same nonbroadcast channels they offer without subscribing to their service. Several bills were suggested to Congress, but they essentially side stepped the issue and left it to the open marketplace. It probably makes good sense to scramble a channel that requires subscription fees to cover its production and transmission costs. But if the channel in question carries advertising, then those advertisers are already paying the costs of operation. Demanding a subscriber fee in addition to advertising dollars is just plain greedy. It is surprising that more advertisers have not complained about such practices. After all, an advertiser wants his commercials seen by as many people as possible. Scrambling and requiring viewer subscriptions, by definition, limits the number of viewers.

There will inevitably be some hot court cases stemming from unauthorized descrambling of satellite channels in the near future. The legal issues are still too foggy and imprecise. There is also more of a problem in enforcing regulations against unauthorized descrambling in a TVRO system. A cable company can use sophisticated equipment to tell what is hooked up to their transmission cables. But a satellite just beams out its signals blindly, and there is no way to determine who is receiving those signals, much less how.

The only reason that this issue hasn't broken out into major court cases and a media circus is that the inherent expense of TVRO systems automatically limits the number of people involved. Because the costs of TVRO systems are hardly likely to come down to popular pricing in the foreseeable future, the legal and moral issues will almost certainly remain fuzzy and unresolved for some time to come.

A lot of the high cost of TVRO systems is built into them almost on the theoretical level. With any new technology, an initial high cost pays for the research and development costs, then prices drop drastically. The early commercial electronic calculators cost $150 and up. Today you can buy a pocket calculator that can do more than those early models for under five bucks. VCRs, CD players, and personal computers all started out prohibitively expensive for most of us, and now they are quite affordable, and more powerful and versatile than the earlier (and now obsolete), much more expensive models. You will probably see this effect to some degree with TVRO systems, but the precise mechanical mechanisms of the dish antenna and its drive system, coupled with the tight specifications of the required microwave circuitry will almost certainly keep construction costs quite high, even after all the basic research and development is paid for. It is very unlikely that TVRO systems will ever be economically competitive with ordinary broadcast television receivers, not even to the extent that color television became cost effective with respect to black-and-white television. TVROs will probably remain a high-ticket luxury item.

Of course, there is always the possibility of an unexpected technological breakthrough that throws all predictions into a cocked hat.

High-definition television

There has been a lot of talk in the media and in technically oriented circles about the coming high-definition television (*HDTV*) system. HDTV provides a larger picture area with much greater detail than ordinary television. There is no question that it is technically superior in all respects.

The question no one seems to be asking is will the general public particularly care about the improvements? There is a good chance that HDTV will lay an egg in the marketplace. It will hold enormous appeal for a small group of technophiles and video connoisseurs, but the general public has not shown much concern over picture quality in the past. Although there have been many complaints about the inferior signals presented on some cable systems, the vast majority of subscribers to these same systems never complain, or even to notice the problem in the first place. Most people, it seems, don't just sit and watch television. They go on about their business with the television on. They do not focus their full attention on the picture, so a super high quality image just isn't all that important.

When VCRs started to become enormously popular, there was a major market war between competing (and incompatible) formats, primarily Beta and VHS. Despite the fact that most technical experts agreed that the Beta format offered better picture quality, VHS won the format wars hands down. VHS won, it seems, because of slightly longer recording times. The Beta videotape format is essentially obsolete today. It is becoming increasingly difficult even to find blank tapes for old Beta machines. (VCRs are discussed in the next chapter.) Clearly, the best possible video image was not a high priority for the general public.

If HDTV does manage to get off the ground at all in this country, it will remain a high-end luxury variation. There is almost no chance that it will ever phase out and replace the current standard television system. (Some other development might do this, but not HDTV).

Despite all the hoopla, HDTV looks like a commercial dead end in this country. Demonstrations of HDTV have been very impressive. It questionable whether the general public will be inclined to pay extra for the advantages of the HDTV system, so the market will be severely limited.

Self-test

1. What is the purpose of the yoke around a television picture tube?

A To correct for inaccurate color reproduction
B To deflect the electron beam to the desired point on the screen
C To hold the picture tube in place
D To generate high-voltage pulses
E None of the above

2. How many complete lines make up a full frame in the U.S.?

A 525
B 60
C 262.5
D 15,750
E None of the above

3. How many frames are displayed per second?

A 1
B 60
C 262
D 30
E None of the above

4. What is the horizontal frequency in a television set?

A 262.5 Hz
B 15,750 Hz
C 60 Hz
D 3.58 MHz
E None of the above

5. How many electron guns does a color television set have?

A One
B Two
C Three
D Four
E None of the above

6. What is the approximate bandwidth of a broadcast television signal?

A 15,750 Hz
B 6 MHz
C 3.58 MHz
D 4.75 MHz
E None of the above

7. What is the highest available channel in modern televisions in the U.S.?

A 82
B 69
C 128
D 99
E None of the above

8. What is represented by the most negative voltage in a television signal?

A The whitest portions of the displayed image
B The blackest portions of the displayed image
C The sync pulses
D The program audio
E None of the above

9. What is the approximate frequency of the color subcarrier?

A 6 MHz
B 4.75 MHz
C 3.58 Mhz
D 1 MHz
E None of the above

10. What is the impedance of coaxial cable used with television aerials?

A 100 Ω
B 300 Ω
C 50 Ω
D 75 Ω
E None of the above

40

VCRs

One of the big success stories in modern commercial electronics is the *VCR*, or *video-cassette recorder.* This device is a special tape recorder that can record the audio and video signals from a television program so they can be played back at a later time. It is a wonderful invention for people who enjoy the old movies that are usually only broadcast at 3:00 in the morning. You can program your VCR to record the late, late, late show while you're sleeping, and then you can enjoy the movie later at some more convenient time. Many VCR owners also enjoy renting (or sometimes buying) recent movies and other programming on prerecorded videotapes and then watching them whenever they feel like it.

The VCR frees you from anybody else's schedule. You are no longer at the mercy of the program directors at the broadcast stations. The VCR can watch whatever he wants to watch, when he wants to watch it. The VCR is clearly an idea with great appeal to the public. According to recent statistics, more than half of all homes with television sets (which accounts for the vast majority of American homes today) also have at least one VCR. Here is a case of a new technological gadget that met a definite need for the general public.

VCRs actually weren't quite the overnight success they might have appeared to be. Videotape recorders had been around for some time, although in somewhat different form. Of course, the first videotape recorders were professional machines for use in broadcast television stations. With videotape, programs no longer need to be aired live (risky, inconvenient, and difficult to repeat) or on film (expensive and time consuming). A videotaped program can be aired at any convenient time and can be repeated as often as the station desires.

There were a few early attempts to market home videotape recorders, but they didn't catch on. Open-reel tapes were used in these early videotape recorders, and several controls had to be precisely adjusted. Operating such a machine was difficult and awkward at best. They were also outrageously expensive, and they bombed on the marketplace. These attempts were quickly forgotten.

Meanwhile, the audio cassette tape had been developed and was doing very well in the marketplace. The general public seemed to love the convenience of the enclosed cassette, freeing them from the minor nuisance of threading the tape onto reels. Several manufacturers more or less simultaneously came up with the brilliant idea of enclosing the videotape in an easy-to-use cassette housing. Adding the cassette housing was the little touch seemingly required to make the home videotape recorder a viable commercial product. It also didn't hurt that recent advances in technology had lowered the costs on the components used in the fairly complicated electronic circuits required to record and playback video signals.

The first commercial VCRs were crude and expensive by today's standards. A very basic machine sold for about $1000. Today you can buy a VCR for less than $200, and it will probably have a number of features that were not included in the older, more-expensive models. The high initial price tag of any new product is necessary to cover the costs of research and development. Once all the preliminary research work has been paid for, and the item is selling in quantities that make large-scale manufacturing worthwhile, the costs can plummet. The same thing has happened with personal computers and CD players in recent years.

The VCR format wars

The basic technical concepts behind all VCRs are pretty much the same, although there are differences in the actual circuitry used, of course. But several VCR formats were offered, and they were mutually incompatible—not on the theoretical level, but on the mechanical level. They used different sizes and shapes for the tape cassettes, different drive systems, different tape widths, different speeds, and so forth. A videotape recorded on a machine of one format simply can't be played back on a VCR of a different format. As a matter of fact, the cassette would not fit into the slot on a machine of a different format.

The first major VCR format was called *Umatic*, and you've probably never heard of it. The Umatic machine was marketed mainly for semiprofessional use, such as in schools and industrial shows. The Umatic format is now long gone, but it played a major role in paving the way for more popular VCR formats.

The two major contenders in the VCR format wars were Beta and VHS. Beta got a head-start on VHS, and for a long time it looked like it would be the eventual winner. Sony developed the Beta, or Betamax format, and their VCRs appeared on the market well before any VHS VCRs were sold. The VHS format was developed by JVC (the Victor Company of Japan). The technical specifications for the Beta system were somewhat better than for the VHS system. For equal quality machines, a Beta VCR would reproduce a better picture than a comparable VHS unit. But the improved video images did not seem to be all that important to the general buying public. Tape running times seemed to be the most critical feature, and VHS always managed to have an edge here—not an enormous difference, but a definite difference did exist.

A standard Beta videocassette held about 500 feet of videotape and could hold a one-hour program. The short tape was certainly a lot better than no recording capability at all, but it was decidedly limited for many programs, such as full-length movies. The VHS cassette is a little larger physically than the Beta cassette, so it can hold more tape—about 800 feet. The basic VHS speed permitted up to two hours of program time on a single cassette. This timing advantage was obvious, and Beta manufacturers quickly came out with a second Beta speed that permitted up to two hours of programming on the same basic Beta cassette. The original speed was called *Beta I*, and the new speed was called *Beta II*. The two formats were now on equal ground as far as tape running times went.

But then the VHS manufacturers used the same basic techniques to develop a second speed of their own. The original VHS speed was called *SP* (standard play), and virtually all prerecorded VHS tapes are recorded at this speed, which produces the best video image quality. The new, slower speed was dubbed LP (long play), and it permitted up to four hours of programming on the basic VHS cassette. Once again, Beta was at a serious marketing disadvantage. So they came up with a Beta III speed, which could record about three hours of programming on the same 500-foot video-cassette. VHS was still in the lead with a maximum LP recording time of four hours to Beta III's three hours, but the VHS manufacturers went ahead and added a third, slower speed of their own anyway, to clinch the issue. This slowest VHS speed is called *EP* (extended play) or SLP (super long play). The only difference here is the name. EP and SLP are the exact same speed. Up to 6 hours of programming can be recorded on a basic VHS videocassette using this speed.

In practice, the VHS LP speed has fallen into disuse, because it really has no particular advantage of its own. Users who want the best recording quality use the fastest SP speed, and users more concerned with tape economy (running time) use the slowest EP speed, which does not give a noticeable degradation in performance from the LP speed. (Although there is some noticeable image degradation when a LP or EP recording is compared with a SP recording of the same program source.)

Both Beta and VHS formats eventually offered extra-length tapes. The cassettes were the same external size, but slightly thinner tape was used, so more tape (a longer running time) could be held inside the standard cassette housing. The standard Beta video cassette was called L-500 and held about 500 feet of tape. A longer L-750 cassette had about half again the running time of the L-500 cassette. That is, on the Beta I speed, the L-500 could record about one hour, and the L-750 cassette could hold about one and a half hours. At the slowest Beta III speed, the standard L-500 cassette could hold up to three hours. The L-750 cassette could record up to four hours. An even longer L-830 was manufactured for a while. It could hold up to five hours at the Beta III speed, still falling short of the VHS format maximum recording time.

The Beta videocassette, of whatever tape length, is about the size of a paperback book, measuring $6.1 \times 3.8 \times 1$ inches. The VHS videocassette is a little larger than the Beta cassette, measuring $7.4 \times 4.1 \times 1$ inches. The standard VHS videocassette is labelled T-120 for its 120-minute (two-hour) recording time at the fastest (SP) recording speed. This same tape could record up to six hours (360 minutes) at the slowest EP/SLP speed. A VHS cassette with a longer (thinner) tape is also available.

This longer tape is the T-180 cassette. As you might suspect, it can record up to 180 minutes (an hour and a half) on the SP speed. At the slowest EP/SLP speed a T-180 videocassette can hold up to eight full hours of programming. Recently there has appeared some T-130 VHS cassettes that offer a little more recording time than the standard T-120 cassettes, but not as much as the T-180 cassettes.

Beta came up with a new, improved version of its format called *SuperBeta*, which offered better recorded image quality and other operational improvements. The VHS manufacturers countered with similar (but somewhat lesser) improvements of their own in the form of VHS HQ. The HQ stood for high quality. SuperBeta was technically superior in many ways, but the buying public didn't seem to particularly care.

For complex reasons no one really understands, VHS eventually emerged as a clear, hands-down winner over Beta in the format wars. Today, although some Beta equipment is still in use, the format is essentially dead. It is very hard to find prerecorded or blank videotapes in the Beta format.

One thing that is clear is that the longer recording times of the VHS format were an important factor in the marketplace. The general public seemed more interested in tape economy and running times than in technical specifications. Perhaps the VHS manufacturers had a clearer view of the marketplace right from the start.

Technical-oriented Beta labelled their videocassettes by the tape length contained, and VHS cassettes are labelled to directly indicate their running time. T-120 (time = 120 minutes) is more meaningful to the general consumer than L-500 (length = 500 feet).

Because the Beta format is now obsolete, you won't need to pay much attention to its technical details, except to note they were basically similar to the VHS format in general principle. Now look into the VHS VCR format in a little more detail.

The tape-drive system

The videotape in a VCR is entirely enclosed inside the plastic cassette housing. You never touch or contact the tape itself. This isolation helps minimize the possibility of dirt and contamination. In operation, some of the tape is automatically drawn out of the cassette and fed through the VCR drive system, so it can be moved past the record/playback heads at a constant speed. Because of the enormous quantities of information required to store and reproduce a video image adequately, a high tape speed would seem to be required. A videotape system would obviously require much greater speeds than in an audio tape recorder. But this speed would result in either unacceptable tape economy (maximum playing time) or a ridiculously bulky and oversize tape cassette. To solve the problem, the tape itself is made to move at a very slow rate, and the record/playback heads are mounted in a revolving drum. The drum rotates at a high speed in the opposite direction as the tape motion past it. The result is a high relative heads-to-tape speed. The tape doesn't care whether it is moving or the record/playback head is moving. In a VCR, both move.

Inexpensive VCRs usually have two record/playback heads mounted on opposite sides of the revolving head drum, as shown in Fig. 40-1. Better quality machines have four head drums, as shown in Fig. 40-2. Four heads means there is less time

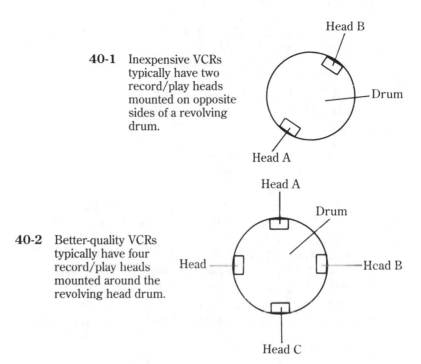

40-1 Inexpensive VCRs typically have two record/play heads mounted on opposite sides of a revolving drum.

40-2 Better-quality VCRs typically have four record/play heads mounted around the revolving head drum.

per frame when no head is in actual contact with the tape, so there is better quality in the reproduced video signal. For good special effects, such as slow motion or freeze frame, a four-head VCR is usually required. When the videocassette is inserted into the VCR and the play or record button is depressed, tiny mechanical fingers open the door to the cassette, and some of the tape is withdrawn and fed through a system of capstans and pinchrollers. It is wound about half-way around the diameter of the head drum, as shown in Fig. 40-3. When the stop button is pushed, the VCR's mechanical drive system winds all the tape back into the cassette and close the door. To remove the cassette from the VCR, the eject button must be used. This button tells the electromechanical system of the VCR to shove the cassette far enough out of the machine so that it can be manually removed by the user.

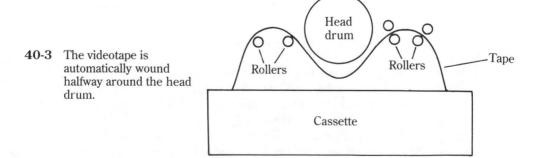

40-3 The videotape is automatically wound halfway around the head drum.

Only the video record/playback heads are mounted in the revolving drum. The audio record/playback head, and the erase head are mounted separately in fixed positions. The slow relative tape-to-head speed doesn't matter as much with these functions, so there is no sense in making the VCR internal mechanisms any more complicated than they have to be by rotating the audio and erase heads. A third stationary head is used to record and playback the control signals along the edge of the tape. These signals are used for synchronization and timing. During playback, the recorded control track also tells the VCR which speed the tape was recorded on. A VCR will not let you play back a tape on the wrong speed. There would be no reason to do so. In an audio tape recorder, it is sometimes amusing to playback a tape at a speed different than the recorded speed. This speed change can also be done to achieve certain special effects. But in a videotape, changing the speed will throw off the horizontal and vertical sync pulses to an unusable degree. Only meaningless garbage would be displayed.

In virtually all modern VCRs, digital circuitry is used to control the basic drive functions—record, play, rewind, fast forward, pause, stop and eject. The tape could be damaged if it was suddenly switched from play to rewind, for example, without momentarily stopping in between. On a VCR, you don't have to worry about forgetting to hit the stop button when changing functions. Depending on the design of the particular VCR in question, either the digital control circuitry will automatically activate the stop function before initiating the new drive function, or (less commonly) new drive commands will be ignored until the current function is stopped.

On many machines, if you hit rewind (or fast forward) during playback, the playback speed will be sped up. The audio is usually turned off in this mode. You can scan through the tape to find a specific point, then return to ordinary playback.

Because digital control circuitry is used in a VCR, it can easily be made programmable. All VCRs have built-in timers or clocks. The VCR can be programmed to turn itself on and activate the record mode at a specific time. It will also automatically adjust its tuner to record the programmed channel. Then the machine will turn itself off at a second preprogrammed time. This way, the VCR can automatically record a program for you when it is not convenient for you to do so yourself, such as in the middle of the night or when you're not at home. With a VCR, you don't ever have to miss your favorite program. Just program the VCR to record it when it is aired, and then you can play back the tape at some later time, at your convenience. This recording and replaying is often called *time shifting*.

Most modern VCRs permit you to preprogram several different recording times over a given period—usually two weeks to one year, depending on the specific design of the VCR in question. Usually there are special provisions for recording daily or weekly programs, without separate programming instructions each time the desired show airs.

There are seemingly countless different approaches to programming a VCR. There are wide variations from manufacturer to manufacturer or even from model to model in a single manufacturer's line. If you've learned how to program VCR A, don't count on the programming VCR B working the same way. This lack of operational standards bothers many consumers who are not technically oriented, and many feel so intimidated; they never even learn how to program their VCRs at all. It is a real shame, because without the programmability features, the VCR owner just isn't getting his or her money's worth. To make matters worse, many manufacturers make the programming process unnecessarily complicated and confusing, and VCR manuals have a well deserved reputation for being badly written (or badly translated from a Japanese original). It is necessary to take your time learning to program a new VCR. Work slowly, step by step. Make sure you understand each step before moving on to the next.

A fairly recent product has appeared to answer such difficulties. It is a universal automated programmer. It is designed to work with the wireless (infrared) remote-control input of most modern VCRs. *TV Guide*, and many other TV program listing publications will give a number for each program. For example, the series *Quantum Leap* airs Tuesday night from 8:00 to 9:00 pm this week. The *TV Guide* gives the programming number as 88505. To program the VCR to record this show with the universal program, all you have to do is punch the number 88505 on a calculator-like keyboard. The universal programmer does the rest. The VCR will automatically tune in the local NBC station at 8:00 pm Tuesday evening. At 9:00 pm, the VCR will automatically turn itself off (unless it has also been programmed to record something else beginning at 9:00 pm). This device should sell very well and help eliminate the intimidation many people feel about learning to program their VCRs.

How signals are recorded on videotape

In an audio tape recording, the signal is laid onto the tape in a perfectly straight-forward and linear manner, as shown in Fig. 40-4. Shown is a full-track recording. It is monophonic (all sound comes from a single speaker), and there is no side 2 to the tape.

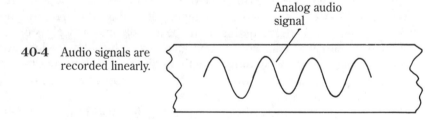

40-4 Audio signals are recorded linearly.

Analog audio signal

Video signals require much greater bandwidths than audio signals. The greater bandwidth means one second of recording time must take up more space on the tape. A high-fidelity audio recording only needs a bandwidth up to about 20 kHz (20,000 Hz), because that is the highest frequency that can be heard by the human ear. Signals with higher frequencies don't need to be recorded, because they would not be audible on playback anyway. (Some audio enthusiasts insist that ultra-audible frequencies can influence the perceived sound quality under some conditions. Double-blind tests have not done much to support such claims, but even if they were true, there still probably wouldn't be much point in an audio tape recorder recording frequencies much above 50 kHz or so). But, as you saw in the previous chapter, a full video signal requires a minimum bandwidth of at least 6 MHz (6,000,000 Hz). This bandwidth is a considerably larger bandwidth than any audio recording will ever encompass. It's not reasonable to expect to cram a 6 MHz bandwidth (video) signal into the same physical space as a 20 kHz bandwidth (audio) signal. More tape area is required for a video recording than for an audio recording of comparable duration.

In less critical audio recording applications, it is often acceptable to use a slower tape speed, which by definition results in less space for the signal (of a given duration) on the tape. Less space reduces the bandwidth of the reproduced signal. Many people can't hear the high frequencies that are lost with the slower tape speed and even people who can hear the difference might not care in certain applications, such as voice recording or background music. But to restrict the bandwidth of a video signal will result in severe degradation of the reproduced image. A minor reduction in the video bandwidth will result in a loss of picture detail. A moderate reduction in the video bandwidth could easily reduce the entire picture to meaningless mush.

Because a video signal requires more physical tape space than an audio recording of comparable duration, a practical videotaping system must find some way to provide this extra required tape area.

There are two obvious approaches that could theoretically work, but neither of them would be a very practical choice in a real-world product, at least not by themselves. You could use a wider tape and/or a faster tape speed. A wider tape sounds good until you think of the degree of increase required.

If a quarter-inch full-track tape at 15 ips (inches per second) adequately records a full 20 kHz bandwidth, how wide or how fast must a comparable videotape be? The video bandwidth is 6 MHz, which is 300 times the 20 kHz bandwidth of the audio system. Theoretically, if you just increased the width of the tape, you would need a tape that is 75 inches (6.25 feet) wide. Obviously, this would not be at all practical. If you used 1/4 inch tape, the video bandwidth would require a tape speed of at least 4,500 ips, or 375 feet per second. A one minute recording would require no less than 22,500 feet of tape. Once again, this would be a wildly impractical recording system.

Even if you try to compromise and increase both the tape width and the tape speed, you would end up needing something like a 37.5 inch wide tape running at 2250 inches per second. This would hardly be an improvement as far as designing a practical video recording system goes.

Well, the problem was obviously solvable, because practical VCRs do exist. Moreover, the tape used in the VHS system is only one-half inch wide, and it runs at a very slow speed. At the fastest SP speed, the tape moves at a rate of 1.31 ips. The second LP speed moves the tape at 0.66 ips. On the slowest EP or SLP speed, the tape creeps along at a speed of only 0.44 ips. How can this possibly work? How can the required 6 MHz video bandwidth possibly be squeezed onto so little tape?

Two tricks are used in VCRs to cram more actual signal into less space on the tape. One of these tricks involves moving the heads. The signal doesn't care if the tape is moving and the record/playback head is stationary, or if the tape is stationary and the record/playback head is moving. All that counts is the relative tape-to-head speed—that is, how fast are these two components moving with respect to one another? It wouldn't be very practical to have a totally motionless tape and a head that moves along its full length, but on the theoretical level, it would work just as well for recording and playing back the signals.

In a VCR, both the tape and the heads are moving. The tape itself moves at a fairly slow absolute speed. Two or four record/playback heads are mounted in a drum assembly, which revolves very rapidly. The drum moves the head with respect to the tape. The actual speed of the tape motion is effectively added to the head drum rotation speed, so the signal being recorded or being played back thinks the tape is moving at a much, much higher speed than it actually is.

To further conserve space on the tape, the video signal is not recorded linearly, as in an audio tape recorder. Instead, it is laid down in a series of discrete, angled strips or tracks, as shown in Fig. 40-5. This drawing shows a video signal recorded at the fastest available speed (SP). The record/playback head has a width of 38 microns, so each individual track is 38 microns wide. Adjacent tracks are separated by a narrow guard band. Each track contains the video data for one field.

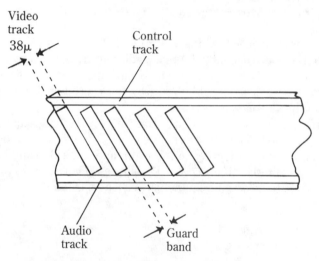

40-5 Video signals are recorded in discrete tracks. At the SP speed, there are guard bands, or narrow blank spaces, between adjacent tracks.

A narrow strip along the bottom edge of the tape carries the audio signal for the recorded program. The audio signal is recorded linearly, as in a standard audio-only tape recorder. Because this audio band is very narrow, the bandwidth of the reproduced audio signals is restricted, but this generally is no particular problem. Broadcast television signals also restrict the bandwidth of the audio signal, and most television receivers are far from high fidelity devices. Usually just a small speaker is used to reproduce the audio portion of a television program. In most cases, you are mainly concerned with reproducing spoken voices, so a restricted audio bandwidth works out just fine. High fidelity sound would rarely add to the enjoyment of most television programs. There are some exceptions, and a few high-fidelity television sets have appeared on the market. A few VCRs use special recording and playback tricks to improve the audio quality. Such devices are pretty rare, and are unlikely to ever become dominant in the general marketplace, so you don't need to get into specific details.

A second linearly recorded track is placed along the upper edge of the video-tape. This track is the control track and it contains the synchronization and timing signals the VCR needs to reproduce the video image properly. The playback heads have to be correctly aligned with the recorded tracks. If the heads contact the tape on the separating guard bands, most of the recorded signal won't be retrieved for reproduction of the picture.

The control band signals are pretty simple, and don't require much bandwidth at all. There is room for the later addition of special control features. Some VCRs include data on the control track for indexing, so a specific point on the tape can be quickly and easily located.

There are also narrow guard bands between the control and audio edge tracks and the angled video tracks in the center of the tape. The guard bands prevent interference problems between the various types of recorded signals (video, audio, and control).

At the slower tape speeds (LP and EP or SLP), the guard bands between the adjacent video tracks are eliminated. In fact, adjacent tracks now overlap slightly, as shown in Fig. 40-6. The overlap is 9 microns, leaving 29 microns of clean track. The effectively narrowed video track width is why the reproduced image is less sharp and detailed on the slower speeds.

To help minimize distortion effects from the overlapping adjacent tracks, azimuth recording is used. Alternate tracks are offset 6 degrees in opposite directions. That is, track A is angled 6 degrees to the right, and track B is angled 6 degrees to the left, for a total angular difference of 12 degrees. There is also an inevitable increase in the signal-to-noise ratio on the slower tape speeds.

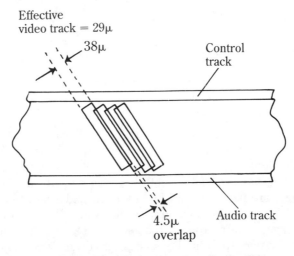

40-6 At the LP or EP speeds, the adjacent video
tracks overlap slightly, with a difference in
the azimuth angles.

VHS-HQ

The original VHS format as designed primarily by JVC was good, but hardly perfect. Beta always had a slight edge in image quality over the VHS format. When Sony came out with the improved SuperBeta subformat, the VHS camp felt a need to respond.

There were growing numbers of videophiles who demanded the best possible picture quality. Although far from the majority of consumers, if the serious videophiles defected en masse to Beta it would have been a significant blow against VHS in the format wars. In addition, as advertisers know, "New and Improved" sells in the general marketplace, even when it isn't too clear to the general consumer just what the improvement is or why it matters. It's largely the appeal of having the latest and what is officially dubbed "the best." JVC's response to Sony's SuperBeta subformat was a subformat of their own called *VHS-HQ*. The HQ, of course, stood for high quality.

An important feature of a successful subformat was that it be compatible with older machines and tapes. The format wars were already bad enough. Neither Sony nor JVC (or any of the manufacturers licensed to use their formats) wanted to leave their existing customers out in the cold by making their own formats obsolete.

The VHS-HQ system was much more flexible than SuperBeta. In fact, the definition of the subformat bordered on being vague or even meaningless. One VHS-HQ machine might be very different in its improvements than another VHS-HQ machine. In other words, the label VHS-HQ didn't exactly tell consumers what they were getting, although some improvement was promised.

Overall, the VHS-HQ subformat offered improved picture quality as a result of as many as five separate modifications in the circuitry. Not all of these circuit modifications are required for a VCR to be licensed as a VHS-HQ unit.

As a minimum definition, JVC defines HQ as a package of three improvements in the reproduced video signal. The goal is reduce the visible video noise and increased the apparent contrast of the reproduced image. Specifically, these HQ improvements are:

Luminance signal (video) noise reduction
Chrominance signal (color) noise reduction
Extension of the white clip level (WCL)

To be licensed to carry the HQ symbol, machines should have the circuitry to carry out all three of these improvements. Matters were complicated somewhat because some models that were released just before the official announcement of the VHS-HQ subformat might contain some or all of the HQ circuits, without being labelled as such. Today this isn't a real problem. Except for a few low-end models, most VHS VCRs on the market today are VHS-HQ machines.

The VHS-HQ system is designed to be fully compatible with the original VHS format. A tape recorded on a standard VHS machine will play back just fine on a VHS-HQ unit, and vice versa. Of course, to achieve the promised HQ improvements, the tape must be both recorded and played back on a HQ machine.

The VHS-HQ subformat also defines two additional video noise-reduction circuits, which are optional. A machine marketed as a VHS-HQ unit might or might not have these extra improvement circuits. The purpose of the required noise-reduction circuits in the VHS-HQ subformat is pretty straightforward. One circuit minimizes noise (unwanted and usually random signal components) in the basic black-and-white (or luminance) signal. The second HQ circuit minimizes noise components in the color signal. Reducing the noise results in a sharper and cleaner image, with fewer stray streaks and speckles.

The recorded video bandwidth is not extended, as was done in SuperBeta. This limits the actual improvement in image detail that is possible in the system, but it ensures full compatibility with older non-HQ VHS equipment and tapes.

The third basic HQ improvement—extension of the white clip level (WCL)—might require a little more explanation. The white clip level is boosted by about 20% in HQ machines. The recording signal is amplified or boosted to sharpen the edges of the image in the picture. There is less bleedover when there is a sudden dark-to-light (or light-to-dark) transition in the picture.

In a standard VCR, the video signals are *preemphasized* (that is, high frequencies are boosted) before modulating the luminance (black-and-white) carrier. Complementary deemphasis during playback will theoretically reduce any noise components introduced by the recording process. (There will always be some noise added in any analog recording system.) However, on a sharp full-black to full-white transition, the preemphasis can cause the signal to overshoot the full-white level corresponding to 100% video modulation. More than 100% modulation of any signal will cause severe distortion of course. To prevent overmodulation of the video FM carrier, a clipping circuit trims off the overshoot. As a result of this clipping, the picture no longer appears as it originally did, after playback demodulation and deemphasis. The transition from black to white starts off rapidly, but before reaching the full-white level it slows down, which slightly smears the edges of the reproduced images. Raising the white clip level, as in the HQ system, reduces the high frequency loss and sharpens the black-to-white transitions. The higher white clip level enhances the apparent contrast and detail of the reproduced picture.

For purposes of video noise reduction, the HQ system basically capitalizes on the presence of redundant information in a normal video picture signal. As you should recall from the previous chapter, in a television monitor the video image is formed by scanning an electron beam across the inner phosphor-coated surface of the picture tube screen. If you look at the screen very closely, you can see that the picture is composed of a series of almost (but not quite) horizontal lines stacked up vertically to give the illusion of a continuous image.

JVC's engineers realized that in most scenes the information changes very little from one horizontal scan line to the next, at least not when the adjacent lines are compared point by point. That is, if a given point on a given line is dark, the points directly above and below it will probably be dark too.

In the VHS-HQ subformat, a CCD (charge-coupled device) is used to delay the video playback information by one horizontal scan line. This delayed data is added to the data from the next incoming horizontal line. At all points where the two lines are the same (correlated), the video signals add directly and double in level. A doubling in level is an increase of 6 decibels (dB). Any noise content of the signal is random (uncorrelated), by definition. Therefore, after the addition, the noise signal will increase by only 3 dB (the square root of the sum of the squares of the instantaneous amplitudes of the signals). Adding adjacent horizontal scan lines in this manner theoretically produces a net improvement of about 3 dB in the luminance S/N (signal-to-noise) ratio. Actually, the procedure is somewhat more complicated than this, but this simplified explanation should give you a good general idea of how the system works. According to JVC, the actual improvement of the S/N ratio is even better than 3 dB.

In a similar procedure, the VHS-HQ chrominance (color) vertical processor correlates the color information line by line to create a 3 dB improvement in the chroma S/N ratio.

A VHS-HQ VCR will reproduce more apparent detail in the picture than a comparable standard VHS machine. To a large extent, the improved detail in the HQ picture is essentially an illusion. The recorded bandwidth is not recorded. Both standard VHS and VHS-HQ have the exact same horizontal resolution—240 lines. But in visual terms, there is a definite improvement in the images reproduced through the VHS-HQ system, and the HQ subformat is fully compatible with all standard VHS machines and tapes. Many viewers even notice a slight improvement in the pictures from a standard VHS tape played back on a VHS-HQ machine.

Camcorders

VCRs are not only used to record television programs (either from over-the-air broadcasts or from cable) or to play prerecorded tapes. They are also used to make original videos and home movies. To do this, you obviously need some sort of video camera. Initially, separate video cameras were used with existing VCRs. The output of the video camera is not connected to the VCR antenna inputs, but to special line input connectors to bypass the VCR internal tuner circuitry. There is no need or point in tuning to a camera "channel." The line input signal from the video camera electrically looks like the output signal from the tuner. It is modulated onto the VCR usual IF carrier frequency.

Of course, this separate camera and VCR setup was fine in a small studio or in the home, but it was a real pain and nuisance for any sort of location recording. Most standard VCRs require an ac power line to operate, so outdoor or remote locations were very inconvenient and awkward to record.

Portable VCRs were soon developed to eliminate some of these problems. These machines could operate for limited periods from an internal battery pack. The battery pack was an improvement, but there were still some severe limits to the system. The portable VCR/video camera combination was rather bulky and had to be set up, often with multiple connections. The big disadvantage was that camera movement was inherently restricted by the cable connecting it to the VCR. The camera can't move any farther than the connecting cable allows. Usually the absolute maximum cable length was about 85 feet.

For most simple recording situations, the cable length was no particular problem, but if a lot of camera movement was required, the connecting cable could become a major limitation. There were essentially just two choices. The VCR could be placed in a fixed location, limiting the operator's movement, or the operator could continuously carry the VCR about with a shoulder harness or a waist-mounted carrying case. Even though portable VCRs were designed to be relatively light, they can still get excessively heavy and tiring after you lug them around for a while as you attempt to shoot your video.

Manufacturers soon came up with a dandy solution to such limitations, and it has caught on like wildfire. A relatively new combination unit called a *camcorder* (camera recorder) is available at fairly reasonable prices. A camcorder has a VCR built right into the housing of the video camera itself. The built-in VCR omits the tuner circuitry, because it is not needed when you're using the camera.

The first camcorders used standard VHS or Beta cassettes. Standard VHS camcorders are still available. They are convenient because the tape can be directly played on your home VCR. But for camcorder applications, the standard VHS video cassette is excessively bulky. It makes the camcorder rather awkward and heavy.

To reduce camcorder bulk two smaller video cassette formats were developed. They are VHS-C and 8 mm (millimeter). These formats are discussed below. Both are used almost exclusively in camcorders.

The camcorder has enjoyed great popularity. Camcorders have been so popular in the marketplace, they have apparently killed off the separate home video camera altogether. The loss of the video camera is a bit of a shame, because these simpler and less expensive video cameras might still be perfectly adequate for many popular recording applications.

Except for special shots, it is usually a good idea to mount a video camera or a camcorder on a tripod. The tripod restricts movement of the camera, but it ensures a rock-steady picture. Tapes made with hand-held cameras and camcorders all too often have an amateurish, wobbly appearance. If the camera is mounted on a tripod, a cable connecting it to a separate VCR is no particular restriction. If you plan to do only fixed-location, tripod recording with your video camera, it is hardly fair that you are required to buy an extra, unneeded VCR built into a camcorder. An inexpensive camcorder costs from $700 to $1000. Good-quality separate video cameras ran for $200 to $350 when they were available, so the increased cost is not exactly a trivial consideration.

However, camcorders have proven immensely popular with the general public, and the sales of video cameras plummeted, so video equipment manufacturers stopped making them. Probably, if separate, less expensive video cameras were still available that they'd be selling a lot better now that the initial novelty and "sex appeal" of the camcorder concept has worn off. Camcorders are still popular, and they will continue to sell well. They are an unique and useful product. But not everyone needs a full camcorder. Many consumers would find their needs more than adequately filled with a separate video camera added to their existing VCR. Unfortunately, this option is no longer available in today's marketplace.

VHS-C

The VHS-C format was developed by JVC specifically for use in camcorders. The C in the name stands for compact. As the name clearly indicates, VHS-C is a compact version of the standard VHS format. Actually, the VHS-C format is really just a sub-format of VHS.

The VHS-C is about one third the size of the standard VHS cassette. Naturally there is less space inside the cassette, so it can't hold nearly as much tape. At the basic SP speed, this smaller cassette permits up to 20 minutes of recording time. Until a few years ago, VHS-C equipment offered only the SP mode. Many newer VHS-C camcorders also feature the slower EP or SLP speed that permits up to one hour of recording time on the same cassette. The in-between LP speed is rarely used in VHS-C. In fact, it is virtually obsolete, and in many modern standard VHS machines it is offered as a playback-only speed, so older, pre-existing tapes can still be played back.

A VHS-C cassette can be inserted into a special adapter that permits it to be played in a standard VHS machine. Compatibility with existing equipment was an important design consideration. Electronically, and in terms of the signals involved, the VHS-C subformat is identical to the standard VHS format. Some VHS-C camcorders even include the VHS-HQ circuitry discussed earlier in this chapter. The only difference between VHS and VHS-C is the physical size of the cassette itself.

A VHS-C camcorder can be made up to 60% smaller than a comparable standard VHS camcorder. The much lighter weight of the VHS-C camcorder makes it much easier and pleasant to use. Lugging a full-size VHS camcorder about can become very tiring very quickly. The chief limitations in the popularity of this subformat are the short recording times and the necessity of using an adaptor to play back the tapes in the home VCR machine.

The 8 mm format

The dust had barely settled on the format wars between VHS and Beta (with VHS as the clear winner) when yet another, totally incompatible VCR format appeared on the market. This format is the 8 mm format, and it is by far the most compact. Improved technology allows 8 mm machines to use a narrower tape than the older formats. VHS and Beta both use half-inch wide tape. You've probably already guessed that the tape width in the 8 mm format is 8 millimeters, or about half the width used in the other home VCR formats.

The thinner tape permits a much smaller cassette housing. An 8 mm video cassette measures only 95 mm $\times$ 62.5 mm $\times$ 15 mm, and weights a mere 1.5 ounce. It is closer in size and weight to an audio tape cassette than to a cassette from any other videotape format. Six 8 mm cassettes could fit into the space taken up by a single VHS cassette. VHS and Beta cassettes are not just larger than 8 mm cassettes, they are also considerably heavier. A Beta videocassette weighs in at 6.5 ounces and a VHS videocassette weighs 8.5 ounces, or over half a pound.

The smaller, lighter 8 mm format was designed specifically for use in camcorders. In fact, few if any 8 mm home VCRs are the market. A few television sets with built-in 8 mm VCRs were marketed, but apparently they didn't sell very well. There are very few of these still in use. The video camera can serve as a tape deck, and play back its prerecorded signal through your television set, or the video camera line outputs can be fed into the line inputs of a VHS VCR to make a copy of your 8 mm tape in the larger home format.

The 8 mm format is very well suited for camcorders. An 8 mm camcorder can be noticeably smaller and considerably lighter than a VHS camcorder, or even a VHS-C camcorder. The 8 mm camcorder isn't as limited in its recording time as a VHS-C camcorder. A standard 8 mm videocassette can hold about 1 and a half hours of recording.

The 8 mm format also uses a somewhat simpler tape drive system than the older videotape formats. The simpler system helps manufacturers to design and build smaller, lighter weight camcorders. Special electronic circuitry is used to compensate for the smaller tape surface area that can be used to hold the recorded signal. The 8 mm format also uses a slower writing speed than the other VCR formats, and the slower speed limits the recorded signal somewhat. Again, special circuitry helps disguise such signal limitations. The writing speed in the VHS format is 5.8 meters per second. Beta VCRs use a writing speed of 7 meters per second. The writing speed in an 8 mm machine is only 3.8 meters per second.

High-grade metal-powdered and metal-evaporated tapes are used in the 8 mm format. Use of these tapes is necessary to achieve an acceptable quality signal with the slower writing speed and reduced tape area.

In practical terms, the electronics in an 8 mm machine usually do a very good job of compensating for these limitations. The reproduced picture quality is usually quite good, and many consumers who are not technically oriented can't tell the difference between an 8 mm picture, and a VHS or Beta picture.

Beta, as you will recall, was designed by Sony, and VHS was the brainchild of JVC. But no one manufacturer was responsible for the 8 mm format. Instead, over 120 electronic companies from around the world agreed upon the standards for this new format. The standards were signed in March of 1983.

The 8 mm format has not caught on nearly as well as many people had hoped. Because it is less convenient for home use, and there has been a real shortage of prerecorded tapes in this format, most consumers stuck with VHS for their home VCR. Most were then reluctant to use a second format in their camcorder and have to make copies to watch their recordings on their home deck. 8 mm camcorders are alive and well in today's marketplace, but they have not stolen the field. Again, despite its inherent inferiority in some ways (in this case, the larger size and weight in camcorders), VHS is emerging as the overall winner in yet another set of format wars.

VCPs

A playback-only videotape machine would seem to be a natural product. According to many surveys, a surprisingly large proportion of home VCR decks are used only to play back prerecorded (primarily rented) tapes. There are many people who don't even know how to record a tape on their VCR, much less use the programmed timer. For such consumers, a VCP (video-cassette player) would save them quite a bit of money, because it would not need to contain a tuner or any input or recording circuitry. However, VCPs have been slow to appear on the market in significant numbers. Several are available, but they do not seem to have had impressive sales records to date. The main market for VCPs has been rental centers that can then rent out the playback machine, along with the prerecorded videotapes.

A VCP is essentially nothing but a stripped-down VCR. It usually has no special effects, and usually only the SP speed is offered, because SP is the speed used on most prerecorded videotape releases. However, there are some exceptions in some budget-priced tapes. Such a VCP could not play back a tape recorded at the LP or EP (SLP) speeds. The result of such an attempt would be chipmunk sounding audio, and a totally incoherent, garbage image. Several newer VCPs also offer the LP and EP (SLP) modes.

Of course, a VCP does not have the capability of recording programs from any source. There are no inputs to a VCP. The main cost reduction in a VCP over a VCR is the elimination of the built-in television tuner. Even so, the cost savings are not as great as might be expected. Often a VCP offers only limited economic advantage over a comparable VCR, which is probably why they have not become a hot product. Rather curiously, many consumers apparently are quite willing to pay a little bit extra for recording capabilities even if they will never take advantage of them.

Still VCPs do offer certain advantages over full VCRs. VCPs tend to be noticeably lighter and easier to hook up and operate than full VCRs. Currently, all commercial VCPs are in the VHS format only.

Videodiscs

Another video playback-only format is the videodisc. A VCR is the video equivalent of a tape recorder in an audio system; a videodisc player is the equivalent to a record player, or somewhat more accurately, a CD (compact-disc) player. There is no way to record your own videodiscs. The advantage is that a prerecorded videodisc will usually have much better image quality than a prerecorded videotape of the same program. Moreover, videodiscs can offer special features, such as still frames, and added materials that are not available on tape.

Several videodisc formats have appeared on the market over the years. Most were stunning failures in terms of sales. The videodisc format that has survived is the laser disc, which has the best specifications of any modern home video format. Videodisc players don't sell nearly as well as VCRs, but they do turn up respectable sales figures overall. The primary market for the videodisc player is the serious videophile and collector who wants the best possible quality. It is basically the same situation as with high-end audio components that are very expensive, and sell only to serious audiophiles, and most general consumers are entirely satisfied with less expensive systems, especially if convenience features are offered.

The laser disc or videodisc is technically similar to the CD, which is discussed in another chapter of this book. The videodisc is larger than a standard CD. The standard size for a videodisc is 12 inches. An average movie usually fits onto one videodisc. Some longer features might require a two videodisc set.

As in a CD, a small laser is used to read the encoded pits and islands in the videodisc. In the audio CD these pits and islands were digitally encoded, but a videodisc uses analog encoded data for the video information. You don't need to get into the technical details of the encoding process here.

Commercially, videodiscs are doing well as a specialized niche in the much larger, general home video marketplace.

41
Troubleshooting and servicing

An important part of any electronics technician's battery of skills is the ability to locate and correct any problems that might show up in a circuit. A newly designed prototype circuit might not function as the designer intended. The technician must determine what went wrong and correct the error in a new prototype. An experimenter might build a presumedly debugged circuit from a magazine, but it doesn't work. Now the experimenter will need to determine if the error is in the published schematic or if there is an error in construction. In either case, the problem must be found and corrected, or the previous efforts will have been wasted. A piece of electronic equipment that once worked properly might start malfunctioning. A technician must locate and repair the portion of the circuit that has become defective.

The process of locating the problem in a malfunctioning electronic circuit is known as *troubleshooting*. Repairing the defect is known as *servicing*. Troubleshooting a complex piece of equipment might seem like a hopeless task at first. But a little common sense can reduce the job to manageable levels. A color television set, for example, might have several hundred discrete components. Do you check each one to make sure that it is within the manufacturer's specifications? Presumedly it would be possible to locate most defects that way, but it would be extremely inefficient. It could take hours, or even days to find the bad component.

Most complex electronic devices are made up of many less complex stages. By analyzing the functions of these stages, you can narrow down the potential cause of the defect. For example, assume the picture on a color TV is breaking up into broad diagonal stripes. you can see that objects are being displayed, although they are distorted and broken up because of the striping. Colors appear normal. The sound is fine. Where is the problem? As a first step, you look at the block diagram for the set. A typical color TV block diagram is shown in Fig. 41-1. Consider each stage and see if it is the problem.

Because the set is operating, you can assume the power supply circuits are

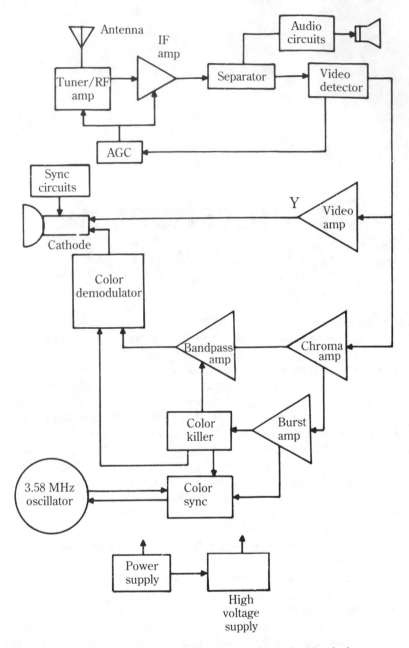

41-1 A good block diagram can help a technician pinpoint faulty
stages in a defective circuit.

working. They might be putting out incorrect voltages, but in this particular case,
that doesn't seem a likely cause of the problem at hand. If you are unsure, it would
only take a few moments to measure the supply voltages. If they are reasonably
close to their specified values, you can move on. Some technicians always measure

supply voltages for all problems. Too high and too low voltages can result in some very odd problems. In this example, you will assume the supply voltages are correct.

You are getting a raster (the CRT is lighting up) so you can eliminate the high-voltage supply and the CRT as suspects. A signal is getting through, so the tuner, RF amplifiers, IF amplifiers, AGC, and separator circuits are probably okay. You have correct sound, so the audio circuits and speaker are not involved. You do have a picture, but it is broken up. This symptom indicates that there isn't a problem in the video detector or video amplifier.

Because the color seems to be all right, the defect is probably not in any of the color circuits (chroma amplifier, bandpass amplifier, burst amplifier, color killer, color sync, and the 3.58 MHz oscillator). What does that leave? There is only one section of the block diagram you haven't at least tentatively ruled out. An experienced technician would probably immediately suspect the sync circuits. The problem you have described is a loss of horizontal synchronization.

In some tough cases, the circuit you suspect might check out fine. There could be a problem in another portion of the circuitry that is causing a surprising symptom. For instance, the sync pulses might be distorted or attenuated in some manner in the video amplifier stage. This problem isn't as likely, but it could happen.

If a set has multiple symptoms, try to isolate the single stage that could be responsible for all or most of the symptoms. There might be defects in more than one stage, but the odds are that there is a common problem, or one component has faded, causing an overload to damage another component.

Consider another problem. The screen is completely dark, and there is no sound at all. The set just sits there and does absolutely nothing. The average person might consider this a more serious problem. After all, the set isn't working at all. But an experienced technician will be unconcerned. It will probably be a relatively easy repair.

What could cause the symptoms described? There could be several combinations of multiple causes. For example, the CRT and one of the audio circuits could be defective. But it's more likely that all of the symptoms are resulting from a problem in a single stage. What stage could account for all of the symptoms (dark screen and no sound)? There is one stage common to everything in the set—the power supply. The other stages probably aren't getting any power. The culprit could be a blown fuse, a burned-out rectifier diode, a shorted filter capacitor, a broken wire, or any of many other possibilities. (Assume the set is plugged in, of course. It saves time to always check out the simple and the obvious first. Dumb mistakes do happen.)

In this chapter, you can briefly consider some of the more common types of defects. But first, examine some of the important weapons in the technician's troubleshooting arsenal.

Test equipment

To determine the source of a problem in an electronic circuit, the technician uses one or more pieces of test equipment. There is no single all-purpose tester that can solve all problems. Each piece of test equipment is designed for specific purposes.

The most basic pieces of test equipment are the technician's eyes, ears, and brain. This comment is not meant to be facetious. The technician observes the symptoms with eyes and ears and analyzes them with his brain. The brain tells the technician what stage or stages are likely culprits.

The technician's eyes can be used to locate many defects. Does a component look burned? Is a wire broken? The technician's nose can also be helpful. Does something smell burned? The technician can also use the tip of the finger to determine if a component is running hot. Of course, the technician's brain must also analyze the data obtained from the various tests performed. What could cause any unusual values found in the circuit? Generally, when speaking of test equipment, however, you are referring to external electronic circuits designed to measure various electrical characteristics and display their results in some manner.

The VOM

The most basic and versatile piece of equipment available to the electronics technician is the VOM, or *volt-ohm-milliammeter*. This device is discussed in chapter 10. A VOM measures dc and ac voltages, dc resistance, and dc current. A few VOMs also measure ac current.

Closely related to the VOM is the VTVM, or *vacuum-tube voltmeter*. Most VTVMs measure only voltage and resistance, but not current. One advantage of the VTVM over the VOM is in measuring high-frequency ac voltages. A VOM reading is not usually reliable for frequencies above a few hundred hertz. A VTVM has a far wider frequency response. The VTVM can also usually measure a wider range of resistances.

The average VTVM is well protected against overloading the meter movement and burning out component parts from excessive test voltages. The meter in a VOM can be damaged if, for example, 120 V was applied when the meter was set to a 10 V range. In a VTVM the meter itself is isolated from the input by the vacuum-tube amplifier and scaling circuitry.

A VTVM usually runs on 120 Vac, which means it can only be used where an electrical socket is available. Most standard VOMs have a small battery for resistance measurements and require no further power source. VTVMs are also more bulky than most VOMs. In addition, the tube circuitry in the VTVM gives off heat.

Because a VOM is not plugged into the electrical socket along with the equipment being tested, there is better isolation, ensuring greater safety for the technician. The use of a battery also tends to minimize interference problems from stray RF signals and ac hum.

The tubes in a VTVM need a certain amount of time to warm up before correct readings can be obtained. In addition, readings might tend to drift slightly, especially as the tubes age. Drift is not a problem with a VOM, which is basically a passive device. Because the VOM involves simpler circuitry than the VTVM, it also tends to be more reliable.

The main difference between meters is sensitivity or input impedance. The higher the input impedance of the meter, the less it will load down the circuit being tested. This is because the meter is placed in parallel with part of the circuit to measure voltage.

A typical VOM has a sensitivity of approximately 20,000 Ω/V (ohms per volt). Some go as high as 50,000 Ω/V. Cheap VOMs with an input sensitivity of only 1000 Ω/V are also available. These can be useful to an electrician, but an electronics technician should steer clear of them—the sensitivity is far too poor for electronics circuitry.

A typical VTVM will have much greater sensitivity than a VOM. Input impedances of 11 MΩ (11,000,000 Ω) are common for VTVMs.

A cross-breed type meter is the FET VOM. The FET version is just a regular VOM with a FET amplifier input stage. The FET offers the advantage of isolating the meter movement from the input signal, but a power source is required for voltage and current measurements as well as for resistance measurements. Power can be supplied by a small internal battery, so portability is not compromised. The sensitivity of a FET VOM is around 1 MΩ/V (1,000,000 Ω/V).

Examine why input sensitivity is so important. Take a look at the simple circuit shown in Fig. 41-2. Assume that the battery is putting out 9 V, and the three resistors have the following values:

$$R1 = 22 \text{ k}\Omega \text{ (22,000 } \Omega)$$
$$R2 = 47 \text{ k}\Omega \text{ (47,000 } \Omega)$$
$$R3 = 33 \text{ k}\Omega \text{ (33,000 } \Omega)$$

41-2   A voltmeter can affect the circuit being tested by acting as a parallel resistance.

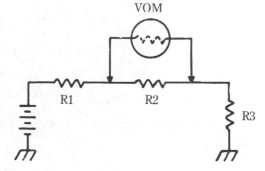

Total circuit resistance is 102,000 Ω. According to Ohm's law, the current drawn through the circuit works out to:

$$I = \frac{E}{R} = \frac{9}{102000} \approx 0.0000882 \text{ A} = 0.0882 \text{ mA}$$

The same amount of current flows through each resistor. You can now calculate the nominal voltage drop across resistor R2:

$$E = IR = 0.0000882 \times 47000 \approx 4.15 \text{ V}$$

Now, assume you hook up a 1000 Ω/V VOM in parallel with R2 to measure the voltage drop. The input resistance of the VOM acts as a parallel resistance with R2, which makes the apparent value of R2 drop to:

$$\frac{1}{R_t} = \frac{1}{R_2} + \frac{1}{R_m} = \frac{1}{47000} + \frac{1}{1000}$$

$$= 0.0000213 + 0.01 = 0.0010213 = \frac{1}{R_t}$$

$$R_t = \frac{1}{0.0010213} \approx 979 \ \Omega$$

That's quite a change. The total circuit resistance is now:

$$R_t = 22000 + 979 + 33000 = 55{,}979 \ \Omega$$

This resistance changes the current flowing through the circuit to:

$$I = \frac{E}{R} = \frac{9}{55979} \approx 0.00016 \ \text{A} = 0.16 \ \text{mA}$$

The measured voltage drop across R2 works out to:

$$E = IR = 0.00016 \times 979 \approx 0.157 \ \text{V}$$

That is certainly an unacceptable error when you are expecting a reading of 4.15 V! If the input resistance of the VOM is 20,000 Ω, things will look a little better, although the error will still be significant. First, solving for the effective resistance of the R_2/meter combination:

$$\frac{1}{R_t} = \frac{1}{R_2} + \frac{1}{R_m} = \frac{1}{4700} + \frac{1}{20{,}000}$$

$$= 0.0000213 + 0.00005 = 0.0000713$$

$$= \frac{1}{R_t} = \frac{1}{0.0000713} \approx 14000 \ \Omega$$

Total effective circuit resistance this time is about 69,000 Ω, making the current flow equal to:

$$I = \frac{9}{69000} \approx 0.00013 \ \text{A} = 0.13 \ \text{mA}$$

The measured voltage drop across R2 using a 20,000 Ω meter works out to:

$$E = 0.00013 \times 14000 \approx 1.8 \ \text{V}$$

Better, but still not all that good. Now, assume you are using a VTVM with an input sensitivity of 11 MΩ (11,000,000). The parallel combination of the meter and R2 in this case becomes:

$$\frac{1}{R_t} = \frac{1}{47000} + \frac{1}{11000000}$$

$$= 0.0000213 + 0.00000009091$$

$$\approx 0.0000214 = \frac{1}{R_t}$$

$$R_t = \frac{1}{0.0000214} \approx 46{,}800 \ \Omega$$

Notice that this value is very close to the original nominal value of R2 alone (47,000 Ω). As you might suspect, there is only a small change in the current flowing through the circuit:

$$I = \frac{9}{101800} = 0.0000884 \ \text{amps} = 0.0884 \ \text{mA}$$

The measured voltage drop across R2 this time would work out to:

$$E = 0.0000884 \times 46800 = 4.14 \ \text{V}$$

That is very close to the calculated nominal value of 4.15 V dropped across R2.

The larger the meter input resistance is in comparison with that of the component being measured, the smaller the error in the reading. The 20,000 Ω VOM didn't fare too well in that example. It would work better in a circuit with smaller resistances. For example, change the resistance values in Fig. 41-2 as follows:

$$R1 = 680 \ \Omega$$
$$R2 = 220 \ \Omega$$
$$R3 = 390 \ \Omega$$

The total circuit resistance is 1290 Ω, so the unmetered current flow through the circuit is:

$$I = \frac{E}{R} = \frac{9}{1290} = 0.00698 \ \text{A} = 6.98 \ \text{mA}$$

This makes the nominal voltage drop across R2 equal to:

$$E = IR = 0.00698 \times 220 \approx 1.53 \ \text{V}$$

If a 20,000 Ω input VOM is placed in parallel with R2 to measure its voltage drop, the effective resistance changes according to the following calculations:

$$\frac{1}{R_t} = \frac{1}{R2} + \frac{1}{R_m} = \frac{1}{220} + \frac{1}{20000}$$

$$= 0.0045455 + 0.00005 = 0.0045955 = \frac{1}{R_t}$$

$$R_t = \frac{1}{0.0045955} \approx 218 \ \Omega$$

The meter resistance is so large in comparison to that of R2, it only has a small effect on the value of the parallel combination. The total effective circuit resistance is 1288 Ω, instead of the nominal 1290 Ω calculated above. This 2 Ω difference won't change the current flowing through the circuit by much:

$$I = \frac{19}{1288} = 0.00699 \text{ A} = 6.99 \text{ mA}$$

The difference is a mere 0.01 mA. The measured voltage drop across R2 would therefore work out to be equal to:

$$E = IR = 0.00699 \times 218 \approx 1.52 \text{ V}$$

Certainly that should be close enough for most purposes.

Many servicing schematics have voltages marked that were made with a 20,000 Ω/V meter. In this case, a meter with higher sensitivity would give poorer results. Be sure you know what is being used as the standard.

The digital voltmeter

In recent years, the VOM has been joined, and in some cases replaced, by the *digital voltmeter*, or *DVM*. Instead of the measured quantity being indicated by the position of a mechanical pointer on a dial face, the values are read out directly in numbers using seven-segment LED or LCD displays, like those used on a calculator. A DVM can measure over a much wider range than a typical VOM. The input sensitivity can be as high as that of a VTVM, but the instrument remains compact, lightweight and portable.

A DVM usually offers good protection against input overloads, and often has built-in polarity indication. If the leads are hooked up backwards (the red lead is made more negative than the black lead), a minus sign will be included on the display.

The better DVMs often include a number of special functions not found on standard VOMs. These include automatic range selection and indication, and the capability for measuring additional electrical quantities, such as conductance (the reciprocal of resistance) and capacitance. Some sort of indicator is generally included to warn you when the battery voltage is getting low.

Some of the newest DVMs incorporate computer circuitry to become extremely powerful pieces of test equipment. Not only are measurements displayed digitally, alphanumeric characters are also displayed to help explain what the numbers mean.

The oscilloscope

If a VOM is figuratively a technician's right arm, the left arm would be the oscilloscope. The VOM and oscilloscope are the two pieces of equipment that are mandatory for anyone doing anything beyond casual puttering with electronics.

The oscilloscope is briefly described in chapter 17. The oscilloscope is an instrument that can display the waveshape of a signal on a CRT screen. Some typical waveforms, as they appear on an oscilloscope display are shown in Fig. 41-3. The frequency of any ac waveform can be found by comparing the number of cycles displayed with the oscilloscope vertical sweep frequency (which can be set to any of a number of values).

The controls on an oscilloscope vary somewhat with the model, but they will usually include the following. (A few manufacturers might use different terms for some of these functions.)

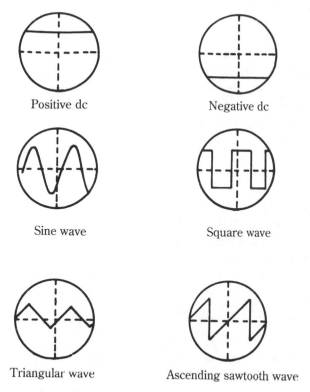

Positive dc   Negative dc

Sine wave   Square wave

Triangular wave   Ascending sawtooth wave

41-3   Various waveforms can be displayed on an oscilloscope.

- Vertical gain—How much deflection from the base line a given input voltage will cause.
- Vertical position—The vertical position control allows the base line to be moved up and down on the CRT screen.
- Horizontal gain—How much deflection will be caused in the right/left dimension when an external signal is used to control the sweep frequency rather than the built-in oscillator.
- Horizontal position—Moves the displayed waveform from right to left on the CRT screen.
- Coarse frequency—Sets the range for the built-in sweep oscillator.
- Fine frequency—Sets the exact frequency of the built-in sweep oscillator.
- Sync—Synchronizes the display with another signal source.
- Intensity—Determines how bright the displayed line(s) will be.
- Focus—Adjusts the clarity of the displayed line(s).

Some oscilloscopes can simultaneously display two signals. This type of machine is called a *dual-trace oscilloscope*. Dual trace is extremely useful for comparing input and output signals or signals from different parts of the circuit being tested.

Signal tracer

A signal tracer is basically an amplifier and output device (such as a speaker) that is used to indicate whether or not a signal is present at the stage of the circuit currently being tested.

Signal injector

A signal injector is the reverse of a signal tracer. It allows you to insert a known signal into a given circuit stage to determine if it is working properly. Signal injectors are available for both audio (AF) and radio (RF) frequencies.

Function generator

A function generator is a multiple-waveform oscillator or signal generator (see chapter 34). It can produce a clean, known signal at a wide variety of frequencies. If you feed a known signal to the input of a suspected circuit, you can examine the output signal with an oscilloscope to determine what if any distortion has occurred in the circuit being tested.

Frequency counter

A frequency counter is used to measure the frequency of an ac signal. It usually has an input stage that converts the input waveform to a clean rectangle wave that can be reliably recognized by the digital circuitry within the frequency counter.

Capacitance meter

A capacitance meter is used to determine the value of an unknown capacitor. It can also be used to check old capacitors for leakage and other problems, or to locate stray capacitances within a circuit. Most capacitance meters work by charging the unknown capacitor to a specific level, and then measure the time it takes for the capacitor to discharge. For more information on the time constant of a capacitor, refer to chapter 6.

Other test equipment

Other pieces of test equipment are manufactured for various special purposes. The ones discussed in this chapter are the most widely available and generally useful. A technician's workbench should also include a good variable-output power supply.

Common problems

Every problem that crops up in electronic circuitry is more or less unique. Nevertheless, there are some fairly common sources of problems that you should be familiar with.

A resistor can change its value with age. This problem is likely to occur if an overvoltage or physical jolt has damaged the resistor. If you suspect a resistor problem, check for resistors that are discolored, or look burnt. Also watch out for tiny hairline cracks. These cracks might not be readily visible. Often a light tap with a screwdriver will cause a cracked resistor to break apart, but a good resistor won't be harmed.

Cracks can also cause problems in capacitors. Especially watch out for ceramic disc capacitors that often tend to crack near their leads. Electrolytic capacitors often become leaky or their electrolyte might dry up as they age. If you see a whitish powder around an electrolytic capacitor, the powder is probably dried electrolyte. Usually, however, this kind of fault will not be visible to the naked eye. Test equipment will have to be brought in to locate the culprit.

Of course, if the insulation between windings of a coil gets damaged or chips off, the inductor will appear to change the number of turns in the coil as adjacent turns short together. A short will cause the inductance to change.

If a PC (printed circuit) board is used, a tiny hairline crack might cause an open circuit. Similarly, because the copper traces are closely spaced, a scrap of wire or solder could land between adjacent paths and cause a short. Stray capacitances, inductances, and resistances might crop up when a new circuit is built, or they might appear in a formerly working circuit when one or more components goes bad or changes value.

Of course, active components can also go bad. Tubes often need to be replaced. Semiconductors are more reliable, but they can be damaged by hard physical jolts, runaway current, over voltages, or excessive heat. Special-purpose transistor testers are widely available. Basic semiconductor junctions (in diodes and bipolar transistors) can also easily be tested with a VOM. The resistance section of the VOM is used for this. Remember that an ohmmeter puts out a specific test voltage and measures the drop across the unknown resistance. If the polarity is applied to the semiconductor junction in one direction the ohmmeter will indicate a very low resistance. Reversing the polarity will cause a much larger resistance reading. For a bipolar transistor, measure between the emitter and the base and between the base and the collector.

In most cases the semiconductor component being tested will have to be removed from the circuit because the rest of the circuit might have additional resistances in parallel with the junction, which could significantly change the readings.

ICs are treated as black boxes. If the signals at all of the input pins are correct, but one or more of the output signals is wrong, it might be reasonable to assume that the IC is bad. Watch out though—some circuits can really fool you. If the input impedance of the next stage has changed due to some defect, the difference in the loading could cause the output being measured to be misread by the meter or oscilloscope.

Especially watch out for signs of any previous repair attempts. A lot of people who don't really know what they are doing often attempt to repair electronic equipment, and their repair might have caused additional problems. A component can be physically damaged or installed with the wrong polarity. A thin lead or wire might be broken. PC board traces might be chipped or shorted.

A major headache is the cold solder joint. A cold solder joint happens when an air bubble has formed under the solder, making an unreliable connection. If a solder joint does not look smooth or shiny, suspect a cold solder joint. Unfortunately, the visual test is not always helpful. Many cold solder joints look just fine, but the electrical connection might be poor, exhibiting an open circuit or a high resistance, or (most frustrating of all) an intermittent connection. See Fig. 41-4.

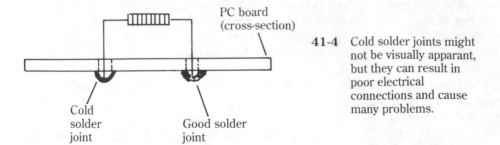

PC board (cross-section)

Cold solder joint

Good solder joint

41-4   Cold solder joints might not be visually apparant, but they can result in poor electrical connections and cause many problems.

A cold solder joint will usually cause problems right away, so the problem most commonly comes up right after a circuit is constructed or repaired. If you suspect a cold solder joint, it might be a good idea to simply reheat all of the solder joints in the area, just to be sure.

Occasionally a cold solder joint will not cause problems until the equipment has been in use for months or even years. They are not terribly common in commercially available equipment, but they do occur from time to time.

The bane of any technician's existence is the intermittent defect. An intermittent defect comes and goes. In some cases you can cause the problem to appear so you can troubleshoot it by simulating specific environmental conditions (high or low ambient temperatures are frequent culprits). All too often, the equipment will refuse to misbehave for the technician, and you can't very well locate a problem that isn't there when you're looking.

Often a technician will operate the equipment under extreme stress to force the intermittent fault to become permanent so that it can be easily pinpointed. This practice is rather risky unless you know what you are doing and are very careful. However, the component causing the intermittent problem is presumedly weaker than any of the good components, so it should fail completely before any other components are damaged.

Flowcharting

If you are servicing an electronic circuit with just a dozen components or so, you can take a brute-force approach and just test everything. But this approach would be highly impractical for any circuit with any degree of complexity. A television set, or a computer, or a VCR (to suggest just a few examples) can contain hundreds, perhaps even thousands of components. It would not be at all reasonable to attempt to test everything in such a device.

If you are attempting to service a given piece of equipment, it must be exhibiting symptoms of some sort. That is, it is presumedly not functioning properly in some way. The malfunction might be simple (a specific indicator light isn't going on), or it might be global (the entire device doesn't work at all). Often there are multiple symptoms that might or might not be related (for example, a TV set might have no sound and a compressed picture). Whatever the symptoms might be, they always provide important clues for servicing the equipment in question.

An electronics technician thinks about the symptoms logically, and tries to imagine what could possibly be causing the observed symptoms. This reasoning will tell him or her what types of tests are most likely to provide worthwhile results. Often many different causes could be behind a given symptom (or group of symptoms), so, if possible, try to list these potential causes in order of probability. Make the most probable tests first, you are less likely to waste time with fruitless tests that way.

When there are multiple symptoms, they might be due to entire separate causes, but it is relatively unlikely for a circuit to simultaneously develop two (or more) entirely unrelated defects. It can happen, but it is more likely that all the observed symptoms stem, directly or indirectly, from a single root cause. Use logical thinking to determine what could be causing all the observed symptoms and test for that first. Only if you run into a dead end on the common cause should you start troubleshooting the observed symptoms independently. Again, you will save a lot of wasted time in the long run.

A very useful troubleshooting technique that is often ignored is the *flowchart*. A *flowchart* is a simple block diagram graph of the equipment to be serviced. The functions of the various subcircuits and their interconnections are shown. This information helps the technician determine what subcircuit(s) is most likely to be causing the observed symptoms. A very simple flowchart for a typical color television set is shown in Fig. 41-1.

Some experienced technicians might be able to hold a full flowchart entirely in their heads, but for most, actually drawing out a flowchart will be helpful, especially in more difficult servicing jobs. There is no reason to be concerned about precise dimensions or perfectly straight lines. Your flowchart is not for publication, it is for your own convenience. If it is neat enough for you to read it, it's good enough. It often helps just to see the circuit functions broken down in visual form like this.

As a rule of thumb, it is a good idea to always consider the power supply a possible culprit for virtually all symptoms. Incorrect supply voltages, or noise on the power lines can cause a lot of surprising results in many circuits.

The shapes used in your flowchart don't really matter, although it is probably useful to use different shapes for certain common types of subcircuits. For example, using a circle for a signal source (such as an oscillator), or a triangle for an amplification stage can make the flowchart easier to read at a glance.

Often each section in the original flowchart will include dozens of components, and can often be broken down to multiple subcircuits. It will usually be helpful to break down any suspected section into more detailed subsections. For example, if you are working on a television set with problems in its sound, but the picture is okay, you'll probably only want to look closer at the audio amplifier section. Draw a more detailed sub-flowchart for this section of circuitry, as shown in Fig. 41- 5. You don't need to identify the subcircuits in any flowchart section where no defect is

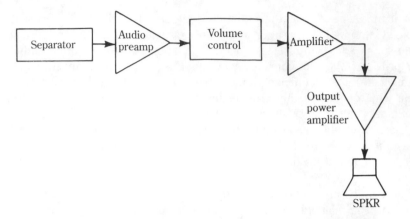

41-5   In narrowing down the problem, it is often helpful to draw a more detailed block diagram of the suspected stage(s).

suspected or likely. You don't expect to do any testing in that part of the equipments circuitry.

The next step in creating a troubleshooting flowchart is to identify the key components in any relevant stages, especially ICs, transistors, transformers, relays, etc. The identification will help focus you in on that components and connections you will want to test, and which probably aren't worth bothering with, because they are unlikely to be defective, given the symptoms observed. For example, in a TV set with sound problems (and no other symptoms), there would be little point in testing the picture tube or the color burst oscillator.

Write the appropriate component identifications in flowchart, as shown in Fig. 41-6. This will be easiest to do by comparing your flowchart with the schematic included in the servicing data. It is very difficult to service a complex electronic

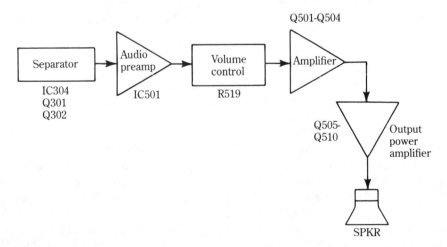

41-6   The next step is to identify the key components in the suspected stage.

circuit without a schematic unless you have a great deal of experience with that particular type of equipment.

You don't have to list every single resistor or capacitor. Just list the key components, especially active components, and controls. Any passive component in the immediate vicinity of these key components is obviously part of the same subcircuit. You might occasionally test a resistor or two that isn't really part of the subcircuit you thought it was, but this will be fairly rare, and won't waste as much time as getting overly detailed in the flowchart.

Notice that there is little point in worrying about the key components in any subcircuits you have already eliminated in the troubleshooting process. If you know the defect must be somewhere in the audio amplifier of the TV, and the video amplifier, and the tuner are probably okay, so you don't care about locating their key components. A little logic and a good flowchart will tell you what tests you should perform to track down the defect(s) causing the observed symptoms.

Servicing without a schematic

All too often it is necessary to service a piece of electronic equipment for which no schematic diagram or other relevant technical literature is available. Clearly this problem increases in seriousness with the complexity of the equipment in question.

The key to solving such problems is to first mentally break up the circuitry into functional blocks. For moderate to complex equipment, it would probably be a good idea to actually draw out a block diagram. For some relatively simple circuits, you might be able to hold the information in your head. Of course, skill in this type of task comes only with experience. Although it will get easier as you gain practical servicing experience, and theoretical knowledge, it will never be truly easy, which is why you should first make every attempt to find a schematic or other service literature if at all possible.

In many cases, there will be some things you simply will not be able to figure out on your own, so if you don't have a schematic, you may be out of luck on some special features and calibration procedures, and the like. Still, there is much you can do with most electronic equipment without a schematic. Just remember, this approach is one of last resort. A schematic will make your job 1000% simpler.

Now, it is difficult to look at a section of circuitry and determine what it is supposed to do. Exotic, unfamiliar, or unmarked ICs further complicates the analysis. An amplifier IC looks just like a timer IC, which looks just like a digital flip-flop IC. Obviously, you won't get very far with your block diagram from just staring at the circuit board. You have to use some logical thinking.

Do not start from the circuit itself, but from the intended function of the equipment. What circuit stages would be required to do the job? In other words, you have to think like a circuit designer. Make a note of any and all controls on the equipment. The controls will provide valuable clues to circuitry functions. Also, components near the control are presumedly part of the same subcircuit.

Don't forget "obvious" stages. Almost any piece of electronic equipment will have a power supply subcircuit of some sort. For some portable equipment, the power supply will be nothing more than some batteries and perhaps a diode or two to protect against possible incorrect installation of the batteries. All you really have to check is the battery voltage and the condition of the diodes, mainly that they are not shorted out.

AC powered equipment will generally have a more complex power supply stage. First check for blown fuses or circuit breakers. Of course, you should make sure the equipment is plugged in. In practice, field technicians often get a service call when there was nothing wrong except the equipment had somehow been un-plugged.

Most ac power supplies have a power transformer of some sort. The secondary winding (or windings) of this transformer will be connected to either one or more rectifier diodes or a voltage regulator IC. (In some circuits, both can be encountered). In either case, there will probably be some filter capacitors too. Usually at least some of these filter capacitors will be fairly large electrolytics, which not only have large capacitance values, but also are physically large. The transformer and the large filter capacitors usually make it relatively easy to identify and isolate the power supply circuitry.

A voltage regulator IC usually resembles a slightly oversize power transistor with three leads (in, out, and common). Sometimes a bridge rectifier is used instead of four separate rectifier diodes. This component will look like a square block of plastic with four leads. Internally, it is simply four matched rectifier diodes encapsulated in a single convenient housing.

Locating the power supply will also provide you valuable clues in tracing the remainder of the circuitry. Find the ground or common connections. Identifying the supply voltage and common connections to any ICs will help cut down the possibilities for identifying the inputs and outputs.

In most electronic equipment, some sort of signal will probably pass through the system. Attempt to locate the point of origin for the main signal. If the equipment uses some sort of external signal source, this process should be relatively easy. Just find the input jacks (or other connectors). For example, an amplifier in a stereo system might take its input signal from a tape deck or a CD player. In a TV or a radio, the main signal originates at the antenna terminals. In other equipment, the main signal can be internally generated. More often than not, the signal comes from an oscillator stage of some sort. An oscilloscope would be a big help in locating a the point of origin of a signal and tracing its passage through the circuitry.

If the equipment you are working on uses signals in the audible range, you might be able to use a signal tracer probe, which can be a simple audio amplifier and speaker with a suitable probe at its input. Refer to chapter 9 for more information on signal tracing.

Pay close attention to any controls, as they will give you clues as to the function of that section of the circuitry. For example, a volume control, or gain control is almost certainly part of an amplifier stage of some sort. (But always remember, that stage might perform additional functions besides simple amplification.)

Servicing electronic equipment without a schematic is not easy, and should be avoided if at all possible. But sometimes it is necessary. Just remember, the task is not truly impossible. Work as slowly and logically as possible. Take as many measurements as you can. It is hard to tell if any single measurement is correct or not, when you look at it by itself. But if you take many measurements throughout a circuit, one or two might stick out like a proverbial sore thumb. Try to relate the odd seeming measurement(s) with the symptoms exhibited by the malfunctioning equipment. If, for example, the problem is excessive distortion, if one stage of the circuit suddenly shows a lot of mysterious, and seemingly unnecessary signals, that might possibly be the source of the distortion. Try to think if there could be any possible reason for the odd signals to be part of the intended design of the circuit. If not, what would have to be done to eliminate them? Does that clear up the misfunction, or does it create new problems?

For obvious reasons, servicing without a schematic or other relevant technical literature or data is not at all recommended for inexperienced electronics technicians. To get some experience, try your hand on some noncritical equipment, especially equipment someone else has thrown out because of some defect.

Another way to learn to do such "data-blind" servicing would be to first attempt diagnosis on a piece of equipment before you look at the schematic and other troubleshooting data. Don't actually do anything to the circuitry at this point—just attempt to figure out how it is supposed to work. Then refer to the available servicing data and see how close you came. The first few times you try, you will almost certainly be way off more often than not, but you will become more skilled with time.

Remember that servicing electronic equipment without a schematic or other appropriate servicing data is always a course of last resort. Only a very foolish technician would ever consider attempting a repair without consulting any and all schematics or troubleshooting materials. The only possible exception would be if he has worked on one particular model of equipment so often that he has the relevant servicing data memorized. Even then, it is probably unnecessarily risky to rely too heavily on human memory. Keep the schematic and other materials handy so you glance at them occasionally, just to be sure.

Self-test

1. How can a block diagram be used in troubleshooting?

A To determine specific operating voltages
B By visually comparing it with the circuit under test
C By narrowing down the stages that could cause the fault in question
D By allowing the technician to systematically check out each stage of the circuit
E None of the above

2. If a television set doesn't work at all (dark screen, no sound) which stage is the most likely source of the problem?

A The power supply
B The video amplifier
C The tuner
D The color killer
E None of the above

3. Which of the following voltmeters would give the most accurate readings?

A 20,000 Ω/V VOM
B 11 MΩ VTVM
C 1000 Ω/V VOM
D 50,000 Ω/V VOM
E None of the above

4. Which of the following can not be measured with a standard VOM?

A Current
B Inductance
C Resistance
D Voltage
E None of the above

5. Which of the following is an advantage of a VTVM over a VOM?

A More portable
B Measures current
C Less expensive
D Operates with less heat
E None of the above

6. What should be the relationship between a meter input resistance (R_m) to the resistance the voltage drop is being measured across (R_x)?

A R must be much larger than R_m
B They should be approximately equal
C R_x must be much smaller than R_m
D None of the above

7. Which of the following is the second most important piece of test equipment for the average electronics technician?

A Oscilloscope
B Frequency counter
C Signal injector
D Capacitance counter
E None of the above

8. Which control on an oscilloscope is used to calibrate the displayed input voltage a specific distance over or below the baseline?

A Coarse frequency
B Vertical gain
C Horizontal gain
D Horizontal position
E None of the above

9. Which of the following is not a common defect in a resistor?

A Changed value
B Burnt
C Cracked
D Leaky
E None of the above

10. If an ohmmeter measures a small resistance when it is hooked up across a semi-conductor diode, what would happen if the polarity of the ohmmeter leads are reversed?

A A higher resistance will be indicated on the meter
B The same resistance will be measured
C The semiconductor junction will probably be damaged
D It doesn't matter—the diode must be bad
E None of the above.

Appendix
Answers to self-tests

Chapter 2

1. *A*

   The atom is made up of electrons, protons, and neutrons, but not coulombs.

2. *D*

   Electrons flow through a conductor to create an electrical current.

3. *C*

   The basic unit for measuring current flow is the ampere.

4. *A*

   The basic unit for measuring electrical power is the watt.

5. *B*

   An ordinary flashlight battery is a dry cell.

6. *A*

   Zinc and copper would make good electrodes in a wet-cell battery.

7. *C*

   Connecting battery cells in series increases the voltage. The total series voltage is equal to the sum of the individual cell voltages.

8. *A*

   Connecting battery cells in parallel increases the current capacity. The total series current is equal to the sum of the individual cell currents.

9. *B*

   If the voltage applied to the circuit is 15 V, and the current flow is 2 A, the power consumed by the circuit is 30 W.

$$P = EI$$

10. **D**
     If 12 V is applied to a circuit that consumes 78 W, the current flow through the circuit equals 6.5 A.

$$I = P/E$$

Chapter 3

1. **A**
    The basic unit of resistance is the ohm.

2. **C**
    If the bands on a resistor are yellow, violet, red, and gold, the value is 4700 Ω, 5%.

3. **D**
    If the bands on a resistor are red, red, orange, and silver, the value is 22,000 Ω, 10%.

4. **D**
    Ohm's law states that $E = IR$. This equation can be rewritten as $I = E/R$ or $R = E/I$, but $R = E^2/I$ is not a correct equivalent.

5. **C**
    For resistances in series, the values add. That is, $R_t = R_1 + R_2$. Therefore, if a 3300 Ω resistor and a 22,000 Ω resistor are connected in series, the total resistance is 25,300 Ω.

6. **A**
    If three 560 Ω resistors are connected in parallel, the total resistance of the combination is 187 Ω.

$$
\begin{aligned}
1/R_t &= 1/R_1 + R_2 + R_3 \\
&= 1/560 + 1/560 + 1/560 \\
&= 0.0017857 + 0.0017857\ 0.0017857 \\
1/R_t &= 0.0053571
\end{aligned}
$$

$$
\begin{aligned}
R_t &= 1/0.0053571 \\
&= 187\ \Omega
\end{aligned}
$$

7. **B**
    If 15 V is applied to a simple series circuit consisting of a 2200 Ω resistor and a 4700 Ω resistor, the value of the current flowing through this circuit is 0.0022 A. The total resistance (R_t) is the sum of the two series resistances:

$$
\begin{aligned}
R_t &= R_1 + R_2 \\
&= 2200 + 4700 \\
&= 6900\ \Omega
\end{aligned}
$$

You can now find the current using Ohm's law:

$$I = E/R$$
$$= 15/6900$$
$$= 0.0022 \text{ A}$$

8. *C*

A 33 kΩ resistor is connected in series with a parallel combination made up of a 56 kΩ resistor and a 6.8 kΩ resistor. First you must find the total effective parallel resistance:

$$1/R_{tp} = 1/56000 + 1/6800$$
$$= 0.0000178 + 0.000147$$
$$1/R_{tp} = 0.0001648$$
$$R_{tp} = 1/0.0001648$$
$$= 6068 \text{ } \Omega$$

This resistance is in series with the 33 kΩ resistor. Resistances in series add, so the total effective resistance of this combination is equal to:

$$R_t = 33000 + 6068$$
$$= 39,067 \text{ } \Omega$$

9. *C*

When resistors are connected in series, the effective resistance is increased.

$$R_t = R_1 + R_2$$

10. *B*

When resistors are connected in parallel, the effective resistance is decreased.

$$1/R_t = 1/R_1 + 1/R_2$$

Chapter 4

1. *B*

Kirchhoff formulated a voltage law and a current law, but not a resistance law.

2. *D*

For Kirchhoff's voltage law, circuits are divided into loops.

3. *C*

A node is a connection point between two or more conductors.

4. *A*

According to Kirchhoff's voltage law, the algebraic sum of all the voltage sources in a loop equals the algebraic sum of all voltage drops in that loop.

5. *B*

Current sources are not included in a loop for Kirchhoff's voltage law.

6. *D*

    To completely analyze a circuit according to Kirchhoff's current law, you need to look at one less than the total number of nodes in the circuit.

7. *C*

    Loop currents can be assumed to flow in either direction, as long as you are consistent throughout the entire circuit.

8. *D*

    According to Kirchhoff's current law, the algebraic sum of all currents entering and exiting a node is always zero.

9. *C*

    If a resistance element is part of two loops, two voltage drops must be calculated for that component.

10. *A*

    In Kirchhoff's current law, the terminal where the current enters the resistance element is assumed to be at a higher potential (more positive) than the other?

Chapter 5

1. *B*

    Alternating-current (ac) frequencies can be measured in cycles per second, hertz, kilocycles, or megahertz, but *waves per second* is meaningless.

2. *C*

    If an ac signal has an rms voltage of 75 V, the peak voltage is 105.75 V.

    $$PEAK = 1.41 \times rms$$

3. *A*

    If an ac signal has a peak voltage of 55 V, the average voltage is 34.98 V.

    $$AVERAGE = 0.636 \times PEAK$$

4. *D*

    If an ac signal has an average voltage of 18 V, the rms voltage is 19.98 V

    $$rms = 1.11 \times AVERAGE$$

5. *C*

    There 360 degrees in one complete wave cycle.

6. *B*

    Assuming sine waves, the rms voltage is always greater than the average voltage.

7. *A*

    If two equal frequency ac signals of equal voltages are combined when one is 180 degrees out of phase with each other, they will cancel each other out, and the resultant voltage will always be zero.

8. *C*

    If the combination of an ac voltage and a dc voltage has an instantaneous volt-age that varies through a range from -2 V to $+10$ V, the peak ac voltage is 6 V. (The peak-to-peak voltage is 12 V.)

9. *B*

    The rms value of the ac voltage in question 8 is 4.242 V.

    $$rms = 0.707 \times PEAK$$

10. *D*

    In question 7, the superimposed dc voltage must have a value of $+4$ V. For a peak voltage of 6 V, the ac zero line has been raised 4 V above true zero.

Chapter 6

1. *D*

    A dielectric is an insulator between the two metal plates in a capacitor.

2. *D*

    The primary function of a capacitor is to store electrical energy. There is some dc resistance in any practical capacitor, but the resistance is incidental.

3. *A*

    Low-frequency signals experience the greatest resistance through a capacitor. Capacitive reactance increases as the signal frequency decreases.

4. *E*

    An ideal capacitor, with no leakage always has a theoretical capacitive re-actance of infinity at dc (0 Hz). The capacitor value does not matter. (Notice that because infinity was not given as a choice for this question, the correct answer is "None of the above."

5. *B*

    The capacitive reactance of a 33 µF capacitor at 500 Hz is 9.55 Ω.

    $$X_c = 1/6.28FC$$
    $$= 1/(6.28 \times 500 \times 0.000033)$$
    $$= 1/0.10362$$
    $$= 9.55 \ \Omega$$

6. *A*

    The capacitive reactance of a 33 µF capacitor at 6500 Hz is 0.74 Ω.

    $$X_c = 1/6.28FC$$
    $$= 1/(6.28 \times 6500 \times 0.000033)$$
    $$= 1/1.34706$$
    $$= 0.74 \ \Omega$$

7. *C*

   The capacitive reactance of a 0.47 µF capacitor at 4300 Hz is 79 Ω.

   $$X_c = 1/6.28FC$$
   $$= 1/(6.28 \times 4300 \times 0.00000047)$$
   $$= 1/0.0126918$$
   $$= 79 \ \Omega$$

8. *C*

   If two 0.25 µF capacitors are connected in series, the total effective capacitance will be 0.125 µF.

   $$1/C_t = 1/C_1 + 1/C_2$$
   $$= 1/0.25 + 1/0.25$$
   $$= 4 + 4$$
   $$1/C_t = 8$$
   $$C_t = 1/8$$
   $$= 0.125 \ \mu F$$

9. *A*

   If two 0.25 µF capacitors are connected in parallel, the total effective capacitance will be 0.50 µF.

   $$C_t = C_1 + C_2$$
   $$= 0.25 + 0.25$$
   $$= 0.50 \ \mu F$$

10. *A*

    The cutoff frequency of a passive lowpass filter made up of a 0.56 µF capacitor and a 330 kΩ resistor is approximately 860 kHz.

    $$F = 159.000/RC$$
    $$= 159,000/(330,000 \times 0.00000056)$$
    $$= 159,000/0.1848$$
    $$= 860,389.61 \ Hz$$
    $$= 860 \ kHz$$

Chapter 7

1. *C*

   Carbon can not be used to make a magnet.

2. *A*

   The ends of a magnet are called poles.

3. *B*

   Another name for magnetic lines of force is *flux*.

4. *C*
   If like poles of two magnets are brought near each other, they will repel each other.

5. *B*
   The magnetic equivalent to electrical voltage is magnetomotive force.

6. *A*
   The magnetic equivalent to electrical current is flux.

7. *C*
   The magnetic equivalent to electrical resistance is reluctance.

8. *D*
   An electrical voltage be induced with a coil and a magnet by moving either the magnet or the coil.

9. *B*
   Rotating an armature in a magnetic field produces ac electricity.

10. *A*
    The frequency range for an ELF field is $0-100$ Hz.

Chapter 8

1. *B*
   An inductance tends to oppose changes in current.

2. *C*
   The reactance of a 25 mH coil at 5000 Hz is 785 Ω.
   $$X_1 = 6.28FL$$
   $$= 6.28 \times 5000 \times 0.025$$
   $$= 785\ \Omega$$

3. *B*
   The reactance of a 25 mH coil at 600 Hz is 785 Ω.
   $$X_1 = 6.28FL$$
   $$= 6.28 \times 600 \times 0.025$$
   $$= 94.2\ \Omega$$
   $$= 94\ \Omega$$

4. *A*
   The inductance of a single-layer coil on a 0.8 inch diameter nonmagnetic core with a length of 1.25 inch, and 320 turns of wire is 960 mH (0.96 H).
   $$L = (0.2d^2N^2)/(3d + 91)$$
   $$= (0.2 \times 0.8^2 \times 320^2)/(3 \times 0.8 + 9 \times 1.25)$$
   $$= (0.2 \times 0.64 \times 102{,}400)/(2.4 + 11.25)$$

$$= 13107.2/13.65$$
$$= 960 \text{ mH}$$

5. **C**

To make a 150 mH coil on a 0.75 nonmagnetic diameter core 1 inch long, you would need 122-½ turns of wire.

$$N = \sqrt{(L(3d + 91))/(0.2d^2)}$$
$$= \sqrt{(105 \times (3 \times 0.75 + 9 \times 1))/(0.2 \times 0.75^2)}$$
$$= \sqrt{(150 \times (2.25 + 9))/(0.2 \times 0.5625)}$$
$$= \sqrt{(150 \times 11.25)/0.1125}$$
$$= \sqrt{1687.5/0.1125}$$
$$= \sqrt{15,000}$$
$$= 122.5 \text{ turns}$$

6. **B**

The inductance of a 1.5 inch coil that is 500 turns on a magnetic core with a cross-sectional area of 0.35 inch, and a permeability rating of 750 is.

$$L = (4.06N^2\mu A)/(0.27 \times 10^8 \times 1)$$
$$= (4.06 \times 500^2 \times 750 \times 0.35)/(0.27 \times 10^8 \times 1.5)$$
$$= (4.06 \times 250,000 \times 262.5)/40,500,000$$
$$= 2,664,375,000/40,500,000$$
$$= 6.6 \text{ mH}$$

7. **D**

A 50-turn torroidal coil 0.15 inch high, with an outside diameter of 0.8 inch, an inside diameter of 0.3 inch and a permeability of 400 has an inductance of approximately 725 mH.

$$L = 0.011684N^2\mu h \times \log_{10}(OD/ID)$$
$$= 0.011684 \times 50^2 \times 400 \times 0.15 \times \log_{10}(0.8/0.3)$$
$$= 0.011684 \times 2500 \times 60 \times \log_{10}(2.666)$$
$$= 1753 \times 0.4150$$
$$= 727 \text{ mH}$$

This value can be rounded off to the 725 mH for convenience.

8. **A**

To double the inductance of the coil described in question 7, you would need 71 turns.

$$L = 0.011684N^2\mu h \times \log_{10}(OD/ID)$$
$$= 0.011684 \times 71^2 \times 400 \times 0.15 \times \log_{10}(0.8/0.3)$$
$$= 0.011684 \times 5041 \times 60 \times \log_{10}(2.666)$$
$$= 3533 \times 0.4150$$
$$= 1466 \text{ mH}$$

This value can be rounded off to the 1,450 mH for convenience.

9. *B*

The time constant of a 500 mH coil and a 3,300 Ω resistor in series is 0.0015 second.

$$T = L/R$$
$$= 0.500/3300$$
$$= 0.0015 \text{ second}$$

10. *C*

In a RL circuit, the time constant is the time required for the induced voltage to reach 63% of its full value.

Chapter 9

1. *B*

Three shielded 100 mII coils connected in series have a total effective inductance of 300 mH. Inductances in series add.

$$I_t = L_1 + L_2 + L_3$$
$$= 100 + 100 + 100$$
$$= 300 \text{ mH}$$

2. *A*

Three shielded 100 mH coils connected in parallel have a total effective inductance of 33.333 mH. The reciprocal of the total parallel inductance equals the sum of the reciprocals of the individual inductances.

$$1/L_t = 1/L_1 + 1/L_2 + 1/L_3$$
$$= 1/100 + 1/100 + 1/100$$
$$= 0.01 + 0.01 + 0.01$$
$$1/L_t = 0.03$$
$$L_t = 1/0.03$$
$$= 33.333 \text{ mH}$$

3. *D*

A transformer consists of two coils wound on a common core.

4. *B*

A transformer with 100 turns in the primary winding and 25 turns in the secondary winding is a step-down transformer.

5. *E*

If 250 V ac is fed to the primary coil of the transformer described in question 4, 62.5 V will be induced in the secondary winding (assuming 100% coupling).

$$INPUT\ VOLTAGE/PRIMARY\ TURNS = VOLTS\ PER\ WINDING$$
$$250/100 = 2.5$$
$$VOLTS\ PER\ WINDING \times SECONDARY\ TURNS = OUTPUT\ VOLTAGE$$
$$2.5 \times 25 = 62.5 \text{ V}$$

6. *D*

    In a step-up transformer, the current in the secondary is less than the current drawn by the primary.

7. *C*

    If the center tap of a transformer is grounded, the two ends of the secondary winding will be 180 degrees out of phase with each other.

8. *B*

    An autotransformer consists of just one coil.

9. *C*

    The effect of some of the magnetic field leaking off due to less than 100% coupling is called *leakage resistance*.

10. *A*

    Iron losses, stray capacitance, and leakage resistance can all contribute to losses in a transformer, but not self-inductance.

Chapter 10

1. *C*

    Impedance is the combination of capacitive reactance, inductive reactance, and dc resistance.

2. *D*

    Ignoring capacitive effects, the impedance of a 250 mH coil with an internal resistance of 55 Ω at 60 Hz is 109 Ω.

$$X_1 = 6.28FL$$
$$= 6.28 \times 60 \times 0.25$$
$$= 94.2 \ \Omega$$

$$Z = \sqrt{R^2 + X_1^2}$$
$$= \sqrt{55^2 + 94.2^2}$$
$$= \sqrt{3025 + 8874}$$
$$= \sqrt{11899}$$
$$= 109 \ \Omega$$

3. *C*

    Ignoring capacitive effects, a 100 mH coil (with an internal resistance of 45 Ω) in parallel with a 4700 Ω resistor has an impedance of 237 Ω at a frequency of 500 Hz.

$$X_1 = 6.28FL$$
$$= 6.28 \times 500 \times 0.1$$
$$= 314 \ \Omega$$

$$Z_a \text{ (coil only)} = \sqrt{R^2 + X_1^2}$$
$$= \sqrt{45^2 + 314^2}$$
$$= \sqrt{2025 + 98,596}$$
$$= \sqrt{100,621}$$
$$= 317 \ \Omega$$

$$Z_b \text{ (resistor only)} = R$$
$$= 4700 \ \Omega$$
$$1/Z_t = 1/Z_a + 1/Z_b$$
$$= 1/317 + 1/4700$$
$$= 0.003154 + 0.000213$$
$$1/Z_t = 0.003367$$
$$Z_t = 237 \ \Omega$$

4. *B*

Ignoring any inductive effects, the impedance of a RC series capacitor made up of a 56K resistor and a 0.033 μF capacitor at a signal frequency of 450 Hz is 57,019 Ω.

$$X_c = 1/(6.28FC)$$
$$= 1/(6.28 \times 450 \times 0.000000033)$$
$$= 1/0.0000932$$
$$= 10,730 \ \Omega$$
$$Z = \sqrt{R^2 + X_c^2}$$
$$= \sqrt{56000^2 + 10730^2}$$
$$= \sqrt{3136000000 + 115132900}$$
$$= \sqrt{3251132900}$$
$$= 57019 \ \Omega$$

5. *A*

Ignoring any effects of dc resistance, the total reactance of a 250 mH coil in series with a 4.7 μF capacitor at a signal frequency of 1000 Hz is 35 Ω.

$$X_1 = 6.28FL$$
$$= 6.28 \times 1000 \times 0.25$$
$$= 1570 \ \Omega$$
$$X_c = 1/(6.28FC)$$
$$= 1/(6.28 \times 1000 \times 0.0000047)$$
$$= 1/0.029516$$
$$= 34 \ \Omega$$
$$X_t = (X_1 \times X_c)/(X_1 - X_c)$$
$$= (1570 \times 34)/(1570 - 34)$$
$$= 53,380/1536$$
$$= 35 \ \Omega$$

6. *B*

Ignoring any effects of dc resistance, the total reactance of a 250 mH coil in series with a 4.7 μF capacitor at a signal frequency of 450 Hz is 84 Ω.

$$X_1 = 6.28FL$$
$$= 6.28 \times 450 \times 0.25$$
$$= 706\ \Omega$$
$$X_c = 1/\ (6.29FC)$$
$$= 1/(6.28 \times 450 \times 0.0000047)$$
$$= 1/0.01328$$
$$= 75\ \Omega$$
$$X_t = (X_1 \times X_c)/(X_1 - X_c)$$
$$= (706 \times 75)/(706 - 75)$$
$$= 52950/631$$
$$= 84\ \Omega$$

7. *D*

Ignoring any effects of dc resistance, the total reactance of a 250 mH coil in series with a 4.7 μF capacitor at a signal frequency of 60 Hz is −113 Ω. (The negative sign simply indicates that the capacitive reactance is greater than the inductive reactance.)

$$X_1 = 6.28FL$$
$$= 6.28 \times 60 \times 0.25$$
$$= 94\ \Omega$$
$$X_c = 1/(6.28FC)$$
$$= 1/(6.28 \times 60 \times 0.0000047)$$
$$= 1/0.00177$$
$$= 565\ \Omega$$
$$X_t = (X_1 \times X_c)/(X_1 - X_c)$$
$$= (94 \times 565)/(94 - 565)$$
$$= 53110/-471$$
$$= -113\ \Omega$$

8. *A*

In a series resonant LC circuit, the impedance at the resonant frequency is determined solely by the dc resistance.

$$Z = \sqrt{(X_1 - X_c)^2 + R^2}$$
$$= \sqrt{0 + R^2}$$
$$= \sqrt{R^2}$$
$$= R$$

9. *D*

In a parallel resonant LC circuit, the impedance at the resonant frequency is nominally infinite.

$$X_t = (X_1 \times X_c)/(X_1 - X_c)$$
$$= (X_1 \times X_c)/0$$
$$= \infty \text{ (theoretical infinity)}$$

10. *B*

If you need an LC circuit to be resonant at 2500 Hz, and use a 150 mH coil, then the capacitance value should be 0.027 μF.

$$C = 1/(39.44F^2L)$$
$$= 1/(39.44 \times 2500^2 \times 0.15)$$
$$= 1/(39.44 \times 6250000 \times 0.15)$$
$$= 1/36975000$$
$$= 0.000000027 \text{ F}$$
$$= 0.027 \text{ μF}$$

In a practical circuit, you would probably use either a 0.022 μF capacitor or a 0.033 μF capacitor. These are the two closest standard values.

Chapter 11

1. *B*

Crystals used in electronics are made of quartz.

2. *D*

A crystal uses a hermetically sealed housing primarily to keep out moisture and dust.

3. *A*

The principle behind the operation of crystal is the piezoelectric effect.

4. *A*

When a mechanical stress is placed along the Y axis of a crystal, an electrical voltage will appear across the X axis.

5. *D*

A crystal can be used in place of either a series resonant or a parallel resonant circuit.

6. *C*

The resonant frequency of a crystal is determined primarily by the size and thickness of the crystal material.

7. *B*

The third harmonic of a crystal with a resonant frequency of 47,000 Hz is 141,000 Hz.

The *N*th harmonic equals $N \times$ the fundamental frequency.
$$3 \times 47,000 = 141,000 \text{ Hz}$$

8. *C*

    A crystal oven should be used when extremely high accuracy is required to prevent any thermal drift effects.

9. *C*

    The biggest disadvantage of using crystals in resonant circuits is that the resonant frequency can not be easily changed, especially during operation.

10. *B*

    The biggest advantage of using crystals in resonant circuits is greater accuracy and stability than with resonant networks made up of capacitors and coils.

Chapter 12

1. *C*

    Impedance can not be easily measured with a simple meter circuit.

2. *A*

    The most common type of meter movement is the D'Arsonval meter.

3. *D*

    An ammeter is used to measure current.

4. *C*

    An ammeter should have as little internal resistance as possible.

5. *B*

    To increase the range of an ammeter, a shunt resistor in parallel with the meter should be added to the circuit.

6. *A*

    A voltmeter is placed in parallel across the component being measured.

7. *B*

    An ohmmeter requires its own power source.

8. *C*

    A VOM is a combination voltmeter, ohmmeter, and milliammeter.

9. *C*

    For the greatest accuracy, the input impedance of a VOM should be as large as possible. The 1,000,000 Ω/V rating suggested as answer E would be very good, but an even higher input impedance would be even more accurate.

10. *D*

    Alternating-current (ac) voltmeters measure the rms voltage of a sine wave.

Chapter 13

1. *C*

    The simplest type of switch is the knife switch.

2. *B*
A DPDT can simultaneously control two separate circuits.

3. *E*
SPST, SPDT, DPST, and DPDT are all standard switch types (although DPST switches are uncommon), so the correct answer is "None of the above."

4. *D*
In Figure 13-6, if R_1 is 2200 Ω & R_2 is 3900 Ω, the circuit resistance will be 1407 Ω when the switch is closed.

$$
\begin{aligned}
R_t &= (R_1 \times R_2)/(R_1 + R_2) \\
&= (2200 \times 3900)/(2200 + 3900) \\
&= 8580000/6100 \\
&= 1407 \ \Omega
\end{aligned}
$$

5. *B*
A switch that disconnects the circuit completely at one position before the connection at the next position is made is called a shorting switch.

6. *C*
When a momentary action switch is released, the switch contacts return to their normal rest position.

7. *D*
A relay contains a coil and a set of switch contacts.

8. *C*
A coil is placed across the coil portion of a relay to protect it from high-voltage transients when the magnetic field collapses.

9. *D*
A device that protects a load circuit from excessive current flow is called a *fuse*.

10. *B*
A fuse should never be replaced with a higher rated unit under any circumstances. Doing so would defeat the purpose of the fuse.

Chapter 17

1. *C*
The active parts of a diode tube are the anode and the cathode.

2. *A*
The simplest type of tube capable of amplification is the triode.

3. *B*
A diode will conduct only when it is forward biased.

4. *B*
The grid in a tube can be described as a metallic mesh.

5. *A*

    If the grid is made more positive than the saturation point, the electrons from the cathode will be drawn to the grid, and will not reach the plate.

6. *D*

    The maximum gain a tube is capable of is defined as μ (amplification factor).

7. *B*

    The purpose of the screen grid is to reduce the effect of interelectrode capacitances.

8. *D*

    A pentode has five electrodes.

9. *C*

    The electrode included in a pentode, but not in a tetrode, is called the suppressor grid.

10. *B*

    A cathode-ray tube is used to display signals in an oscilloscope.

Chapter 18

1. *C*

    Germanium is a semiconductor. Copper is a conductor. Rubber and carbon are insulators.

2. *C*

    A P-type semiconductor has a shortage of holes.

3. *B*

    There is only one junction in a semiconductor diode.

4. *A*

    The arrow in the schematic symbol points toward the cathode.

5. *C*

    The most important specification for a semiconductor diode is the peak inverse voltage, or PIV. (This specification is sometimes called *peak reverse voltage* or *PRV*.

6. *A*

    When the voltage applied to a zener diode exceeds its avalanche point, the current through the diode abruptly rises to a very high value.

7. *D*

    A zener diode is most likely to be used in a voltage regulator circuit.

8. *B*

    A varactor diode is also known as a voltage-controlled capacitor.

9. *D*

"Peak inverse voltage" is not another name for "trigger voltage."

10. *C*

Another name for a four-layer diode is a Shockley diode.

Chapter 19

1. *C*

The correct operating voltage for a typical LED is +3 to +6 V.

2. *B*

An LED glows when it is forward biased.

3. *A*

Current through an LED is limited to a safe value by a small value series resistor.

4. *C*

Seven segments are needed to display any digit.

5. *C*

The purpose of an optoisolator is to provide isolation between a controlling and a controlled circuit.

6. *D*

The normal flash rate of a flasher LED is about three times a second.

7. *A*

LCD stands for *liquid-crystal display.*

8. *D*

The visibility of an LED is affected by bright ambient light, so this is not an advantage of the component.

9. *A*

Another name for a laser diode is injection laser.

10. *B*

Laser light is coherent, focused, and monochromatic, but not diffuse.

Chapter 20

1. *B*

The most basic type of transistor is the bipolar transistor.

2. *A*

The three leads to a standard bipolar transistor are called the emitter, base, and collector.

3. *D*

   To ensure maximum heat transfer from a transistor, smear the transistor surface with silicon grease.

4. *B*

   The three basic transistor circuit types are the common-collector, common-emitter, and common-base circuits. There is no such thing as a common-cathode transistor circuit, because a transistor does not have a cathode.

5. *C*

   About 95% of the current through a transistor will flow through the collector.

6. *C*

   The input and output of a common-emitter amplifier are 180 degrees out of phase with each other.

7. *A*

   The input impedance in a common-base amplifier is very low.

8. *B*

   For any practical transistor, the beta value is considerably larger than the alpha value.

9. *D*

   The correct formula for determining the beta of a transistor is $B = a/(1 - a)$.

10. *B*

    The biggest difference between an NPN transistor circuit and a PNP transistor circuit is that all polarities are reversed.

Chapter 21

1. *B*

   A Darlington pair is made up of two similar bipolar transistors in series.

2. *A*

   The three leads on a UJT are labelled emitter, base 1, and base 2.

3. *B*

   There are two PN junctions in a FET.

4. *D*

   The type of transistor that most closely resembles a vacuum tube in its operating characteristics is the FET.

5. *D*

   JFETs, MOSFETs, and IGFETs are all variations on the basic FET. An EEFET does not exist.

6. *A*

   The best description of an SCR is a diode with an electrically operated switch.

7. *B*
   A diac is a dual-polarity trigger diode.

8. *A*
   The three leads of an SCR are called the anode, the cathode, and the gate.

9. *C*
   The primary advantage of advantage of a VMOSFET over a MOSFET is that the VMOSFET can operate at much higher power levels.

10. *B*
    The only component from those listed that can operate at microwave frequencies of the Gunn diode.

Chapter 22

1. *B*
   A component that changes its resistance in response to light intensity is called a photoresistor.

2. *A*
   A typical photovoltaic cell has a nominal voltage of 0.5 V.

3. *D*
   Light sensitive devices depend upon the photoelectric effect for their operation.

4. *D*
   A photodiode, a LASCR, and a photoresistor are all light-sensitive transducers, but a SCR is not.

5. *B*
   Semiconductors are more likely to be photosensitive than either conductors or insulators.

6. *C*
   The base lead is often omitted in a phototransistor.

7. *C*
   Ordinary transistors are not sensitive to light because they are normally enclosed in light-tight housings.

8. *D*
   A photoresistor does not have any PN junctions at all. It is a junctionless device. Therefore, it has no particular polarity.

9. *A*
   Most (but not all) photovoltaic cells are made up primarily of silicon.

10. *B*
    A photovoltaic cell performs a function similar to that of a dc battery.

Chapter 24

1. *A*

   SSI (small-scale integration) ICs are, by definition, the least complex.

2. *C*

   The words *linear* and *analog* are essentially synonymous for electronics work.

3. *A*

   The inverting input of an op amp will shift a signal 180 degrees.

4. *D*

   If an inverting amplifier circuit has a value of 2.2 kΩ for R_1 and 18 kΩ for R_2, the gain will be approximately -8.2.

$$G = -R_2/R_1$$
$$= -18000/2200$$
$$= -8.1818$$
$$= -8.2$$

5. *B*

   Assuming a gain of 1, if 0.27 is fed to the inverting input of an op amp and -0.33 is fed to the noninverting input, the output will be equal to -0.60 V

$$V_o = (V_{in} + \times G) - (V_{in} - \times G)$$
$$= (-0.33 \times 1) - (0.27 \times 1)$$
$$= -0.33 - 0.27$$
$$= -0.60 \text{ V}$$

6. *B*

   The 741, the 709, and the 748 are all common op amp ICs. The 555 is a timer IC (see chapter 25).

7. *D*

   The 324 IC contains four independent op amps.

8. *B*

   An op amp circuit using a resistor as the input component and a capacitor as the feedback component is called an integrator.

9. *D*

   The XR-2208 is an operational multiplier IC.

10. *B*

    In a differentiator circuit with a 22 kΩ (22,000 Ω) resistor, a 0.01 μF (0.00000001 F) and an input frequency of 375 Hz, the gain is approximately -0.5.

$$G = -6.28FCR_2$$
$$= -6.28 \times 375 \times 0.00000001 \times 22000$$

$$= -0.5181$$
$$= -0.5$$

Chapter 25

1. *B*

   A monostable multivibrator has one stable output state.

2. *A*

   The name of the multivibrator circuit with no stable output states is astable.

3. *B*

   The output pulse of a 555 monostable multivibrator circuit when the timing resistor is 560 kΩ and the timing capacitor is 0.22 μF is about 0.13

   $$T = 1.1RC$$
   $$= 1.1 \times 560000 \times 0.00000022$$
   $$= 0.13552 \text{ second}$$
   $$= 0.13 \text{ second}$$

4. *D*

   Approximately the longest timing period that can be achieved with a standard single-stage 555 monostable multivibrator circuit is 11,000 seconds.

5. *C*

   Assume an astable multivibrator circuit built around a 555 IC uses the following component values:

   $$R_a = 33 \text{ k}\Omega$$
   $$R_b = 27 \text{ k}\Omega$$
   $$C_t = 0.1 \text{ μF}$$

   The output frequency of this circuit is about 166 Hz.

   $$F = 1.44/((R_a + 2R_b)C_t)$$
   $$= 1.44/((33000 + 2 \times 27000) \times 0.0000001)$$
   $$= 1.44/((33000 + 54000) \times 0.0000001)$$
   $$= 1.44/(87000 \times 0.0000001)$$
   $$= 1.44/0.0087$$
   $$= 166 \text{ Hz}$$

6. *B*

   The correct formula for finding the output frequency of a standard 555 astable multivibrator circuit is:

   $$F = 1.44/((R_a + 2R_b)C_t)$$

7. *A*

   The wave shape at the output of a standard 555 astable multivibrator circuit is a rectangular wave. A square wave is a special form of the rectangular wave, but it is impossible to achieve a true square wave in a standard 555 timer circuit.

8. *C*

There are four timer stages in the 558 chip.

9. *C*

The 3905 is not a variation on the basic 555 timer. The 558 is a quad 555 timer, the 556 is a dual 555 timer, and the 7555 is a CMOS version of the basic 555 timer.

10. *B*

In a monostable multivibrator circuit built around the 322 precision timer IC, using a timing resistor of 270 kΩ and a timing capacitor of 33 μF, the timing period is approximately 8.9 seconds.

$$T = RC$$
$$= 270000 \times 0.000033$$
$$= 8.9 \text{ seconds}$$

Chapter 26

1. *B*

1001 0111 binary in decimal is 151.

2. *C*

169 expressed in binary is 1010 1001.

3. *A*

The action of an OR gate can be described as the output goes HIGH when one or more of the inputs is HIGH.

4. *B*

An NAND gate be made from an AND gate with inverters at the inputs.

5. *A*

If the input to an inverter is a logic 1, the output is 0. Remember, the output of an inverter is always at the opposite logic state as its input.

6. *B*

The base of the octal numbering system is eight.

7. *C*

An exclusive-OR gate does not produce a 1 at the output when all inputs are 1s. All of the other gate types suggested as possible answers to this question do have a 1 output under these conditions.

8. *D*

The only input combinations that will produce a logic 1 output from a three input NOR gate is 000.

9. *A*

If both inputs of an OR gate are inverted, the result will be the same as an NAND gate.

10. *C*

A majority logic gate might have any odd number of inputs. (One does not count as an odd number in this case.) If there were an even number of inputs, there could be an equal tie, with no majority condition to control the output.

Chapter 27

1. *C*

Another name for a flip-flop circuit is a bistable multivibrator.

2. *A*

When a D-type flip-flop is triggered, the outputs reverse states.

3. *C*

JK, RS, and D flip-flops are all standard and common. There is no such thing as a QS flip-flop.

4. *C*

A multistage counter can be built up from several flip-flop stages.

5. *B*

The maximum count of a five stage counter in binary is 11111.

6. *D*

The maximum count of a five-stage counter in decimal is 31. There are 32 counting steps, but that includes a count of 0, so the maximum count value is always one less than the number of counting steps.

7. *A*

If the Q output of a counter stage is at logic 1, then the $\overline{Q}$ output must be at logic 0. The $\overline{Q}$ output is always at the opposite state as the Q output, by definition.

8. *D*

A counter stage can divide an input frequency by a factor of 2. There is one complete output pulse cycles for every two input pulse cycles.

9. *D*

1101 is a disallowed state for a BCD circuit, because it has a decimal value of 11.

10. *A*

A six-stage PISO shift register has one output. *SO* means *serial output*, implying a single input, regardless of the number of stages.

Chapter 28

1. *A*

   VMOS is not a standard logic family but a type of semiconductor construction (see chapter 21).

2. *C*

   The DTL logic family has been obsolete for a long time now and is no longer commercially available.

3. *B*

   Of the choices given, TTL is the one most commonly used by electronics hobbyists, although in recent years, TTL has been clearly over-taken by CMOS in popularity.

4. *D*

   TTL devices require a well-regulated +5 V power supply.

5. *B*

   CMOS devices can use power supply voltages ranging from 3 to 15 V. +12 V will usually give the best overall performance.

6. *A*

   A TTL device that uses Schottky diodes in its internal circuitry always has an *S* in the middle of its identification number, so from the choices given, the 74LS04. (The *L* stands for low power.)

7. *D*

   Of the choices given, the ECL logic family has the highest switching speed capability.

8. *A*

   CMOS devices can be easily damaged by static electricity.

9. *C*

   The 7404 is not a CMOS device. It is a TTL chip.

10. *B*

    The CMOS logic family can stand the greatest degree of variation in the supply voltage. TTL and its subfamilies can tolerate little or no supply voltage fluctuation. The supply voltage for TTL must be a well-regulated +5 V.

Chapter 29

1. *A*

   The CPU section of the computer is the brain, executing the actual program instructions.

2. *D*

   There is no inherent difference between a data byte and a command byte. The only difference is in how they are used by the CPU, which is determined by the software itself.

3. *C*

There are eight bits are there in a standard byte.

4. *A*

A program is a series of commands to perform a specific task.

5. *B*

A CPU can understand only machine language directly. All other programming languages require an interpreter or compiler.

6. *A*

A bus is a digital signal line.

7. *B*

ROM is the type of memory that is not user programmable.

8. *E*

When power is interrupted the data stored in ordinary RAM is lost.

9. *D*

A modem is not a type of long-term data storage.

10. *B*

Assuming a simple, direct 16-bit address bus, the maximum normal memory size is 64 kB. In modern personal computers bank switching or some other special technique is used to achieve much larger usable memory sizes.

Chapter 30

1. *A*

A sensor that reacts to light is photosensitive.

2. *C*

Use only low-power dc for a touch-switch circuit. Never use ac power in such a circuit.

3. *A*

The piezoelectric effect can be described as a mechanical stress along the X axis produces an electrical signal along the Y axis.

4. *A*

A condenser microphone requires its own dc voltage supply (usually a small battery).

5. *A*

Skin resistance is used in lie detectors.

6. *D*

A typical application of hand capacitance is in touch switches.

7. *C*

A crystal can be used as a pressure sensor.

8. *B*

In a thermistor resistance varies with temperature.

9. *E*

Carbon, ribbon, condenser, and dynamic are all common microphone types, so "None of the above" is the correct answer.

10. *C*

A woofer is a speaker specifically designed to be good at reproducing low frequencies.

Chapter 32

1. *B*

Batteries supply dc power only.

2. *D*

A diode can be used as a half-wave rectifier. In fact, diodes are often simply called *rectifiers*.

3. *B*

The biggest disadvantage of the half-wave rectifier circuit is that it wastes half of each input ac cycle.

4. *A*

The purpose of the output capacitor in a power supply circuit is to filter out ac ripple.

5. *B*

A minimum of two diodes are needed for a full-wave rectifier circuit.

6. *D*

A bridge rectifier consists of four diodes.

7. *C*

The advantage of using a bridge rectifier for full-wave rectification is that no center tap is required on the transformer.

8. *C*

A typical voltage regulator IC has three leads—input, output, and common.

9. *C*

If four 33 kΩ resistors are used in the circuit of Fig. 32-15, and V+ equals 18 V, then the voltage reading on meter M2 will be ¾ of the full supply voltage, or 13.5 V.

10. *B*

The purpose of a zener diode in a power supply circuit is to provide voltage regulation.

Chapter 33

1. *C*
   An amplifier alters the level of an input signal.

2. *C*
   Class A, Class B, Class C, and Class AB are all standard amplifiers, but Class BC is not.

3. *B*
   A push-pull amplifier circuit is subject to crossover distortion because the negative and positive portions of the signal are amplified separately.

4. *A*
   A buffer amplifier has a gain of unity, or one. The amplitude of the output signal is the same as the amplitude of the input signal.

5. *D*
   A Class A exhibits both low efficiency and low distortion.

6. *B*
   A Class C exhibits both high efficiency and high distortion.

7. *C*
   A Class B amplifier is biased so it is cut off for half of each cycle.

8. *D*
   A Class C amplifier offers the highest efficiency.

9. *A*
   A Class A amplifier offers the lowest distortion.

10. *C*
    In a push-pull amplifier circuit, you should use one NPN transistor and one PNP transistor.

Chapter 34

1. *B*
   The simplest, purest waveform is the sine wave.

2. *C*
   Hartley oscillators, crystal oscillators, and Colpitts oscillators all produce sine waves, but a multivibrator generates a rectangular wave.

3. *A*
   A Hartley oscillator has a split inductance.

4. *C*
   If the two capacitors in the LC tank of a Colpitts oscillator have values of 0.01 μF and 0.0047 μF, the total effective capacitance is 0.0032 μF.

$$1/C_t = 1/C_1 + 1/C_2$$
$$= 1/0.01 + 1/0.0047$$
$$= 100 + 213$$
$$1/C_t = 313$$
$$C_t = 1/313$$
$$= 0.0032 \ \mu F$$

5. *A*

A triangular wave is sometimes called a delta wave.

6. *D*

In a rectangular wave with a duty cycle of 1:4, every fourth harmonic is omitted, so the lower harmonic content of this signal includes the fundamental, and the second, third, fifth, sixth, seventh, ninth, and tenth harmonics.

7. *A*

A square wave has a duty cycle of 1:2.

8. *B*

All harmonics are stronger in amplitude in a square wave than the comparable harmonic in a triangle wave of the same frequency. The fifth harmonic is stronger in a square wave.

9. *D*

The square-wave oscillator made up of four inverter stages and a capacitor (Fig. 34-22) has a frequency range of 1 Hz to 1 MHz.

10. *D*

A waveform that begins at a low level, smoothly builds up to a maximum level, then quickly snaps back to the minimum and starts over is an ascending sawtooth wave.

Chapter 35

1. *A*

A passive lowpass filter is made up of a series resistor and a capacitor to ground.

2. *C*

A bandpass filter rejects all but a specified continuous range of frequencies.

3. *C*

The nominal cutoff frequency of a highpass passive filter made up of a 680 Ω resistor and a 0.5 μF capacitor is approximately 468 Hz.

$$F = 159000/RC$$
$$= 159000/(680 \times 0.5)$$
$$= 159000/340$$
$$= 468 \ Hz$$

Remember, for this equation, the frequency is in Hertz, the resistance is in ohms, and the capacitance is in microfarads.

4. *A*

To make a passive lowpass filter with a cutoff frequency of 1000 Hz, using a 470 Ω resistor, the capacitor value should be 0.33 μF

$$
\begin{aligned}
C &= 159000/FR \\
&= 159000/(1000 \times 470) \\
&= 159000/470000 \\
&= 0.33~\mu\text{F}
\end{aligned}
$$

Remember, for this equation, the frequency is in Hertz, the resistance is in ohms, and the capacitance is in microfarads.

5. *B*

In the active lowpass filter circuit of Fig. 35-4, you are assuming the following component values:

R_1 1000 Ω
R_2 2200 Ω
R_3 10000 Ω
C_1 0.5 μF
C_2 0.01 μF

The nominal cutoff frequency of this circuit is approximately equal to 480 Hz.

$$
\begin{aligned}
F &= 1/(6.28~\sqrt{(R_2\,R_3\,C_1\,C_2)} \\
&= {}^1/(6.28~\sqrt{(2200 \times 10000 \times 0.0000005 \times 0.00000001)}) \\
&= 1/(6.28 \times \sqrt{0.0000011}) \\
&= 1/(6.28 \times 0.000332) \\
&= 1/0.00208 \\
&= 480~\text{Hz}
\end{aligned}
$$

6. *D*

A passive bandpass filter can be created by adding an inductor in parallel with the capacitor in a passive lowpass filter circuit.

7. *B*

The bandwidth of a passive bandpass filter is defined as:

$$BW = 159000/RC$$

8. *A*

The Q of a bandpass filter is equal to the center frequency divided by the bandwidth.

$$Q = F_r/BW$$

9. *A*

    A band-reject filter is also known as a notch filter because of the appearance of
    its filter response graph.

10. *C*

    In a passive lowpass filter made up of a 390 Ω resistor and a 0.1 μF capacitor,
    a 1.8 mH inductor should be added for a center frequency of about 12,000 Hz.

    $$L = 1/(4\pi^2 F^2\, C)$$
    $$= 1/(4 \times 3.14^2 \times 12000^2 \times 0.0000001)$$
    $$= 1/(4 \times 9.86 \times 144000000 \times 0.0000001)$$
    $$= 1/568$$
    $$= 0.0018 \text{ H}$$
    $$= 1.8 \text{ mH}$$

Chapter 36

1. *D*

    Amplitude modulation, frequency modulation, and phase modulation are all
    commonly used for transmission or storage of information. Ring modulation is
    not normally used for transmission or storage.

2. *A*

    The frequency of the upper sideband produced when a 50,000 Hz carrier signal
    is amplitude modulated by a 4500 Hz program signal is 54,500 Hz.

    $$F_{us} = F_c + F_p$$

3. *B*

    Two sidebands are produced in an AM signal (assuming sine waves).

4. *C*

    When a 2500 Hz carrier is frequency modulated by a 300 Hz program signal
    so the carrier signal varies between 1900 Hz and 3100 Hz produces four side-
    bands.

    $$\Delta F = 3100 - 1900$$
    $$= 1200 \text{ Hz}$$

    $$\text{MI} = \Delta F / F_p$$
    $$= 1200/300$$
    $$= 4$$

5. *D*

    In the previous question, raising the program signal amplitude so the carrier
    signal is varied between 1000 Hz and 4000 Hz, will increase the number of
    sidebands generated to:

    $$\Delta F = 4000 - 1000$$
    $$= 3000 \text{ Hz}$$

$$\begin{aligned}
\text{MI} &= \Delta F/F_{\text{p}} \\
&= 3000/300 \\
&= 10
\end{aligned}$$

6. *C*
   Pulse-code modulation is used to transmit analog data in digital equipment.

7. *D*
   A digital signal can be converted into an analog signal with a D/A converter. (*D/A* stands for *digital to analog*.)

8. *A*
   Assuming sine waves, if a 15,000 Hz carrier signal is amplitude modulated by a 2000 Hz program signal, only two sidebands will be produced. The frequencies do not matter in AM sideband generation, only the number of harmonics included in the carrier and program signals. There are two sidebands generated for each pair of frequency components in the input signals.

9. *A*
   Two signals whose frequencies are close together are superimposed and mixed into a single tone with amplitude fluctuations produces an effect known as heterodyning.

10. *B*
    The modulation index determines the number of sidebands in an FM signal.

Chapter 37

1. *B*
   Electromagnetic waves are used for radio transmission. (In some cases they might be sine waves, but not necessarily. The transmitted waveforms can be either symmetrical or asymmetrical.)

2. *B*
   500 kHz is included in the MF portion of the radio spectrum.

3. *D*
   The simplest form of radio transmission uses Morse code, in which the carrier is simply turned on and off. There is no modulating program signal.

4. *A*
   A phase-locked loop (or PLL) is used to hold a VCO on frequency.

5. *C*
   Reflected waves are not a basic type of signal path for radio transmission. Radio transmissions can be received by reflected waves, but reflected waves are usually a distortion or interference problem, not an intentional means of transmission.

6. *B*
   A half-wave antenna to receive signals at 65,000,000 Hz should be 7.2 feet long.

$$X = 468/F$$
$$= 468/65$$
$$= 7.2 \text{ feet}$$

The frequency must be in Megahertz for this equation.

7. *B*

A half-wave dipole exhibits two lobes.

8. *C*

A reflector for a half-wave antenna for 100 MHz signals should be 4.91 feet long. (About 5% longer than the half-wave dipole itself.)

$$X = (468/F) \times 1.05$$
$$= (468/100) \times 1.05$$
$$= 4.68 \times 1.05$$
$$= 4.91 \text{ feet}$$

9. *C*

SWR is usually caused by mismatched impedances.

10. *D*

AM signals can be demodulated with a diode.

Chapter 38

1. *B*

A system with two sound sources to allow more realistic effects is called stereophonic.

2. *A*

The highest frequency that can be recorded on a magnetic tape recorder is determined primarily by the recording head gap.

3. *B*

3.75 ips and 15 ips are standard speeds for reel-to-reel tape recorders. 1.875 is the standard tape speed for cassette tapes. 33.33 ips is not a standard speed for tape recorders.

4. *B*

Two channels are used on a four-track stereo tape. There are two complete programs (one forward and one backwards), each with two tracks for the right and left channels.

5. *D*

A stereo FM station transmits (L + R) and (L − R) signals.

6. *C*

The pilot signal in an FM stereo broadcast signal has a frequency of 19 kHz.

7. *A*

   A magnetic cartridge requires a preamplifier because of its low output amplitude.

8. *C*

   When a monaural FM receiver is tuned to a stereo FM station, it receives only
   the (L + R) signal.

9. *B*

   SQ is a matrixing system for quadraphonic sound.

10. *E*

    There is now no legal standard has been set for AM stereo broadcast.

Chapter 39

1. *B*

   The purpose of the yoke is to deflect the electron beam to the desired point on
   the screen.

2. *A*

   In the U.S., 525 complete lines make up a full frame.

3. *D*

   There are 30 frames a second, each made up of two interlaced fields (60 fields
   a second).

4. *B*

   The horizontal frequency in a television set is 15,750 Hz.

5. *C*

   There are three electron guns in a color television set.

6. *B*

   A broadcast television signal has a bandwidth of about 6 MHz.

7. *B*

   In modern U.S. televisions, the highest available channel is 69.

8. *A*

   The most negative voltage in a television signal represents the whitest portions
   of the displayed image.

9. *C*

   The color subcarrier frequency is approximately 3.58 MHz.

10. *D*

    Coaxial cable used with television aerials has an impedance of 75 Ω.

Chapter 41

1. *C*
   A block diagram is used in troubleshooting to aid the technician in narrowing down the stages that could cause the fault in question.

2. *A*
   If a television set doesn't work at all, the power supply is a likely source of the problem, because it affects all other stages.

3. *B*
   The higher the input impedance, the more accurate the voltmeter will be, so the 11 MΩ VTVM is the most accurate of the choices offered.

4. *B*
   Inductance can not be directly measured with a standard VOM.

5. *E*
   The only real advantage of a VTVM over a VOM is the higher input impedance, which is not one of the offered choices, so the answer is "None of the above."

6. *C*
   The resistance the voltage drop is being measured across (Rx) must be much smaller than the meter input resistance (R_m) for maximum accuracy and minimum loading.

7. *A*
   The oscilloscope is the second most important piece of test equipment for the average electronics technician. The multimeter is the most important.

8. *B*
   The vertical gain on an oscilloscope is used to calibrate the displayed voltage a specific distance above or below the baseline.

9. *D*
   Defective resistors might be changed in value, burned, or cracked, but not leaky. A leaky component is one that acts like a resistor when it isn't supposed to function that way.

10. *A*
    If an ohmmeter measures a small resistance across a semiconductor, reversing the polarity of the ohmmeter leads will result in a much higher resistance reading on the meter.

Index

A

absolute value of modulation signal, 515
active devices, 218
alkaline batteries, 12-13
alternating current (ac), 48-62
  adaptors, 52
  adding out of phase voltage, 56, 61
  average voltage, 54-55, 60, 61
  capacitor in ac circuit, 65-66
  common ground, 52-53
  current flow, 48-50
  dc combined with ac, 59-62
  degrees of cycle, 56, 61
  direct current, 59-62
  grounding, 52-53
  in phase, 50, 55
  meters, 159-160
  out of phase, 55
  peak voltage, 54-55, 60
  peak-to-peak voltage, 54-55, 60
  plugs, 52
  polarity, 48-50
  root mean square (rms), 54-55, 60, 61
  sine wave, 50, 60-61
  sources of ac voltage, 50-53
  transformers, 51
  vector diagrams, 57-59
  voltage, polarity, 48-50
ammeters, 153-155, 161
ampere, 8
amplification, 218
amplification factor of triodes, 226, 237
amplifiers (*see also* op amps), 2, 467-477
  asymmetrical clipping, 473
  audio, 350, 546-547
  audio amplifier, 546-547
  biasing, 468
  buffers or buffer amplifiers, 467
  classes of amplifiers, 467-468, 475-476
  clipping, asymmetrical, 473
  common-base amplifier, 281, 330-332
  common-collector amplifier, 282, 333
  common-emitter amplifier, 280-281, 332-333
  conduction, 474
  crossover distortion, 472, 476

difference amplifier, 347
distortion, 477
inverting op amps, 342-343, 357
noninverting op amps, 343-344
Norton amplifiers, 349-350
output symmetry, 473
push-pull amplifiers, 471-472, 477
RF, 350
transistor amplifier configurations, 280-283
voltage-controlled amplifiers (VCA), 515
amplitude modulation (AM), 514-516, 521, 522
  stereo, 560-563, 567
  transmitter circuit, 529-530, 549
analog delay, SAD 1024, 355-356
analog-to-digital (A/D) conversion, 518-521
AND gates, 374-375
anode, diode, 219
antennas, 525, 528, 530-536, 594-600
  balun, 599
  coaxial cabling, 598-599, 610
  dipole, 533, 549, 596-597
  direct waves, 530-531
  directional (movable), 595
  directors, 536-537
  fading area, 531
  filtering action of, 534
  gain, 536
  ghosting in television reception, 594-595
  ground waves, 530-531, 549
  half-wave, 533-534, 549
  hand capacitance/resistance, 428-430
  hertz, 532-533
  lobes in reception pattern, 535
  Marconi antenna, 532-533
  mounting, safety guidelines, 599-600
  nulls in reception pattern, 535
  parasitic elements, 536-537
  propagation of radio waves, 531
  quarter-wave, 532
  reception patterns for, 534-535
  reflectors, 536-537, 549
  service areas, radio stations, 531
  sky waves, 530-532, 549
  standing wave ratio (SWR), 536-537, 549

twinlead, 300-ohm , 598
Yagi antennas, 536, 597-598
arithmetic logic unit (ALU), 418
armature of meters, 151
assembly-language programming, 419
asymmetrical clipping in amplifiers, 473
atomic theory, electrons and electricity, 4-7, 9, 17, 240-241
attenuation, 77-78, 218
automatic gain control (AGC) in television, 579
autotransformers, 124-125, 129
avalanche point of zener diodes, 247
axes of crystals, 144, 149

B

balun, antennas, 599
bar graph displays, 257-260
BARITT diodes, 310-311
  punch-through voltage, 311
batteries, 9-15, 17-18, 465
biasing amplifiers, 468
biasing semiconductors, 244-245
binary system, 372-373, 382-383
biorhythms and electromagnetic fields, 97-102
bits, 397, 419, 426
bleeder resistors, 464
block diagrams, 184-185, 629-630
  troubleshooting and repair, 645
breadboarding techniques, 188-190
breakdown voltage of capacitors, 70
buffering in op amps, 343
buffers, 377-378
  buffer amplifiers, 467
bus, 419, 427
bytes, 397, 426

C

cable TV, 595-596, 600-604
camcorders (*see also* VCRs), 623-624
capacitive reactance, 66-68, 82, 83
capacitors and capacitance, 1, 63-83
  ac circuit use of capacitor, 65, 66
  attenuation in filters, 77-78
  axial leads on capacitor, 74
  breakdown voltage, 70